VOLUME SIXTY ONE

ADVANCES IN
MARINE BIOLOGY

Advances in Sponge Science: Phylogeny, Systematics, Ecology

Advances in MARINE BIOLOGY

VOLUME SIXTY ONE

ADVANCES IN
MARINE BIOLOGY

Advances in Sponge Science: Phylogeny, Systematics, Ecology

Edited by

MIKEL A. BECERRO*, MARIA J. URIZ,
MANUEL MALDONADO AND XAVIER TURON
Center for Advanced Studies (CEAB-CSIC)
Acc Cala S Francesc 14
17300 Blanes, Girona, Spain
*Current address: Natural Products and Agrobiology Institute
(IPNA-CSIC)
Avda Astrofisico Francisco Sanchez 3
38206 La Laguna, Tenerife, Canary Islands, Spain

AMSTERDAM • BOSTON • HEIDELBERG • LONDON
NEW YORK • OXFORD • PARIS • SAN DIEGO
SAN FRANCISCO • SINGAPORE • SYDNEY • TOKYO
Academic Press is an imprint of Elsevier

ELSEVIER

Academic Press is an imprint of Elsevier
32 Jamestown Road, London NW1 7BY, UK
Radarweg 29, PO Box 211, 1000 AE Amsterdam, The Netherlands
Linacre House, Jordan Hill, Oxford OX2 8DP, UK
225 Wyman Street, Waltham, MA 02451, USA
525 B Street, Suite 1900, San Diego, CA 92101-4495, USA

First edition 2012

ISBN: 978-0-12-387787-1
ISSN: 0065-2881

For information on all Academic Press publications
visit our website at elsevierdirect.com

Printed and bound in UK

12 13 14 15   10 9 8 7 6 5 4 3 2 1

Contributors to Volume 61

C. Borchiellini
Institut Méditerranéen de Biodiversité et d'Ecologie marine et continentale, UMR 7263 IMBE, Station Marine d'Endoume, Chemin de la Batterie des Lions, Marseille, France

N. Boury-Esnault
IMBE-UMR7263 CNRS, Université d'Aix-Marseille, Station Marine d'Endoume, Marseille, France

P. Cárdenas
Département Milieux et Peuplements Aquatiques, Muséum National d'Histoire Naturelle, UMR 7208 "BOrEA", Paris, France

M. Dohrmann
Department of Invertebrate Zoology, Smithsonian National Museum of Natural History, Washington, DC, USA

D. Erpenbeck
Department of Earth and Environmental Sciences, Palaeontology & Geobiology, and GeoBio-Center, Ludwig-Maximilians-Universität München, München, Germany

C. Larroux
Department of Earth and Environmental Sciences, Palaeontology & Geobiology, Ludwig-Maximilians-Universität München, München, Germany

D. V. Lavrov
Department of Ecology, Evolution, and Organismal Biology, Iowa State University, Ames, IA, USA

M. Maldonado
Department of Marine Ecology, Centro de Estudios Avanzados de Blanes (CEAB-CSIC), Blanes, Girona, Spain

T. Pérez
IMBE-UMR7263 CNRS, Université d'Aix-Marseille, Station Marine d'Endoume, Marseille, France

Klaus Rützler
Department of Invertebrate Zoology, National Museum of Natural History, Smithsonian Institution, Washington D.C., USA

Xavier Turon
Department of Marine Ecology, Centre d'Estudis Avançats de Blanes (CEAB-CSIC), Blanes, Girona, Spain

Maria J. Uriz
Department of Marine Ecology, Centre d'Estudis Avançats de Blanes (CEAB-CSIC), Blanes, Girona, Spain

O. Voigt
Department of Earth and Environmental Sciences, Palaeontology & Geobiology, Ludwig-Maximilians-Universität München, München, Germany

G. Wörheide
Department of Earth and Environmental Sciences, Palaeontology & Geobiology; GeoBio-Center, Ludwig-Maximilians-Universität München, and Bayerische Staatssammlung für Paläontologie und Geologie, München, Germany

Janie Wulff
Department of Biological Science, Florida State University, Tallahassee, FL, USA

CONTENTS

PREFACE

The idea of this special contribution reviewing the latest advances in sponge science was conceived during the World Sponge Conference held in Girona (Spain) in September 2010. Dr. Michael Lesser, editor of the Advances in Marine Biology book series, first suggested the production of a sponge-dedicated monograph. As organizers of the conference, we realized that the amount of information available had increased exponentially in recent years. As this overwhelming new information is scattered over an enormous volume of scientific papers published in journals of very different disciplines, we agreed that a thorough compilation and comprehensive review would be appropriate and useful. The monograph could convey the latest advances in sponge science to sponge specialists besides providing a comprehensive overview to a wider audience with interest in invertebrate biology, marine ecology, molecular ecology, or phylogeny among others. This contribution is timely because we lack reviews in some topics, while in other aspects, reviews were either too old or have become outdated because significant progress has been achieved in the past years. So we took the bait and you have in your hands the results of our efforts to sum up the most relevant and up-to-date scientific literature on the Phylum Porifera.

Sponges are extraordinary animals. With over 8000 extant described species, these organisms are major players in many scientific disciplines. Sponges have relevant roles in shaping the ecological functioning of many marine benthic communities, hold a strategic position for understanding the evolutionary origin of animals, and produce a great variety of secondary metabolites and skeletal structures that have made them preferred targets in biotechnological research. This contribution, split in two thematic volumes, comprises a representative selection of the most active fields of sponge research. Even if not exhaustive, this multiauthor blend of visions offers a wide portrait of the state of the art in sponge science. We have intended the volumes to highlight recent developments in multiple scientific fields, while identifying current limitations and knowledge gaps and delineating challenges and foreseeable future directions.

More specifically, the contributions include an overview of the titanic research work performed on taxonomy and ecology of Caribbean sponges over the past decades. The amazing array of ecological interactions in which sponges engage, with special emphasis on the diversity and functionality of their associated microbiomes, are dealt with in other chapters. The revolution that new molecular tools have represented in ecological studies is also covered in a dedicated chapter. The role of sponges in biogeochemical nutrient cycling is reviewed for the first time. The cell and molecular biology of sponges is a rocketing field, which gets its most recent advances and insights discussed from a modern perspective. Some chapters deal with sponge systematics and phylogeny, which are being hotly debated from several points of view, including a variety of hypotheses to interpret the relationships between sponge groups, other basal invertebrates, and early bilaterian animals. The rich chemical warfare featured by sponges, which has made this group a prolific source of new active natural products, has also been addressed, as well as the sponge machinery for processing and accumulating silica and its implications in tissue engineering. Although some of these chapters provide a good balance between basic and applied research, more biotechnologically oriented issues related to the culture of sponges, sponge cells, or symbionts for the production of chemicals have also found its place in the monograph. The chapters have been organized in two volumes: one covering the topics of phylogeny, systematics, and ecology, and the other dealing with physiology, chemical and microbial diversity, and biotechnology.

We address these volumes to both sponge specialists and nonspecialists, pursuing a twofold goal. We have intended to make the forefront of sponge research easily accessible to the nonspecialist, illustrating the state of the art of the field, and presenting current controversial issues. For the specialist, we wanted this monograph to be a handy, valuable update on the most recent advances in sponge science. We hope we have achieved our goals, at least partially. It goes without saying that the value of the volumes is mostly the merit of the contributing authors and the willing reviewers who altruistically devoted much time to read and make useful suggestions on the manuscripts. Our warmest thanks to all of them as well as to the AMB editorial staff who took care of editing and producing these books. We also thank you, the reader, for your interest in sponges and sponge science. We hope this collection of reviews is entertaining, useful, and inspiring for you all.

Mikel A. Becerro, Maria J. Uriz,
Manuel Maldonado and Xavier Turon

SERIES CONTENTS FOR LAST FIFTEEN YEARS*

---

* The full list of contents for volumes 1–37 can be found in volume 38

CHAPTER ONE

Deep Phylogeny and Evolution of Sponges (Phylum Porifera)

G. Wörheide[*,†,‡,1], M. Dohrmann[§], D. Erpenbeck[*,†], C. Larroux[*],
M. Maldonado[¶], O. Voigt[*], C. Borchiellini[||] and D. V. Lavrov[#]

Contents

[*] Department of Earth and Environmental Sciences, Palaeontology & Geobiology, Ludwig-Maximilians-Universität München, München, Germany
[†] GeoBio-Center, Ludwig-Maximilians-Universität München, München, Germany
[‡] Bayerische Staatssammlung für Paläontologie und Geologie, München, Germany
[§] Department of Invertebrate Zoology, Smithsonian National Museum of Natural History, Washington, DC, USA
[¶] Department of Marine Ecology, Centro de Estudios Avanzados de Blanes (CEAB-CSIC), Blanes, Girona, Spain
[||] Institut Méditerranéen de Biodiversité et d'Ecologie marine et continentale, UMR 7263 IMBE, Station Marine d'Endoume, Chemin de la Batterie des Lions, Marseille, France
[#] Department of Ecology, Evolution, and Organismal Biology, Iowa State University, Ames, IA, USA
[1] Corresponding author: Email: woerheide@lmu.de

Advances in Marine Biology, Volume 61
ISSN 0065-2881, DOI: 10.1016/B978-0-12-387787-1.00007-6

Abstract

Sponges (phylum Porifera) are a diverse taxon of benthic aquatic animals of great ecological, commercial, and biopharmaceutical importance. They are arguably the earliest-branching metazoan taxon, and therefore, they have great significance in the reconstruction of early metazoan evolution. Yet, the phylogeny and systematics of sponges are to some extent still unresolved, and there is an on-going debate about the exact branching pattern of their main clades and their relationships to the other non-bilaterian animals. Here, we review the current state of the deep phylogeny of sponges. Several studies have suggested that sponges are paraphyletic. However, based on recent phylogenomic analyses, we suggest that the phylum Porifera could well be monophyletic, in accordance with cladistic analyses based on morphology. This finding has many implications for the evolutionary interpretation of early animal traits and sponge development. We further review the contribution that mitochondrial genes and genomes have made to sponge phylogenetics and explore the current state of the molecular phylogenies of the four main sponge lineages (Classes), that is, Demospongiae, Hexactinellida, Calcarea, and Homoscleromorpha, in detail. While classical systematic systems are largely congruent with molecular phylogenies in the class Hexactinellida and in certain parts of Demospongiae and Homoscleromorpha, the high degree of incongruence in the class Calcarea still represents a challenge. We highlight future areas of research to fill existing gaps in our knowledge. By reviewing sponge development in an evolutionary and phylogenetic context, we support previous suggestions that sponge larvae share traits and complexity with eumetazoans and that the simple sedentary adult lifestyle of sponges probably reflects some degree of secondary simplification. In summary, while deep sponge phylogenetics has made many advances in the past years, considerable efforts are still required to achieve a comprehensive understanding of the relationships among and within the main sponge lineages to fully appreciate the evolution of this extraordinary metazoan phylum.

Key Words: sponges; Porifera; non-Bilateria; phylogeny; evolution; evo-devo; Demospongiae; Hexactinellida; Homoscleromorpha; Calcarea

1. INTRODUCTION

Sponges are sessile aquatic organisms that inhabit most marine and many freshwater habitats. Adult sponges are of large ecological importance as, for example, filter-feeders and bioeroders (Bell, 2008) and have considerable commercial/biopharmaceutical value (Faulkner, 2002). Their systematics, phylogeny, evolution, and taxonomy have often been proven difficult to reconstruct because many sponges possess only a few systematically/phylogenetically informative morphological characters, and some skeletal traits, which for a long time served as the sole basis for sponge systematics, are prone to homoplasies (reviewed in Erpenbeck and Wörheide, 2007) and relatively variable as a function of local environmental conditions (Maldonado *et al.*, 1999). Nevertheless, significant progress has been achieved in recent years (e.g. Cárdenas *et al.*, 2009, 2011; Dohrmann *et al.*, 2011, 2012; Morrow *et al.*, 2012; Voigt *et al.*, 2012b).

Because of their early-branching position in the animal tree of life (Philippe *et al.*, 2009; Pick *et al.*, 2010), sponges are instrumental in the on-going efforts to better understand the main trajectories of early animal evolution and to decipher the paleogenomics of the last common ancestor of animals (Taylor *et al.*, 2007). Additionally, other non-bilaterian taxa (i.e. Placozoa, Cnidaria, and Ctenophora) and their relationships to each other and to the Bilateria have gained substantial interest as they are of great importance for understanding the evolution of key metazoan traits (Miller, 2009). The statement *"Nothing in biology makes sense except in the light of (a) phylogeny"* (modified after Dobzhansky, 1973) is especially true for the non-bilaterian part of the animal tree of life.

This review is intended to summarize the current state of the debate on the phylogenetic relationships within and among the main sponge lineages and their relationships to other non-bilaterian animals. Erpenbeck and Wörheide (2007) reviewed the then current status of the molecular phylogeny of sponges. They concluded with the statement that "*Coming years will bring the science of sponge systematics closer to its long-awaited goal of a fully consistent phylogeny*". Since then, numerous phylogenies have been published, and the reconstruction of deep-level animal relationships has shifted from the analyses of single or a small number of genes to phylogenomic approaches analyzing dozens to hundreds of genes (e.g. Hejnol *et al.*, 2009; Philippe *et al.*, 2009) and complete mitochondrial genomes (e.g. Lavrov *et al.*, 2008)—we might now ask the question: are we there yet?

2. HIGHER-LEVEL NON-BILATERIAN RELATIONSHIPS

In recent years, several contradicting hypotheses about higher-level non-bilaterian relationships have been published (reviewed by Edgecombe *et al.*, 2011; Philippe *et al.*, 2011). Conflicting results among studies addressing non-bilaterian relationships are not completely unexpected because such studies attempt to reconstruct cladogenetic events that occurred hundreds of millions of years ago (Ma), possibly as early as the Cryogenian ($\sim$650 Ma, Peterson *et al.*, 2008; Erwin *et al.*, 2011). Resolving such ancient splits with molecular sequence data is always difficult because of phylogenetic signal erosion along long terminal branches caused by multiple substitutions (saturation) ("non-phylogenetic signal", see Philippe *et al.*, 2011), which is often combined with short internal branches along which little phylogenetic signal has accumulated (see Rokas and Carroll, 2006). In such cases, the phylogenetic signal along those short internal branches is too low to achieve high statistical support (see Felsenstein, 1985). As a consequence of these difficulties, the relationships of the non-bilaterian taxa, including the origin of Porifera, remain among the most important open questions concerning the higher-level relationships of the Metazoa (Edgecombe *et al.*, 2011; Telford and Copley, 2011).

Due to reductions in sequencing costs, increasing amounts of DNA sequence data have been generated in genome and transcriptome sequencing projects in recent years, and then included in "phylogenomic" analyses. Phylogenomics, described by Eisen and Fraser (2003) as the "intersection of evolution and genomics", currently uses either data from fully sequenced genomes or more commonly, due to the lower resource demands, from expressed sequence tag (EST)/transcriptome sequencing projects to build large alignments (supermatrices) (Philippe and Telford, 2006). Phylogenomics should be distinguished from multi-gene analyses (e.g. Sperling *et al.*, 2009), which typically include fewer than 30 genes that are selected before rather than after sequencing.

In an early phylogenomic study, Rokas *et al.* (2005) used 50 protein-coding genes to reconstruct animal evolution and found that non-bilaterian relationships were unresolved. The authors concluded that these cladogeneses, which most likely occurred several million years before the Bilateria diversified during the "Cambrian Explosion" (Peterson *et al.*, 2008), happened so fast (possibly within about 20 million years) that it is very difficult, if not impossible, to resolve these relationships with sequence data from extant organisms (Rokas and Carroll, 2006).

Dunn *et al.* (2008) applied a much broader phylogenomic approach by analyzing 150 genes, focusing on the relationships within Bilateria. Here, they added a large amount of new EST data for many taxa, which led to increased resolution in this part of the tree. However, non-bilaterian taxa were not extensively sampled, and only a few representatives of Porifera, Cnidaria, and Ctenophora were included. The most publicized result of this study was the position of the Ctenophora as a sister group to the remaining Metazoa, including sponges. A follow-up study from the same group added some additional non-bilaterian taxa (including Placozoa) and reconstructed very similar relationships (Hejnol *et al.*, 2009). Pick *et al.* (2010) significantly improved the taxon sampling of the Dunn *et al.* (2008) dataset. They added EST data from 18 additional non-bilaterian species, including previously unsampled placozoans and sponges but used the same genes and phylogenetic reconstruction methods. In contrast to the findings of Dunn *et al.* (2008), they found that monophyletic sponges branched off first, followed by Ctenophora as the sister group to the remaining metazoans (Cnidaria, Placozoa, Bilateria). The non-bilaterian relationships, although not highly supported by posterior probabilities, were stable regardless of whether only choanoflagellates (the closest living relatives of Metazoa, see e.g. Carr *et al.*, 2008), or the full set of outgroups used by Dunn *et al.* (2008) that included the more distantly related Fungi, were included in the analyses. Pick *et al.* (2010) thus concluded that the early-branching position of Ctenophora found by Dunn *et al.* (2008) was an artefact of insufficient taxon sampling leading to long branch attraction (LBA, *sensu* Felsenstein, 1978, a phenomenon where taxa with long branches are attracted to each other in phylogenetic analyses without being truly related). This conclusion was further corroborated by Philippe *et al.* (2011).

Schierwater *et al.* (2009) published a combined analysis of nuclear protein-coding and mitochondrial genes with morphological characters. They recovered a clade of diploblastic (i.e. non-bilaterian) animals (within which Placozoa branched off first) as the sister group to the triploblastic Bilateria, which led the authors to derive far-reaching conclusions about the evolution of characters such as the nervous system and to propose a "modernized Urmetazoa hypothesis". However, according to topology tests (Shimodaira and Hasegawa, 1999), their preferred tree was not significantly better than competing hypotheses. Furthermore, it was recently shown that their supermatrix contained genes with questionable orthology, frameshift errors, point mutations, as well as biological and *in silico* contaminations (Philippe *et al.*, 2011). An analysis of the same dataset after correction of errors resulted in a different tree topology in which the diploblastic clade was no longer supported (Philippe *et al.*, 2011).

Finally, Philippe *et al.* (2009) published a study based on 128 genes and the most comprehensive sampling of non-bilaterian taxa at that date. Their results revived traditional views on deep animal relationships in that they recovered a highly supported monophyletic Porifera (discussed in more detail below) as sister to the remaining Metazoa, as well as a clade uniting Ctenophora and Cnidaria (=Coelenterata) as sister to the Bilateria, with Placozoa the sister to the Eumetazoa (Coelenterata+Bilateria) clade (see Fig. 1.1). From the morphological perspective, this tree is plausible because if ctenophores or placozoans would have branched off prior to sponges (i.e. form the sister group to the remaining metazoans including sponges),

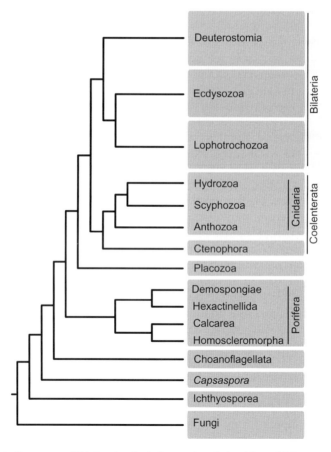

Figure 1.1 Consensus of higher-level phylogenetic relationships of Metazoa, including relationships to non-metazoan relatives (redrawn from Philippe *et al.*, 2009).

cytological features of, for example, sponges—such as choanocyte-like cells, presumably shared by choanoflagellate protists and the hypothetic animal ancestor—would have been lost in Ctenophora or Placozoa and independently in the ancestor of Cnidaria and Bilateria, being only retained by the common ancestor of sponges.

2.1. The status of phylum Porifera: Monophyletic or paraphyletic?

Porifera is morphologically well supported as a monophylum within the Metazoa as judged by the main biphasic life cycle, filter-feeding habits in combination with a sessile adult form, pinacocytes, choanocytes, and aquiferous system (e.g. Böger, 1983; Ax, 1996; Reitner and Mehl, 1996), although exceptions to the classical sponge bauplan exist (e.g. some sponges lack a larval stage and/or mineral skeleton, carnivorous sponges lack choanocyte chambers and an aquiferous system, see Erpenbeck and Wörheide, 2007). The extant classes within the phylum Porifera are also morphologically well defined—Calcarea (calcareous sponges or calcisponges) produce extracellular calcite spicules, Hexactinellida (glass sponges) are characterized by triaxonic silica spicules and adult tissues largely formed by syncytia, and Demospongiae possess monaxonic, tetraxonic, and/or polyaxonic silica spicules, and/or collagen-derived skeletal structures (e.g. spongin fibres and filaments, masses of collagen fibrils) (Hooper and Van Soest, 2002d). Recently, the Homoscleromorpha, which were considered as a subgroup of demosponges at different taxonomic levels (Hooper and Van Soest, 2002d), have received special attention (see below) and are now regarded as a separate class (e.g. Gazave et al., 2010b, 2012).

The monophyly of the extant sponge classes is generally supported by molecular data (Erpenbeck and Wörheide, 2007; Gazave et al., 2012), but the phylogenetic relationships among them and to other metazoan taxa are still regarded as contentious. Morphological analyses have supported different scenarios of relationships within a monophyletic Porifera (see Erpenbeck and Wörheide, 2007 for a summary), while molecular studies, beginning in the early 1990s, predominantly suggested that sponges might be a paraphyletic assemblage sharing a grade of construction rather than common ancestry (see Table 1.1).

Lafay et al. (1992) were among the first to investigate higher-level sponge phylogeny with molecular sequence data, using about 400 bp of 28S rDNA. Their analyses suggested that sponges are paraphyletic (with high support) and they also found paraphyletic demosponges. Calcarea was reconstructed as the sister group to Ctenophora, but this inference was not robustly supported. Furthermore, Hexactinellida and Bilateria were not included. Cavalier-Smith et al. (1996) analyzed 450 near-complete eukaryotic 18S rDNA sequences, including about 100 animal species, and they also found

Table 1.1 Non-exhaustive summary of molecular phylogenetic studies that include statements about sponge mono- versus paraphyly

Author	Molecular marker	Inference method (Model)	Sponge lineages included (number of taxa)				Bilateria included	Monophyly, support	Paraphyly, support
			Demospongiae	Homoscleromorpha	Hexactinellida	Calcarea			
Lafay et al. (1992)	Partial 28S rDNA	NJ, MP, ML	9	–	–	2	No		Yes, high
Cavalier-Smith et al. (1996)	18S rDNA	NJ, MP, ML	4	–	1	3	Yes		Yes, low
Van de Peer and de Wachter (1997)	18S rDNA (secondary structure)	Distance	2	–	–	2	Yes		Yes, low (BS 67)
Zrzavy et al. (1998)	18S rDNA	MP	3	–	–	3	Yes		Yes, low
Collins (1998)	18S rDNA	MP, ME, ML	5	1	1	3	Yes		Yes, low
Kruse et al. (1998)	cPKC	NJ	2	–	1	1	Yes		Yes, low
Schütze et al. (1999)	Hsp70, cPKC, calmodulin, tubulin	NJ	2	–	1	1	Yes		Only with Hsp70 and cPKC, low
Kim et al. (1999)	18S rDNA		3	–	–	3	Yes	Unresolved	Unresolved
Adams et al. (1999)	18S rDNA (secondary structure)	MP, ME, ML	7	–	2	4	No		Yes, low
Medina et al. (2001)	18S rDNA, 28S rDNA	ML with KH tests	2	–	1	1	Yes	Equivocal	Equivocal
Peterson and Eernisse (2001)	18S rDNA	MP	8	–	2	4	Yes		Yes, low
Rokas et al. (2003)	a-tubulin, b-tubulin, EF-2, HSP90, HSP70	ML	4	–	1	2	Yes	Unresolved	Unresolved
Manuel et al. (2003)	18S rDNA	MP, ML	9	–	2	17	Yes	Equivocal	Equivocal
Rokas et al. (2005)	50 genes	ML, MP, BI	1	–	1	1	Yes	Unresolved	Unresolved
Borchiellini et al. (2001)	18S rDNA	NJ, MP	12	–	5	2	No		Yes, high (BS: MP 83/Nj 85)
Peterson and Butterfield (2005)	Seven nuclear housekeeping genes	MP, ML, distance	3	–	–	2	Yes		Yes, medium (BS 76)
Peterson et al. (2005)	Seven nuclear housekeeping genes, mtDNA COI, 18S rDNA	MP	3	–	–	2	yes		Yes, low (BS 62)

Study	Data	Method						Support
Sperling et al. (2007)	Seven nuclear housekeeping genes	Partitioned BI	9	1	–	2	Yes	Yes, high
Dohrmann et al. (2008)	18S rDNA, 28S rDNA, 16S rDNA (mt)	Partitioned BI, secondary structure	6	2	32	4	No	Yes, low (PP 0.6/0.7, BS 74)
Dohrmann et al. (2009)	18S rDNA, 28S rDNA, 16S rDNA (mt)	Partitioned BI, secondary structure	6	2	43	4	No	Yes, low-moderate (PP 0.59/0.84)
Sperling et al. (2009)	Seven nuclear housekeeping genes	BI (CAT-GTR)	20	2	3	4	Yes	Yes, low (PP 0.65/0.71)
Sperling et al. (2010)	Seven nuclear housekeeping genes	BI (CAT-GTR)	20	2	3	5	Yes	Yes, moderate to low (PP 0.92/0.75)
Philippe et al. (2009)	128 genes	BI (CAT)	4	1	2	2	Yes	Yes, high (Bayesian BS 96)
Pick et al. (2010)	150 genes	BI (CAT)	6	2	3	2	Yes	Yes, moderate (PP 0.91)
Erwin et al. (2011)	Seven nuclear housekeeping genes, 18S rDNA, 28S rDNA	Partitioned BI (GTR)	14	2	–	5	Yes	Yes, high

Abbreviations: MP, maximum parsimony; ML, maximum likelihood; BI, Bayesian inference; NJ, neighbour-joining; KH, Kishino–Hasegawa test; PP, posterior probability; BS, bootstrap support; mt, mitochondrial; CAT, CAT model; GTR, general time-reversible model; rDNA, ribosomal DNA; KH, Kishino–Hasegawa.

sponge paraphyly with Calcarea identified as a sister to Ctenophora; however, this was again poorly supported. Van de Peer and de Wachter (1997) were the first to use RNA secondary structure to guide alignment of 18S rDNA sequences, and they investigated about 500 eukaryote species. The authors did not focus on animal phylogeny, although they also found sponges to be paraphyletic with Calcarea as a sister group to Ctenophora, but with low support; no hexactinellids were included. Koziol *et al.* (1997) analyzed a protein-coding gene, *Hsp*70, from three sponges and one bacterium and recovered a completely unresolved tree. Similarly, using the same gene, Borchiellini *et al.* (1998) were unable to resolve the branching order between Cnidaria, Ctenophora, and the three classes of Porifera with convincing support, as their results were highly dependent on the tree reconstruction method. Subsequent studies using protein-coding genes (Kruse *et al.*, 1998; Schütze *et al.*, 1999) have generally provided low support for paraphyletic sponges, but these studies suffered from poor taxon sampling, and they only used simple distance methods.

Zrzavy *et al.* (1998) recovered sponge paraphyly from a combined analysis of 18S rDNA and morphology. The authors reported that siliceous sponges diverged early (although hexactinellids were not included). They used maximum parsimony for tree reconstruction and provided no statistical support measures; sponges were recovered as monophyletic when the morphological data were analyzed alone. Adams *et al.* (1999) also analyzed 18S rDNA and again recovered a weakly supported sister group relationship between Calcarea and Ctenophora, while Cnidaria was identified as sister to a siliceous sponge clade (Demospongiae + Hexactinellida). Collins (1998), also using 18S rDNA, found Demospongiae + Hexactinellida as the sister group to the remaining Metazoa but could not resolve the position of Calcarea and Ctenophora with convincing support. Further rDNA analyses by Kim *et al.* (1999) and Medina *et al.* (2001), the latter including 28S sequences in addition to 18S sequences, likewise found no unambiguous support for either sponge mono- or paraphyly. Another 18S rDNA analysis (Borchiellini *et al.*, 2001) supported paraphyletic sponges with Calcarea as the sister group of eumetazoans, followed by Demospongiae and then Hexactinellida. Their preferred topology received relatively high bootstrap support, but the authors only used simple distance and parsimony algorithms, and Bilateria were not included. Their proposition to elevate Calcarea to the phylum level did not find wide acceptance.

Peterson and Eernisse (2001) conducted a similar study to that of Zrzavy *et al.* (1998) using maximum parsimony to analyze 18S rDNA and morphology, this time including hexactinellid sequences. Their results were similar to those of Zrzavy *et al.* (1998), but again, statistical support for the paraphyly hypothesis was not assessed. In another 18S rDNA study, which focused on the phylogeny of Calcarea, Manuel *et al.* (2003) found no convincing support for either sponge mono- or paraphyly and concluded that "18S rRNA alone is inefficient for resolving sponge [. . .] monophyly". In summary, early molecular studies produced many conflicting hypotheses regarding sponge interrelationships (summarized in Table 1.1) while analyses based on morphology consistently support sponge monophyly (see above). However, those molecular studies were based on only one or a few genes (often partial) with little phylogenetic signal, often missed some important in-group taxa, and frequently suffered from systematic biases.

Rokas *et al.* (2003) were among the first to analyze multiple protein-coding genes from non-bilaterian animals but failed to resolve their relationships, including sponge mono- or paraphyly. They concluded that none of these genes contains sufficient phylogenetic signal to resolve deep metazoan phylogeny.

Peterson *et al.* (2004) used seven nuclear housekeeping genes to investigate metazoan evolution, and this was followed by a series of studies that steadily increased taxon sampling of the same set of genes and always found sponges to be paraphyletic (Peterson and Butterfield, 2005; Peterson *et al.*, 2005; Sperling *et al.*, 2007, 2009, 2010). These authors are among the strongest proponents of the sponge paraphyly hypothesis, and they derived far-reaching conclusions about early animal evolution from it. Peterson *et al.* (2005) and Peterson and Butterfield (2005) only included Calcarea and Demospongiae, and paraphyly (with Calcarea closer to Eumetazoa) was not strongly supported (Table 1.1). Sperling *et al.* (2007) added more demosponges and a homoscleromorph and found the latter as sister to Eumetazoa with good support (see also below for further discussion of this grouping). However, hexactinellids were still missing from their data set, preventing a relevant test of sponge monophyly. Also, very distantly related outgroups (one plant, one fungus) were used, which might have introduced a bias (see Philippe *et al.*, 2011).

Dohrmann *et al.* (2008, 2009) investigated the phylogeny of Hexactinellida using 18S and 28S rDNA (also 16S rDNA, but no outgroups were included for this partition) and found monophyletic sponges (albeit with low support) and a highly supported sister group relationship between Calcarea and Homoscleromorpha (see below for further discussion of this grouping). Hexactinellida were shown to be closely related to demosponges; although the latter were paraphyletic with respect to the former, this was attributed to insufficient taxon sampling of Demospongiae.

Sperling *et al.* (2009) again increased the taxon sampling of their housekeeping gene dataset by including among others another homoscleromorph and three hexactinellids. Although their topology was similar to that of their previous study (Sperling *et al.*, 2007), with Hexactinellida recovered as sister to Demospongiae, critical nodes were not well supported under their best-fitting substitution model. In particular, the nodes responsible for sponge paraphyly, that is, the positions of Calcarea and Homoscleromorpha as successive sister groups to Eumetazoa, only had 0.65 and 0.71 Bayesian posterior probability. Sperling *et al.* (2010) added another calcareous sponge, which resulted in increased support for sponge paraphyly (0.92 for the position of Calcarea). The same gene-sampling was used by Erwin *et al.* (2011) in combination with rDNA sequences, and their analysis recovered a Calcarea + Homoscleromorpha clade (consistent with Dohrmann *et al.*, 2008; Philippe *et al.*, 2009) as sister to Eumetazoa. However, Erwin *et al.* (2011) did not use the substitution model identified by Sperling *et al.* (2009) as best-fitting for the nuclear housekeeping genes and, even more surprisingly, included less demosponges than in Sperling *et al.* (2010) and removed the Hexactinellida altogether.

Sperling *et al.* (2010) also analyzed sponge micro RNAs (miRNAs), a set of novel molecular markers that have been proven valuable in studies of bilaterian relationships (Sperling and Peterson, 2009). None of seven (out of eight) demosponge-specific miRNAs were found in any of the hexactinellid, calcarean, or homoscleromorph small RNA libraries, so miRNAs could not contribute to resolving the mono- versus paraphyly issue. However, the presence of miR-2019 only in the Hexactinellida and Demospongiae supports their sister group relationship (Sperling *et al.*, 2010).

Philippe *et al.* (2009) were the first to apply a phylogenomic approach to the problem of sponge paraphyly. They included the most comprehensive sampling of non-bilaterian taxa to date, including all four extant sponge classes, for a set of 128 genes and recovered sponge monophyly with high support. Within Porifera, they found a sister group relationship of Hexactinellida and Demospongiae, and this "Silicea *sensu stricto*" clade was sister to a Homoscleromorpha + Calcarea clade (see Fig. 1.1), although the latter was less well supported than in Dohrmann *et al.* (2008). Pick *et al.* (2010) recovered a similar topology from their extended dataset from the Dunn *et al.* (2008) study (see above), although sponge monophyly was not as highly supported.

Due to the present lack of complete mitochondrial genome data from Calcarea and unique modes of mtDNA evolution in Calcarea and Hexactinellida (see below), mitogenomics could not yet contribute significantly to evaluations of the interrelationships of the four sponge classes. Studies based on mitochondrial genomes consistently find a sister group relationship between Homoscleromorpha and Demospongiae (Wang and Lavrov, 2007; Lavrov *et al.*, 2008), but resolving higher-level relationships of non-bilaterians using mitochondrial (genome) data has proven to be generally difficult (Lavrov, 2007).

In summary, the majority of studies that have suggested sponge paraphyly provide non-significant support for this hypothesis and/or are hampered by insufficient data (particularly taxon sampling) and/or methodological shortcomings (e.g. simple distance methods for phylogeny reconstruction). Although the final verdict is still open, the congruence of phylogenetic hypotheses derived from independent data types represents the strongest evidence to support one of these alternatives (Pisani *et al.*,

2007). Consequently, the reconstruction of poriferan relationships provided by Philippe *et al.* (2009) and corroborated by Pick *et al.* (2010) and Philippe *et al.* (2011) represents—with respect to the monophyly of sponges—the working hypothesis that is at present preferred by us (Fig. 1.1). Sponge monophyly is (a) supported by currently the largest amount of phylogenomic data (in terms of amino acid positions and in-group taxon sampling) (Philippe *et al.*, 2009; Pick *et al.*, 2010) and (b) is congruent with cladistic analyses of morphological characters (e.g. Böger, 1983; Ax, 1996; Reitner and Mehl, 1996). It should be noted, however, that the alleged sister group relationship of Calcarea and Homoscleromorpha, as reported from recent molecular studies (Dohrmann *et al.*, 2008; Philippe *et al.*, 2009; Pick *et al.*, 2010; Erwin *et al.*, 2011) is presently difficult to support by morphological synapomorphies (see discussion below).

2.2. Why is the phylogenetic status of sponges important for understanding early animal evolution?

In the sponge paraphyly scenario (Fig. 1.2), either Calcarea or Homoscleromorpha or both are more closely related to the rest of the metazoans than to Demospongiae and Hexactinellida. The possible position of Homoscleromorpha as a sister to Eumetazoa has received special attention because these are the only sponges which possess a basement membrane (Fig. 1.3) with evidence of the presence of type-IV collagen in this layer (Boute *et al.*, 1996), which is traditionally considered to define "true" epithelia and might then be interpreted as a synapomorphy of Homoscleromorpha and Eumetazoa (Sperling *et al.*, 2007). Consequently, Homoscleromorpha were included in the Epitheliozoa (a clade combining Eumetazoa and Placozoa, Ax, 1996) by Sperling *et al.* (2009). This scenario opens the possibility that "true" epithelia and developmental mechanisms involved in epithelial patterning and morphogenesis would have appeared before the emergence of Eumetazoa, which is consistent with a conserved function of *Wnt* signalling in epithelial morphogenesis in Homoscleromorpha and Eumetazoa (Ereskovsky *et al.*, 2009; Windsor and Leys, 2010).

The most remarkable feature of the sponge body plan is the aquiferous system, a system of internal canals in which water is pumped from the external medium to chambers lined by choanocytes (flagellate filtering cells). In the paraphyly scenario, the aquiferous system and the choanocytes would have to be interpreted most parsimoniously as ancestral features of Metazoa, implying that the most recent common ancestor of all extant animals was a sponge-like organism (Fig. 1.2). In this scenario, non-sponge metazoans are derived from a sponge-like ancestor through loss of poriferan attributes (Fig. 1.2). Maldonado (2004) proposed that such a step could have involved a neotenic evolution from a poriferan-like larval stage and Nielsen

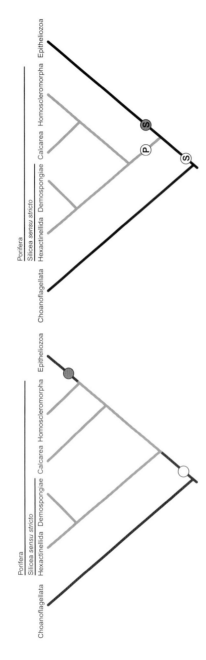

Figure 1.2 Alternative scenarios for the higher-level relationships of extant sponges. *Left*: sponge paraphyly (e.g. Sperling *et al.*, 2009). According to this scenario, a sponge-like body plan (white circle) was acquired in the last common ancestor of Porifera and Epitheliozoa (*sensu* Ax, 1996; = Cnidaria, Ctenophora, Placozoa, Bilateria) and subsequently lost (red circle) from the last common ancestor of Epitheliozoa. Alternative paraphyly scenarios exist mainly in earlier studies, where homoscleromorphs were often not included (see text for details). *Right*: sponge monophyly (e.g. Philippe *et al.*, 2009). According to this scenario, the sponge-like body plan (white circle) was acquired either in the stem lineage of Porifera (P) or, if choanocytes are considered homologous to choanoflagellate cells as judged by outgroup comparison to the well-established sister group of the Metazoa, the Choanoflagellata (see text for details), in stem-group metazoans (S). The latter scenario would require one gain and one loss (indicated by white/red dots marked with S), as in the paraphyly hypothesis. (For interpretation of the references to colour in this figure legend, the reader is referred to the Web version of this chapter.)

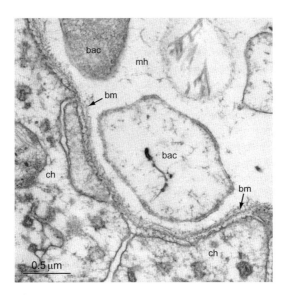

Figure 1.3 Details of the basement membrane (bm) reinforcing the proximal side of the choanocyte layer (ch) in the homoscleromorph *Corticium candelabrum*. Note the abundant intercellular bacteria (bac) and collagen fibrils in the sponge mesohyl (mh) adjacent to the choanocytes (ch).

(2008) suggested it to be a homoscleromorph-like larva that became sexually mature.

In contrast, in the monophyly hypothesis, with sponges as the sister group of all other metazoans, there are two options to make inferences about the metazoan ancestor. These depend on whether sponge choanocytes are considered as homologous to choanoflagellate cells or not. Based on the well-established sister group relationship of Choanoflagellata and Metazoa (King *et al.*, 2008) homology appears most likely, but seemingly different functional properties and ultrastructural differences led some authors to consider that gross morphological similarities could be rather convergences (see discussion in Woollacott and Pinto, 1995; Karpov and Leadbeater, 1998; Philippe *et al.*, 2009). If the latter is true, then the poriferan body plan with, for example, its aquiferous system was at a minimum present only in the stem group of extant sponges (Fig. 1.2) and no immediate inferences can be made about the metazoan ancestor. If sponge choanocytes are indeed homologous to choanoflagellate cells, then the stem group of extant metazoans could well have been a filter-feeding sponge-like organism.

Furthermore, the sponge monophyly scenario either implies that a basement membrane and, consequently, "true" epithelia as classically recognized were present in the last common ancestor of the Metazoa and were subsequently lost in sponge lineages other than Homoscleromorpha

(Fig. 1.2; see also discussion in Lavrov, 2007) or evolved convergently in Homoscleromorpha and Eumetazoa (see Fig. 3 in Philippe *et al.*, 2009). Loss of a basement membrane has been described in some Turbellaria (Plathyhelminthes, Brusca and Brusca, 2003), indicating that such a loss is indeed possible. Recent findings by Leys and Riesgo (2012) indicated that type-IV collagen, a major constituent of a basement membrane, is more ubiquitously distributed in different sponge lineages (found also in demosponges and calcareans) than previously appreciated. Type-IV collagen thus has likely been acquired in stem-group metazoans (as suggested by Aouacheria *et al.*, 2006). Whether the basement membrane then is a more recent independent innovation of homoscleromorphs and eumetazoans, in both cases involving the co-option of type-IV collagen, or symplesiomorphic for the Metazoa is unsolved at present and more research is needed to address these issues.

3. MITOCHONDRIAL DNA IN SPONGE PHYLOGENETICS

Mitochondria—the energy-producing organelles present in most eukaryotic cells—contain their own genome (mt-genome or mtDNA), which is separate from that of the nucleus. For technical and historical reasons, mtDNA has been one of the favourite molecular markers in animal phylogenetic, population genetic, and biogeographic studies as it provides convenient access to a set of orthologous genes with few or no introns, little or no recombination, usually uniparental inheritance, and high evolutionary rates (for a review see Moritz *et al.*, 1987). Although complete sequences of animal mtDNA have been determined since the early 1980s (e.g. Anderson *et al.*, 1981), the first complete mitochondrial genomes of sponges were only published in 2005 (Lavrov *et al.*, 2005). Since then, complete mitochondrial genome sequences have been determined for ∼30 sponges, and current projects aim to bring this number into the 100s. Here, we describe the general organization of mtDNA in sponges and we review a few studies that inferred phylogenies based on mitochondrial sequences. Our focused attention on mtDNA is due to its unique role in animal phylogenetics and our advanced knowledge of its genomic organization in sponges.

3.1. The mitochondrial genomes of sponges

Studies of the mitochondrial genomes of sponges have produced two main unexpected outcomes. First, the study of mtDNA from a few species of demosponges revealed its unique organization, which is different from that in bilaterian animals (Lavrov *et al.*, 2005). Second, a sampling of additional mtDNA from Demospongiae as well as from Hexactinellida, Calcarea, and

Homoscleromorpha showed distinct modes and rates of mitochondrial genome evolution in each of these groups.

So far, most mitochondrial genomes have been determined for the class Demospongiae, including at least one genome for each traditionally recognized order in this group (Lavrov *et al.*, 2005; Erpenbeck *et al.*, 2007d, 2009; Belinky *et al.*, 2008; Lukic-Bilela *et al.*, 2008; Wang and Lavrov, 2008; Ereskovsky *et al.*, 2011). Mitochondrial genomes in Demospongiae are characterized by the retention of several ancestral features (e.g. shared with non-metazoan eukaryotes), including a minimally modified genetic code, the presence of extra genes, conserved structures of tRNA genes, and the existence of multiple non-coding regions (Lavrov *et al.*, 2005). At the same time, some variation has been found in their size, gene content, and gene order (Wang and Lavrov, 2008). The rate of nucleotide substitutions in demosponges is low, although a significant acceleration in evolutionary rates occurred in the Keratosa (G1) lineage (Wang and Lavrov, 2008). However, it is unclear whether this acceleration was restricted to a certain period in the history of the Keratosa or represents an on-going process, as several species with very different morphologies are separated by small mitochondrial genetic distances (Erpenbeck *et al.*, 2009).

Although most of the determined mitochondrial genomes come from the class Demospongiae, the best sampled group is the Homoscleromorpha, considering the ratio of published complete mitochondrial genomes (14) to the number of described species (< 100) (Wang and Lavrov, 2007, 2008; Gazave *et al.*, 2010b). The mitochondrial genomes of homoscleromorphs are similar overall to those of demosponges and retain the same ancestral genomic features. However, two different mitochondrial organizations have been found within this group (Gazave *et al.*, 2010b) corresponding to the families of spiculate and aspiculate homoscleromorphs (Plakinidae and Oscarellidae, respectively; see below). Interestingly, one or two introns are present in the gene for cytochrome oxidase subunit 1 (*cox1* or COI) of several species of Plakinidae (Gazave *et al.*, 2010b), in the same positions where introns have also been found in the demosponge family Tetillidae (Szitenberg *et al.*, 2010). Their location in the standard DNA bar coding primer sites in *cox1* greatly complicates the amplification of this gene as a marker for sponge species identification.

Hexactinellida is currently represented by mitochondrial genomes of three species: *Iphiteon panicea*, *Sympagella nux* (Haen *et al.*, 2007), and *Aphrocallistes vastus* (Rosengarten *et al.*, 2008), although several additional genomes are forthcoming. Mitochondrial genomes in this group show a distinctly different organization that is superficially similar to that of bilaterian animals (Haen *et al.*, 2007). In particular, Bilateria and Hexactinellida share a change in the mitochondrial genetic code and unusual tRNA structures that are unknown outside these groups. Additionally, glass sponges are characterized by phylogenetically diverse and extensive usage of translational frameshifting in mitochondrial translation (Haen *et al.*, unpublished data).

The mitochondrial genome of calcareous sponges remains poorly character-ized. However, a partial mitochondrial genome of the calcinean sponge *Cla-thrina clathrus* has been reported but has not yet been published (Lavrov *et al.*, 2006; Kayal *et al.*, 2010). Preliminary data indicate that this genome is highly unusual and exhibits a very high rate of sequence evolution. In addition to this genomic sequence, several mitochondrial genes of the calcinean *Leucetta chago-sensis* were obtained from its cDNA library. The rate of mitochondrial sequence evolution in this group appears to be much higher than in other sponges (Voigt *et al.*, 2012a). Consequently, a first study on the intraspecific variation of the *cox3* gene in *Leucetta chagosensis* suggests that mitochondrial genes are very useful for phylogeographic studies of Calcarea (Voigt *et al.*, 2012a).

3.2. Inferring sponge phylogeny from mtDNA

Although poriferan mtDNA likely evolves as a single locus, its individual genes display different rates of sequence evolution (Wang and Lavrov, 2008) and so may be more or less appropriate for a specific phylogenetic inference. However, many studies utilized only *cox1* in sponge phylogenetics (e.g. Erpenbeck *et al.*, 2002, 2007a; Nichols, 2005; Cárdenas *et al.*, 2011). We note, however, that other genes, in particular *cob*, have been shown to be more phylogenetically informative (Farias *et al.*, 2001; Lavrov *et al.*, 2008) and some regions in the mtDNA genome may appear to be more informa-tive than others for a particular group (e.g. Rua *et al.*, 2011).

To date, several studies have used complete mtDNA sequences to study the phylogenetic relationships of sponges. In particular, Lavrov *et al.* (2008) investigated demosponge relationships using 21 complete mt genomes representing all recognized orders in the group and Gazave *et al.* (2010b) used 14 complete and 2 partial mitochondrial genomes to study the rela-tionships within Homoscleromorpha. In addition, individual mitochondrial genomes have been used in several other studies (Haen *et al.*, 2007; Belinky *et al.*, 2008; Lavrov *et al.*, 2008; Lukic-Bilela *et al.*, 2008; Erpenbeck *et al.*, 2009; Ereskovsky *et al.*, 2011). Phylogenetic results from these studies are described in the following sections of this review.

4. THE CURRENT STATUS OF THE MOLECULAR PHYLOGENY OF DEMOSPONGIAE

4.1. Introduction to Demospongiae

Demosponges inhabit most aquatic habitats, including all oceans from the intertidal to the abyss, from the tropics to the polar seas, and (almost) all types of freshwater habitats. This diversity in habitats is reflected in their taxonomic diversity. Demosponges are by far the most diverse group of Porifera, comprising about 85% of all extant sponge species.

Demosponges comprise cellular (i.e. not syncytial) Porifera possessing spongin (sometimes greatly reduced) whose mineral skeleton (if present) consists of either monaxonic, tetraxonic, or polyaxonic, -but never triaxonic-, siliceous spicules, and/or occasionally a calcareous basal skeleton. The mineral skeleton can be partially or entirely replaced by an organic skeleton consisting of spongin; alternatively, the skeleton may be reduced to its minimal expression in some demosponges, which only contain abundant collagen fibrils in their mesohyl. So far, there has been no evidence of the presence of a basal lamina as reported for Homoscleromorpha. Molecular data indicate that the definition of Demospongiae in the Systema Porifera (Hooper and Van Soest, 2002d) is now outdated because the inclusion of homoscleromorph sponges within Demospongiae has been rejected based on molecular and cytological data (see other parts of this article).

4.2. Taxonomic overview

In Boury-Esnault's (2006) review of the literature on the evolution of demosponges, the transition from an emphasis on morphology to an emphasis on genetics is described and how different data sets, analytical methods, and interpretations over the decades have resulted in many different classifications is discussed. With the advent of molecular techniques in sponge systematics (Kelly-Borges et al., 1991), one of the first molecular phylogenies (e.g. Lafay et al., 1992) indicated that the then-accepted classifications, which were based on morphology (e.g. Halichondrida, Van Soest et al., 1990), lacked significant support from the molecular data.

Among the most important recent contributions to our understanding of the relationships between the demosponge taxa is the congruence between nuclear and mitochondrial gene trees. As both are reconstructed from independent loci, which have potentially different evolutionary histories and substitution patterns (see, e.g. Moore, 1995), congruent topologies provide strong evidence for accuracy. This congruence should be regarded as more important in phylogenies than high bootstrap support values, which indicate only the support from the underlying data of a single dataset (e.g. Felsenstein, 1985).

One of the first phylogenies directly targeting the deeper demosponge relationships was the work of Borchiellini et al. (2004), who expanded the 18S and 28S rDNA data set of Manuel et al. (2003) to include representatives of almost all accepted demosponge orders (sensu, Hooper and Van Soest, 2002d). Subsequent analyses with complete mitochondrial genomes of selected demosponge taxa (Lavrov et al., 2008; Wang and Lavrov, 2008) provided important support for the new understanding of the deeper phylogenetic splits in demosponges as it revealed a high level of congruence with the nuclear gene trees. However, the number of taxa analyzed for these studies was relatively low (1–2 species per order) compared to the diversity of demosponges and the uncertain monophyly of many orders. In addition, nuclear housekeeping gene data contradict some aspects of these new nuclear ribosomal and mitochondrial phylogenies (Sperling *et al.*, 2009) (see below).

Despite this finding, surprisingly little new insight into the deeper splits among demosponge lineages has been published since the last review (Erpenbeck and Wörheide, 2007) until the work of Morrow *et al.* (2012), who unravelled phylogenetic relationships of the "G4" clade (see below; see also Fig. 1.4, and for a new definition of the higher demosponge clades

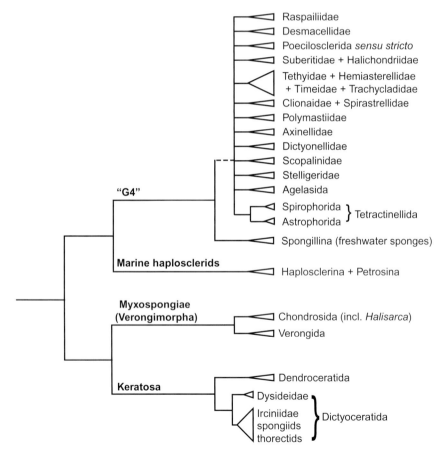

Figure 1.4 Overview of the current phylogenetic relationships of Demospongiae as evident from nuclear ribosomal and mitochondrial gene trees (see the text for details). The names of the deeper clades are adopted from Borchiellini *et al.* (2004), Boury-Esnault (2006) and Erpenbeck *et al.*, (2012b).

see the contribution of Cárdenas *et al.,* Chapter 2, this volume). Most new publications focused particularly on the phylogenetic relationships on shallower levels including species (e.g. López-Legentil and Pawlik, 2008), genera (e.g. Pöppe *et al.,* 2010), families (e.g. Gazave *et al.,* 2010a), and orders (e.g. Cárdenas *et al.,* 2011), leaving many questions about deep demosponge phylogeny unaddressed.

Among the (mostly) aspiculate demosponges, the orders Verongida, Halisarcida, and Chondrosida (with a single genus containing siliceous spicule elements) form a clade termed Myxospongiae ("G2", Borchiellini *et al.,* 2004 also termed "Verongimorpha", (Erpenbeck *et al.,* 2012b)), as revealed by ribosomal (Addis and Peterson, 2005; Nichols, 2005; Schmitt *et al.,* 2005; Holmes and Blanch, 2007; Redmond *et al.,* 2007), mitochondrial (Nichols, 2005; Rot *et al.,* 2006; Lavrov *et al.,* 2008; Wang and Lavrov, 2008), and nuclear housekeeping gene data (Sperling *et al.,* 2009; Fig. 1.4). The monogeneric Halisarcida do not possess any skeletal elements besides collagenous fibrils. Such askeletal taxa are likewise found in Verongida (*Hexadella*), which includes sponges possessing spongin skeletons, and in Chondrosida, which include askeletal (*Chondrosia*), spiculose (*Chondrilla*), and spongin skeleton-possessing (e.g. *Thymosia*) taxa.

The orders Dictyoceratida and Dendroceratida form a clade termed Keratosa (clade "G1", Borchiellini *et al.,* 2004; Addis and Peterson, 2005; Schmitt *et al.,* 2005; Lavrov *et al.,* 2008; Wang and Lavrov, 2008; Fig. 1.4). Both orders include sponges with a spongin skeleton lacking authigenic (produced by the organism itself) mineral elements (with the exception of the coralline sphinctozoan *Vaceletia*; see below).

Molecular phylogenies, based on both nuclear ribosomal and mitochondrial data, suggest a deep split between (mostly) spiculose and (mostly) aspiculose demosponges (Borchiellini *et al.,* 2004; Addis and Peterson, 2005; Nichols, 2005; Schmitt *et al.,* 2005; Holmes and Blanch, 2007; Redmond *et al.,* 2007; Lavrov *et al.,* 2008; Wang and Lavrov, 2008). While ribosomal RNA data initially could not unambiguously resolve whether Keratosa and Myxospongiae are sister taxa (as suggested by 18S phylogenies) or Keratosa are sister group to all other demosponges (as suggested by 28S phylogenies) (Borchiellini *et al.,* 2004), subsequent trees reconstructed from mitochondrial genomes suggested the former, although support values were low (Lavrov, 2007; Lavrov *et al.,* 2008). Consequently, sponges (mostly) without a mineral skeleton might form a sister group to the (mostly) spiculose sponges (Fig. 1.4). However, analyses of nuclear housekeeping genes suggested that Myxospongiae are the sister group to all other Demospongiae (Sperling *et al.,* 2009), albeit with low support. Therefore, the deepest split among the demosponge lineages currently remains unresolved.

The clade of (mostly) spiculose sponges contains the majority of the demosponge taxa. Nuclear ribosomal and mitochondrial data univocally suggest that marine species of the order Haplosclerida (i.e. two of its suborders, Petrosina and Haplosclerina) form a sister group to all other spiculose sponges (Fig. 1.4). This marine Haplosclerida clade corresponds to a clade termed "G3" in the nuclear ribosomal analyses of Borchiellini *et al.* (2004) and is congruent with several other nuclear and mitochondrial gene trees (Schmitt *et al.*, 2005; Holmes and Blanch, 2007; Lavrov *et al.*, 2008; Wang and Lavrov, 2008). DNA data repeatedly demonstrated that marine Haplosclerida do not form a clade with their freshwater counterparts (suborder Spongillina), which form a clade with all remaining demosponge taxa subsequently termed "G4" that is the sister group to all other (mostly spiculose) demosponges (see Borchiellini *et al.*, 2004; Addis and Peterson, 2005; Nichols, 2005; Itskovich *et al.*, 2007; Redmond *et al.*, 2007; Lavrov *et al.*, 2008; Wang and Lavrov, 2008). In contrast, marine and freshwater Haplosclerida form a well-supported clade in phylogenetic reconstructions based on nuclear housekeeping genes (Sperling *et al.*, 2009).

The "G4" clade is the by far most diverse among the higher taxa of demosponges and comprises besides the freshwater sponges a wide range of taxa that are morphologically classified into the orders Astrophorida, Spirophorida, Poecilosclerida, Hadromerida, Halichondrida, and Agelasida (Fig. 1.4). A plethora of molecular gene trees (discussed below) indicates the non-monophyly of several of these orders, and the resolution of the "G4" clade into a new classification and subsequent re-interpretation of morphological characters is the focus of several recent studies (e.g. Erpenbeck *et al.*, 2012a; Morrow *et al.*, 2012).

Molecular data corroborated the earlier views, particularly after introducing cladistic character analyses in sponge systematics (Van Soest, 1990), that a division of demosponges into the subclasses "Ceractinomorpha" and "Tetractinomorpha" is invalid (Hooper, 1984; Van Soest, 1984). Those two subclasses were based primarily on reproductive features (see, e.g. Van Soest, 1991) and were consequently disregarded in the last major classification of sponge genera (Systema Porifera, Hooper and Van Soest, 2002d). They were subsequently abandoned because of a lack of molecular support (Boury-Esnault, 2006).

4.3. Molecular phylogenetics

4.3.1. Keratosa
Dictyoceratida and Dendroceratida are sister groups that form the Keratosa. In dendroceratids, the fibre skeleton is frequently (but not always) dendritic and arises from a basal plate, and it is generally less dense than in dictyoceratids. Dictyoceratids possess anastomosing and, in comparison to dentroceratids, mostly denser fibre networks.

4.3.1.1. Dendroceratida
All dendroceratid sponges possess eurypylous choanocyte chambers. Eurypylous choanocyte chambers connect directly with inhalant and exhalant canals, whereas diplodal choanocyte chambers connect only via a so-called

prosodus or aphodus. Morphologically, Dendrocertida are divided into two families, Darwinellidae, which possess a mostly dendritic skeleton (i.e. the skeletal elements branch, but do not rejoin) and Dictyodendrillidae, which possess a mostly anastomose skeleton (i.e. the skeletal elements branch and rejoin and might form a network [reticulum]) (Bergquist and Cook, 2002b). Published molecular data on dendroceratids are scarce (see, e.g. Borchiellini *et al.*, 2004), but recent results based on nuclear (28S rDNA) and mitochondrial (*cox1*) sequences support the monophyly of Dendroceratida while rejecting the monophyly of the two traditionally recognized families (Erpenbeck *et al.*, unpublished data).

4.3.1.2. Dictyoceratida

Dictyoceratida is composed of the families Dysideidae, Irciniidae, Thorectidae, and Spongiidae. Thorectidae and Spongiidae are mostly distinguishable by the presence of laminated (Thorectidae) or homogeneous (Spongiidae) spongin fibre bark, while Irciniidae have characteristic collagenous filaments, which impart a rubber-like consistency (Cook and Bergquist, 2002). All of these three families have diplodal choanocyte chambers, while members of the fourth family, the Dysideidae, have eurypylous choanocyte chambers, which has led to speculations about a dendroceratid origin of the Dysideidae (see, e.g. Bergquist, 1980). However, recent research based on 28S rDNA and *cox1* sequences supports the dictyoceratid origin of dysideid sponges (Erpenbeck *et al.*, 2012b). Most molecular phylogenies published so far are consistent with monophyly of Dictyoceratida (e.g. Borchiellini *et al.*, 2004; Nichols, 2005; Holmes and Blanch, 2007; Redmond *et al.*, 2007; Lavrov *et al.*, 2008; Wang and Lavrov, 2008; Wörheide, 2008), although taxon sampling is still somewhat limited (but see also Erpenbeck *et al.*, 2012b). Dysideidae are probably a sister group to the remaining dictyoceratids (congruent with Borchiellini *et al.*, 2004; but see also Sperling *et al.*, 2009), indicating monophyly of dictyoceratids with diplodal choanocyte chambers. Within this clade, Irciniidae form a monophyletic group, while the validity of Spongiidae and Thorectidae is still to be resolved (Erpenbeck *et al.*, 2012b). A striking discovery was that *Vaceletia*, a coralline sponge with sphinctozoan bauplan that had until then been placed in its own order Verticillitida, also falls within the Dictyoceratida based on 18S and 28S rDNA data (Wörheide, 2008). This finding was subsequently corroborated by complete mtDNA data (Wang and Lavrov, 2008). Thus, *Vaceletia* appears to be the only recent keratose sponge with an authigenic, although secondary, mineral skeleton.

4.3.2. Myxospongiae (Verongimorpha)

Myxospongiae comprise taxa of the orders Chondrosida, Halisarcida, and Verongida. Myxospongid synapomorphies are mainly cytological, for example, the orientation and position of the accessory centriole, the nuclear apex, and the Golgi apparatus relative to each other, as observed in the ultrastructure of epithelial and larval cells (Maldonado, 2009). Molecular data indicate a sister group relationship between Chondrosida and Halisarcida (Boury-Esnault, 2006), with the Verongida more distantly related. However, the exact branching pattern is dependent on the uncertain taxonomic status of the Chondrosida (see below). On grounds of the cytological and molecular congruencies, Maldonado (2009) elevated the clade formed by these three orders to the subclass level (Myxospongia).

4.3.2.1. Chondrosida

Among the myxospongid taxa, chondrosids possess a marked cortex with fibrillar collagen (Boury-Esnault, 2002). Its four genera display a wide range of skeletal features, such as the possession of siliceous spicules (*Chondrilla*), an irregular, sparse network of small nodal fibres (e.g. *Thymosia*), or only collagen fibrils in the mesohyl (*Thymosiopsis*). Nevertheless, Chondrosida do not appear to be monophyletic in molecular trees. While the molecular data support a close relationship between *Chondrilla*, *Thymosia*, and *Thymosiopsis* (Vacelet *et al.*, 2000), neither nuclear ribosomal (Borchiellini *et al.*, 2004) nor mitochondrial data (Erpenbeck *et al.*, 2007a) group these three genera with *Chondrosia* (see also Nichols, 2005).

4.3.2.2. Halisarcida

The monogeneric Halisarcida (*Halisarca*) include sponges without a skeleton but with a highly organized ectosomal and subectosomal collagen as well as tubular and branched choanocyte chambers (Bergquist and Cook, 2002c). According to recent molecular data, Halisarcida form a clade with *Chondrilla*, *Thymosia*, and *Thymosiopsis*, indicating the non-monophyly of chondrosids (Erpenbeck *et al.*, unpublished data). Recently the Halisarcida have been merged with the Chondrosida (Ereskovsky *et al.*, 2011).

4.3.2.3. Verongida

Verongida is the largest order within the Myxospongiae. Verongid sponges are characterized by the presence of spongin fibres in all but one genus, with a generally well-laminated bark, a dark cellular pith (=fine inclusions) (Bergquist and Cook, 2002d), and the production of bromotyrosines (see Erpenbeck and Van Soest, 2007 for a discussion). Verongida are classified into four families, of which the Ianthellidae possess eurypylous choanocyte chambers while the other three families Aplysinidae, Aplysinellidae, and Pseudoceratinidae (with diplodal choanocyte chambers) differ based on their skeletal characteristics (Bergquist and Cook, 2002d).

Monophyly of Verongida has been demonstrated in a series of analyses (Borchiellini *et al.*, 2004; Nichols, 2005; Holmes and Blanch, 2007; Kober and Nichols, 2007; Redmond *et al.*, 2007). Current rDNA internal transcribed spacer (ITS) data indicate the non-monophyly of several verongid families as the monogeneric Pseudoceratinidae (*Pseudoceratina*) form the sister group to *Verongula* (Aplysinidae), while the other Aplysinidae branch earlier (*Aplysina*) (Erwin and Thacker, 2007). Other Aplysinidae (Aiolochroia, although Aplysinidae insertae sedis, Bergquist and Cook, 2002a) form a sister group to Ianthellidae (Erwin and Thacker, 2007). Mitochondrial (*cox1*) and 28S rDNA data support the findings of Erwin and Thacker (2007), indicating Ianthellidae monophyly and the non-monophyly of aplysinid and aplysinellid sponges (Erpenbeck *et al.*, 2012b).

4.3.3. Marine Haplosclerida

Haplosclerida is regarded as an evolutionarily successful taxon with respect to diversity and habitat (e.g. Van Soest and Hooper, 2002b). The skeleton of marine haplosclerids displays a (partial) isodictyal reticulation (i.e. triangular meshes with sides of one spicule length) of diactinal (two rayed) spicules. Marine Haplosclerida are currently classified into the suborders Haplosclerina and Petrosina with three families each. Their molecular phylogeny is still one of the largest mysteries in demosponge systematics. Molecular data have so far been unable to confirm the morphological classification, including the monophyly of the marine haplosclerid suborders, families, and even genera (particularly the species-rich genera *Haliclona* and *Callyspongia*, see, e.g. McCormack *et al.*, 2002; Erpenbeck *et al.*, 2004; Itskovich *et al.*, 2007; Raleigh *et al.*, 2007; Redmond *et al.*, 2007; Redmond and McCormack, 2008; Voigt *et al.*, 2008). Reasons for these discrepancies are still unknown, although an elevated substitution rate in comparison to the other demosponge orders has been detected for the nuclear ribosomal genes (Erpenbeck *et al.*, 2004) and occasionally in mitochondrial genes (Erpenbeck *et al.*, 2007d), which can cause tree reconstruction artefacts but may not entirely explain the incongruent branching patterns. In recent years, attempts to resolve marine haplosclerid phylogeny found congruency of topologies reconstructed from ribosomal RNA (including a study on the suitability of ITS, Redmond and McCormack, 2009) and mitochondrial markers. These congruencies diminish the possibility of reconstruction artefacts as a source of the contradictions to morphology and strengthen the need for a revised marine haplosclerid classification. An analysis of 28S rDNA (McCormack *et al.*, 2002), 18S rDNA (Redmond *et al.*, 2007), including secondary structure analyses (Redmond and McCormack, 2008; Voigt *et al.*, 2008), and two different fragments of *cox1* (Itskovich *et al.*, 2007; Raleigh *et al.*, 2007) suggest the presence of a large clade including several intermixed *Callyspongia* (Callyspongiidae) and *Haliclona* (Chalinidae) species, while most petrosids, niphatids, and phloeodictyids branch earlier.

4.3.4. The "G4" clade

The remaining demosponge taxa form a clade designated as "G4" by Borchiellini *et al.* (2004) (also termed "Democlavia" by Sperling *et al.*, 2009). It comprises by far the largest taxonomic diversity of demosponges. Molecular data suggest that most of the morphologically defined orders are not monophyletic and a recent study of Morrow *et al.* (2012) led to a new classification of the "G4" clade based on analyses of mitochondrial and nuclear DNA sequences.

The order Halichondrida occupies a pivotal position in the history of demosponge phylogeny (for an overview see Erpenbeck *et al.*, 2012a). After Van Soest and Hooper reported inconsistencies in the current classifications of Poecilosclerida and Axinellida, respectively (Hooper, 1984; Van Soest, 1984), and cladistic character analyses were introduced in sponge systematics (Van Soest, 1990), both authors independently concluded that the distinction between Ceractinomorpha and Tetractinomorpha is unparsimonious and suggested the re-merging of the order "Axinellida" with Halichondrida. Monophyly of the re-defined order Halichondrida and its five families (Van Soest and Hooper, 2002a) could not be demonstrated in morphological (see Erpenbeck, 2004), biochemical (Erpenbeck and Van Soest, 2005), or molecular data sets (see, e.g. Morrow *et al.*, 2012). In fact, halichondrid polyphyly has been repeatedly demonstrated, since both ribosomal RNA (Lafay *et al.*, 1992) and biochemical data (Van Soest and Braekman, 1999) suggested a close relationship between Agelasida and axinellids (later corroborated with several independent molecular data sets, see Erpenbeck *et al.*, 2006). Molecular data also demonstrated that the family Dictyonellidae (Van Soest *et al.*, 1990), which was mostly defined based on the absence of specific characters, consisted of unrelated taxa (Nichols, 2005), and its nominal genus *Dictyonella* did not form a clade with Halichondriidae. Axinellidae has been reported as polyphyletic in molecular phylogenies and this is also the case of its nominal genus *Axinella* (Alvarez *et al.*, 2000; see Gazave *et al.*, 2010a for a recent review). Similar, other taxa included in Axinellidae, such as *Reniochalina*, *Ptilocaulis*, and *Phakellia*, do not form a monophyletic group with *Axinella* in all molecular phylogenies (Erpenbeck *et al.*, 2007b,c, 2012; Holmes and Blanch, 2007; Morrow *et al.*, 2012). Halichondrida are also polyphyletic due to a close relationship between Halichondriidae and the hadromerid Suberitidae repeatedly that emerged from molecular analyses (e.g. Chombard and Boury-Esnault, 1999; McCormack and Kelly, 2002; Erpenbeck *et al.*, 2004, 2005b, 2012; Morrow *et al.*, 2012).

The order Hadromerida, which has been frequently targeted for molecular analyses but is often too weakly represented with respect to the diverse spicule and skeletal shape (Kelly-Borges *et al.*, 1991; Chombard and Boury-Esnault, 1999; Borchiellini *et al.*, 2004), was eventually shown to be paraphyletic with respect to Poecilosclerida (Nichols and Barnes, 2005; Kober and Nichols, 2007; Morrow *et al.*, 2012). Likewise, Poecilosclerida (Hooper and Van Soest, 2002c) itself has been found to be polyphyletic based on molecular markers. This taxon was established on the basis of chelae microscleres, which are present in most of the Poecilosclerida genera. Other non-chelae bearing taxa, such as Raspailiidae or Desmacellidae, are assigned to Poecilosclerida due to skeletal similarities other than chelae (see Hooper and Van Soest, 2002c for details). However, mitochondrial data revealed that chelae-bearing Poecilosclerida are unrelated to chelae-lacking Raspailiidae, some Desmacellidae and several microcionid taxa (Erpenbeck *et al.*, 2007a), which was later supported by ribosomal RNA analyses (Erpenbeck *et al.*, 2007b,c; Morrow *et al.*, 2012).

The polyphyly of lithistid demosponges has been accepted for a longer time. Lithistid sponges are characterized by the presence of irregular articulated choanosomal siliceous spicules called desmas that interlock and form a rigid skeleton in most fossil and recent genera. Based on this feature, they were grouped together as Order Lithistida Schmidt, 1870. However, polyphyly of this order had been suspected for about a century (see, e.g. Pisera and Lévi, 2002), which has been supported by molecular data (e.g. Kelly-Borges and Pomponi, 1994) and is accepted in the most recent morphological classification (Hooper and Van Soest, 2002d). Lithistid sponges are currently divided into 13 extant families (Pisera and Lévi, 2002). While the phylogenetic relationships of all extant lithistid taxa have yet to be fully resolved, molecular data demonstrated that several triaene-bearing lithistid sponges fall into the Tetractinellida (see below) (Cárdenas *et al.*, 2011).

4.3.4.1. Spongillina (freshwater sponges)

The monophyly of freshwater sponges has been supported by molecular data in several analyses (e.g. Addis and Peterson, 2005; Itskovich *et al.*, 2007; Redmond *et al.*, 2007; Voigt *et al.*, 2008). However, its nominal family Spongillidae was found to be paraphyletic (see below), particularly with respect to Lubomirskiidae, the Lake-Baikal endemic family (Itskovich *et al.*, 1999, 2008; Addis and Peterson, 2005; Meixner *et al.*, 2007; Redmond *et al.*, 2007). Monophyly of Lubomirskiidae has been suggested based on *cox1* and tubulin intron analyses (but see also Schröder *et al.*, 2003; Itskovich *et al.*, 2006), but more recent *cox1* and 18S rDNA data contradict this hypothesis (e.g. Itskovich *et al.*, 2007; Meixner *et al.*, 2007). Nevertheless, support in the latter analysis is rather low and newer ITS2 data again strongly support Lubomirskiidae monophyly (Itskovich *et al.*, 2008).

Many members of the family Spongillidae (e.g. Itskovich *et al.*, 2008) and several of its genera, such as *Ephydatia*, have been found to be paraphyletic (e.g. Addis and Peterson, 2005; Meixner *et al.*, 2007). This indicates the need for a revised taxonomy of freshwater sponges (Addis and Peterson, 2005; Harcet *et al.*, 2010a). Several endemic taxa are thought to have been derived from widespread Spongillidae, such as *Spongilla* or *Ephydatia* (or Erpenbeck *et al.*, 2011 for Lake Tanganyika sponges; see, e.g. Meixner *et al.*, 2007 for Lake Baikal).

The phylogenetic position of the remaining freshwater sponge families is unresolved as they are clearly underrepresented in current gene trees. In most analyses, the Metaniidae *Corvomeyenia* splits first from all Spongillina (e.g. Addis and Peterson, 2005; Itskovich *et al.*, 2007; Redmond *et al.*, 2007; Voigt *et al.*, 2008), and in several analyses, the lithistid *Vetulina* (Vetulinidae) is a sister group to freshwater sponges (see, e.g. Addis and Peterson, 2005; Itskovich *et al.*, 2007).

4.3.4.2. Tetractinellida

Molecular data support the monophyly of Tetractinellida, which include the orders Astrophorida and Spirophorida (Vacelet *et al.*, 2000; Borchiellini *et al.*, 2004; Addis and Peterson, 2005; Nichols, 2005; Erpenbeck *et al.*, 2007a; Holmes and Blanch, 2007; Itskovich *et al.*, 2007; Redmond *et al.*, 2007; Lavrov *et al.*, 2008; Wang and Lavrov, 2008; Morrow *et al.*, 2012). Morphologically, the Tetractinellida are distinguished by the possession of tetractine (four rayed) megascleres, which have one ray clearly prolonged and the remaining three approximately evenly short (triaenes). Several lithistid sponges have been found to fall into the Tetractinellida, among them *Aciulites* sp. (Scleritodermidae), *Theonella* sp. and *Discodermia dissoluta* (Theonellidae), and *Corallistes* sp. (Corallistidae) (e.g. Addis and Peterson, 2005; Nichols, 2005; Itskovich *et al.*, 2007).

4.3.4.2.1. Astrophorida

Astrophorida are conventionally (leaving apart the lithistids) divided into five families (Hooper and Van Soest, 2002b). The most comprehensive molecular phylogeny of Astrophorida was recently published by Cárdenas *et al.* (2011), who extended an earlier study on the taxonomic status of the family Geodiidae (Cárdenas *et al.*, 2009) by additional astrophorid representatives. They found that the astrophorid suborders Euastrophorida and Streptosclerophorida are polyphyletic (as indicated earlier by Chombard *et al.*, 1998); likewise, the families Geodiidae, Ancorinidae, and Pachastrellidae as well as many genera are polyphyletic (Cárdenas *et al.*, 2011). The combined analysis of 28S and *cox1* results in the following phylogenetic hypothesis, which is significantly congruent with earlier analyses based on a much lower taxon sampling (Chombard *et al.*, 1998; Nichols, 2005):

A well-supported geodinid clade has been recovered, including three *Geodia* subclades termed "*Geodia*", "*Depressogeodia*", and "*Cydonium*" (following the PhyloCode) as well as *Ecionemia*, *Rhabdastrella*, and *Stelletta* species, which are currently classified as Ancorinidae. This geodinid clade is a sister group to a calthropellid clade (*Calthropella*) and an erylinid clade (including Erylus, Penares, and Pachymatisma, see also Cárdenas *et al.*, 2007). Together, the geodinid and erylinid + calthropellid clade form a (poorly supported) geodiid clade, which is a sister to Pachastrellidae (*Poecillastra* + *Pachastrella* + *Triptolemma*).

Sister to this geodiid + pachastrellid clade are Ancorinidae and several lithistid families, such as Corallistidae, Phymaraphiniidae, and Theonellidae, as well as the pachastrellid genera, *Characella*, and *Dercitus*. The remainder of pachastrellids branch earlier in the Astrophorida phylogeny. For these, Cárdenas *et al.* (2011) suggest a new family, designated as Vulcanellidae (for *Vulcanella* and *Poecillastra*), and the resurrection of the families Theneidae (for *Thenea* spp.) and Thoosidae for *Alectona millari*, which is currently classified in Alectonidae (Hadromerida). Maldonado (2004) suggested that the occurrence of discotriaenes (triaenes in which the short rays form a single disc) in the larva of *Thoosa* and *Alectona* indicate that those genera did not belong to the family Clionaidae (Order Hadromerida) and suggested transferring them to the order Astrophorida. A subsequent molecular study (Borchiellini *et al.*, 2004) corroborated the suggestion that *Alectona millari* was more closely related to members of the order Astrophorida than to representatives of Hadromerida. Another alectonid, *Neamphius*, also falls into Astrophorida (Cárdenas *et al.*, 2011).

Although the deeper splits of this phylogeny are weakly supported or unsupported, it provides important clues about demosponge character evolution. It also reminds us that even taxa that are relatively rich in complex characters compared to other demosponges are prone to character misinterpretations resulting in unrecognized homoplasies.

4.3.4.2.2. Spirophorida
Even 5 years after the last review of the field (Erpenbeck and Wörheide, 2007), Tetillidae is still the only Spirophorida family with published data for molecular phylogenetics. This is probably due to the encrusting or excavating habit of Samidae and Spirasigmidae, which are more prone to DNA contamination than the more massive tetillids. Therefore, the monophyly of Spirophorida lacks confirmation from molecular data, but as their sigmaspire microscleres are unique among Demospongiae, this hypothesis might remain unchallenged (see Hooper and Van Soest, 2002a). Tetillidae so far appear to be monophyletic, as *Tetilla* and *Cinachyrella* form a clade in several larger phylogenies (e.g. Nichols, 2005; Redmond *et al.*, 2007). The largest phylogenetic contribution to Tetillidae is based on an analysis of a mitochondrial intron in the Tetillidae (Szitenberg *et al.*, 2010). The tree derived from the corresponding *cox1* fragment displays the genera *Cinachyrella*, *Tetilla*, and *Craniella* as non-monophyletic. However, additional data are necessary to verify and explain these outcomes.

4.3.4.3. Agelasids + axinellids + raspailids + dictyonellids + heteroxyids
Agelasida possess spicules with spines arranged into verticills. They contain the Astroscleridae, which have a calcareous basal skeleton (Wörheide, 1998), and the soft-bodied monogeneric (*Agelas*) Agelasidae (see also Parra-Velandia, 2011 for internal relationships of Caribbean species of this family). The close relationship of the families Astroscleridae *s.s.* (*Astrosclera*) and Agelasidae was repeatedly demonstrated with molecular (Chombard *et al.*, 1997; Alvarez *et al.*, 2000; Nichols, 2005) and biochemical data (see

review in Wörheide, 1998). Nevertheless, recent molecular data indicate paraphyly of the Astroscleridae, suggesting an *Agelas* + *Astrosclera* clade to which other Astroscleridae (*Stromatospongia vermicola* + *Ceratoporella nicholsoni*) form a sister group (Parra-Velandia, 2011).

Molecular data have repeatedly indicated a close relationship between Agelasida and axinellid taxa, especially several *Axinella* species and *Stylissa* (see Erpenbeck *et al.*, 2006 for details). Polyphyly of *Axinella* was first demonstrated with 28S rDNA (Alvarez *et al.*, 1998), and additional 18S data indicate at least three separate clades of *Axinella* with *A. damicornis*, *A. verrucosa*, and *A. corrugata* in a sister group to Agelasida together with *Cymbastela cantharella* (Gazave *et al.*, 2010a, who subsequently termed this clade "Cymbaxinella"). However, *Cymbastela* has been demonstrated to be polyphyletic (Alvarez *et al.*, 2000; Alvarez and Hooper, 2010; Erpenbeck *et al.*, 2012a), and *C. cantharella* might be unrelated to the type species *C. stipitata* for which a close relationship to Agelasida has never been shown.

In addition, molecular analyses also group some raspailid taxa (Poecilosclerida) with this Agelasida/axinellid assemblage. The 28S rDNA sequences of *Amphinomia* are almost identical with their agelasid sequences of the clade (Erpenbeck *et al.*, 2007c); furthermore, molecular data found "*Eurypon* cf. *clavatum*" closely related to Agelasida/Axinellidae (besides *Prosuberites laughlini*; Hadromerida: Suberitidae and *Hymerhabdia typica*; Halichondrida: Bubaridae) (Nichols, 2005; Itskovich *et al.*, 2007; Morrow *et al.*, 2012). Agelasida were re-defined based on the new taxon composition (Morrow *et al.*, 2012).

All raspailiid taxa so far investigated with molecular markers, including its nominal genus *Raspailia*, are unrelated to Poecilosclerida *s.s.* and form a clade with the axinellids *Ptilocaulis* and *Reniochalina* (Erpenbeck *et al.*, 2007a; Holmes and Blanch, 2007), with the heteroxyid halichondrid *Didiscus* (Erpenbeck *et al.*, 2007b), and with the former hadromerid (*incertae sedis*) family Sollasellidae (Van Soest *et al.*, 2006; Erpenbeck *et al.*, 2007b). Molecular analyses show that *Raspailia* (*s.s.*), *Eurypon*, *Sollasella*, *Aulospongus*, and *Ectyoplasia* form a Raspailiinae clade, while several other *Raspailia* subgenera, for example, *Parasyringella*, do not appear to be monophyletic.

Gazave *et al.* (2010a) recovered two additional *Axinella* spp. clades: one clade, subsequently termed "Axinellidae", including the type species *Axinella polypoides*, *Dragmacidon*, and other *Axinella* (including *Axinella dissimilis* and *Axinella aruensis*); and the other, subsequently termed "*Acanthella*", including *Axinella cannabia* and the dictyonellids *Acanthella acuta* and *Dictyonella*. In previous molecular analyses, *Acanthella* was the only dictyonellid with close relationships to the nominal genus *Dictyonella*, and it formed a clade with the axinellid *Cymbastela* (including the type species *C. stipitata*) and the halichondriid *Axinyssa* (Alvarez *et al.*, 2000; Erpenbeck *et al.*, 2005b). This clade now forms with *Phakellia* and the lithistid *Desmanthus* a re-defined family Dictyonellidae (Morrow *et al.*, 2012).

4.3.4.4. Halichondridae + Suberitidae

Ribosomal and mitochondrial genes indicate a close relationship between Halichondriidae (order Halichondrida) and Suberitidae (order Hadromerida) (Chombard and Boury-Esnault, 1999; McCormack and Kelly, 2002; Erpenbeck *et al.*, 2004, 2005b, 2012), although this lacks support from evaluation of Elongation Factor 1 alpha and biochemical analyses (Erpenbeck and Van Soest, 2005; Erpenbeck *et al.*, 2005a). *Axinyssa* is the only halichondrid without molecular phylogenetic affinities with this Halichondriidae + Suberitidae clade. *Axinyssa* is also the only halichondriid without an ectosomal skeleton—a feature shared by *Acanthella* and *Dictyonella*, which are close relatives based on molecular phylogenies (see above). Molecular analyses also support the morphological distinction of the genus *Johannesia* from *Vosmaeria* (Gerasimova *et al.*, 2008).

4.3.4.5. Polymastiidae

Polymastiidae, albeit so far only represented by *Polymastia* spp., form a monophyletic group in several 28S phylogenies (Nichols, 2005), sometimes in the form of a sister group to the Halichondridae + Suberitidae clade (Kober and Nichols, 2007; Morrow *et al.*, 2012).

4.3.4.6. Clionaidae + Spirastrellidae

Nichols' (2005) 28S analysis resulted in a spiraster-bearing Clionaidae + - Spirastrellidae clade including *Spirastrella*, *Diplastrella* (both Spirastrellidae), *Cliona*, *Pione*, and *Cervicornia* (all Clionaidae). Neither of these two families was found to be monophyletic. Another 28S analysis based on the D2 fragment corroborated these results, finding the genera *Cliona* and *Spheciospongia* to be non-monophyletic and *Cliothosa* nested within *Cliona* (Barucca *et al.*, 2007). Nevertheless, this fragment could not support Clionaidae as monophyletic because *Diplastrella* falls in this clade (see also Kober and Nichols, 2007; Morrow *et al.*, 2012). Finally, 28S analyses placed *Placospongia* (Placospongidae) in an unsupported sister group relationship to Clionaidae (Nichols, 2005; Kober and Nichols, 2007).

4.3.4.7. Tethyidae + hemiasterellids

Analyses of 28S support a clade combining Tethyidae with several hemiasterellid species (*Axos cliftoni*, *Adreus* spp.), although neither family has been found to be monophyletic (Nichols, 2005; Kober and Nichols, 2007; see Heim *et al.*, 2007a,b,c also for other tethyid species phylogenies). Other hemiasterellid taxa, such as *Stelligera* and *Paratimea*, fall outside this clade (Nichols, 2005; Morrow *et al.*, 2012) and form with the heteroxyid *Halicnemia* a re-erected Family Stelligeridae (Morrow *et al.*, 2012). As hemiasterellid taxa show similarities to several other demosponge families (Hooper, 2002), the polyphyletic status of this group is not surprising. Nichols (2005), Kober and Nichols (2007), and Morrow *et al.* (2012)

recovered a close relationship of Timeidae to Tethyidae + hemiasterellids (which is morphologically supported by the presence of asterose microscleres) to which *Trachycladus* (Trachycladidae) is the sister group.

4.3.4.8. Poecilosclerida *sensu stricto* (primary chelae-bearing poecilosclerids)

Poecilosclerida is the largest order of sponges with respect to the numbers of families and genera (Hooper and Van Soest, 2002d), but it is the least studied by means of molecular systematics. Nuclear ribosomal and mitochondrial analyses recently revealed that several taxa classified as poecilosclerids, all of them lacking the Poecilosclerida-characteristic chelae microscleres, do not form a clade with most non-chelae-bearing poecilosclerids (Erpenbeck *et al.*, 2007a,b,c). However, these Poecilosclerida *sensu stricto* may contain taxa with an assumed secondary loss of chelae, such as *Tedania* (Tedaniidae), which groups within chelate poecilosclerids (Erpenbeck *et al.*, 2007a).

Nevertheless, the suborders of the chelae-bearing Poecilosclerida, Mycalina, Microcionina, and Myxillina (see Hooper and Van Soest, 2002c) could not be supported by molecular data (e.g. Nichols, 2005). Podospongiidae are Mycalina *incertae sedis* based on an interpretation that the protorhabd of spinorhabds is potentially a sigmancistra derivative (Kelly and Samaai, 2002), and the sequences of the podospongiids *Negombata* and *Diacarnus* form a monophyletic group in *cox1* analyses. However, molecular data so far do not support a clade combining Podospongiidae with other Mycalidae, but rather with myxillids (Nichols, 2005; Schmitt *et al.*, 2005; Rot *et al.*, 2006; Itskovich *et al.*, 2007; Morrow *et al.*, 2012).

4.4. Future work

It is evident that in recent years, our understanding of the phylogenetic relationships of demosponges has greatly improved, particularly due to congruence between mitochondrial and nuclear data. However, recruitment of additional, independent markers is clearly needed, especially to contribute to the resolution of the deeper splits. Nevertheless, many details of morphological character evolution remain unclear. The widespread inconsistency between gene trees and morphology-based taxonomy in marine Haplosclerida is probably among the most difficult issues to solve in demosponge phylogeny.

Additionally, there are many taxa for which the traditional placement has been rejected based on molecular data, and most of them currently await a new assignment, including dictyonellids, such as the relationships of *Svenzea* and *Scopalina* to other "G4" sponges. These taxa were studied intensively on the species level (e.g. Blanquer *et al.*, 2005; Blanquer and Uriz, 2008, 2010) and are currently placed in a newly erected Family Scopalinidae (Morrow *et al.*, 2012). Likewise, the phylogenetic position of *Biemna* and other desmacellids is unresolved (see also Mitchell *et al.*, 2011; Morrow *et al.*, 2012), as well as most taxa of Poeciloscerida *sensu stricto*.

Increased taxon sampling of Axinellidae, Raspailiidae, Agelasida, Dictyonellidae, and (former) Heteroxyidae are needed to fully appreciate the emerging classification schemes from molecular data. Distinguishing between raspailids and axinellid species is difficult and subjective (Alvarez and Hooper, 2010) and might be complicated by hybridization (Alvarez *et al.*, 2007). Furthermore, the analysis of Morrow *et al.* (2012) provides a new view on the classification, but mostly fails to provide a robust resolution of the phylogenetic relationships of the major clades; additional analyses using slow-evolving molecular markers are desirable.

5. THE CURRENT STATUS OF THE MOLECULAR PHYLOGENY OF HEXACTINELLIDA

5.1. Introduction to Hexactinellida

Hexactinellida (glass sponges) are exclusively marine and siliceous sponges largely restricted to the deep sea, with a few notable exceptions, such as massive glass sponge reefs found in SCUBA-accessible depths off the Canadian west coast (e.g. Conway *et al.*, 2001; Krautter *et al.*, 2001; Cook *et al.*, 2008) and population of sublittoral caves in the Mediterranean by one species (*Oopsacas minuta*: Vacelet *et al.*, 1994; Bakran-Petricioli *et al.*, 2007). Currently, 623 extant species are considered valid according to the World Porifera Database (Van Soest *et al.*, 2011), but because the deep sea is still to a large extent unexplored and vast museum collections await revision by a limited number of experts, this is probably a gross underestimate of the actual diversity of this group (Reiswig, 2002). Glass sponges are remarkably distinct from other sponges in many aspects of their biology (reviewed in Leys *et al.*, 2007). In particular, their syncytial tissue organization and triaxonic spicule symmetry clearly distinguish them from the other three major sponge groups and make them one of the best-supported higher-level metazoan monophyla (Mehl, 1992). They also differ from other sponges because they generally have a richer set of morphological characters, displaying a complex skeletal anatomy and a vast array of different spicule types that provide a wealth of information for the taxonomy of the group.

5.2. Taxonomic overview

The current classification of extant Hexactinellida (Dohrmann *et al.*, 2011; Hooper *et al.*, 2011) recognizes two subclasses, characterized by distinct types of microscleres: Amphidiscophora (with amphidiscs) and Hexasterophora (with hexasters). Amphidiscophora contains a single extant order, Amphidiscosida, with three families: Hyalonematidae, Pheronematidae, and the monogeneric Monorhaphididae. With 16 families in four orders, Hexasterophora is much more diverse. Within this subclass, two main types of skeletal organization are distinguished: lyssacine, that is, mainly composed of unfused spicules (which is also the sole type of skeletal organization found in Amphidiscophora), and dictyonine, that is, with rigid skeletons (dictyonal frameworks) composed of fused six-rayed (hexactine) megascleres in addition to loose spiculation. The lyssacine hexasterophorans (Rossellidae, Euplectellidae, and Leucopsacidae) are placed in a single order Lyssacinosida. The dictyonine taxa are divided into three orders: Aulocalycoida (Aulocalycidae and Uncinateridae) and Lychniscosida (Aulocystidae and the monogeneric Diapleuridae) are rare, species-poor groups; the majority of dictyonine genera are placed in the Hexactinosida, including Euretidae, Tretodictyidae, Farreidae, Dactylocalycidae, Aphrocallistidae, and the monogeneric Auloplacidae, Fieldingiidae, Craticulariidae, and Cribrospongiidae. While most genera and families of Hexactinellida are morphologically well-defined taxa, order-level relationships, relationships between the families and intrafamilial relationships (e.g. division of larger families into subfamilies) are difficult to resolve with morphological data (Dohrmann *et al.*, 2008).

5.3. Molecular phylogenetics

5.3.1. Current status

Although nuclear and mitochondrial sequences of a few glass sponge species have become available since the early 1990s (see Table 1.1), the internal phylogenetic relationships of this group were only recently investigated with molecular data. The first molecular phylogenetic study of Hexactinellida (Dohrmann *et al.*, 2008) included 34 species from 27 genera, 9 families, and 3 orders (Amphidiscosida, Hexactinosida, and Lyssacinosida) and was based on nuclear 18S, partial nuclear 28S, and partial mitochondrial 16S rDNA sequences. As expected from morphological predictions, monophyly of Hexactinellida and of its two subclasses was highly supported. Furthermore, and in contrast to the molecular phylogenies of Demospongiae and Calcarea (see the respective sections above and below), the reconstructed relationships within these groups are also remarkably congruent with the taxonomic classification—all but one genera and all families with more than one species included were found to be monophyletic. Also, almost all of the included species of Hexactinosida formed a highly supported clade corresponding to the Sceptrulophora (Mehl, 1992), a taxon that was only recently formally introduced (Dohrmann *et al.*, 2011). As the name suggests, its members are characterized by the possession of sceptrules, a distinct class of spicules that is regarded as synapomorphic and that occurs in different variations, most commonly scopules or clavules (see Dohrmann *et al.*, 2011). In contrast, the Hexactinosida as a whole were found to be non-monophyletic because the only included sceptrule-lacking species (*Iphiteon panicea*; Dactylocalycidae) either formed the sister group to the Lyssacinosida or was nested within that group, depending on the substitution models employed. However, these results were not totally unexpected given that monophyly of Hexactinosida and Lyssacinosida had been called into question before on purely morphological grounds (Mehl, 1992).

In a follow-up study based on increased taxon sampling of the same markers, Dohrmann *et al.* (2009) resolved the position of *Iphiteon* as sister to Lyssacinosida, supporting monophyly of the latter (it should be noted, however, that additional, so far unsampled, dictyonal taxa might be nested within Lyssacinosida [see below]). Since then, the taxonomic sampling of Hexactinellida has been increased to 50 species (38 genera, 10 families, 3 orders), and the rDNA dataset was supplemented with an additional marker, *cox1* (Dohrmann *et al.*, 2011, 2012). Below we discuss the relationships within Sceptrulophora and Lyssacinosida based on the combined analysis of rDNA and *cox1* sequences from these two studies (Fig. 1.5).

The subdivision of Sceptrulophora into Scopularia and Clavularia (Mehl, 1992), based on the presence of scopules (most taxa) or clavules

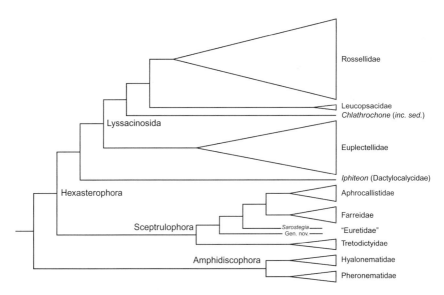

Figure 1.5 Overview of current knowledge about higher-level phylogeny of Hexactinellida. Based on maximum likelihood analysis of combined 18S, 28S, 16S rDNA, and *cox1* sequences (~4600 bp). See Dohrmann *et al.* (2011, 2012) for details. Gen. nov., yet-to-be described new genus of "Euretidae"; *inc. sed.*, Lyssacinosida *incertae sedis*.

(restricted to Farreidae) is strongly rejected by the molecular data. Instead, Farreidae (=Clavularia) is nested within a paraphyletic "Scopularia" as sister group to the Aphrocallistidae (see Dohrmann *et al.*, 2011). Monophyly of "typical" Farreidae is highly supported, but the monospecific genus *Sarostegia*, which appeared somewhat misplaced in this family, clearly groups outside this clade and was consequently moved to Euretidae, consistent with earlier classifications (Dohrmann *et al.*, 2011). Although the exact position of *Sarostegia* remains unclear due to low bootstrap support, a grouping with the other included euretid is consistent with the molecular data (Dohrmann *et al.*, 2011). However, because Euretidae is particularly speciose and morphologically diverse and because there seem to be no potentially apomorphic characters uniting its genera, more taxa need to be sampled to test the monophyly of this family. Monophyly of the similarly species-rich Tretodictyidae, which is so far resolved as the sister group to all remaining sceptrulophorans, is currently only moderately supported by molecular data. However, this family is morphologically well characterized, so it can be expected that molecular support will solidify with inclusion of additional genera. Finally, monophyly of Aphrocallistidae, a species-poor but highly abundant family that includes the only extant examples of reef-building sponge species (see above), is highly supported by both morphological and molecular evidence. However, reciprocal monophyly of its two constituent genera could not be demonstrated by the combined molecular data, a result that is somewhat puzzling and might be related to gene-tree—species-tree conflict (Dohrmann *et al.*, 2011).

Among the Lyssacinosida, monophyly of all three families is highly supported by the combined DNA sequence data. While for the Euplectellidae (the "venus-flower basket" family) this result was expected from morphology, in case of the Rossellidae (the most speciose family of Hexactinellida) and the small family Leucopsacidae (three genera) this can be viewed as a positive surprise because morphological autapomorphies of these taxa are hard to pin down. In contrast, at the intrafamilial level, the situation is more "typical" for sponges: molecular data do not support any of the currently recognized subfamilies of Rossellidae (Rossellinae, Lanuginellinae) or Euplectellidae (Euplectellinae, Corbitellinae, Bolosominae). However, with the exception of Lanuginellinae (see Dohrmann *et al.*, 2012), these taxa are either negatively defined (Rossellinae=non-Lanuginellinae) or defined based on homoplasy-prone characters (Dohrmann *et al.*, 2009, 2012). On the interfamilial level, Euplectellidae has been identified as the sister group of the remaining lyssacinosidans, among which the unplaced monospecific *Clathrochone* is sister to a Rossellidae+Leucopsacidae clade. Although Tabachnick (2002) apparently favours a closer relationship of Leucopsacidae to Euplectellidae, the reasons for this proposal are unclear; it remains to be shown what (if any) morphological characters would support or contradict the higher-level molecular phylogeny of Lyssacinosida.

5.4. Future work

Although the taxon sampling achieved so far is already fairly comprehensive, it is heavily skewed towards Hexasterophora. Relationships within Hyalonematidae and Pheronematidae (Amphidiscophora) should be further investigated by incorporating additional taxa, and the phylogenetic position of *Monorhaphis* (Monorhaphididae), which is famous for its up to 3 m long giant anchor spicule, remains to be determined.

Within Hexasterophora, taxon sampling of the dictyonal groups still needs improvement. Of special importance are the sceptrule-lacking Lychniscosida, Aulocalycoida, and Dactylocalycidae. These taxa are crucial for understanding skeletal evolution because their dictyonal frameworks differ considerably from those found in Sceptrulophora. Mehl (1992) rejected a closer relationship of Lychniscosida—a relict group that was highly diverse and reef-building in the Mesozoic—to other dictyonal sponges, instead proposing a position within Lyssacinosida, which remains to be tested with molecular data. While Lychniscosida is morphologically well supported, this is not the case for Aulocalycoida—although members of this group display a similar type of framework, constructional differences between the families (Hooper and Van Soest, 2002d; Janussen and Reiswig, 2003; Leys *et al.*, 2007; Reiswig and Kelly, 2011) raise doubts about their homology, and even monophyly of the families is not well established. In Uncinateridae, the presence of scopules in *Tretopleura* has been confirmed (Dohrmann, personal observation), and molecular data indicate a nested position of this genus within Sceptrulophora (Dohrmann, unpublished data). Therefore, at least the Uncinateridae, or parts thereof, belong in Sceptrulophora; the position and status of Aulocalycidae still remain elusive. Dactylocalycidae only consists of the already sampled *Iphiteon* (see above) and the type genus *Dactylocalyx*; if molecular data can confirm monophyly of this family and its position as sister to Lyssacinosida remains stable, Dactylocalycidae should best be classified in a separate order. Finally, within Sceptrulophora, monophyly and intergeneric relationships of Euretidae and Tretodictyidae need to be further investigated (see above), and the positions of the four monogeneric families remain to be resolved.

Within Lyssacinosida, intrafamilial relationships of Rossellidae and Euplectellidae are in need of further clarification. A dense taxon sampling comprising the majority of genera will be required to determine if the molecular phylogeny supports morphologically diagnosable clades that could be classified as subfamilies; if this is not the case, subfamilies should be abandoned among Lyssacinosida. Finally, inclusion of *Hyaloplacoida* (*incertae sedis*) might support the designation of a fourth family, if this taxon groups with *Clathrochone* (see above), which can be predicted from their similar spiculation (see Hooper and Van Soest, 2002d).

6. THE CURRENT STATUS OF THE MOLECULAR PHYLOGENY OF HOMOSCLEROMORPHA

6.1. Introduction to Homoscleromorpha

Homoscleromorpha is a small group of marine sponges (<100 described species), the monophyly of which is well accepted on the basis of their general organization and the shared features of their cytology and embryology. Their affinities to other sponges, however, are less clear and have recently been questioned. Traditionally, homoscleromorph sponges were considered as a family or a suborder of the subclass Tetractinellida of the class Demospongiae mainly due to the shared presence of siliceous tetractinal-like calthrops (Lévi, 1956). This small group, however, progressively appeared to be problematic. In recent molecular phylogenetic studies that recovered sponges as monophyletic, Homoscleromorpha appears to be most closely related to Calcarea (Dohrmann *et al.*, 2008; Philippe *et al.*, 2009; Pick *et al.*, 2010), although support for this is only moderate to low in the latter two studies (see Part 2 of this chapter and Table 1.1 for alternative relationships that have been proposed). This grouping has been claimed earlier to be consistent with similarities of spicule shape and gross larval morphology in these two groups (Grothe, 1989; Van Soest, 1990), but these morphological similarities are rather superficial (and therefore of limited phylogenetic value); so far, no clear-cut morpho-anatomical characters appear to support this clade (Gazave *et al.*, 2010b). Recently, Gazave *et al.* (2012) considered sponges to be monophyletic, formally raised the Homoscleromorpha to class-level and proposed the presence of cross-striated rootlets in larval ciliated cells of both cinctoblastula (Homoscleromorpha) (Boury-Esnault *et al.*, 2003), amphiblastula (Calcaronea), and calciblastula (Calcinea) (Gallissian and Vacelet, 1992; Ereskovsky and Willenz, 2008) as a possible synapomorphy of Homoscleromorpha and Calcarea.

Homoscleromorpha are often encrusting or lobate with a smooth surface, and they usually occur at shallow depths, but a few have been recovered from abyssal depths. Homoscleromorph sponges display a large number of characters that distinguish them from Demospongiae (Muricy and Diaz, 2002; Uriz *et al.*, 2003; Uriz, 2006; Maldonado and Riesgo, 2007; 2008b; Ereskovsky *et al.*, 2009; Ereskovsky, 2010; Gazave *et al.*, 2010b). They are characterized by an aquiferous system with sylleibid-like or leuconoid organization with eurypylous, diplodal, or aphodal choanocyte chambers. These sponges possess a unique type of tetractine spicules (calthrops), distinguishable from calthrops of the Demospongiae and their derivatives by their small size, ramification (lophose calthrops), and/or reduction (diods and triods) of one to all four actines. These spicules are secreted not only within sclerocytes (as in the demosponges) but also within epithelial cells, showing a unique

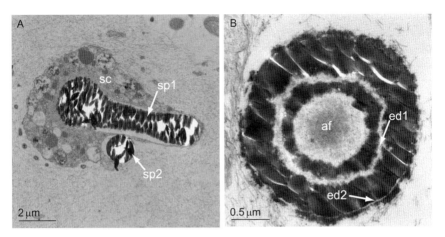

Figure 1.6 (A) A sclerocyte (sc) of the homosclerophorid *Corticium candelabrum,* showing an intracellular spicule (sp1) and another (sp2) that appears to be in the process of extrusion to the surrounding mesohyl. (B) Cross-section of a spicule belonging to *C. candelabrum* in an early stage of silicification. This still growing intercellular spicule has an axial filament (af) and two concentric extra-axial organic deposits (ed1, ed2) between the silica layers. Modified from Maldonado and Riesgo (2007).

silicification process characterized by amorphous axial filaments and two concentric extra-axial organic layers (Maldonado and Riesgo, 2007; Fig. 1.6). These spicules do not form a well-organized skeleton. Homoscleromorpha possess flagellated exopinacocytes and endopinacocytes, unique flagellated apopylar cells, an incubated cinctoblastula larva with cross-striated ciliary rootlets that are surprisingly derived from the accessory centriole (a unique feature in Porifera), and asynchronous spermatogenesis that occurs inside of spermatic cysts. Another feature of the Homoscleromorpha is that they are the only sponge lineage in which adult cell layers are underlain by a basement membrane containing type-IV collagen and *zonula adherens* cell junctions. However, whether the epithelium of the larval stage, although reported (Boury-Esnault *et al.*, 2003), always has a basement membrane remains discussed (see discussion in Maldonado and Riesgo, 2008b).

6.2. Taxonomic overview

Since 1995, Homoscleromorpha has been composed of a single order (Homosclerophorida) with a single family (Plakinidae) and seven genera (*Oscarella, Plakina, Plakortis, Plakinastrella, Corticium, Pseudocorticium,* and *Placinolopha*) (Boury-Esnault *et al.*, 1995; Hooper *et al.*, 2002; Van Soest *et al.*, 2011). The genera have been distinguished based on four morphological characters (Diaz and Soest, 1994; Muricy and Diaz, 2002): the presence or absence of a siliceous skeleton; the presence or absence of a cortex associated with the architecture of the aquiferous system and the type of choanocyte chambers; if spicules are present, they are characterized based on the number of spicule size classes; and the presence and type of ramification in the actins of the calthrops. Molecular phylogenetics have recently changed the taxonomic system (see below), now two families (Plakinidae, Oscarellidae) are accepted, with five genera (68 species) and two genera (17 species), respectively (Hooper *et al.*, 2011).

6.3. Molecular phylogenetics

The internal relationships within this group have recently been investigated using molecular data for six of the seven valid genera (Gazave *et al.*, 2010b), resulting in a revision of the suprageneric classification in the World Porifera Database (Van Soest *et al.*, 2011). Based on the congruence of the results from mitochondrial, nuclear, and chemical markers (Gazave *et al.*, 2010b; Ivaniševic *et al.*, 2010), it has been proposed that the subdivision of Homoscleromorpha, which was abandoned in 1995 (Boury-Esnault *et al.*, 1995), into Oscarellidae (aspiculate genera, including the genus *Oscarella*) and Plakinidae (spiculate genera) should be restored (see Fig. 1.7). It was only after the designation of a new genus, *Pseudocorticium*, which is similar in histological traits to the spiculate genus *Corticium* but devoid of a mineral skeleton like *Oscarella* (Solé-Cava *et al.*, 1992), that it was proposed to merge

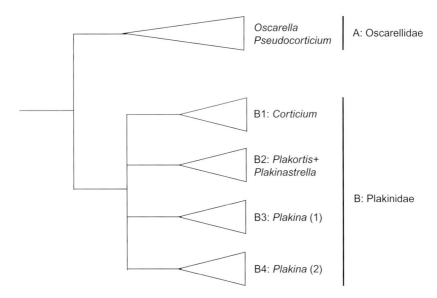

Figure 1.7 Internal relationships of Homoscleromorpha.

all the homoscleromorphs into a single family, the Plakinidae, with *Pseudo-corticium* as an aspiculate morph of *Corticium*. Thus, the absence of a skeleton in *Oscarella* and *Pseudocorticium* was phylogenetically non-informative (Boury-Esnault *et al.*, 1995). Recent molecular phylogenetic results (Gazave *et al.*, 2010b) have challenged this view, as they supported a sister group relationship of *Pseudocorticium* and *Oscarella*. This hypothesis implies that the cortex, aquiferous system organization, and external morphological similarities of *Corticium* and *Pseudocorticium*, previously interpreted as syna-pomorphies, represent convergent characters. In contrast, the presence or absence of spicules in these two genera can be considered as diagnostic. In addition, mitochondrial gene arrangement consistently gives strong support for this scenario (Wang and Lavrov, 2007, 2008; Gazave *et al.*, 2010b). Indeed, the mitochondrial genomes of the Oscarellidae species share a specific gene order, the presence of *tatC*, as well as genes for 27 tRNAs (Wang and Lavrov, 2007; Gazave *et al.*, 2010b), whereas the species included in the Plakinidae clade share the lack of *tatC* as well as the lack of 20 of the 25 tRNA genes typically found in demosponges (Wang and Lavrov, 2008; Gazave *et al.*, 2010b). In Oscarellidae, the monophyly of the genus *Oscarella* has not been confirmed by all molecular analyses. However, the hypothesis of a paraphyletic *Oscarella* as suggested by 18S and mitochondrial data sets (Gazave *et al.*, 2010b) needs further testing with the inclusion of more *Oscarella* species. Within the family Plakinidae, a more robust hypothesis is obtained based on 28S data, congruent with the morphologically well-defined genera *Plakortis*, *Corticium*, and *Plaki-nastrella* and validating morphological characters as diagnostic for these clades (Muricy and Diaz, 2002). In contrast, regardless of the genetic marker and analytical method used, the genus *Plakina* appears paraphyletic with two of the four *Plakina* species being more closely related to *Corticium*. This scenario is not surprising based on the lack of clear apomorphic characters, which has already led several authors to question the monophyly of this genus (Muricy *et al.*, 1996, 1998; Muricy and Diaz, 2002; Gazave *et al.*, 2010b). Other molecular and morphological analyses of extant species are needed to resolve this issue and propose a subdivision into several genera. Yet, the presence of several characters (i.e. well-developed mesohyl, well-differentiated ectosome, large subectosomal cavities, and tetralophose calthrops) has been proposed to sup-port a clade uniting *Plakina jani* and *Plakina trilopha* (Gazave *et al.*, 2010b). At a higher taxonomic level, molecular analyses support the grouping of *Plakortis* and *Plakinastrella*. A synapomorphy of this clade could be the absence of lophose spicules, which are present in all the other spiculate genera.

6.4. Future work

Molecular analyses reject the monophyly of *Plakina*, which should be tested using a larger taxon sampling. Additional data are also needed to resolve the question of the phylogenetic status of the genus *Oscarella*. Also, more detailed studies of *Pseudocorticium* and *Oscarella* species are needed, and the phylogenetic position of *Placinolopha*, the only genus not yet included in any dataset, should be determined.

7. THE CURRENT STATUS OF THE MOLECULAR PHYLOGENY OF CALCAREA

7.1. Introduction to Calcarea

Calcareous sponges (Class Calcarea) include about 675 accepted extant species (Van Soest *et al.*, 2011), which are exclusively marine. They occur mostly in shallow waters; only a few species are known from the deep sea (for an overview see, e.g. Rapp *et al.*, 2011). In contrast to the intracellularly formed siliceous spicules found in the other sponge classes, Calcarea are characterized by calcium carbonate spicules that are excreted to the extracellular space (Manuel *et al.*, 2002; Sethmann and Wörheide, 2008). In most Calcarea, the skeleton is exclusively composed of free spicules, but some species additionally possess a rigid basal skeleton of fused or cemented spicules (Manuel *et al.*, 2002). Three basic spicule types can be distinguished depending on the numbers of actines: diactines, triactines, and tetractines. Variation in spicule morphology is limited compared to other sponges (Manuel, 2006). Four different types of aquiferous systems occur in Calcarea. In asconoid Calcarea, all internal cavities are lined with choanocytes (this organization is referred to as homocoel). In syconoid, sylleibid and leuconoid Calcarea, choanocytes occur in choanocyte chambers, and parts of the internal cavities (inhalant and exhalant canals or the atrium) are lined with pinacocytes (heterocoel organization). In the traditional taxonomy, the arrangement of the spicules and the organization of the aquiferous system are important characters (Manuel, 2006). All species of Calcarea are viviparous (Manuel *et al.*, 2002).

7.2. Taxonomic overview

Calcarea is divided into two subclasses: Calcinea and Calcaronea. This subdivision is supported by several characters: the position of the nucleus in the choanocytes (basal in Calcinea, apical in Calcaronea), development (eversion of stomoblastula in Calcaronea), larval types (coeloblastula in Calcinea, amphiblastula in Calcaronea), the spicule type that is built first during ontogenesis (Calcinea: triactines; Calcaronea: diactines) (Bidder, 1898; Hartman, 1958; Manuel *et al.*, 2002; Manuel, 2006), and different values of $\delta^{13}C$ isotopes in the spicules (Reitner, 1992; Wörheide and Hooper, 1999). Several autapomorphies for each subclass can also be found in the secondary structure of the 18S rRNA (Voigt *et al.*, 2008).

The definition of orders, families, and genera is based on characters of skeletal architecture and the aquiferous system (Manuel, 2006). The classification of Calcarea is mainly typological and not based on phylogenetic analyses (Erpenbeck and Wörheide, 2007). Unsurprisingly then, the first phylogenetic analysis of morphological characters showed only little resolution below the subclass level, suggesting a high level of homoplasy (Manuel et al., 2003).

In the following, we refer to the latest taxonomic revisions at the subclass level for Calcinea (Borojevic et al., 1990) and Calcaronea (Borojevic et al., 2000). Importantly, the classification is based on the idea of gradual evolution and that extant Calcarea represent different evolutionary "steps", from sponges with a simple, asconoid, and olynthus-like organization to more complex forms through several intermediate stages on different evolutionary paths (reviewed and illustrated by Manuel, 2006).

Calcinea contains two orders, Murrayonida and Clathrinida. The order Murrayonida comprises Calcinea with a reinforced calcite skeleton, calcareous plates, or spicule tracts. Only a few species belong to this order (three families, three genera, four species, Van Soest et al., 2011). Order Clathrinida includes the majority of calcinean species (6 families, 16 genera, 160 species, Van Soest et al., 2011), with skeletons that are only composed of free spicules.

In Calcaronea, three orders are recognized: Leucosolenida, Lithonida, and Baerida. Leucosolenida contains the majority of calcaronean species (9 families, 43 genera, 467 species, Van Soest et al., 2011). Their skeleton is composed of free spicules without calcified non-spicular reinforcements (Borojevic et al., 2000). Lithonida comprises a small number of calcaronean species with reinforced skeletons (2 families, 6 genera, 20 species, Van Soest et al., 2011). Baerida is a similarly small group (3 families, 8 genera, 17 species, Van Soest et al., 2011). Sponges of this order have skeletons formed exclusively or in substantial parts by microdiactines (Borojevic et al., 2000).

7.3. Molecular phylogenetics

7.3.1. Current status

Only a few molecular studies have aimed at resolving relationships of the entire class by analysis of small and large subunit ribosomal RNA genes (Manuel et al., 2003, 2004; Dohrmann et al., 2006; Voigt et al., 2012b). An overview of the relationships according to Voigt et al. (2012b) is shown in Fig. 1.8. A common outcome of these studies is the monophyly of Calcarea and its subclasses Calcinea and Calcaronea, while relationships below the subclass level strongly conflict with the classification system described above. Many of the supraspecific taxa cannot be recovered as monophyletic (e.g. Dohrmann et al., 2006; Voigt et al., 2012b), and the phylogenetic

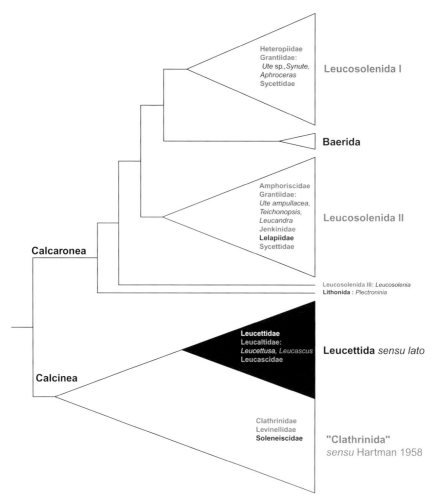

Figure 1.8 Relationships of Calcarea inferred from ribosomal RNA-gene sequences (see Voigt *et al.*, 2012b for details). Taxa that are not monophyletic are shown in grey. When other members of a family also occur in a separate clade, the genus or species names are given.

hypotheses that were brought forward contradict the scenarios of morphological evolution that are the foundations of the current taxonomic system (see above).

7.3.1.1. Calcinea

In Calcinea, the orders Clathrinida and Murrayonida are not monophyletic. Instead, homocoel (asconoid) genera without a cortex form a paraphyletic grade leading to a clade containing all included Calcinea with a cortex

(Voigt *et al.*, 2012b). The included species of Murrayonida do not group together and are nested within the clade of cortex-bearing Clathrinida (Dohrmann *et al.*, 2006; Voigt *et al.*, 2012b), which includes all sampled species from families Leucettidae and Leucascidae, as well as the sampled heterocoel members of Leucaltidae (*Leucaltis* and *Leucettusa*). The genus *Leucetta* (Leucettidae) is not monophyletic, and *Ascandra*, a homocoel member of Leucaltidae, is more closely related to other homocoel Calcinea than to *Leucaltis* or *Leucettusa* (Voigt *et al.*, 2012b). Relationships among homocoel Calcinea are not resolved, as many nodes are poorly supported (Dohrmann *et al.*, 2006; Voigt *et al.*, 2012b). Within this paraphyletic group, the family Clathrinidae and the genus *Clathrina* are not monophyletic (Dohrmann *et al.*, 2006; Voigt *et al.*, 2012b).

The clade of cortex-bearing Calcinea can be classified by broadening the definition of the order Leucettida (Hartman, 1958) to include Calcinea with a cortex and heterocoel organization. This order was rejected by Borojevic *et al.* (1990) and merged with Clathrinida. These authors instead suggested independent gains of a cortex in Leucaltidae, Leucettidae + Leucascidae, and Murrayonida. However, molecular phylogenies reject the monophyly of Leucaltidae, Leucascidae, and Murrayonida (Dohrmann *et al.*, 2006; Voigt *et al.*, 2012b), thereby contradicting this evolutionary scenario. Instead, Leucettida *sensu lato* can be defined as Calcinea with a cortex, which would also include Murrayonida. The asconoid aquiferous system of *Ascaltis* may be interpreted as a secondary simplification within this clade, a hypothesis that needs to be tested further (Voigt *et al.*, 2012b).

In summary, despite discrepancies with the classification of Borojevic *et al.*, an evolution from simple to more complex forms in Calcinea is supported by molecular phylogenies (Manuel *et al.*, 2003; Dohrmann *et al.*, 2006). However, the suggested independent evolutionary paths in Leucaltidae and Murrayonida are rejected (Voigt *et al.*, 2012b).

7.3.1.2. Calcaronea

In Calcaronea, the order Leucosolenida is paraphyletic because it includes species of the order Baerida (Dohrmann *et al.*, 2006; Voigt *et al.*, 2012b). Baerida is also paraphyletic as far as the classical taxonomy is concerned, as it includes the hyper-calcified sponge *Petrobiona massiliana*, which is currently classified in the order Lithonida (Manuel *et al.*, 2003; Dohrmann *et al.*, 2006; Voigt *et al.*, 2012b). However, the grouping in Baerida is also supported by morphological characters, indicating misclassification of this genus (Manuel *et al.*, 2003). The only other included lithonid (*Plectroninia neocaledoniense*) is the sister taxon to all other Calcaronea (Dohrmann *et al.*, 2006), which has led to the speculation that the rigid basal skeleton of fused spicules in this species might be an ancestral character of Calcaronea (Dohrmann *et al.*, 2006). However, this hypothesis needs further testing by inclusion of more species of Lithonida (Dohrmann *et al.*, 2006). The asconoid species *Leucosolenia* sp. branches off after *Plectroninia* (Dohrmann *et al.*, 2006), which calls into question the primitive state of the asconoid aquiferous system in this subclass because *Plectroninia* has a leuconoid aquiferous system.

The remaining Calcaronea form the sister clades (Leucosolenida I + Baerida) and Leucosolenida II (Voigt *et al.*, 2012b). Leucosolenida I includes all sampled Heteropiidae (*Sycettusa, Syconessa, Grantessa*), two *Sycon* species (*S. capricorn* and *S. ciliatum*), and some species from Grantiidae with giant cortical diactines (*Ute* sp., *Synute* and *Aphroceras*, Voigt *et al.*, 2012b). Within this clade, Heteropiidae and *Sycettusa* are not monophyletic (Dohrmann *et al.*, 2006; Voigt *et al.*, 2012b). Giant cortical diactines also occur in the heteropiid genera *Heteropia* and *Paraheteropia* (Borojevic *et al.*, 2000), which were not included in molecular analyses. A closer relationship between such Grantiidae and Heteropiidae and between *Sycon* and Grantiidae has been suggested before (e.g. Borojevic, 1965; Borojevic *et al.*, 2000). However, both *Sycon* and Grantiidae are polyphyletic according to molecular data (Manuel *et al.*, 2003; Dohrmann *et al.*, 2006; Voigt *et al.*, 2012b). Other *Sycon* species and *Ute ampullacea* are included in Leucosolenida II (Voigt *et al.*, 2012b). Leucosolenida II also includes species from the families Amphoriscidae, Jenkinidae, and Lelapiidae and from some additional genera of Grantiidae (*Grantia, Teichonopsis,* and *Leucandra*). Besides Lelapiidae, which is only represented by the genus *Grantiopsis*, these families are not monophyletic (Manuel *et al.*, 2003, 2004; Dohrmann *et al.*, 2006; Voigt *et al.*, 2012b).

The morphological evolution in Calcaronea is poorly understood. As mentioned above, the early-branching position of *Plectroninia* might imply that the common ancestor of the subclass was not asconoid as suggested before (e.g. Borojevic *et al.*, 2000; Manuel, 2006), but was leuconoid with a rigid skeleton of fused spicules (Dohrmann *et al.*, 2006). The syconoid aquiferous system is the most frequent in the included taxa (see, e.g. Voigt *et al.*, 2012b). Ancestral character state reconstruction suggests that leuconoid aquiferous systems evolved several times independently (Manuel *et al.*, 2003, 2004; Voigt *et al.*, 2012b). A cortex might have evolved early in Calcaronea, possibly before or after the split of *Leucosolenia*, and several syconoid taxa lacking a cortex (e.g. *Sycon, Syconessa*) might have lost it secondarily (e.g. Voigt *et al.*, 2012b). However, these inferences have to be treated with caution, as inclusion of additional taxa might result in a different conclusion.

7.4. Future work

In summary, molecular data suggest that morphological evolution in this taxonomically difficult class of sponges is even more complex than anticipated based on previous studies. Approaches to resolve the phylogeny of Calcarea will be more problematic than in other sponges because the classical taxonomy is of limited value as a framework to guide taxon sampling. Additionally, taxon-specific revisions (e.g. Klautau and Valentine, 2003) need to be treated with caution because they possibly do not consider monophyletic groups, which in turn can hamper the recognition of potential morphological synapomorphies.

An alternative phylogenetic classification of Calcarea cannot yet be established from molecular phylogenies, although the recognition of monophyletic Leucettida *sensu lato* in Calcinea may provide a starting point. Until a classification based on a better understanding of morphological character evolution is available, it appears crucial to include DNA data in any taxonomic study and to include all available taxa of the subclass of the target species. Future molecular phylogenetic studies should include many more species, but not only from the still unsampled families and genera. It would also be desirable to extend and test the results obtained from ribosomal RNA data with independent phylogenetic markers, such as mitochondrial genes.

With such additional data at hand, remaining questions will have to be addressed: In Calcinea, the validity of Leucettida *sensu lato* must be tested, and the relationships of the asconoid Clathrinida remain to be resolved. The positions of *Burtonulla*, a heterocoel genus of Levinellidae, and *Paramurrayona* (Murrayonida) with respect to Leucettida *sensu lato* should be determined. In Calcaronea, inclusion of members of the families Lepidoleuconidae and Trichogypsiidae is needed to further test the monophyly of Baerida (*sensu* Manuel *et al.*, 2003), and among Leucosolenida, the position and monophyly of the still unsampled Achramorphidae and Sycanthidae needs to be tested. Additional taxa are also required to shed light on the phylogenetic affinities of Heteropiidae, certain *Sycon* species, and Grantiidae of the genera *Ute*, *Synute*, and *Aphroceras*. In this context, the inclusion of Heteropiidae with giant cortical diactines would be especially interesting, as the resemblance in skeletal architecture between certain Heteropiidae and Grantiidae has been recognized before (Borojevic, 1965; Borojevic *et al.*, 2000, 2002). In Leucosolenida II (Voigt *et al.*, 2012b), the connections between species of Jenkinidae, Amphoriscidae, Grantiidae, Sycettidae, and Lelapiidae need to be clarified. Finally, the monophyly of Minchinellidae (Lithonida) needs to be tested.

8. THE EVOLUTION OF SPONGE DEVELOPMENT

With a phylogeny mostly based on molecular markers that are independent of morphological characters, it is now possible to map traits, trace their origin, and define shared ancestral features of Porifera and, more generally, Metazoa. In particular, the analysis of development in a phylogenetic framework may identify some of the key innovations that accompanied the origin of the Metazoa. The use of embryonic development to reconstruct early animal evolution dates back to Haeckel's Gastraea theory (Haeckel, 1874), which was largely inspired by embryonic and adult sponges. As multicellular animals evolved from a protist ancestor, cells had to acquire different identities, specialize in particular functions and become organized into tissues and organs to form a macroscopic, coordinated organism. Such crucial attributes of multicellularity evolved through the assembly of a primordial metazoan developmental program, which was then modified to produce the large diversity of body plans found across the Metazoa. By comparing embryonic development in the different branches of the metazoan tree, we can attempt to reconstruct the first animal developmental program and understand the core traits that underpin multicellularity in animals. In this endeavour, it is crucial to examine the arguably earliest-branching extant metazoan taxon—Porifera.

Animal embryonic development progresses through three major steps: (1) blastulation or cleavage—from the zygote, cell divisions produce a multicellular embryo of generally undifferentiated cells called blastomeres; (2) gastrulation—spatial redistribution and initial differentiation of the blastomeres delineate embryonic germ layers and symmetry; and (3) organogenesis—differentiation and patterning of the germ layers into organs and along one or two axes of symmetry. The reproductive process in the phylum Porifera shows astonishing complexity and diversity. Development in sponges seems to occur similarly to other metazoans, which can be illustrated by examining the model demosponge *Amphimedon queenslandica* (Haplosclerida) (Degnan *et al.*, 2009). After a period of cleavage, segregation of the primary cell layers (termed gastrulation), patterning along an anterior–posterior (AP) axis, and cell differentiation give rise to a typical parenchymella larva with an obvious axis of symmetry and at least eleven differentiated cell types, apparently organized into three concentric layers in *A. queenslandica* (Leys and Degnan, 2001; 2002). As an example of embryonic patterning, pigment cells scattered throughout the outer layer migrate to the posterior pole and are organized into a photosensory ring. The competent *A. queenslandica* larva responds to light and biochemical settlement cues, settles on its anterior end and, during metamorphosis, the aquiferous system of the juvenile sponge is formed. Despite the similarities between sponge and eumetazoan development, the extent to which the processes are homologous has been long debated—in particular, regarding gastrulation, germ layers, and symmetry. Sponges have long been interpreted as having no true tissues or organs and hence representing a primitive animal body plan.

While there are excellent recent reviews analyzing the large diversity of embryogenesis in sponges (Leys, 2004; Maldonado, 2004; Leys and Ereskovsky, 2006; Ereskovsky, 2010), our purpose here will be to focus on developmental traits that are informative in a phylogenetic framework in order to gain insight into the ancestral sponge developmental program. We will point out certain reproductive traits whose phylogenetic value has been revised, including the mode of reproduction and spermatozoon ultrastructure. Additionally, we will discuss other traits that are comparable to other animals and phylogenetically informative, including larval form and gastrulation, we will briefly review relevant molecular analyses of sponge embryogenesis, and we will discuss what these features can tell us about ancestral sponge development.

8.1. Differences in the mode of reproduction and spermatozoon ultrastructure are not synapomorphies of higher-level sponge clades

For many years, the externally developing oviparous condition versus the brooding viviparous condition was assumed to represent a strong phylogenetic signal. Therefore, by finding relative congruence between these reproductive features and some skeletal features, Lévi (1957, 1973) established the first modern taxonomic classification of Demospongiae, discriminating three large lineages: Homosclerophorida (or Homoscleromorpha, brooding sponges with minute tetractinal to diactinal spicules), Tetractinomorpha (with tetraxonic spicules and derived forms, without spongin, typically oviparous), and Ceractinomorpha (without tetraxonic spicules, with variable levels of spongin, typically viviparous). As previously discussed in this chapter, the advent of molecular methods has revealed that Tetractinomorpha and Ceractinomorpha are not monophyletic, suggesting that oviparity evolved independently multiple times from viviparous ancestors (Borchiellini *et al.*, 2004).

It was thought until recently that the absence of a "true acrosome" was the rule in sponge spermatozoa with the notable exception of the homoscleromorph sponges, which have rounded or C-shaped simple acrosomes (reviewed in Reiswig, 1983; Boury-Esnault and Jamieson, 1999; Riesgo and Maldonado, 2009). This feature has often been proposed to support a closer relationship between eumetazoans and homoscleromorph sponges relative to that of the other three major sponge lineages. However, it is not as phylogenetically informative as once thought. Indeed, a large conical acrosome has also been documented in the calcaronean *Sycon calcaravis* (Nakamura and Okada, 1998), and the most atypical and complex spermatozoon known in demosponges so far belongs to the poecilosclerid *Crambe crambe* (Riesgo and Maldonado, 2009). The elongated V-shaped spermatozoon has a sophisticated acrosomal complex with an associated organelle called a perforatorium, which is far more complex than homoscleromorph acrosomes. The prevailing idea that the organization of the spermatozoon would have increased in complexity in the animal lineage (e.g. Franzén, 1987; Reunov, 2001) has hence been disproved by the discovery of both "simple" and "complex" spermatozoa in Porifera. The absence of an acrosome in most sponges might be a derived condition related to particular mechanisms mediating the process of oocyte fertilization.

8.2. Diversity in sponge larval types: The parenchymella larva may be ancestral to Demospongiae

Sponge embryonic development typically gives rise to a larval stage, with up to eight major larval types clearly identified to date (see Fig. 1.9), in addition to three other described larvae that are difficult to categorize (e.g. Maldonado and Riesgo, 2009). These major larval types are defined according to not only differences in their final morphology and cytology but also a distinctive embryogenesis (reviewed in Maldonado and Bergquist, 2002;

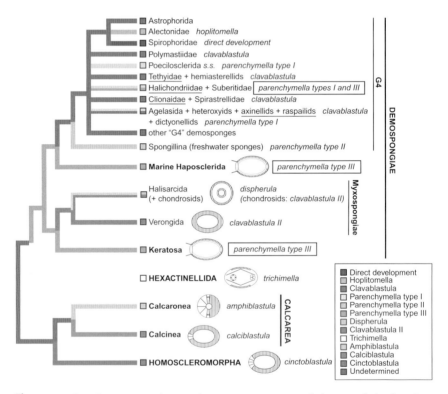

Figure 1.9 Larval types traced onto the consensus sponge phylogeny. Striped regions indicate ciliated cells in the larval schematics. Character states at ancestral nodes were reconstructed using Mesquite (Maddison and Maddison, 2011), and a parsimony criterion of optimality with multistate coding of all the larval types in a single character. Although they have the same name, clavablastulae from the G4 clade and those from the Myxospongiae clade were treated as separate characters (Myxospongiae: clavablastula II) as these hollow entirely ciliated larvae are most likely not homologous (Maldonado and Bergquist, 2002). The resulting tree features type III parenchymella as the ancestral demosponge larval type. Where more than one taxon is included at the tips of the branches, the group where the indicated larval type is found is underlined. (For colour version of this figure, the reader is referred to the Web version of this chapter.)

Maldonado, 2004; Ereskovsky, 2010). The larva of certain invertebrates with divergent adult body plans, such as echinoderms or ascidians, display a core set of fundamental animal synapomorphic traits and gene expression patterns that are lacking in the adult form. Similarly, although sponges have little in common with other animals as adults, their larvae are more readily comparable with eumetazoans. Hence, sponge larval development may be the only stage that is evolutionarily conserved with other animals, illustrating the importance of analyzing larval evolution in this group.

In hexactinellid sponges, an elongated and ciliated trichimella larva has been described that is highly differentiated along an AP axis including in its ciliation (Fig. 1.9). It contains a large syncytium formed by cell fusion (Okada, 1928; Boury-Esnault et al., 1999; Leys et al., 2006). The ovate and ciliated cinctoblastula larva of homoscleromorphs is in essence a monolayered epithelium differentiated into three distinct regions along an AP axis with at least five cell types (Boury-Esnault et al., 2003; Ereskovsky, 2010). In the subclass Calcinea of calcareous sponges, an ovate and ciliated coeloblastula (i.e. hollow) called a calciblastula consisting of one cell layer with one or two cell types is released into the water column (Minchin, 1900; Johnson, 1979; Amano and Hori, 2001; Maldonado and Bergquist, 2002). Calcaronean sponges (the other subclass of Calcarea) have an amphiblastula larva with anterior ciliated micromeres, posterior macromeres, and four "cellules en croix" (cross cells) that might be phototactic (Tuzet and Grassé, 1973; Franzen, 1988; Amano and Hori, 1992; Leys and Eerkes-Medrano, 2005).

In contrast to the other three major sponge clades, there is great larval diversity among demosponges, but most members of this group release a highly differentiated parenchymella larva (described above for A. queenslandica) (Harrison and De Vos, 1991; Maldonado and Bergquist, 2002; Maldonado and Riesgo, 2008a). The parenchymella type shows some morphological variability regarding ciliation and cytology, the phylogenetic significance of which remains unexplored. Some parenchymellae are entirely and homogeneously covered by equally long cilia or have a small region at the posterior pole devoid of cilia (herein considered to be type I). In freshwater sponges, the parenchymella contains a large cavity probably involved in osmoregulation (type II). Other parenchymellae have a bare posterior pole surrounded by a ring of pigmented cells with long cilia, which functions as an organ-like photoreceptory structure (as in A. queenslandica, type III).

Demosponge subclasses, as well as some orders and genera, appear to be paraphyletic, but four major clades have been detected: Keratosa (G1), Myxospongiae (G2), marine Haposclerida (G3), and a large unnamed clade termed G4, with G1 + G2 and G3 + G4 being relatively well supported (Borchiellini et al., 2004 and discussed earlier in this chapter). Type III parenchymella larvae are well documented in the Keratosa clade both among dictyoceratids and dendroceratids (Woollacott and Hadfield, 1989; Maldonado et al., 2003; Ereskovsky and Tokina, 2004; Mariani et al., 2005) and in the marine haplosclerid clade (including *A. queenslandica*) (Woollacott, 1993; Fromont, 1994; Maldonado and Young, 1999; Leys and Degnan, 2001; Mariani *et al.*, 2005; Fig. 1.9). In the Myxospongiae clade, both verongids and chondrosids have ciliated coeloblastula (hollow) larvae (termed clavablastulae) that develop externally (Usher and Ereskovsky, 2005; Maldonado, 2009; Fig. 1.9), while halisarcids release a ciliated dispherula larva (Lévi, 1956). Depending on the level of cell ingression into the blastocoel, the dispherula larva may be coeloblastula-like or parenchymella-like (Gonobobleva and Ereskovsky, 2004; Ereskovsky, 2010). In the large G4 clade, which includes many paraphyletic orders, poecilosclerids mainly have type I parenchymella (Bergquist *et al.*, 1970, 1977; Wapstra and van Soest, 1987; Mariani *et al.*, 2005), freshwater sponges have type II (Brien, 1973; Saller, 1988), and halichondrids have types I and III (Woollacott, 1990; Maldonado and Young, 1996; Fig. 1.9). Clavablastula and hoplitomella larvae and direct development are also found in this clade; these clavablastulae are unlikely to be homologous to those found in the Myxospongiae clade (Maldonado and Bergquist, 2002; Maldonado and Riesgo, 2008a). Thus, as the very distinctive type III parenchymella is definitely present in three of the four demosponge clades (and the Myxospongiae are sister to the Keratosa; Fig. 1.9), it is most parsimonious to propose that it is the ancestral form for Demospongiae, with other larval types derived from it (e.g. types I and II parenchymella, dispherula, clavablastula). The analysis shown in Fig. 1.9 suggests a tentative pattern of phylogenetic relationships for these larval forms and supports type III parenchymella as the ancestral demosponge larval type.

Parenchymella sub-epithelial layers have been described in dictyoceratids (Keratosa; e.g. Ereskovsky and Tokina, 2004), halichondrids (G4; Brien, 1973), freshwater sponges (G4; Brien, 1973), and marine haplosclerids (e.g. Woollacott, 1993; Leys and Degnan, 2001). In poecilosclerids (G4), three layers are described, but the intermediate layer is particularly wide (Boury-Esnault, 1976; Bergquist and Green, 1977). Thus, as it is present in the three demosponge clades with parenchymella larvae, it is likely that the intermediate layer was present in the ancestral parenchymella. As it arises long after gastrulation and a third cell layer is absent from other sponge classes, it is unlikely that the third layer is related to the mesoderm germ layer of bilaterians. It probably represents a patterning event that arose in this lineage.

8.3. Sponge gastrulation as the morphogenetic movements during embryogenesis

Gastrulation can be defined as the movement of cells in the embryo to form the primary germ layers (Brusca *et al.*, 1997). It occurs after cleavage and is a key step in development because the multicellular animal becomes organized into two or three cell layers and along one or two axes of symmetry. Eumetazoans become either "diploblastic", with two germ layers (ectoderm and endoderm), or "triploblastic", with three germ layers (ectoderm, mesoderm, endoderm). It is increasingly evident from molecular data that gastrulation, germ layer formation, and axial patterning were associated in the cnidarian–bilaterian ancestor (Lee *et al.*, 2006). As endoderm gives rise to the gut, gastrulation is also associated with gut formation. This part of the definition has made it problematic to define gastrulation in sponges or to agree on whether they undergo gastrulation at all (Rasmont, 1979; Ereskovsky, 2010) as sponges feed using a specialized aquiferous system with no known homology to the eumetazoan gut. Furthermore, there are two phases of reorganization of cell layers that have been documented in all sponge lineages except Hexactinellida: during embryogenesis and during metamorphosis. In the latter phase, an "inversion of germ layers" results in the formation of the aquiferous system, which is the analogue of a sponge "gut" (Amano and Hori, 1996; Leys and Degnan, 2002), and some authors argue that this is gastrulation (Brien, 1973; Simpson, 1984). Other authors, however, have described gastrulation as the earlier cellular movements that follow cleavage and result in the embryonic "germ" layers in certain sponges (Lévi, 1956; Efremova, 1997; Boury-Esnault *et al.*, 1999; Leys and Degnan, 2002; Maldonado and Bergquist, 2002; Leys, 2004; Maldonado, 2004). We favour the latter interpretation, based on the association of cell movements during embryogenesis with the formation of primary cell layers and axial patterning as well as developmental timing (Leys and Degnan, 2002; Maldonado, 2004). We do not, however, argue that gastrulation occurs during embryogenesis in every sponge lineage but rather that this was the case in the ancestral sponge, with some lineages possibly conserving this trait and modifications in other lineages.

In the context of the demosponge common ancestor having a type III parenchymella larva (a plausible possibility discussed above), formation of the primary cell layers would have likely occurred through the migration of cells resulting in an outer layer of micromeres and a central core of macromeres (Borojevic and Lévi, 1965; Leys and Degnan, 2002). It is unclear whether micromeres migrate outwards, macromeres migrate inwards, or both. In the hexactinellid sponge *Oopsacas minuta*, cellular reorganization occurs by cellular delamination—oriented unequal cleavage resulting in micromeres outside and macromeres inside—a gastrulation mode described in hydrozoans (Cnidaria) (Okada, 1928; Boury-Esnault *et al.*, 1999; Leys *et al.*, 2006). These are the strongest cases for gastrulation in sponges, as these processes occur at the end of cleavage and coincide with the formation of two cell layers and the appearance of polarity in the embryo. In the case of demosponges, molecular expression data from *A. queenslandica* provide additional evidence that this is true gastrulation (discussed below).

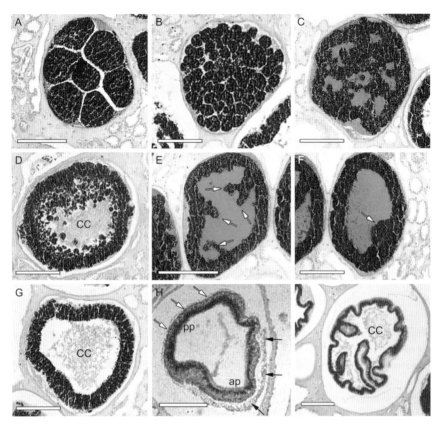

Figure 1.10 Blastomere reorganization during the gastrulation process of the homosclero-phorid *Corticium candelabrum*, through which the solid blastula becomes a hollow embryo by multipolar outward cell migration (i.e. multipolar egression). (cc) larval cavity, filled with symbiotic bacteria and collagen fibrils. (pp) posterior and (ap) anterior embryo pole, which will become the posterior and anterior larval pole, respectively, relative to the direction of swimming. Scale bars: 100 μm. Modified from Maldonado and Riesgo (2008b). (For colour version of this figure, the reader is referred to the Web version of this chapter.)

In homoscleromorphs, cells of the solid blastula migrate to the outer region of the embryo to form one cell layer during a unique process called multipolar egression (Boury-Esnault *et al.*, 2003; Maldonado and Riesgo, 2007; Fig. 1.10). These cell movements follow cleavage. For some authors, this process differs from gastrulation in that the resulting embryo apparently consists of one uniform cell layer and lacks polarity (Ereskovsky, 2010). However, this remarkable reorganization of the embryo marks the onset of polarization and regionalization processes in the embryo, suggesting that it is akin to gastrulation (Maldonado and Riesgo, 2007).

A highly unusual morphogenetic phenomenon occurs during the embryogenesis of calcaronean sponges when the coeloblastula with cilia facing inwards everts. This process maintains one cell layer and occurs at the end of cleavage when the embryo is already polarized (Franzen, 1988; Leys and Eerkes-Medrano, 2005). As neither of these types of morphogenetic movements have a parallel elsewhere in the Metazoa and/or result in two cell layers, it is difficult to equate them with gastrulation at this point in time.

The calcinean calciblastula larvae released from the adult sponge can be interpreted as the blastula stage with gastrulation occurring later because these larvae appear to be less differentiated than in other lineages (Maldonado, 2004). They are primarily composed of a uniform layer of ciliated cells. No morphogenetic movement occurs during embryogenesis, but cells ingress into the hollow larva while it is free-swimming before metamorphosis begins (Borojevic, 1969). This process is akin to eumetazoan gastrulation by multipolar ingression. Such putative gastrulation after larval release is reminiscent of the continuing "gastrulation" of the swimming planula larva of the cnidarian *Nematostella vectensis* (Magie *et al.*, 2007).

8.4. Molecular evidence for homology between sponge and eumetazoan development

Although sponge embryology has been studied since the nineteenth century, a concerted effort to understand the genetic mechanisms underlying sponge development, and hence to determine the homology or lack thereof of sponge and eumetazoan developmental mechanisms, has begun in detail only in the last decade. In particular, sequencing of the genome of the haplosclerid demosponge *A. queenslandica* and comparison with data from other sponges and early-branching phyla have enabled a large leap in our understanding of the nature of the ancestral metazoan genome (Srivastava *et al.*, 2010b).

Embryogenesis in well-studied bilaterian model organisms, such as vertebrates or *Drosophila*, is governed by a common set of genetic tools, primarily transcription factors and signalling pathways, which are found at all levels of the developmental program. Transcription factors directly switch genes on or off in a specific manner while signalling pathways transmit signals between cells. Comparative genomic analyses have shown that the large majority of gene classes encoding developmental proteins arose with animals (Larroux *et al.*, 2007, 2008; Simionato *et al.*, 2007; Gazave *et al.*, 2009; Richards and Degnan, 2009; Adamska *et al.*, 2010; Bridgham *et al.*, 2010; Srivastava *et al.*, 2010a,b). However, the *A. queenslandica* genome only has a fraction of the genes that are shared by most eumetazoans. This simpler genetic toolkit may represent secondary loss in this lineage or it may reflect a simpler developmental program in the animal ancestor. The presence of these animal developmental genes in sponges strongly supports homology between the embryogeneses of all metazoans.

Regardless, in order to determine the ancestral sponge developmental gene content, more data from all four classes are needed. Fortunately, ESTs from other sponges have already been sequenced (Nichols *et al.*, 2006; Gazave *et al.*, 2009; Labepie *et al.*, 2009; Philippe *et al.*, 2009; Harcet *et al.*, 2010b; Pick *et al.*, 2010).

The expression of transcription factors and signalling pathway components in *A. queenslandica* embryogenesis suggests that sponge and eumetazoan development are homologous. These genetic tools appear to be used in a similar manner in this sponge as they are in other animals. In some cases, conservation of gene function between sponges and bilaterians (*Drosophila* or vertebrate) has been shown (Coutinho *et al.*, 2003; Richards *et al.*, 2008; Bridgham *et al.*, 2010; Hill *et al.*, 2010). The localized expression of the signalling molecule *WntA* just prior to the segregation of cell layers in *A. queenslandica* suggests that early morphogenetic movements in demosponges may be homologous to eumetazoan gastrulation (Adamska *et al.*, 2007a). Expression patterns of *WntA* and *TGF-beta* suggest a role for these signalling ligands in axial patterning during gastrulation (Adamska *et al.*, 2007a, 2011). This proposed role, shared with cnidarians and bilaterians, awaits confirmation by functional gene studies but does suggest that the primary body axes of sponge and eumetazoan larvae are homologous (Adamska *et al.*, 2007a, 2011). Similarly, the canonical Wnt signalling pathway as well as the TGF-beta and Hedgehog-like pathways appear to pattern the photosensory ring, the only organ-like structure in the larva (Adamska *et al.*, 2007a,b, 2010, 2011). Expression analyses in the homoscleromorph sponge *Oscarella lobularis* suggest that the Wnt signalling pathway also has a conserved function in metazoan epithelial patterning and morphogenesis (Labepie *et al.*, 2009; see also Windsor and Leys, 2010). The Notch pathway also seems to fulfil a similar role in *A. queenslandica* as it does in eumetazoans, determining different cell fates of daughter cells during cell division (Richards *et al.*, 2008). The expression of transcription factors in certain cell lineages suggests that they contribute to the gene regulatory networks that govern cell fate determination and differentiation, as they do in eumetazoans (e.g. Larroux *et al.*, 2006; Fahey *et al.*, 2008; Gauthier and Degnan, 2008; Richards *et al.*, 2008; Bridgham *et al.*, 2010; Holstien *et al.*, 2010; Srivastava *et al.*, 2010a; Larroux, unpublished data).

These data come with certain caveats. It is often difficult to make sense of expression data because we know little about the functions of different larval cells and have no embryonic cell lineage data. Additionally, there have been no studies demonstrating the function of genes in sponge embryogenesis. However, advances with pharmacological disruption of signalling pathways and RNA inhibition in sponge adults and juveniles (Lapébie *et al.*, 2009; Windsor and Leys, 2010; Rivera *et al.*, 2011) are promising and suggest we may have success in applying these tools to the study of sponge embryogenesis in the near future.

8.5. Ancestral sponge development

Despite the important differences between the embryogeneses of different sponge groups, development in the Porifera essentially follows similar steps to those in eumetazoans and can thus point us towards a reconstruction of ancestral animal development. Larval types and "gastrulation" modes vary greatly between Hexactinellida, the two subclasses of Calcarea, Homoscleromorpha, and Demospongiae. Within the first four clades, however, different species seem to develop largely in the same manner (although data are limited to one of the five orders for Hexactinellida). In contrast, there is a great deal of diversity within the Demospongiae, but we proposed in the second section that parenchymella-type development could be ancestral. Hence, by comparing hexactinellids, calcaroneans, calcineans, homoscleromorphs, and demosponges with parenchymellae, we can propose some hypotheses regarding development in the poriferan common ancestor.

Cell movement in development is by no means exclusive to Metazoa and does not entail homology of animal developmental traits. For example, *Volvox* spp., multicellular algae, "gastrulate" by inverting their cell layer using cytoplasmic bridges (Viamontes and Kirk, 1977), a process resembling the inversion of calcaronean sponges. Nonetheless, gastrulation is a central step in eumetazoan embryogenesis. The debate regarding gastrulation in sponges must be considered within the context of a sponge developmental program that incorporates a number of eumetazoan attributes, and it thus seems most parsimonious to infer homology of gastrulation across Metazoa. While there is some evidence for gastrulation in demosponges and hexactinellids, the homology of morphogenetic movements in homoscleromorph and calcareous sponges with gastrulation remains a matter of discussion. Determining whether sponges truly gastrulate awaits further testing and molecular data. It is worth noting, however, that in contrast to the low diversity in the modes of gastrulation in bilaterians, which mainly gastrulate by invagination, cnidarians display a large variety of gastrulation modes, some of which are unique to the phylum (Byrum and Martindale, 2004). This could also be the case in sponges, which have had a longer time to evolve than cnidarians, and inversion and multipolar egression may one day be accepted in textbooks as modes of gastrulation. It could also be that more plastic embryogenesis in sponges (with less developmental constraints than other animals) enabled certain lineages to lose the gastrulation step or shift its timing. If sponge embryonic cell movements are revealed to be homologous to eumetazoan gastrulation, the ancestral poriferan and metazoan mode of gastrulation would have probably been through cell migration rather than invagination, based on sponge and cnidarian gastrulation (Price and Patel, 2004). Furthermore, as the process of gastrulation is intimately linked to the formation of germ layers in eumetazoans, demosponge, and hexactinellid germ layers would likely correspond to endoderm and ectoderm.

If homoscleromorph and calcareous sponges are sister groups, as suggested by the molecular phylogenies discussed above, it is interesting to note that the larvae from both of these groups are hollow and single layered, although their development and state of differentiation are quite different. Likewise, in Silicea *sensu stricto* (the sister group of the Calcarea + Homoscleromorpha clade, see above), the hexactinellid- and parenchymella-type demosponge larvae are similar because they are solid and highly differentiated. The trichimella of hexactinellids, the parenchymella proposed to be ancestral to Demospongiae, the cinctoblastula of homoscleromorphs, and the calciblastula or amphiblastula of calcareous sponges are all non-feeding (i.e. lecithotrophic) and ciliated larvae with a clear AP axis. Along with the similar nature of the planula larva of cnidarians (although the planulae of some anthozoans are planktotrophic), it is most parsimonious to postulate that both the poriferan and metazoan common ancestors had such a larva in their life history.

The multiplication of sponge developmental models, with for example *Oscarella* (Homoscleromorpha) (Nichols *et al.*, 2006; Ereskovsky *et al.*, 2009), *Sycon* (Calcarea) (Manuel and Le Parco, 2000; Adamska *et al.*, 2011), and *Ephydatia* (Demospongiae) (Elliott and Leys, 2003; Funayama *et al.*, 2005) species, promises to advance our understanding of ancestral sponge embryogenesis. Although we have not discussed it in this review, most of the molecular research on sponge developmental mechanisms nowadays is actually undertaken on sponge adults, juveniles, or cell culture (e.g. Adell *et al.*, 2003; Perovic *et al.*, 2003; Funayama *et al.*, 2005, 2010; Gazave *et al.*, 2008; Wiens *et al.*, 2008; Labepie *et al.*, 2009; Windsor and Leys, 2010). This body of research has considerably advanced our understanding of sponge and ancestral metazoan development. In conjunction, we propose that efforts to study the molecular basis of sponge embryogenesis should be renewed in order to make significant progress towards understanding the fundamental characters of sponge and animal development.

9. CONCLUSIONS AND OUTLOOK

Based on the discussion in this chapter, it is clear that deep sponge phylogenetics has come a long way in recent years. Large-scale phylogenomic analyses have so far rejected the hypothesis that sponges are paraphyletic; instead, several studies are consistent with the notion of monophyletic Porifera. It has also become clear from evolutionary developmental studies of sponges that sponge larvae share traits and complexity with eumetazoans and that the simple sedentary adult lifestyle of sponges probably reflects some degree of secondary simplification.

An unexpected sister-group relationship between the former demosponge group Homoscleromorpha—now considered the fourth extant sponge class—and the Calcarea within monophyletic Porifera has been suggested in a few studies. Although this relationship has not yet received unequivocal support, and clear morphological synapomorphies remain to be identified, this would shed some new light on the evolution of some of the key traits of sponges as well as on the early evolution of the Metazoa. Type-IV collagen, previously only thought to occur in Homoscleromorpha and Eumetazoa, now appears to be plesiomorphic for the Metazoa because it has recently been found in Calcarea and Demospongiae too (its presence in the Hexactinellida remains to be detected). Monophyletic sponges with a Calcarea + Homoscleromorpha clade would either suggest that the basement membrane is also plesiomorphic for the Metazoa and is now found in the Homoscleromorpha and Eumetazoa but lost from the other sponge lineages or that it convergently evolved in Homoscleromorpha and Eumetazoa. In either case, a basement membrane would not be synapomorphic for an "Epitheliozoa" clade (Homoscleromorpha + Placozoa + Eumetazoa). A Calcarea + Homoscleromorpha clade also has important implications for the evolution of spiculogenesis in sponges. It would either imply that silica spiculogenesis is plesiomorphic for Porifera and was lost in Calcarea or that it evolved several times independently in sponges (see also Maldonado and Riesgo, 2007). Both Demospongiae and Hexactinellida produce their spicules around an axial filament, which in demosponges contains silicatein. However, while silicatein was apparently characterized in a single hexactinellid species, *Crateromorpha meyeri* (Müller *et al.*, 2008), other studies have failed to demonstrate that classical silicateins are ubiquitously involved in spiculogenesis in other hexactinellids (Ehrlich *et al.*, 2010; Veremeichik *et al.*, 2011). Clearly, more work is needed, but the results have so far called into question the homology of spiculogenesis in Silicea *sensu stricto*. Additionally, the Homoscleromorpha appear to secrete their silica spicules differently than Demospongiae, but their spiculogenesis awaits more detailed study.

While most of the higher-level relationships in Demospongiae appear resolved and corroborated by independent molecular markers, the "mixed-bag" "G4" clade still represents a serious challenge, as many relationships within this clade await robust resolution (but see Morrow *et al.*, 2012). Higher-level relationships in Hexactinellida appear largely congruent with previous morphological systematics, but some critical taxa (such as Lychniscosida and Aulocalycidae) await to be included in molecular studies. The Homoscleromorpha are clearly distinct from the demosponges, and their internal phylogeny is largely resolved, although taxon sampling could be improved. The phylogeny of Calcarea remains largely unresolved because molecular phylogenies are highly incongruent with the taxonomic system based on morphological characters. Here, probably the most work is needed to fully understand the basis for this incongruence. Calcarea are also among the few non-bilaterian taxa where no complete mitochondrial genome has yet been sequenced.

As discussed above, we have made great progress in deep sponge phylogenetics, but we still have a long way to go to achieve a comprehensive understanding of the relationships among and within the main sponge lineages, which will be crucial to fully appreciate the evolution of this extraordinary metazoan phylum.

ACKNOWLEDGEMENTS

G.W. acknowledges funding from the German Science Foundation (DFG, Projects Wo896/ 3, 5, 6, 7, 9), partly through the Priority Program 1174 "Deep Metazoan Phylogeny". M.D. acknowledges the Smithsonian Institution for a Postdoctoral Fellowship. M.M. acknowledges funding from the Spanish Ministry for Innovation and Science (BFU2008-00227/ BMC). C.L. was supported by a Humboldt Postdoctoral Research Fellowship. We would like to thank Gonzalo Giribet and an anonymous reviewer for their valuable comments on an earlier version of the manuscript.

REFERENCES

Adams, C. L., McInerney, J. O., and Kelly, M. (1999). Indications of relationships between poriferan classes using full-length 18S rRNA gene sequences. *Memoirs of the Queensland Museum* **44**, 33–44.
Adamska, M., Degnan, S. M., Green, K. M., Adamski, M., Craigie, A., Larroux, C., and Degnan, B. M. (2007a). Wnt and TGF-b expression in the sponge *Amphimedon queenslandica* and the origin of metazoan embryonic patterning. *PLoS One* **2**, e1031.
Adamska, M., Matus, D. Q., Adamski, M., Green, K., Rokhsar, D. S., Martindale, M. Q., and Degnan, B. M. (2007b). The evolutionary origin of hedgehog proteins. *Current Biology* **17**, R836–R837.
Adamska, M., Larroux, C., Adamski, M., Green, K., Lovas, E., Koop, D., Richards, G. S., Zwafink, C., and Degnan, B. M. (2010). Structure and expression of conserved Wnt pathway components in the demosponge *Amphimedon queenslandica*. *Evolution & Development* **12**, 494–518.
Adamska, M., Degnan, B. M., Green, K., and Zwafink, C. (2011). What sponges can tell us about the evolution of developmental processes. *Zoology* **114**, 1–10.
Addis, J. S., and Peterson, K. J. (2005). Phylogenetic relationships of freshwater sponges (Porifera, Spongillina) inferred from analyses of 18S rDNA, COI mtDNA, and ITS2 rDNA sequences. *Zoologica Scripta* **34**, 549–557.
Adell, T., Nefkens, I., and Müller, W. E. G. (2003). Polarity factor 'Frizzled' in the demosponge *Suberites domuncula*: Identification, expression and localization of the receptor in the epithelium/pinacoderm. *FEBS Letters* **554**, 363–368.
Alvarez, B., and Hooper, J. N. A. (2010). Taxonomic revision of the order Halichondrida (Porifera: Demospongiae) of northern Australia. Family Dictyonellidae. *The Beagle* **26**, 13–36.
Alvarez, B., van Soest, R. W. M., and Rützler, K. (1998). A revision of Axinellidae (Porifera: Demospongiae) of the Central West Atlanic Region. *Smithsonian Contributions to Zoology* **598**, 1–47.
Alvarez, B., Crisp, M. D., Driver, F., Hooper, J. N. A., and Van Soest, R. W. M. (2000). Phylogenetic relationships of the family Axinellidae (Porifera: Demospongiae) using morphological and molecular data. *Zoologica Scripta* **29**, 169–198.

Alvarez, B., Krishnan, M., and Gibb, K. (2007). Analysis of intragenomic variation of the rDNA internal transcribed spacers (ITS) in Halichondrida (Porifera: Demospongiae). *Journal of the Marine Biological Association of the United Kingdom* **87**, 1599–1605.

Amano, S., and Hori, I. (1992). Metamorphosis of calcareous sponges. 1. Ultrastructure of free-swimming larvae. *Invertebrate Reproduction and Development* **21**, 81–90.

Amano, S., and Hori, I. (1996). Transdifferentiation of larval flagellated cells to choanocytes in the metamorphosis of the demosponge *Haliclona permollis*. *The Biological Bulletin* **190**, 161–172.

Amano, S., and Hori, I. (2001). Metamorphosis of coeloblastula performed by multipotential larval flagellated cells in the calcareous sponge *Leucosolenia laxa*. *The Biological Bulletin* **200**, 20–32.

Anderson, S., Bankier, A., Barrell, B., de Bruijn, M., Coulson, A., Drouin, J., Eperon, I., Nierlich, D., Roe, B., Sanger, F., Schreier, P., Smith, A., Staden, R., and Young, I. (1981). Sequence and organization of the human mitochondrial genome. *Nature* **290**, 457–465.

Aouacheria, A., Geourjon, C., Aghajari, N., Navratil, V., Deléage, G., Lethias, C., and Exposito, J.-Y. (2006). Insights into early extracellular matrix evolution: Spongin short chain collagen-related proteins are homologous to basement membrane type IV collagens and form a novel family widely distributed in invertebrates. *Molecular Biology and Evolution* **23**, 2288–2302.

Ax, P. (1996). Multicellular animals: A new approach to the phylogenetic order in nature. (vol. 1). Springer Verlag, Berlin.

Bakran-Petricioli, T., Vacelet, J., Zibrowius, H., Petricioli, D., Chevaldonné, P., and Rada, T. (2007). New data on the distribution of the 'deep-sea' sponges *Asbestopluma hypogea* and *Oopsacas minuta* in the Mediterranean Sea. *Marine Ecology* **28**, 10–23.

Barucca, M., Azzini, F., Bavestrello, G., Biscotti, M. A., Calcinai, B., Canapa, A., Cerrano, C., and Olmo, E. (2007). The systematic position of some boring sponges (Demospongiae, Hadromerida) studied by molecular analysis. *Marine Biology* **151**, 529–535.

Belinky, F., Rot, C., Ilan, M., and Huchon, D. (2008). The complete mitochondrial genome of the demosponge *Negombata magnifica* (Poecilosclerida). *Molecular Phylogenetics and Evolution* **47**, 1238–1243.

Bell, J. J. (2008). The functional roles of marine sponges. *Estuarine, Coastal and Shelf Science* **79**, 341–353.

Bergquist, P. R. (1980). A revision of the supraspecific classification of the orders Dictyo-ceratida, Dendroceratida, and Verongida (class Demospongiae). *New Zealand Journal of Zoology* **7**, 443–503.

Bergquist, P. R., and Cook, S. d. C. (2002a). Family Aplysinidae Carter, 1875. In "Systema Porifera. A Guide to the Classification of Sponges" (J. N. A. Hooper and R. W. M. Van Soest, eds), pp. 1082–1085. Kluwer Academic/Plenum Publishers, New York, Boston, Dordrecht, London, Moscow.

Bergquist, P. R., and Cook, S. d. C. (2002b). Order Dendroceratida Minchin, 1900. In "Systema Porifera. A Guide to the Classification of Sponges" (J. N. A. Hooper and R. W. M. Van Soest, eds), p. 1067. Kluwer Academic/Plenum Publishers, New York, Boston, Dordrecht, London, Moscow.

Bergquist, P. R., and Cook, S. d. C. (2002c). Order Halisarcida Bergquist, 1996. In "Systema Porifera. A Guide to the Classification of Sponges" (J. N. A. Hooper and R. W. M. Van Soest, eds), p. 1077. Kluwer Academic/Plenum Publishers, New York, Boston, Dordrecht, London, Moscow.

Bergquist, P. R., and Cook, S. d. C. (2002d). Order Verongida Bergquist, 1978. In "Systema Porifera. A Guide to the Classification of Sponges" (J. N. A. Hooper and R. W. M. Van Soest, eds), p. 1081. Kluwer Academic/Plenum Publishers, New York, Boston, Dordrecht, London, Moscow.

Bergquist, P. R., and Green, C. R. (1977). Ultrastructural-study of settlement and metamorphosis in sponge larvae. *Cahiers De Biologie Marine* **18,** 289–302.

Bergquist, P. R., Sinclair, M. E., and Hogg, J. J. (1970). Adaptation ot intertidal existence: Reproductive cycles and larval behaviours in Demospongiae. *Symposia of the Zoological Society of London* **25,** 247–271.

Bergquist, P. R., Green, C. R., Sinclair, M. E., and Roberts, H. S. (1977). Morphology of cilia in sponge larvae. *Tissue & Cell* **9,** 179–184.

Bidder, G. P. (1898). The skeleton and classification of calcareous sponges. *Proceedings of the Royal Society of London* **6,** 61–76.

Blanquer, A., and Uriz, M. J. (2008). 'A posteriori' searching for phenotypic characters to describe new cryptic species of sponges revealed by molecular markers (Dictyonellidae: *Scopalina*). *Invertebrate Systematics* **22,** 489–502.

Blanquer, A., and Uriz, M. J. (2010). Population genetics at three spatial scales of a rare sponge living in fragmented habitats. *BMC Evolutionary Biology* **10,** 13.

Blanquer, A., Uriz, M. J., and Pascual, M. (2005). Polymorphic microsatellite loci isolated from the marine sponge *Scopalina lophyropoda* (Demospongiae: Halichondrida). *Molecular Ecology Notes* **5,** 466–468.

Böger, H. (1983). Versuch über das phylogenetische System der Porifera. *Meyniana* **40,** 143–154.

Borchiellini, C., Boury-Esnault, N., Vacelet, J., and Le Parco, Y. (1998). Phylogenetic analysis of the *Hsp70* sequences reveals the monophyly of Metazoa and specific phylogenetic relationships between animals and fungi. *Molecular Biology and Evolution* **15,** 647–655.

Borchiellini, C., Manuel, M., Alivon, E., Boury-Esnault, N., Vacelet, J., and Le Parco, Y. (2001). Sponge paraphyly and the origin of Metazoa. *Journal of Evolutionary Biology* **14,** 171–179.

Borchiellini, C., Chombard, C., Manuel, M., Alivon, E., Vacelet, J., and Boury-Esnault, N. (2004). Molecular phylogeny of Demospongiae: Implications for classification and scenarios of character evolution. *Molecular Phylogenetics and Evolution* **32,** 823–837.

Borojevic, R. (1965). Éponges Calcaires des Côtes de France I. *Amphiute paulini* Hanitsch: les generes *Amphiute* Hanitsch et *Paraheteropia* n. gen. *Archives De Zoologie Experimentale Et Generale* **106,** 665–670.

Borojevic, R. (1969). Étude de développement et de la différenciation celullaire d'éponges calcaires calcinéennes (genre *Clathrina* et *Ascandra*). *Annales D'embryologie et de Morphogénese* **2,** 15–36.

Borojevic, R., and Lévi, C. (1965). Morphogenèse expérimentale d'une éponge à partir des cellules de la larve nageante dissociée. *Zeitschrift für Zellforschung* **68,** 57–69.

Borojevic, R., Boury-Esnault, N., and Vacelet, J. (1990). A revision of the supraspecific classification of the subclass Calcinea (Porifera, class Calcarea). *Bulletin du Museum National d'Histoire Naturelle, Section A, Zoologie Biologie et Ecologie Animales* **12,** 243–276.

Borojevic, R., Boury-Esnault, N., and Vacelet, J. (2000). A revision of the supraspecific classification of the subclass Calcaronea (Porifera, class Calcarea). *Zoosystema* **22,** 203–263.

Borojevic, R., Boury-Esnault, N., Manuel, M., and Vacelet, J. (2002). Order Leucosolenida Hartman, 1958. In "Systema Porifera. A Guide to the Classification of Sponges" (J. N. A. Hooper and R. W. M. Van Soest, eds), pp. 1157–1184. Kluwer Academic/ Plenum Publishers, New York, Boston, Dordrecht, London, Moscow.

Boury-Esnault, N. (1976). Ultrastructure de la larve parenchymella d'*Hamigera hamigera* (Schmidt) (Demosponge, Poecilosclerida). Origine des cellules grises. *Cahiers De Biologie Marine* **17,** 9–20.

Boury-Esnault, N. (2002). Order Chondrosida Boury-Esnault & Lopes, 1985. Family Chondrillidae Gray, 1872. In "Systema Porifera. A Guide to the Classification of Sponges" (J. N. A. Hooper and R. W. M. Van Soest, eds), pp. 291–298. Kluwer Academic/Plenum Publishers, New York, Boston, Dordrecht, London, Moscow.

Boury-Esnault, N. (2006). Systematics and evolution of Demospongiae. *Canadian Journal of Zoology-Revue Canadienne De Zoologie* **84,** 205–224.

Boury-Esnault, N., and Jamieson, B. (1999). Porifera. In "Reproductive biology of invertebrates. Progress in male gamete ultrastructure and phylogeny" (K. G. Adiyodi and R. G. Adiyodi, eds), pp. 1–20. John Wiley and Sons, Chichester.

Boury-Esnault, N., Muricy, G., Gallissian, M.-F., and Vacelet, J. (1995). Sponges without skeleton: A new Mediterranean genus of Homoscleromorpha (Porifera, Demospongiae). *Ophelia* **43,** 25–43.

Boury-Esnault, N., Efremova, S., Bezac, C., and Vacelet, J. (1999). Reproduction of a hexactinellid sponge: First description of gastrulation by cellular delamination in the Porifera. *Invertebrate Reproduction and Development* **35,** 187–201.

Boury-Esnault, N., Ereskovsky, A., Bezac, C., and Tokina, D. (2003). Larval development in the Homoscleromorpha (Porifera, Demospongiae). *Invertebrate Biology* **122,** 187–202.

Boute, N., Exposito, J. Y., Boury-Esnault, N., Vacelet, J., Noro, N., Miyazaki, K., Yoshizato, K., and Garrone, R. (1996). Type IV collagen in sponges, the missing link in basement membrane ubiquity. *Biology of the Cell* **88,** 37–44.

Bridgham, J. T., Eick, G. N., Larroux, C., Deshpande, K., Harms, M. J., Gauthier, M. E. A., Ortlund, E. A., Degnan, B. M., and Thornton, J. W. (2010). Protein evolution by molecular tinkering: Diversification of the nuclear receptor superfamily from a ligand-dependent ancestor. *PLoS Biology* **8,** e1000497.

Brien, P. (1973). Les démosponges. In "Traité de Zoologie" (P.-P. Grassé, ed.), pp. 133–461. Masson, Paris.

Brusca, R. C., and Brusca, G. J. (2003). Invertebrates. Sinauer Associates, Sunderland, MA.

Brusca, G. J., Brusca, R. C., and Gilbert, S. F. (1997). Characteristics of metazoan development. In "Embryology, Constructing the Organism" (S. F. Gilbert and A. M. Raunio, eds). Sinauer Associates, Sunderland, MA.

Byrum, C. A., and Martindale, M. Q. (2004). Gastrulation in the Cnidaria and Ctenophora. In "Gastrulation: From Cells to Embryos" (C. D. Stern, ed.), pp. 33–50. Cold Spring Harbor Laboratory Press, Cold Spring Harbor, NY.

Cárdenas, P., Xavier, J., Tendal, O. S., Schander, C., and Rapp, H. T. (2007). Redescription and resurrection of *Pachymatisma normani* (Demospongiae: Geodiidae), with remarks on the genus *Pachymatisma*. *Journal of the Marine Biological Association of the United Kingdom* **87,** 1511–1525.

Cárdenas, P., Rapp, H. T., Schander, C., and Tendal, O. (2009). Molecular taxonomy and phylogeny of the Geodiidae (Porifera, Demospongiae, Astrophorida)—Combining phylogenetic and Linnaean classification. *Zoologica Scripta* 1–18.

Cárdenas, P., Xavier, J., Reveillaud, J., Schander, C., and Rapp, H. (2011). Molecular phylogeny of the Astrophorida (Porifera, Demospongiae) reveals an unexpected high level of spicule homoplasy. *PLoS One* **6,** e18318.

Carr, M., Leadbeater, B. S. C., Hassan, R., Nelson, M., and Baldauf, S. L. (2008). Molecular phylogeny of choanoflagellates, the sister group to Metazoa. *Proceedings of the National Academy of Sciences of the United States of America* **105,** 16641–16646.

Cavalier-Smith, T., Allsopp, M. T. E. P., Chao, E. E., Boury-Esnault, N., and Vacelet, J. (1996). Sponge phylogeny, animal monophyly, and the origin of the nervous system: 18S rRNA evidence. *Canadian Journal of Zoology* **74,** 2031–2045.

Chombard, C., and Boury-Esnault, N. (1999). Good Congruence between morphology and molecular phylogeny of Hadromerida, or how to bother sponge taxonomists. *Memoirs of the Queensland Museum* **44,** 100.

Chombard, C., Boury-Esnault, N., Tillier, A., and Vacelet, J. (1997). Polyphyly of "sclerosponges" (Porifera, Demospongiae) supported by 28S ribosomal sequences. *The Biological Bulletin* **193,** 359–367.

Chombard, C., Boury-Esnault, N., and Tillier, S. (1998). Reassessment of homology of morphological characters in tetractinellid sponges based on molecular data. *Systematic Biology* **47**, 351–366.

Collins, A. G. (1998). Evaluating multiple alternative hypotheses for the origin of Bilateria: An analysis of 18S rRNA molecular evidence. *Proceedings of the National Academy of Sciences of the United States of America* **95**, 15458–15463.

Conway, K. W., Krautter, M., Barrie, J. V., and Neuweiler, M. (2001). Hexactinellid sponge reefs on the Canadian continental shelf: A unique 'living fossil'. *Geoscience Canada* **28**, 71–78.

Cook, S. d. C., and Bergquist, P. R. (2002). Order Dictyoceratida Minchin, 1900. In "Systema Porifera. A Guide to the Classification of Sponges" (J. N. A. Hooper and R. W. M. Van Soest, eds), p. 1021. Kluwer Academic/Plenum Publishers, New York, Boston, Dordrecht, London, Moscow.

Cook, S. E., Conway, K. W., and Burd, B. (2008). Status of the glass sponge reefs in the Georgia Basin. *Marine Environmental Research* **66**, S80–S86.

Coutinho, C. C., Fonseca, R. N., Mansure, J. J. C., and Borojevic, R. (2003). Early steps in the evolution of multicellularity: Deep structural and functional homologies among homeobox genes in sponges and higher metazoans. *Mechanisms of Development* **120**, 429–440.

Degnan, B. M., Adamska, M., Craigie, A., Degnan, S. M., Fahey, B., Gauthier, M., Hooper, J. N. A., Larroux, C., Leys, S. P., Lovas, E., and Richards, G. S. (2009). The demosponge *Amphimedon queenslandica*: Reconstructing the ancestral metazoan genome and deciphering the origin of animal multicellularity. In "Emerging Model Organisms" (D. A. Crotty and A. Gann, eds), pp. 139–165. Cold Spring Harbor Laboratory Press, Cold Spring Harbor, NY.

Diaz, M. C., and van Soest, R. W. M. (1994). The Plakinidae: A systematic review. In "Sponges in Time and Space; Biology, Chemistry, Paleontology" (R. W. M. van Soest, T. G. van Kempen and J. C. Braekman, eds), pp. 93–110. A.A. Balkema, Rotterdam.

Dobzhansky, T. (1973). Nothing in biology makes sense except in the light of evolution. *The American Biology Teacher* **35**, 125–129.

Dohrmann, M., Voigt, O., Erpenbeck, D., and Wörheide, G. (2006). Non-monophyly of most supraspecific taxa of calcareous sponges (Porifera, Calcarea) revealed by increased taxon sampling and partitioned Bayesian analysis of ribosomal DNA. *Molecular Phylogenetics and Evolution* **40**, 830–843.

Dohrmann, M., Janussen, D., Reitner, J., Collins, A. G., and Wörheide, G. (2008). Phylogeny and evolution of glass sponges (Porifera, Hexactinellida). *Systematic Biology* **57**, 388–405.

Dohrmann, M., Collins, A. G., and Wörheide, G. (2009). New insights into the phylogeny of glass sponges (Porifera, Hexactinellida): Monophyly of Lyssacinosida and Euplectellinae, and the phylogenetic position of Euretidae. *Molecular Phylogenetics and Evolution* **52**, 257–262.

Dohrmann, M., Göcke, C., Janussen, D., Reitner, J., Lüter, C., and Wörheide, G. (2011). Systematics and spicule evolution in dictyonal sponges (Hexactinellida: Sceptrulophora) with description of two new species. *Zoological Journal of the Linnean Society* **163**, 1003–1025.

Dohrmann, M., Haen, K. M., Lavrov, D. V., and Wörheide, G. (2012). Molecular phylogeny of glass sponges (Porifera, Hexactinellida): Increased taxon sampling and inclusion of the mitochondrial protein-coding gene, cytochrome oxidase subunit I. *Hydrobiologia* **687**, 11–20.

Dunn, C. W., Hejnol, A., Matus, D. Q., Pang, K., Browne, W. E., Smith, S. A., Seaver, E., Rouse, G. W., Obst, M., Edgecombe, G. D., Sørensen, M. V., Haddock, S. H. D., Schmidt-Rhaesa, A., Okusu, A., Kristensen, R. M., Wheeler, W. C., Martindale, M. Q., and Giribet, G. (2008). Broad phylogenomic sampling improves resolution of the animal tree of life. *Nature* **452**, 745–749.

Edgecombe, G. D., Giribet, G., Dunn, C. W., Hejnol, A., Kristensen, R. M., Neves, R. C., Rouse, G. W., Worsaae, K., and Sørensen, M. V. (2011). Higher-level metazoan relationships: Recent progress and remaining questions. *Organisms, Diversity and Evolution* **11**, 151–172.

Efremova, S. M. (1997). Once more on the position among the Metazoa- gastrulation and grminal layers of sponges. *Berliner Geowissenschaftliche Abhandlungen* **20**, 7–15.

Ehrlich, H., Deutzmann, R., Brunner, E., Cappellini, E., Koon, H., Solazzo, C., Yang, Y., Ashford, D., Thomas-Oates, J., Lubeck, M., Baessmann, C., Langrock, T., Hoffmann, R., Wörheide, G., Reitner, J., Simon, P., Tsurkan, M., Ereskovsky, A. V., Kurek, D., Bazhenov, V. V., Hunoldt, S., Mertig, M., Vyalikh, D. V., Molodtsov, S. L., Kummer, K., Worch, H., Smetacek, V., and Collins, M. J. (2010). Mineralization of the metre-long biosilica structures of glass sponges is templated on hydroxylated collagen. *Nature Chemistry* **2**, 1084–1088.

Eisen, J. A., and Fraser, C. M. (2003). Phylogenomics: intersection of evolution and genomics. *Science* **300**, 1706–1707.

Elliott, G. R. D., and Leys, S. P. (2003). Sponge coughing: Stimulated contractions in a juvenile freshwater sponge, *Ephydatia muelleri*. *Integrative and Comparative Biology* **43**, 817.

Ereskovsky, A. (2010). The Comparative Embryology of Sponges. Springer, Dordrecht, The Netherlands.

Ereskovsky, A. V., and Tokina, D. B. (2004). Morphology and fine structure of the swimming larvae of *Ircinia oros* (Porifera, Demospongiae, Dictyoceratida). *Invertebrate Reproduction and Development* **45**, 137–150.

Ereskovsky, A. V., and Willenz, P. (2008). Larval development in *Guancha arnesenae* (Porifera, Calcispongiae, Calcinea). *Zoomorphology* **127**, 175–187.

Ereskovsky, A. V., Borchiellini, C., Gazave, E., Ivanisevic, J., Lapébie, P., Perez, T., Renard, E., and Vacelet, J. (2009). The Homoscleromorph sponge *Oscarella lobularis*, a promising sponge model in evolutionary and developmental biology. *BioEssays* **31**, 89–97.

Ereskovsky, A., Lavrov, D., Boury-Esnault, N., and Vacelet, J. (2011). Molecular and morphological description of a new species of *Halisarca* (Demospongiae: Halisarcida) from Mediterranean Sea and a redescription of the type species *Halisarca dujardini*. *Zootaxa* **2768**, 5–31.

Erpenbeck, D. (2004). On the phylogeny of halichondrid demosponges. (PhD Thesis, 208 pp., University of Amsterdam, Amsterdam).

Erpenbeck, D., and Van Soest, R. W. M. (2005). A survey for biochemical synapomorphies to reveal phylogenetic relationships of halichondrid demosponges (Metazoa: Porifera). *Biochemical Systematics and Ecology* **33**, 585–616.

Erpenbeck, D., and Van Soest, R. W. M. (2007). Status and perspective of sponge chemosystematics. *Marine Biotechnology* **9**, 2–19.

Erpenbeck, D., and Wörheide, G. (2007). On the molecular phylogeny of sponges (Porifera). *Zootaxa* **1668**, 107–126.

Erpenbeck, D., Breeuwer, J. A. J., van der Velde, H. C., and van Soest, R. W. M. (2002). Unravelling host and symbiont phylogenies of halichondrid sponges (Demospongiae, Porifera) using a mitochondrial marker. *Marine Biology* **141**, 377–386.

Erpenbeck, D., McCormack, G. P., Breeuwer, J. A. J., and Van Soest, R. W. M. (2004). Order level differences in the structure of partial LSU across demosponges (Porifera): New insights into an old taxon. *Molecular Phylogenetics and Evolution* **32**, 388–395.

Erpenbeck, D., Breeuwer, J. A. J., and van Soest, R. W. M. (2005a). Identification, characterization and phylogenetic signal of an elongation factor-1 alpha fragment in demosponges (Metazoa, Porifera, Demospongiae). *Zoologica Scripta* **34**, 437–445.

Erpenbeck, D., Breeuwer, J. A. J., and van Soest, R. W. M. (2005b). Implications from a 28S rRNA gene fragment for the phylogenetic relationships of halichondrid sponges

(Porifera: Demospongiae). *Journal of Zoological Systematics and Evolutionary Research* **43**, 93–99.

Erpenbeck, D., Breeuwer, J. A. J., Parra-Velandia, F. J., and van Soest, R. W. M. (2006). Speculation with spiculation?—Three independent gene fragments and biochemical characters versus morphology in demosponge higher classification. *Molecular Phylogenetics and Evolution* **38**, 293–305.

Erpenbeck, D., Duran, S., Rützler, K., Paul, V., Hooper, J. N. A., and Wörheide, G. (2007a). Towards a DNA taxonomy of Caribbean demosponges: A gene tree reconstructed from partial mitochondrial CO1 gene sequences supports previous rDNA phylogenies and provides a new perspective on the systematics of Demospongiae. *Journal of the Marine Biological Society of the United Kingdom* **87**, 1563–1570.

Erpenbeck, D., Hooper, J. N. A., List-Armitage, S. E., Degnan, B. M., Wörheide, G., and Van Soest, R. W. M. (2007b). Affinities of the family Sollasellidae (Porifera, Demospongiae). II. Molecular evidence. *Contributions to Zoology* **76**, 95–102.

Erpenbeck, D., List-Armitage, S. E., Alvarez, B., Degnan, B. M., Hooper, J. N. A., and Wörheide, G. (2007c). The systematics of Raspailiidae (Demospongiae, Poecilosclerida, Microcionina) reanalysed with a ribosomal marker. *Journal of the Marine Biological Society of the United Kingdom* **87**, 1571–1576.

Erpenbeck, D., Voigt, O., Adamski, M., Adamska, M., Hooper, J. N. A., Wörheide, G., and Degnan, B. M. (2007d). Mitochondrial diversity of early-branching Metazoa is revealed by the complete mt genome of a haplosclerid demosponge. *Molecular Biology and Evolution* **24**, 19–22.

Erpenbeck, D., Voigt, O., Wörheide, G., and Lavrov, D. (2009). The mitochondrial genomes of sponges provide evidence for multiple invasions by Repetitive Hairpin-forming Elements (RHE). *BMC Genomics* **10**, 591.

Erpenbeck, D., Sutcliffe, P., Cook, S. C., Dietzel, A., Maldonado, M., van Soest, R. W. M., Hooper, J. N. A., and Wörheide, G. (2012b). Horny sponges and their affairs: On the phylogenetic relationships of keratose sponges. *Molecular Phylogenetics and Evolution*. online early. doi: 10.1016/j.ympev.2012.02.024.

Erpenbeck, D., Weier, T., de Voogd, N. J., Wörheide, G., Sutcliffe, P., Todd, J. A., and Michel, E. (2011). Insights into the evolution of freshwater sponges (Porifera: Demospongiae: Spongillina): Barcoding and phylogenetic data from Lake Tanganyika endemics indicate multiple invasions and unsettle existing taxonomy. *Molecular Phylogenetics and Evolution* **61**, 231–236.

Erpenbeck, D., Hall, K., Alvarez, B., Büttner, G., Sacher, K., Schätzle, S., Schuster, A., Vargas, S., Hooper, J. N. A., and Wörheide, G. (2012a). The phylogeny of halichondrid demosponges: Past and present re-visited with DNA-barcoding data. *Organisms, Diversity and Evolution* **12**, 57–70.

Erwin, P. M., and Thacker, R. W. (2007). Phylogenetic analyses of marine sponges within the order Verongida: A comparison of morphological and molecular data. *Invertebrate Biology* **126**, 220–234.

Erwin, D. H., Laflamme, M., Tweedt, S. M., Sperling, E. A., Pisani, D., and Peterson, K. J. (2011). The Cambrian conundrum: Early divergence and later ecological success in the early history of animals. *Science* **334**, 1091–1097.

Fahey, B., Larroux, C., Woodcroft, B. J., and Degnan, B. M. (2008). Does the high gene density in the sponge NK homeobox gene cluster reflect limited regulatory capacity? *The Biological Bulletin* **214**, 205–217.

Farias, I., Orti, G., Sampaio, I., Schneider, H., and Meyer, A. (2001). The cytochrome b gene as a phylogenetic marker: The limits of resolution for analyzing relationships among cichlid fishes. *Journal of Molecular Evolution* **53**, 89–103.

Faulkner, J. (2002). Marine natural products. *Natural Products Reports* **19**, 1–48.

Felsenstein, J. (1978). Cases in which parsimony or compatibility methods will be positively misleading. *Systematic Zoology* **27**, 401–410.

Felsenstein, J. (1985). Confidence limits on phylogenies: An approach using the bootstrap. *Evolution* **39**, 783–791.

Franzén, Å. (1987). Spermatogenesis. In "Reproduction in Marine Invertebrates" (A. C. Giese, J. S. Pearse and V. B. Pearse, eds), pp. 1–47. Boxwood Press, Pacific Groove, CA.

Franzen, W. (1988). Oogenesis and larval development of *Scypha ciliata* (Porifera, Calcarea). *Zoomorphology* **107**, 349–357.

Fromont, J. (1994). Reproductive development and timing of tropical sponges (Order Haplosclerida) from the Great-Barrier-Reef, Australia. *Coral Reefs* **13**, 127–133.

Funayama, N., Nakatsukasa, M., Hayashi, T., and Agata, K. (2005). Isolation of the choanocyte in the fresh water sponge, *Ephydatia fluviatilis* and its lineage marker, Ef annexin. *Development, Growth & Differentiation* **47**, 243–253.

Funayama, N., Nakatsukasa, M., Mohri, K., Masuda, Y., and Agata, K. (2010). Piwi expression in archeocytes and choanocytes in demosponges: Insights into the stem cell system in demosponges. *Evolution & Development* **12**, 275–287.

Gallissian, M. F., and Vacelet, J. (1992). Ultrastructure of the oocyte and embryo of the calcified sponge, *Petrobiona massiliana* (Porifera, Calcarea). *Zoomorphology* **112**, 133–141.

Gauthier, M., and Degnan, B. M. (2008). The transcription factor NF-kappa B in the demosponge *Amphimedon queenslandica*: Insights on the evolutionary origin of the Rel homology domain. *Development Genes and Evolution* **218**, 23–32.

Gazave, E., Lapébie, P., Renard, E., Bézac, C., Boury-Esnault, N., Vacelet, J., Pérez, T., Manuel, M., and Borchiellini, C. (2008). NK homeobox genes with choanocyte-specific expression in homoscleromorph sponges. *Development Genes and Evolution* **218**, 479–489.

Gazave, E., Labépie, P., Richards, G. S., Brunet, F., Ereskovsky, A. V., Degnan, B. M., Borchiellini, C., Vervoort, M., and Renard, E. (2009). Origin and evolution of the Notch signalling pathway: An overview from eukaryotic genomes. *BMC Evolutionary Biology* **9**, 249.

Gazave, E., Carteron, S., Chenuil, A., Richelle-Maurer, E., Boury-Esnault, N., and Borchiellini, C. (2010a). Polyphyly of the genus *Axinella* and of the family Axinellidae (Porifera: Demospongiae(p)). *Molecular Phylogenetics and Evolution* **57**, 35–47.

Gazave, E., Lapébie, P., Renard, E., Vacelet, J., Rocher, C., Ereskovsky, A., Lavrov, D., and Borchiellini, C. (2010b). Molecular phylogeny restores the supra-generic subdivision of homoscleromorph sponges (Porifera, Homoscleromorpha). *PLoS One* **5**, e14290.

Gazave, E., Lapébie, P., Ereskovsky, A., Vacelet, J., Renard, E., Cárdenas, P., and Borchiellini, C. (2012). No longer Demospongiae: Homoscleromorpha formal nomination as a fourth class of Porifera. *Hydrobiologia* **687**, 3–10.

Gerasimova, E., Erpenbeck, D., and Plotkin, A. (2008). *Vosmaeria* Fristedt, 1885 (Porifera, Demospongiae, Halichondriidae): Revision of species, phylogenetic reconstruction and evidence for split. *Zootaxa* **1694**, 1–37.

Gonobobleva, E. L., and Ereskovsky, A. V. (2004). Metamorphosis of the larva of *Halisarca dujardini* (Demospongiae, Halisarcida). *Bulletin de l'Institut Royal des Sciences Naturelles de Belgique Biologie* **74**, 101–115.

Grothe, F. (1989). On the phylogeny of homoscleromorphs. *Berliner Geowissenschaftliche Abhandlungen, Reihe A* **106**, 155–164.

Haeckel, E. (1874). Die Gastrea-Theorie, die phylogenetische Klassifikation des Tierreichs, und die Homologie der Keimblätter. *Jenaische Zeitschrift für Naturwissenschaft* **8**, 1–55.

Haen, K. M., Lang, B. F., Pomponi, S. A., and Lavrov, D. V. (2007). Glass sponges and bilaterian animals share derived mitochondrial genomic features: A common ancestry or parallel evolution? *Molecular Biology and Evolution* **24**, 1518–1527.

Harcet, M., Bilandžija, H., Bruvo-Mađarić, B., and Ćetković, H. (2010a). Taxonomic position of *Eunapius subterraneus* (Porifera, Spongillidae) inferred from molecular data—A revised classification needed? *Molecular Phylogenetics and Evolution* **54**, 1021–1027.

Harcet, M., Roller, M., Cetkovic, H., Perina, D., Wiens, M., Müller, W. E. G., and Vlahovicek, K. (2010b). Demosponge EST sequencing reveals a complex genetic toolkit of the simplest metazoans. *Molecular Biology and Evolution* **27**, 2747–2756.

Harrison, F. W., and De Vos, L. (1991). Porifera. In "Microscopic Anatomy of Invertebrates" (F. W. Harrison and J. A. Westfall, eds), pp. 29–89. Wiley-Liss Inc., New York.

Hartman, W. D. (1958). A re-examination of Bidder's classification of the Calcarea. *Systematic Zoology* **7**, 97–110.

Heim, I., Nickel, M., and Brümmer, F. (2007a). Phylogeny of the genus *Tethya* (Tethyidae: Hadromerida: Porifera): Molecular and morphological aspects. *Journal of the Marine Biological Association of the United Kingdom* **87**, 1615–1627.

Heim, I., Nickel, M., and Brümmer, F. (2007b). Molecular markers for species discrimination in poriferans: A case study on species of the genus *Aplysina*. In "Porifera Research: Biodiversity, Innovation and SustainaBility" (M. R. Custódio, G. Lôbo-Hajdu, E. Hajdu and G. Muricy, eds), pp. 361–371. Museu Nacional de Rio de Janiero Book Series, Rio de Janeiro, Brazil.

Heim, I., Nickel, M., Picton, B., and Brümmer, F. (2007c). Description and molecular phylogeny of *Tethya hibernica* sp. nov. (Porifera, Demospongiae) from Northern Ireland with remarks on the European species of the genus *Tethya*. *Zootaxa* **1595**, 1–15.

Hejnol, A., Obst, M., Stamatakis, A., Ott, M., Rouse, G. W., Edgecombe, G. D., Martinez, P., Baguñà, J., Bailly, X., Jondelius, U., Wiens, M., Müller, W. E. G., Seaver, E., Wheeler, W. C., Martindale, M. Q., Giribet, G., and Dunn, C. W. (2009). Assessing the root of bilaterian animals with scalable phylogenomic methods. *Proceedings of the Royal Society Series B Biological Sciences* **276**, 4261–4270.

Hill, A., Boll, W., Ries, C., Warner, L., Osswalt, M., Hill, M., and Noll, M. (2010). Origin of Pax and Six gene families in sponges: Single PaxB and Six1/2 orthologs in *Chalinula loosanoffi*. *Developmental Biology* **343**, 106–123.

Holmes, B., and Blanch, H. (2007). Genus-specific associations of marine sponges with group I crenarchaeotes. *Marine Biology* **150**, 759–772.

Holstien, K., Rivera, A., Windsor, P., Ding, S. Y., Leys, S. P., Hill, M., and Hill, A. (2010). Expansion, diversification, and expression of T-box family genes in Porifera. *Development Genes and Evolution* **220**, 251–262.

Hooper, J. N. A. (1984). *Sigmaxinella soelae* and *Desmacella ithystela*, two new desmacellid sponges (Porifera, Axinellida, Desmacellidae) from the Northwest shelf of Western Australia, with a revision of the family Desmacellidae. *Monograph series of the Northern Territory Museum of Arts and Sciences* **2**, 1–58.

Hooper, J. N. A. (2002). Family Hemiasterellidae Lendenfeld, 1889. In "Systema Porifera. Guide to the Classification of Sponges" (J. N. A. Hooper and R. W. M. Van Soest, eds), pp. 186–195. Kluwer Academic/Plenum Publishers, New York, Boston, Dordrecht, London, Moscow.

Hooper, J. N. A., and Van Soest, R. W. M. (2002a). Family Spirasigmidae Hallmann, 1912. In "Systema Porifera. A Guide to the Classification of Sponges" (J. N. A. Hooper and R. W. M. Van Soest, eds), pp. 102–104. Kluwer Academic/Plenum Publishers, New York, Boston, Dordrecht, London, Moscow.

Hooper, J. N. A., and Van Soest, R. W. M. (2002b). Order Astrophorida Sollas, 1888. In "Systema Porifera. A Guide to the Classification of Sponges" (J. N. A. Hooper and R. W. M. Van Soest, eds), pp. 105–107. Kluwer Academic/Plenum Publishers, New York, Boston, Dordrecht, London, Moscow.

Hooper, J. N. A., and Van Soest, R. W. M. (2002c). Order Poecilosclerida Topsent, 1928. In "Systema Porifera. A Guide to the Classification of Sponges" (J. N. A. Hooper and R. W. M. Van Soest, eds), pp. 403–408. Kluwer Academic/Plenum Publishers, New York, Boston, Dordrecht, London, Moscow.

Hooper, J. N. A., and Van Soest, R. W. M. (2002d). Systema Porifera, A Guide to the Classification of Sponges. KluwerAcademic/Plenum Publishers, New York, Boston, Dordrecht, London, Moscow.

Hooper, J. N. A., Van Soest, R. W. M., and Debrenne, F. (2002). Phylum Porifera Grant, 1836. In "Systema Porifera: A Guide to the Classification of Sponges" (J. N. A. Hooper and R. W. M. van Soest, eds), pp. 9–14. Kluwer Academic/Plenum Publishers, New York, Boston, Dordrecht, London, Moscow.

Hooper, J. N. A., Van Soest, R. W. M., and Pisera, A. (2011). Phylum Porifera Grant, 1826 [sic!]. In "Animal Biodiversity: An Outline of Higher-Level Classification and Survey of Taxonomic Richness" (Z.-Q. Zhang, ed.), pp. 13–18. Zootaxa.

Itskovich, V. B., Belikov, S. I., Efremova, S. M., and Masuda, Y. (1999). Phylogenetic relationships between Lubomirskiidae, Spongillidae and some marine sponges according partial sequences of 18S rDNA. In "Memoirs of the Queensland Museum." Proceedings of the 5th International Sponge Symposium, Brisbane, June-July 1998 (J. N. A. Hooper, ed.), pp. 275–280.

Itskovich, V. B., Belikov, S., Efremova, S. M., Masuda, Y., Krasko, A., Schröder, H. C., and Müller, W. E. G. (2006). Monophyletic origin of freshwater sponges in ancient lakes based on partial structures of COXI gene. Hydrobiologia **568,** 155–159.

Itskovich, V., Belikov, S., Efremova, S., Masuda, Y., Perez, T., Alivon, E., Borchiellini, C., and Boury-Esnault, N. (2007). Phylogenetic relationships between freshwater and marine Haplosclerida (Porifera, Demospongiae) based on the full length 18S rRNA and partial COXI gene sequences. In "Porifera Research: Biodiversity, Innovation and Sustainability" (M. R. Custódio, G. Lôbo-Hajdu, E. Hajdu and G. Muricy, eds), pp. 383–391. Museu Nacional Rio de Janeiro, Rio de Janeiro, Brazil.

Itskovich, V., Gontcharov, A., Masuda, Y., Nohno, T., Belikov, S., Efremova, S., Meixner, M., and Janussen, D. (2008). Ribosomal ITS sequences allow resolution of freshwater sponge phylogeny with alignments guided by secondary structure prediction. Journal of Molecular Evolution **67,** 608–620.

Ivaniševic, J., Thomas, O. P., Lejeusne, C., Chevaldonné, P., and Pérez, T. (2010). Metabolic fingerprinting as an indicator of biodiversity: Towards understanding inter-specific relationships among Homoscleromorpha sponges. Metabolomics **7,** 289–304.

Janussen, D., and Reiswig, H. M. (2003). Re-description of Cyathella lutea Schmidt and formation of a new subfamily Cyathellinae (Hexactinellida, Aulocalycoida, Aulocalycidae). Senckenbergiana Biologica **82,** 1–10.

Johnson, M. (1979). Gametogenesis and embryonic development in the calcareous sponges Clathrina coriacea and C. blanca from Santa Catalina Island, California. Bulletin of the Southern California Academy of Sciences **78,** 183–191.

Karpov, S. A., and Leadbeater, B. S. C. (1998). Cytoskeleton structure and composition in choanoflagellates. The Journal of Eukaryotic Microbiology **45,** 361–367.

Kayal, E., Voigt, O., Wörheide, G., and Lavrov, D. V. (2010). Calcareous sponges redefine the limits of mtDNA evolution in the animal kingdom. In "8th World Sponge Conference: Ancient Animals, New Challenges. Book of Abstracts." p. 76, Girona, Spain.

Kelly-Borges, M., and Pomponi, S. A. (1994). Phylogeny and classification of lithistid sponges (Porifera: Demospongiae): A preliminary assessment using ribosomal DNA sequence comparisons. Molecular Marine Biology and Biotechnology **3,** 87–103.

Kelly-Borges, M., Bergquist, P. R., and Bergquist, P. L. (1991). Phylogenetic relationships within the order Hadromerida (Porifera, Demospongiae, Tetractinomorpha) as indicated by ribosomal RNA sequence comparisons. Biochemical Systematics and Ecology **19,** 117–125.

Kelly, M., and Samaai, T. (2002). Family Podospongiidae de Laubenfels, 1936. In "Systema Porifera. A Guide to the Classification of Sponges" (J. N. A. Hooper and R. W. M. Van Soest, eds), pp. 694–702. Kluwer Academic/Plenum Publishers, New York, Boston, Dordrecht, London, Moscow.

Kim, J., Kim, W., and Cunningham, C. W. (1999). A new perspective on lower metazoan relationships from 18S rDNA sequences. *Molecular Biology and Evolution* **16,** 423–427.

King, N., Westbrook, M., Young, S., Kuo, A., Abedin, M., Chapman, J., Fairclough, S., Hellsten, U., Isogai, Y., Letunic, I., Marr, M., Pincus, D., Putnam, N., Rokas, A., Wright, K., Zuzow, R., Dirks, W., Good, M., Goodstein, D., Lemons, D., Li, W., Lyons, J., Morris, A., Nichols, S., Richter, D., Salamov, A., Sequencing, J., Bork, P., Lim, W., Manning, G., Miller, W., McGinnis, W., Shapiro, H., Tjian, R., Grigoriev, I., and Rokhsar, D. (2008). The genome of the choanoflagellate *Monosiga brevicollis* and the origin of metazoans. *Nature* **451,** 783–788.

Klautau, M., and Valentine, C. (2003). Revision of the genus *Clathrina* (Porifera, Calcarea). *Zoological Journal of the Linnean Society* **139,** 1–62.

Kober, K., and Nichols, S. (2007). On the phylogenetic relationships of hadromerid and poecilosclerid sponges. *Journal of the Marine Biological Association of the United Kingdom* **87,** 14.

Koziol, C., Leys, S. P., Müller, I. M., and Müller, W. E. G. (1997). Cloning of *Hsp*70 genes from the marine sponges *Sycon raphanus* (Clacarea) and *Rhabdocalyptus dawsoni* (Hexactinellida). An approach to solve the phylogeny of sponges. *Biological Journal of the Linnean Society* **62,** 581–592.

Krautter, M., Conway, K. W., Barrie, J. V., and Neuweiler, M. (2001). Discovery of a 'living dinosaur': Globally unique modern hexactinellid sponge reefs off British Columbia, Canada. *Facies* **44,** 265–282.

Kruse, M., Leys, S. P., Müller, I. M., and Müller, W. E. G. (1998). Phylogenetic position of the Hexactinellida within the phylum Porifera based on the amino acid sequence of the protein kinase C from *Rhabdocalyptus dawsoni*. *Journal of Molecular Evolution* **46,** 721–728.

Labépie, P., Gazave, E., Ereskovsky, A., Derelle, R., Bézac, C., Renard, E., Houliston, E., and Borchiellini, C. (2009). WNT/beta-catenin signalling and epithelial patterning in the homoscleromorph sponge *Oscarella*. *PLoS One* **4,** e5823.

Lafay, B., Boury-Esnault, N., Vacelet, J., and Christen, R. (1992). An analysis of partial 28S ribosomal RNA sequences suggests early radiations of sponges. *Biosystems* **28,** 139–151.

Larroux, C., Fahey, B., Liubicich, D., Hinman, V. F., Gauthier, M., Gongora, M., Green, K., Wörheide, G., Leys, S. P., and Degnan, B. M. (2006). Developmental expression of transcription factor genes in a demosponge: Insights into the origin of metazoan multicellularity. *Evolution & Development* **8,** 150–173.

Larroux, C., Fahey, B., Degnan, S. M., Adamski, M., Rokhsar, D. S., and Degnan, B. M. (2007). The NK homeobox gene cluster predates the origin of Hox genes. *Current Biology* **17,** 706–710.

Larroux, C., Luke, G. N., Koopman, P., Rokhsar, D. S., Shimeld, S. M., and Degnan, B. M. (2008). Genesis and expansion of metazoan transcription factor gene classes. *Molecular Biology and Evolution* **25,** 980–996.

Lavrov, D. (2007). Key transitions in animal evolution: A mitochondrial DNA perspective. *Integrative and Comparative Biology* **47,** 734–743.

Lavrov, D. V., Forget, L., Kelly, M., and Lang, B. F. (2005). Mitochondrial genomes of two demosponges provide insights into an early stage of animal evolution. *Molecular Biology and Evolution* **22,** 1231–1239.

Lavrov, D., Frishman, N., Haen, K., Shao, Z., and Wang, X. (2006). Mitochondrial genomics of sponges—Implications for sponge phylogeny and animal mtDNA evolution. In "7th International Sponge Symposium: Biodiversity, Innovation, Sustainability. Book of abstacts" (M. Custodio, G. Lobo-Hajdu, E. Hajdu and G. Muricy, eds), p. 192. Museu Nacional, Rio de Janeiro, Brasil.

Lavrov, D., Wang, X., and Kelly, M. (2008). Reconstructing ordinal relationships in the Demospongiae using mitochondrial genomic data. *Molecular Phylogenetics and Evolution* **49**, 111–124.

Lee, P. N., Pang, K., Matus, D. Q., and Martindale, M. Q. (2006). A WNT of things to come: Evolution of Wnt signaling and polarity in cnidarians. *Seminars in Cell & Developmental Biology* **17**, 157–167.

Lévi, C. (1956). Etudes des *Halisarca* de Roscoff. Embryologie et systématique des Démosponges. *Archives of Zoological Experimental Genetics* **93**, 1–181.

Lévi, C. (1957). Ontogeny and systematics in sponges. *Systematic Zoology* **6**, 174–183.

Lévi, C. (1973). Systématique de la Classe des *Demospongiaria* (Démosponges). In "Spongiaires. Anatomie, Physiologie, Systématique, Ecologie" (P. P. Grassé, ed.), pp. 577–631. Masson et Cíe, Paris.

Leys, S. P. (2004). Gastrulation in sponges. In "Gastrulation: From Cells to Embryos" (C. Stern, ed.), pp. 23–31. Cold Spring Harbor Laboratory Press, Cold Spring Harbor, NY.

Leys, S. P., and Degnan, B. M. (2001). Cytological basis of photoresponsive behavior in a sponge larva. *The Biological Bulletin* **201**, 323–338.

Leys, S. P., and Degnan, B. M. (2002). Embryogenesis and metamorphosis in a haplosclerid demosponge: Gastrulation and transdifferentiation of larval ciliated cells to choanocytes. *Invertebrate Biology* **121**, 171–189.

Leys, S. P., and Eerkes-Medrano, D. (2005). Gastrulation in calcareous sponges: In search of Haeckel's gastraea. *Integrative and Comparative Biology* **45**, 342–351.

Leys, S. P., and Ereskovsky, A. V. (2006). Embryogenesis and larval differentiation in sponges. *Canadian Journal of Zoology* **84**, 262–287.

Leys, S. P., and Riesgo, A. (2011). Epithelia, an evolutionary novelty of metazoans. *Journal of Experimental Zoology. Part B, Molecular and Developmental Evolution* (online early), doi: 10.1002/jez.b.21442.

Leys, S. P., Cheung, E., and Boury-Esnault, N. (2006). Embryogenesis in the glass sponge *Oopsacas minuta*: Formation of syncytia by fusion of blastomeres. *Integrative and Comparative Biology* **46**, 104–117.

Leys, S. P., Mackie, G. O., and Reiswig, H. M. (2007). The biology of glass sponges. *Advances in Marine Biology* **52**, 1–145.

López-Legentil, S., and Pawlik, J. (2008). Genetic structure of the Caribbean giant barrel sponge *Xestospongia muta* using the I3-M11 partition of COI. *Coral Reefs* **28**, 157–165.

Lukic-Bilela, L., Brandt, D., Pojskic, N., Wiens, M., Gamulin, V., and Müller, W. E. G. (2008). Mitochondrial genome of *Suberites domuncula*: Palindromes and inverted repeats are abundant in non-coding regions. *Gene* **412**, 1–11.

Maddison, W. P., and Maddison, D. R. (2011). Mesquite: A Modular System for Evolutionary Analysis. Version 2.75, http://mesquiteproject.org.

Magie, C. R., Daly, M., and Martindale, M. Q. (2007). Gastrulation in the cnidarian *Nematostella vectensis* occurs via invagination not ingression. *Developmental Biology* **305**, 483–497.

Maldonado, M. (2004). Choanoflagellates, choanocytes, and animal multicellularity. *Invertebrate Biology* **123**, 1–22.

Maldonado, M. (2009). Embryonic development of verongid demosponges supports the independent acquisition of spongin skeletons as an alternative to the siliceous skeleton of sponges. *Biological Journal of the Linnean Society* **97**, 427–447.

Maldonado, M., and Bergquist, P. R. (2002). Phylum Porifera. In "Atlas of Marine Invertebrate Larvae" (C. M. Young, M. A. Sewell and M. E. Rice, eds), pp. 21–50. Academic Press, San Diego.

Maldonado, M., and Riesgo, A. (2007). Intra-epithelial spicules in a homosclerophorid sponge. *Cell and Tissue Research* **328,** 639–650.

Maldonado, M., and Riesgo, A. (2008a). Reproduction in the phylum Porifera: A synoptic overview. *Treballs de la SCB* **59,** 29–49.

Maldonado, M., and Riesgo, A. (2008b). Reproductive output in a Mediterranean population of the homosclerophorid *Corticium candelabrum* (Porifera, Demospongiae), with notes on the ultrastructure and behavior of the larva. *Marine Ecology* **29,** 298–316.

Maldonado, M., and Riesgo, A. (2009). Gametogenesis, embryogenesis, and larval features of the oviparous sponge *Petrosia ficiformis* (Haplosclerida, Demospongiae). *Marine Biology* **156,** 2181–2197.

Maldonado, M., and Young, C. M. (1996). Effects of physical factors on larval behavior, settlement and recruitment of four tropical demosponges. *Marine Ecology Progress Series* **138,** 169–180.

Maldonado, M., and Young, C. M. (1999). Effects of the duration of larval life on postlarval stages of the demosponge *Sigmadocia caerulea*. *Journal of Experimental Marine Biology and Ecology* **232,** 9–21.

Maldonado, M., Carmona, M. C., Uriz, M. J., and Cruzado, A. (1999). Decline in Mesozoic reef-building sponges explained by silicon limitation. *Nature* **401,** 785–788.

Maldonado, M., Durfort, M., McCarthy, D. A., and Young, C. M. (2003). The cellular basis of photobehavior in the tufted parenchymella larva of demosponges. *Marine Biology* **143,** 427–441.

Manuel, M. (2006). Phylogeny and evolution of calcareous sponges. *Canadian Journal of Zoology* **84,** 225–241.

Manuel, M., and Le Parco, Y. (2000). Homeobox gene diversification in the calcareous sponge, *Sycon raphanus*. *Molecular Phylogenetics and Evolution* **17,** 97–107.

Manuel, M., Borojevic, R., Boury-Esnault, N., and Vacelet, J. (2002). Class Calcarea Bowerbank,1864. In "Systema Porifera: A Guide to the Classification of Sponges" (N. A. Hooper and R. W. M. van Soest, eds), pp. 1103–1110. Kluwer Academic/ Plenum Publishers, New York, Boston, Dordrecht, London, Moscow.

Manuel, M., Borchiellini, C., Alivon, E., Le Parco, Y., Vacelet, J., and Boury-Esnault, N. (2003). Phylogeny and evolution of calcareous sponges: Monophyly of Calcinea and Calcaronea, high level of morphological homoplasy, and the primitive nature of axial symmetry. *Systematic Biology* **52,** 311–333.

Manuel, M., Borchiellini, C., Alivon, E., and Boury-Esnault, N. (2004). Molecular phylogeny of calcareous sponges using 18S rRNA and 28S rRNA sequences. *Bollettino dei Musei e degli Istituti Biologici* **68,** 449–461.

Mariani, S., Uriz, M. J., and Turon, X. (2005). The dynamics of sponge larvae assemblages from northwestern Mediterranean nearshore bottoms. *Journal of Plankton Research* **27,** 249–262.

McCormack, G. P., and Kelly, M. (2002). New indications of the phylogenetic affinity of *Spongosorites suberitoides* Diaz et al., 1993 (Porifera, Demospongiae) as revealed by 28S ribosomal DNA. *Journal of Natural History* **36,** 1009–1021.

McCormack, G. P., Erpenbeck, D., and Van Soest, R. W. M. (2002). Major discrepancy between phylogenetic hypotheses based on molecular and morphological criteria within the Order Haplosclerida (Phylum Porifera: Class Demospongiae). *Journal of Zoological Systematics and Evolutionary Research* **40,** 237–240.

Medina, M., Collins, A. G., Silberman, J. D., and Sogin, M. L. (2001). Evaluating hypotheses of basal animal phylogeny using complete sequences of large and small subunit rRNA. *Proceedings of the National Academy of Sciences of the United States of America* **98,** 9707–9712.

Mehl, D. (1992). Die Entwicklung der Hexactinellida seit dem Mesozoikum. Paläobiologie, Phylogenie und Evolutionsökologie. *Berliner Geowissenschaftliche Abhandlungen, Reihe (E), Palaeobiologie* **2,** 1–164.

Meixner, M. J., Luter, C., Eckert, C., Itskovich, V., Janussen, D., von Rintelen, T., Bohne, A. V., Meixner, J. M., and Hess, W. R. (2007). Phylogenetic analysis of freshwater sponges provide evidence for endemism and radiation in ancient lakes. *Molecular Phylogenetics and Evolution* **45**, 875–886.

Miller, G. (2009). Origins. On the origin of the nervous system. *Science* **325**, 24–26.

Minchin, E. A. (1900). The Porifera and Coelenterata. In "Treatise on Zoology" (R. Lakenster, ed.), pp. 1–178. A. & Ch. Black, London.

Mitchell, K. D., Hall, K. A., and Hooper, J. N. A. (2011). A new species of *Sigmaxinella* Dendy, 1897 (Demospongiae, Poecilosclerida, Desmacellidae) from the Tasman Sea. *Zootaxa* **2901**, 19–34.

Moore, W. S. (1995). Inferring phylogenies from mtDNA variation—Mitochondrial-gene trees versus nuclear-gene trees. *Evolution* **49**, 718–726.

Moritz, C., Dowling, T. E., and Brown, W. M. (1987). Evolution of the animal mitochondrial DNA: Relevance for population biology and systematics. *Annual Review of Ecology and Systematics* **18**, 269–292.

Morrow, C. C., Picton, B. E., Erpenbeck, D., Boury-Esnault, N., Maggs, C. A., and Allcock, A. L. (2012). Congruence between nuclear and mitochondrial genes in Demospongiae: A new hypothesis for relationships within the G4 clade (Porifera: Demospongiae). *Molecular Phylogenetics and Evolution* **62**, 174–190.

Müller, W. E. G., Wang, X., Kropf, K., Boreiko, A., Schloßmacher, U., Brandt, D., Schröder, H., and Wiens, M. (2008). Silicatein expression in the hexactinellid *Crateromorpha meyeri*: The lead marker gene restricted to siliceous sponges. *Cell and Tissue Research* **333**, 339–351.

Muricy, G., and Diaz, M. (2002). Order Homosclerophorida Dendy, 1905, Family Plakinidae Schulze, 1880. In "Systema Porifera. A Guide to the Classification of Sponges" (J. N. A. Hooper and R. W. M. Van Soest, eds), pp. 71–82. Kluwer Academic/Plenum Publishers, New York, Boston, Dordrecht, London, Moscow.

Muricy, G., Boury-Esnault, N., Bézac, C., and Vacelet, J. (1996). Cytological evidence for cryptic speciation in Mediterranean *Oscarella* species (Porifera, Homoscleromorpha). *Canadian Journal of Zoology* **74**, 881–896.

Muricy, G., Boury-Esnault, N., Bézac, C., and Vacelet, J. (1998). Taxonomic revision of the Mediterranean *Plakina* Schulze (Porifera, Demospongiae, Homoscleromorpha). *Zoological Journal of the Linnean Society* **124**, 169–203.

Nakamura, Y., and Okada, K. (1998). The ultrastructure of spermatozoa and its structural change in the choanocytes of *Sycon calcaravis* Hozawa. In "Sponge Sciences: Multidisciplinary Perspectives" (Y. Watanabe and N. Fusetani, eds), pp. 179–191. Springer, Tokyo.

Nichols, S. A. (2005). An evaluation of support for order-level monophyly and interrelationships within the class Demospongiae using partial data from the large subunit rDNA and cytochrome oxidase subunit I. *Molecular Phylogenetics and Evolution* **34**, 81–96.

Nichols, S., and Barnes, P. (2005). A molecular phylogeny and historical biogeography of the marine sponge genus *Placospongia* (Phylum Porifera) indicate low dispersal capabilities and widespread crypsis. *Journal of Experimental Marine Biology and Ecology* **323**, 1–15.

Nichols, S. A., Dirks, W., Pearse, J. S., and King, N. (2006). Early evolution of animal cell signaling and adhesion genes. *Proceedings of the National Academy of Sciences of the United States of America* **103**, 12451–12456.

Nielsen, C. (2008). Six major steps in animal evolution: Are we derived sponge larvae? *Evolution & Development* **10**, 241–257.

Okada, Y. (1928). On the development of a hexactinellid sponge *Farrea sollasii*. *Journal of the Faculty of Science, Imperial University of Tokyo* **2**, 1–29.

Parra-Velandia, F. J. (2011). Speciation scenarios of sessile organisms in the Caribbean Sea: The genus *Agelas* (Porifera: Demospongiae), a case of high diversity in the area. (Ph.D. Thesis, Universidad Nacional de Colombia, Bogota, 165 pp.).

Perovic, S., Schroder, H. C., Sudek, S., Grebenjuk, V. A., Batel, R., Stifanic, M., Müller, I. M., Müller, W. E. G., Nicholas, K. B., and Nicholas, H. B. (2003). Expression of one sponge Iroquois homeobox gene in primmorphs from *Suberites domuncula* during canal formation. *Evolution & Development* **5,** 240–250.

Peterson, K. J., and Butterfield, N. J. (2005). Origin of the Eumetazoa—Testing ecological predictions of molecular clocks against the Proterozoic fossil record. *Proceedings of the National Academy of Sciences of the United States of America* **102,** 9547–9552.

Peterson, K. J., and Eernisse, D. J. (2001). Animal phylogeny and the ancestry of bilaterians: Inferences from morphology and 18S rDNA gene sequences. *Evolution & Development* **3,** 170–205.

Peterson, K. J., McPeek, M. A., and Evans, D. A. D. (2005). Tempo and mode of early animal evolution: Inferences from rocks, Hox, and molecular clocks. *Paleobiology* **31,** (Suppl. 2), 36–55.

Peterson, K. J., Cotton, J. A., Gehling, J. G., and Pisani, D. (2008). The Ediacaran emergence of bilaterians: Congruence between the genetic and the geological fossil records. *Philosophical Transactions of the Royal Society of London. Series B, Biological Sciences* **363,** 1435–1443.

Peterson, K. J., Lyons, J. B., Nowak, K. S., Takacs, C. M., Wargo, M. J., and McPeek, M. A. (2004). Estimating metazoan divergence times with a molecular clock. *Proceedings of the National Academy of Sciences of the USA* **101,** 6536–6541.

Philippe, H., and Telford, M. J. (2006). Large-scale sequencing and the new animal phylogeny. *Trends in Ecology & Evolution* **21,** 614–620.

Philippe, H., Derelle, R., Lopez, P., Pick, K., Borchiellini, C., Boury-Esnault, N., Vacelet, J., Deniel, E., Houliston, E., Quéinnec, E., Da Silva, C., Wincker, P., Le Guyader, H., Leys, S., Jackson, D. J., Schreiber, F., Erpenbeck, D., Morgenstern, B., Wörheide, G., and Manuel, M. (2009). Phylogenomics revives traditional views on deep animal relationships. *Current Biology* **19,** 706–712.

Philippe, H., Brinkmann, H., Lavrov, D. V., Littlewood, D. T. J., Manuel, M., Wörheide, G., and Baurain, D. (2011). Resolving difficult phylogenetic questions: Why more sequences are not enough. *PLoS Biology* **9,** e1000602.

Pick, K. S., Philippe, H., Schreiber, F., Erpenbeck, D., Jackson, D. J., Wrede, P., Wiens, M., Alié, A., Morgenstern, B., Manuel, M., and Wörheide, G. (2010). Improved phylogenomic taxon sampling noticeably affects non-bilaterian relationships. *Molecular Biology and Evolution* **27,** 1983–1987.

Pisani, D., Benton, M. J., and Wilkinson, M. (2007). Congruence of morphological and molecular phylogenies. *Acta Biotheoretica* **55,** 269–281.

Pisera, A., and Lévi, C. (2002). 'Lithistid' Demospongiae. In "Systema Porifera. A Guide to the Classification of Sponges" (J. N. A. Hooper and R. W. M. Van Soest, eds), pp. 299–301. Kluwer Academic/Plenum Publishers, New York, Boston, Dordrecht, London, Moscow.

Pöppe, J., Sutcliffe, P., Hooper, J. N. A., Wörheide, G., and Erpenbeck, D. (2010). COI Barcoding reveals new clades and radiation patterns of Indo-Pacific sponges of the Family Irciniidae (Demospongiae: Dictyoceratida). *PLoS One* **5,** e9950.

Price, A. L., and Patel, N. H. (2004). The evolution of gastrulation: Cellular and molecular aspects. In "Gastrulation: From Cells to Embryos" (C. D. Stern, ed.), pp. 695–701. Cold Spring Harbor Laboratory Press, Cold Spring Harbor, NY.

Raleigh, J., Redmond, N. E., Delahan, E., Torpey, S., Van Soest, R. W. M., Kelly, M., and McCormack, G. P. (2007). Mitochondrial cytochrome oxidase 1 phylogeny supports alternative taxonomic scheme for the marine Haplosclerida. *Journal of the Marine Biological Association of the United Kingdom* **87,** 1577–1584.

Rapp, H. T., Janussen, D., and Tendal, O. S. (2011). Calcareous sponges from abyssal and bathyal depths in the Weddell Sea, Antarctica. *Deep-Sea Research Part II* **58,** 58–67.

Rasmont, R. (1979). Les eponges: Des metazoaires et des societes de cellules. *Colloques Internationaux du Centre National de la Recherche Scientifique* **291,** 21–29.

Redmond, N. E., and McCormack, G. P. (2008). Large expansion segments in 18S rDNA support a new sponge clade (Class Demospongiae, Order Haplosclerida). *Molecular Phylogenetics and Evolution* **47,** 1090–1099.

Redmond, N. E., and McCormack, G. P. (2009). Ribosomal internal transcribed spacer regions are not suitable for intra- or inter-specific phylogeny reconstruction in haplosclerid sponges (Porifera: Demospongiae). *Journal of the Marine Biological Association of the United Kingdom* **89**, 1251–1256.

Redmond, N. E., Van Soest, R. W. M., Kelly, N., Raleigh, J., Travers, S. A. A., and McCormack, G. P. (2007). Reassessment of the classification of the Order Haplosclerida (Class Demospongiae, Phylum Porifera) using 18S rRNA gene sequence data. *Molecular Phylogenetics and Evolution* **43**, 344–352.

Reiswig, H. M. (1983). Porifera. In "Reproductive Biology of Invertebrates" (K. G. Adiyodi and R. G. Adiyodi, eds), pp. 1–21. Wiley and sons, Chichester.

Reiswig, H. M. (2002). Class Hexactinellida Schmidt, 1870. In "Systema Porifera. A Guide to the Classification of Sponges" (J. N. A. Hooper and R. W. M. Van Soest, eds), pp. 1201–1202. Plenum, New York, Boston, Dordrecht, London, Moscow.

Reiswig, H. M., and Kelly, M. (2011). The marine fauna of New Zealand: Hexasterophoran Glass Sponges of New Zealand (Porifera: Hexactinellida: Hexasterophora): Orders Hexactinosida, Aulocalycoida and Lychniscosida. *NIWA Biodiversity Memoirs* **124**, 1–176.

Reitner, J. (1992). "Coralline Spongien". Der Versuch einer phylogenetisch taxonomischen Analyse. *Berliner Geowissenschaftliche Abhandlungen, Reihe (E), Palaeobiologie* **1**, 1–352.

Reitner, J., and Mehl, D. (1996). Monophyly of the Porifera. *Verhandlungen des natrurwissenschaftlichen Vereins Hamburg* **36**, 5–32.

Reunov, A. (2001). Problem of terminology in characteristics of spermatozoa of Metazoa. *Russian Journal of Developmental Biology* **36**, 335–351.

Richards, G. S., and Degnan, B. M. (2009). The dawn of developmental signaling in the metazoa. *Cold Spring Harbor Symposia on Quantitative Biology* **74**, 81–90.

Richards, G. S., Simionato, E., Perron, M., Adamska, M., Vervoort, M., and Degnan, B. M. (2008). Sponge genes provide new insight into the evolutionary origin of the neurogenic circuit. *Current Biology* **18**, 1156–1161.

Riesgo, A., and Maldonado, M. (2009). An unexpectedly sophisticated, V-shaped spermatozoon in Demospongiae (Porifera): Reproductive and evolutionary implications. *Biological Journal of the Linnean Society* **97**, 413–426.

Rivera, A. S., Hammel, J. U., Haen, K. M., Danka, E. S., Cieniewicz, B., Winters, I. P., Posfai, D., Wörheide, G., Lavrov, D. V., Knight, S. W., Hill, M. S., Hill, A. L., and Nickel, M. (2011). RNA interference in marine and freshwater sponges: Actin knockdown in *Tethya wilhelma* and *Ephydatia muelleri* by ingested dsRNA expressing bacteria. *BMC Biotechnology* **11**, 67.

Rokas, A., and Carroll, S. B. (2006). Bushes in the Tree of Life. *PLoS Biology* **4**, e352.

Rokas, A., King, N., Finnerty, J., and Carroll, S. B. (2003). Conflicting phylogenetic signals at the base of the metazoan tree. *Evolution & Development* **5**, 346–359.

Rokas, A., Krüger, D., and Carroll, S. B. (2005). Animal evolution and the molecular signature of radiations compressed in time. *Science* **310**, 1933–1938.

Rosengarten, R., Sperling, E., Moreno, M., Leys, S., and Dellaporta, S. (2008). The mitochondrial genome of the hexactinellid sponge *Aphrocallistes vastus*: Evidence for programmed translational frameshifting. *BMC Genomics* **9**, 33.

Rot, C., Goldfarb, I., Ilan, M., and Huchon, D. (2006). Putative cross-kingdom horizontal gene transfer in sponge (Porifera) mitochondria. *BMC Evolutionary Biology* **6**, 71.

Rua, C. P. J., Zilberberg, C., and Solé-Cava, A. M. (2011). New polymorphic mitochondrial markers for sponge phylogeography. *Journal of the Marine Biological Association of the United Kingdom* **91**, 1015–1022.

Saller, U. (1988). Oogenesis and larval development of *Ephydatia fluviatilis* (Porifera, Spongillidae). *Zoomorphology* **108**, 23–28.

Schierwater, B., Eitel, M., Jakob, W., Osigus, H.-J., Hadrys, H., Dellaporta, S. L., Kolokotronis, S.-O., DeSalle, R., and Penny, D. (2009). Concatenated analysis sheds light on early metazoan evolution and fuels a modern "Urmetazoon" hypothesis. *PLoS Biology* **7,** e20.

Schmitt, S., Hentschel, U., Zea, S., Dandekar, T., and Wolf, M. (2005). ITS-2 and 18S rRNA gene phylogeny of Aplysinidae (Verongida, Demospongiae). *Journal of Molecular Evolution* **60,** 327–336.

Schröder, H. C., Efremova, S. M., Itskovich, V. B., Belikov, S., Masuda, Y., Krasko, A., Müller, I. M., Müller, W. E. G., Medina, M., Weil, E., and Szmant, A. M. (2003). Molecular phylogeny of the freshwater sponges in Lake Baikal. *Journal of Zoological Systematics and Evolutionary Research* **41,** 80–86.

Schütze, J., Krasko, A., Custodio, M. R., Efremova, S. M., Müller, I. M., and Müller, W. E. G. (1999). Evolutionary relationships of Metazoa within the eukaryotes based on molecular data from Porifera. *Proceedings of the Royal Society Series B Biological Sciences* **266,** 63–73.

Sethmann, I., and Wörheide, G. (2008). Structure and composition of calcareous sponge spicules: A review and comparison to structurally related biominerals. *Micron* **39,** 209–228.

Shimodaira, H., and Hasegawa, M. (1999). Multiple comparisons of log-likelihoods with applications to phylogenetic inference. *Molecular Biology and Evolution* **16,** 1114–1116.

Simionato, E., Ledent, V., Richards, G., Thomas-Chollier, M., Kerner, P., Coornaert, D., Degnan, B. M., and Vervoort, M. (2007). Origin and diversification of the basic helix-loop-helix gene family in metazoans: Insights from comparative genomics. *BMC Evolutionary Biology* **7,** 33.

Simpson, T. L. (1984). The Cell Biology of Sponges. Springer Verlag, New York.

Solé-Cava, A. M., Boury-Esnault, N., Vacelet, J., and Thorpe, J. P. (1992). Biochemical genetic divergence and systematics in sponges of the genera *Corticium* and *Oscarella* (Demospongiae: Homoscleromorpha) in the Mediterranean Sea. *Marine Biology* **113,** 299–304.

Sperling, E. A., and Peterson, K. J. (2009). MicroRNAs and metazoan phylogeny: Big trees from little genes. In "Animal Evolution: Genomes, Fossils, and Trees" (M. J. Telford and D. T. J. Littlewood, eds), pp. 157–170. Oxford University Press, Oxford.

Sperling, E. A., Pisani, D., and Peterson, K. J. (2007). Poriferan paraphyly and its implications for Precambrian paleobiology. *Journal of the Geological Society of London, Special Publications* **286,** 355–368.

Sperling, E. A., Peterson, K. J., and Pisani, D. (2009). Phylogenetic signal dissection of nuclear housekeeping genes supports the paraphyly of sponges and the monophyly of Eumetazoa. *Molecular Biology and Evolution* **26,** 2261–2274.

Sperling, E. A., Robinson, J. M., Pisani, D., and Peterson, K. J. (2010). Where's the glass? Biomarkers, molecular clocks, and microRNAs suggest a 200-Myr missing Precambrian fossil record of siliceous sponge spicules. *Geobiology* **8,** 24–36.

Srivastava, M., Larroux, C., Lu, D. R., Mohanty, K., Chapman, J., Degnan, B. M., and Rokhsar, D. S. (2010a). Early evolution of the LIM homeobox gene family. *BMC Biology* **8,** 4.

Srivastava, M., Simakov, O., Chapman, J., Fahey, B., Gauthier, M. E. A., Mitros, T., Richards, G. S., Conaco, C., Dacre, M., Hellsten, U., Larroux, C., Putnam, N. H., Stanke, M., Adamska, M., Darling, A., Degnan, S. M., Oakley, T. H., Plachetzki, D. C., Zhai, Y. F., Adamski, M., Calcino, A., Cummins, S. F., Goodstein, D. M., Harris, C., Jackson, D. J., Leys, S. P., Shu, S. Q., Woodcroft, B. J., Vervoort, M., Kosik, K. S., Manning, G., Degnan, B. M., and Rokhsar, D. S. (2010b). The *Amphimedon queenslandica* genome and the evolution of animal complexity. *Nature* **466,** 720–727.

Szitenberg, A., Rot, C., Ilan, M., and Huchon, D. (2010). Diversity of sponge mitochondrial introns revealed by *cox1* sequences of Tetillidae. *BMC Evolutionary Biology* **10,** 288.

Tabachnick, K. R. (2002). Family Euplectellidae Gray, 1867. In "Systema Porifera. A Guide to the Classification of Sponges" (J. N. A. Hooper and R. W. M. van Soest, eds), pp. 1388–1434. Plenum, New York, Boston, Dordrecht, London, Moscow.

Taylor, M. W., Thacker, R. W., and Hentschel, U. (2007). Genetics. Evolutionary insights from sponges. *Science* **316,** 1854–1855.

Telford, M. J., and Copley, R. R. (2011). Improving animal phylogenies with genomic data. *Trends in Genetics* **27,** 186–195.

Tuzet, O., and Grassé, P. P. (1973). Éponges calcaires. In "Traité de zoologie. Spongiaires," pp. 7–132. Masson et Cie, Paris.

Uriz, M.-J. (2006). Mineral skeletogenesis in sponges. *Canadian Journal of Zoology* **84,** 322–356.

Uriz, M., Turon, X., Becerro, M., and Agell, G. (2003). Siliceous spicules and skeleton frameworks in sponges: Origin, diversity, ultrastructural patterns, and biological functions. *Microscopy Research and Technique* **62,** 279–299.

Usher, K. M., and Ereskovsky, A. V. (2005). Larval development, ultrastructure and metamorphosis in *Chondrilla australiensis* Carter, 1873 (Demospongiae, Chondrosida, Chondrillidae). *Invertebrate Reproduction and Development* **47,** 51–62.

Vacelet, J., Boury-Esnault, N., and Harmelin, J.-G. (1994). Hexactinellid cave, a unique deep-sea habitat in the SCUBA zone. *Deep-Sea Research Part I* **41,** 965–973.

Vacelet, J., Borchiellini, C., Perez, T., Bultel-Ponce, V., Brouard, J. P., and Guyot, M. (2000). Morphological, chemical and biochemical characterization of a new species of sponge without skeleton (Porifera, Demospongiae) from the Mediterranean Sea. *Zoosystema* **22,** 313–326.

Van de Peer, Y., and de Wachter, R. (1997). Evolutionary relationships among the eukaryotic crown taxa taking into account site-to-site rate variation in 18S rRNA. *Journal of Molecular Evolution* **45,** 619–630.

Van Soest, R. W. M. (1984). Marine sponges from Curacao and other Caribbean Localities. Part III. Poecilosclerida. In "Studies on the Fauna of Curaçao and Other Caribbean Islands" (P. Wagenaar-Hummelinck and L. J. van der Steen, eds). Foundation for Scientific Research in Suriname and the Netherlands Antilles, Utrecht, The Netherlands, (No. 112).

Van Soest, R. W. M. (1990). Towards a phylogenetic classification of sponges. In "New Perspectives in Sponge Biology: Papers Contributed to the Third International Conference on the Biology of Sponges, Convened by Willard D. Hartman and Klaus Rützler, Woods Hole, Massachusetts, 17–23 November 1985." (K. Rützler, V. V. Macintyre and K. P. Smith, eds), pp. 344–348. Smithsonian Institution Press, Washington D.C.

Van Soest, R. W. M. (1991). Demosponge higher taxa classification re-examined. In "Fossil and Recent Sponges" (J. Reitner and H. Keupp, eds), pp. 54–71. Springer Verlag, Berlin, Heidelberg.

Van Soest, R. W. M., and Braekman, J. C. (1999). Chemosystematics of Porifera: A review. *Memoirs of the Queensland Museum* **44,** 569–589.

Van Soest, R. W. M., and Hooper, J. N. A. (2002a). Order Halichondrida Gray, 1867. In "Systema Porifera, A Guide to the Classification of Sponges" (J. N. A. Hooper and R. W. M. Van Soest, eds), pp. 721–723. KluwerAcademic/Plenum Publishers, New York, Boston, Dordrecht, London, Moscow.

Van Soest, R. W. M., and Hooper, J. N. A. (2002b). Order Haplosclerida Topsent, 1928. In "Systema Porifera. A Guide to the Classification of Sponges" (J. N. A. Hooper and R. W. M. Van Soest, eds), pp. 831–832. Kluwer Academic/Plenum Publishers, New York, Boston, Dordrecht, London, Moscow.

Van Soest, R. W. M., Diaz, M. C., and Pomponi, S. A. (1990). Phylogenetic classification of the halichondrids (Porifera, Demospongiae). *Beaufortia* **40,** 15–62.

Van Soest, R. W. M., Hooper, J. N. A., Beglinger, E., and Erpenbeck, D. (2006). Affinities of the family Sollasellidae (Porifera, Demospongiae). I. Morphological evidence. *Contributions to Zoology* **75,** 133–144.

Van Soest, R. W. M., Boury-Esnault, N., Hooper, J. N. A., Rützler, K., de Voogd, N. J., Alvarez de Glasby, B., Hajdu, E., Pisera, A. B., Manconi, R., Schoenberg, C., Janussen, D., Tabachnick, K. R., Klautau, M., Picton, B., and Kelly, M. (2011). World Porifera database. http://www.marinespecies.org/porifera, (accessed 01 May 2011).

Veremeichik, G. N., Shkryl, Y. N., Bulgakov, V. P., Shedko, S. V., Kozhemyako, V. B., Kovalchuk, S. N., Krasokhin, V. B., Zhuravlev, Y. N., and Kulchin, Y. N. (2011). Occurrence of a silicatein gene in glass sponges (Hexactinellida: Porifera). *Marine Biotechnology* **13**, 810–819.

Viamontes, G. I., and Kirk, D. L. (1977). Cell shape change and the mechanism of inversion in *Volvox*. *The Journal of Cell Biology* **75**, 719–730.

Voigt, O., Erpenbeck, D., and Wörheide, G. (2008). Molecular evolution of rDNA in early diverging Metazoa: First comparative analysis and phylogenetic application of complete SSU rRNA secondary structures in Porifera. *BMC Evolutionary Biology* **8**, 69.

Voigt, O., Eichmann, V., and Wörheide, G. (2012a). First evaluation of mitochondrial DNA as a marker for phylogeographic studies of Calcarea: a case study from *Leucetta chagosensis*. *Hydrobiologia* **687**, 101–106.

Voigt, O., Wülfing, E., and Wörheide, G. (2012b). Molecular phylogenetic evaluation of classification and scenarios of character evolution in calcareous sponges (Porifera, Class Calcarea). *PLoS ONE*, doi:10.1371/journal.pone.0033417.

Wang, X., and Lavrov, D. (2007). Mitochondrial genome of the homoscleromorph *Oscarella carmela* (Porifera, Demospongiae) reveals unexpected complexity in the common ancestor of sponges and other animals. *Molecular Biology and Evolution* **24**, 363–373.

Wang, X., and Lavrov, D. (2008). Seventeen new complete mtDNA sequences reveal extensive mitochondrial genome evolution within the Demospongiae. *PLoS One* **3**, e2723.

Wapstra, M., and van Soest, R. W. M. (1987). Sexual reproduction, larval morphology and behaviour in demosponges from the southwest of the Netherlands. In "Taxonomy of Porifera from the N.E. Atlantic and Mediterranean Sea" (J. Vacelet and N. Boury-Esnault, eds), pp. 281–307. Springer Verlag, New York.

Wiens, M., Belikov, S. I., Kaluzhnaya, O. V., Adella, T., Schröder, H. C., Perovic-Ottstadt, S., Kaandorp, J. A., and Müller, W. E. G. (2008). Regional and modular expression of morphogenetic factors in the demosponge *Lubomirskia baicalensis*. *Micron* **39**, 447–460.

Windsor, P. J., and Leys, S. P. (2010). Wnt signaling and induction in the sponge aquiferous system: Evidence for an ancient origin of the organizer. *Evolution & Development* **12**, 484–493.

Woollacott, R. M. (1990). Structure and swimming behavior of the larva of *Halichondria melanadocia* (Porifera, Demospongiae). *Journal of Morphology* **205**, 135–145.

Woollacott, R. M. (1993). Structure and swimming behavior of the larva of *Haliclona tubifera* (Porifera, Demospongiae). *Journal of Morphology* **218**, 301–321.

Woollacott, R. M., and Hadfield, M. G. (1989). Larva of the sponge *Dendrilla cactus* (Demospongiae, Dendroceratida). *Transactions of the American Microscopical Society* **108**, 410–413.

Woollacott, R., and Pinto, R. (1995). Flagellar basal apparatus and its utility in phylogenetic analyses of the Porifera. *Journal of Morphology* **226**, 247–265.

Wörheide, G. (1998). The reef cave dwelling ultraconservative coralline demosponge *Astrosclera willeyana* Lister 1900 from the Indo-Pacific—Micromorphology, ultrastructure, biocalcification, isotope record, taxonomy, biogeography, phylogeny. *Facies* **38**, 1–88.

Wörheide, G. (2008). A hypercalcified sponge with soft relatives: *Vaceletia* is a keratose demosponge. *Molecular Phylogenetics and Evolution* **47**, 433–438.

Wörheide, G., and Hooper, J. N. A. (1999). Calcarea from the Great Barrier Reef. 1: Cryptic Calcinea from Heron Island and Wistari Reef (Capricorn-Bunker Group). *Memoirs of the Queensland Museum* **43**, 859–891.

Zrzavy, J., Mihulka, S., Kepka, P., Bezdek, A., and Tietz, D. (1998). Phylogeny of the Metazoa based on morphological and 18S ribosomal DNA evidence. *Cladistics* **14**, 249–285.

CHAPTER TWO

SPONGE SYSTEMATICS FACING NEW CHALLENGES

P. Cárdenas*, T. Pérez[†] and N. Boury-Esnault[†,1]

Contents

* Département Milieux et Peuplements Aquatiques, Muséum National d'Histoire Naturelle, UMR 7208 "BOrEA", Paris, France
[†] IMBE-UMR7263 CNRS, Université d'Aix-Marseille, Station Marine d'Endoume, Marseille, France
[1] Corresponding author: Email: nicole.boury-esnault@orange.fr

Advances in Marine Biology, Volume 61
ISSN 0065-2881, DOI: 10.1016/B978-0-12-387787-1.00010-6

Abstract

Systematics is nowadays facing new challenges with the introduction of new concepts and new techniques. Compared to most other phyla, phylogenetic relationships among sponges are still largely unresolved. In the past 10 years, the classical taxonomy has been completely overturned and a review of the state of the art appears necessary. The field of taxonomy remains a prominent discipline of sponge research and studies related to sponge systematics were in greater number in the Eighth World Sponge Conference (Girona, Spain, September 2010) than in any previous world sponge conferences. To understand the state of this rapidly growing field, this chapter proposes to review studies, mainly from the past decade, in sponge taxonomy, nomenclature and phylogeny.

In a first part, we analyse the reasons of the current success of this field. In a second part, we establish the current sponge systematics theoretical framework, with the use of (1) cladistics, (2) different codes of nomenclature (*PhyloCode* vs. Linnaean system) and (3) integrative taxonomy. Sponges are infamous for their lack of characters. However, by listing and discussing in a third part all characters available to taxonomists, we show how diverse characters are and that new ones are being used and tested, while old ones should be revisited. We then review the systematics of the four main classes of sponges

(Hexactinellida, Calcispongiae, Homoscleromorpha and Demospongiae), each time focusing on current issues and case studies. We present a review of the taxonomic changes since the publication of the *Systema Porifera* (2002), and point to problems a sponge taxonomist is still faced with nowadays. To conclude, we make a series of proposals for the future of sponge systematics. In the light of recent studies, we establish a series of taxonomic changes that the sponge community may be ready to accept. We also propose a series of sponge new names and definitions following the *PhyloCode*. The issue of phantom species (potential new species revealed by molecular studies) is raised, and we show how they could be dealt with. Finally, we present a general strategy to help us succeed in building a Porifera tree along with the corresponding revised Porifera classification.

Key Words: cladistics; integrative taxonomy; *PhyloCode*; phylogeny; Linnaean system; classification; Demospongiae; Calcispongiae; Hexactinellida; Homoscleromorpha

1. INTRODUCTION

In the special volume of the *Canadian Journal of Zoology* dedicated to "Biology of neglected groups: Porifera (sponges)", Mackie (2006) underlined that "Systematics is not a static science". Plenty of new ideas regarding the evolution and classification of extant and fossils sponges emerged in this beginning of the twenty-first century. In the preface of the book of abstracts of the Eighth World Sponge Conference (Girona, Spain, September 2010), the organizing committee underlined that "contrary to the expectations from general scientific trends, the field of taxonomy has survived Conference after Conference, and remains a prominent discipline of sponge research, bypassing the changing scientific paradigm" (Uriz *et al.*, 2010; Fig. 2.1). In Girona, 92 contributions have been proposed to the section "Taxonomy & Faunistics" representing thus more than 25% of all contributions, and 46 to the section "Phylogeny & Evolution" counting for about 15% of all contributions. Percentages for both fields were higher than in the previous world sponge conferences. The revival of sponge taxonomy is also illustrated by the numerous "special volumes" dedicated to sponges in the past decades (Table 2.1). As underlined in the preface of a Sponge Biodiversity volume dedicated to sponges: "whereas taxonomy as a Science appears to be in decline globally, we may consider ourselves lucky to see an upsurge in α-taxonomy of sponges from younger scientists" van Soest (2007). Indeed, since 2000, 17 PhDs were defended specifically in sponge taxonomy and/or phylogeny.

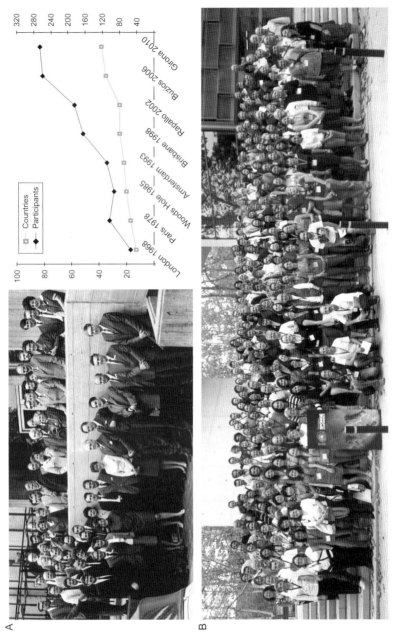

Figure 2.1 Comparison of the number of participants and their countries of origin in the eight international sponge conferences held between 1968 and 2010. (A) Participants to the London Conference in 1968; (B) participants to the Girona Conference in 2010 (photo courtesy Xavier Turon). (For colour version of this figure, the reader is referred to the Web version of this chapter.)

Table 2.1 Numerous "special volumes" dedicated to sponge biology published since 2000

2000_Zoosystema: Volume dedicated to Professor Lévi on sponge taxonomy (Vacelet, 2000)

2002_*Systema Porifera*: A Guide to the Classification of Sponges (Hooper and van Soest, 2002), Kluwer Academic/Plenum Publishers, New York (45 authors from 17 countries)

2003_Microscopy Research and Technique: Volume dedicated to Biology of silica deposition (Uriz, 2003)

2004_Sponge Science in the New Millennium: Proceedings of the VI Symposium (Pansini *et al.*, 2004)

2006_Canadian Journal of Zoology: Biology of neglected groups: Porifera (sponges) (Saleuddin and Fenton, 2006)

2007_Journal of the Marine Biological Association of the United Kingdom: Sponge Biodiversity (van Soest, 2007)

2007_Porifera Research Biodiversity, Innovation and Sustainability: Proceedings of the VII Symposium, (Custódio *et al.*, 2007)

2008_Marine Ecology: Advances in Sponge Research. A tribute to Klaus Rützler (Ott *et al.*, 2008)

2012_Hydrobiologia: Ancient animals, new challenges: Developments in sponge research (Maldonado *et al.*, 2012)

This revival of sponge taxonomy is linked to the emergence of new young spongologists, the huge burst of molecular systematics, the exploration of new geographical areas as well as deep waters thanks to the emergence of new technologies like small submersibles and Remotely Operated Vehicles (ROVs) (Krautter *et al.*, 2006; van Soest *et al.*, 2007a). Even the exploration of supposedly well-known biogeographical areas revealed a higher biodiversity than expected: the Mediterranean Sea (Pansini and Longo, 2008; Coll *et al.*, 2010; Calcinai *et al.*, 2011; Ereskovsky *et al.*, 2011; Pérez *et al.*, 2011), the North-East Atlantic (Rapp, 2006; Picton and Goodwin, 2007; Cárdenas *et al.*, 2011; Goodwin *et al.*, 2011b), the Caribbean Sea (Zea, 2001; Díaz *et al.*, 2005; Rützler *et al.*, 2009; van Soest, 2009), the Southern Ocean (Antarctic) (Calcinai and Pansini, 2000; Ríos *et al.*, 2004; Plotkin and Janussen, 2008; Bertolino *et al.*, 2009; Janussen and Reiswig, 2009; Rapp *et al.*, 2011) and the South Pacific (Fromont *et al.*, 2008; Hooper *et al.*, 2008; Tabachnick *et al.*, 2008; Kelly *et al.*, 2009; Alvarez and Hooper, 2009, 2010; Reiswig and Kelly, 2011).

However, there are regions of the world where the taxonomic effort is, for the time being, very low: Northwest Pacific coasts with the exception of the Korean coasts (Sim *et al.*, 1990; Sim and Shin, 2006; Jeon and Sim, 2008) (see Hooper *et al.*, 2000 Checklist of South China), the littoral zone of West and East Africa, with the exception of South Africa, coasts of the

Indian Peninsula, etc. The emergence of teams of specialized systematicians is needed for these regions.

After the low increase of works dedicated to sponge molecular phylogeny at the end of the twentieth century, we assisted in the past 10 years to a burst of sponge works dedicated to phylogeny (see reviews by Erpenbeck and Wörheide, 2007 and Chapter 1).

The number of new Porifera names per year has been irregular these past 20 years with major peaks due to large taxonomy revisions (Hooper, 1996; Hooper and van Soest, 2002), monographs on campaigns in poorly surveyed areas (Lévi, 1993; Pulitzer-Finali, 1993) or proceedings and special sponge volumes (Fig. 2.2 and Table 2.1). In 2002, the *Systema Porifera* (see Section 2.1) is of course responsible for the peak of new names (other than new species names). Today, large taxonomic monographs on sponges following oceanographic campaigns become scarce, the most recent ones including the hexactinellids from the MAR-ECO campaign in the North-East Atlantic (Tabachnick and Collins, 2008), the Demospongiae from the Brazilian Antarctic Programme (PROANTAR) (Campos *et al.* 2007a,b) or sponges from Clipperton Island (van Soest *et al.*, 2011a). Most taxonomists presently prefer to publish smaller papers focusing on the description of a new species or the revision of a genus or family. For this, two journals are clearly preferred by sponge taxonomists: *Zootaxa* (IF 2010: 0.853) and *Journal of the Marine Biological Association of the United Kingdom* (JMBA)

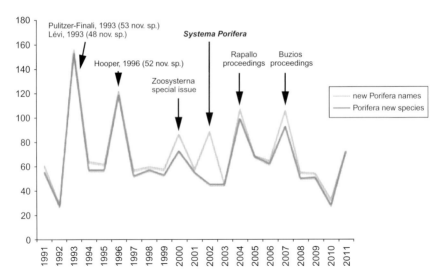

Figure 2.2   Number of new Porifera species (in black) and new Porifera names (in grey) between 1991 and July 2011. Peaks of new names can be explained by large taxonomic revisions, monographs on campaigns in poorly surveyed areas or proceedings and special sponge volumes.

(IF 2010: 0.933) which have, respectively, published 50 and 47 papers on sponge taxonomy between 2006 and September 2011. For sponge phylogeny papers, *Molecular Phylogenetics and Evolution* (IF 2010: 3.889) is clearly preferred (16 papers between 2006 and September 2011). But new online journals are now also being used such as *Zookeys* (IF 2010: 0.514) for taxonomy and *PLoS ONE* (IF 2010: 4.411) for phylogeny.

A total of 8264 extant sponge species are currently loaded in the World Porifera Database (WPD), representing 667 *Calcispongiae*[p], 589 *Hexactinellida*[p], 84 *Homoscleromorpha*[p] and 6922 *Demospongiae*[p] (van Soest *et al.*, 2011b; WPD accessed 15 April 2011). An average of 50 new species/year are described, so if we have 8000 more species to describe as suggested by van Soest (2007), and should the taskforce of taxonomists remain stable, we would need another 160 years to describe all the sponge species of our planet. Among the priorities, many candidate species (cryptic or not) are at the moment only known from sequences (Klautau *et al.*, 1999; Lazoski *et al.*, 2001; Pöppe *et al.*, 2010; Reveillaud *et al.*, 2010, 2011; Xavier *et al.*, 2010a) and awaiting a formal description and the finding of morphological characters to be discriminated (see Section 10.2).

The establishment of a solid theoretical framework through integrative taxonomy, cladistic theory and phylogenetic nomenclature should help to propose solid phylogenetic hypothesis such that we can hope to be at a dawn of a new fruitful era for sponge systematics.

2. Why This Increasing Number of Works on Sponge Taxonomy?

2.1. Development of new tools to share information: The *Systema Porifera* and databases

Among the new tools for sponge taxonomy, it is first and foremost necessary to mention the publication in 2002 of the *Systema Porifera*, a multi-author book edited by Hooper and van Soest where the taxonomic knowledge was summarized and rationalized. The *Systema Porifera* provides a revised taxonomic survey of the Porifera from the genus to the phylum rank. Diagnoses and type species descriptions have been standardized, and terminology made consistent. When available, type material for type species has been redescribed. Keys have been constructed at each rank from class to genus. The *Systema Porifera* is a practical guide to allocate sponges to their proper taxonomic position for non-sponge taxonomists and beginners.

Meanwhile, as in other phyla, sponge taxonomic information has been increasingly digitized, globalized and democratized through internet databases, thus participating to the emergence of the new discipline called "biodiversity informatics" (Sarkar, 2009). A logical follow-up and addition to the *Systema Porifera* is the WPD of all Recent sponges ever described. This database is part of the World Register of Marine Species (WoRMS) (http://www.marinespecies.org/porifera/), a global initiative aiming at registering all marine organisms. This WPD, which is continuously updated by van Soest and 15 co-editors, lists all the described sponge taxa and gives useful information on their potential taxonomic revision, their distribution across the oceans and the most relevant associated literatures (some of which can be downloaded).

The sponge genetree server (www.spongegenetrees.org) is providing a phylogenetic backbone for poriferan evolutionary studies (Erpenbeck *et al.*, 2008). It aims at summarizing the increasing knowledge on sponge molecular systematics and at visualizing DNA sequence information on sponges that has accumulated in the past years in public databases. It is continuously updated automatically. Several additional collaborative web projects are in development, which use the web resources to facilitate cooperation between the different teams like, for example, the Porifera Tree of Life (PorToL) project (http://www.portol.org), LifeDesk (http://porifera. lifedesks.org), the Sponge Barcoding Project (SBP) (http://www. spongebarcoding.org/) and the sponge reef project (http://www.porifera. org/a/ciopen.html). It is important to mention that PorToL and LifeDesk are contributing to the building of an encyclopaedia of life (Wilson, 2003) already well on its way (Encyclopaedia of Life: www.eol.org). Other websites, not specific to sponges, are likely to play important roles in sponge taxonomy: Zoobank (www.zoobank.org/) can store nomenclatural acts and type specimen data, while MorphoBank (www.morphobank.org/) and Morph-D-Base (www.morphdbase.de) are especially designed to store morphological character matrices for morphological phylogenetics and cladistic research. MorphoBank can also be used as a way to store the media of many specimens observed, during a taxonomic revision, for example (Cárdenas *et al.*, 2007). MarinLit is a database of the marine natural product literature (Blunt, 2011). In addition to the usual bibliographic data, this database contains compound information (structures, formulae, molecular mass) and taxonomic data.

Other databases are dedicated to sponge diversity, such as the European Register of Marine Species (van Soest, 2001 for chapter on Porifera) http:// www.marbef.org/data/erms.php, the checklists of Italian sponge fauna (Pansini and Longo, 2003, 2008), http://www.sibm.it/CHECKLIST/02% 20PORIFERA/porifera.htm, an annotated list of NE Pacific sponge species (Austin *et al.*, 2007) http://www.mareco.org/KML/Projects/NEsponges_ content.asp, the Porifera volume of the Zoological Catalogue of Australia (Hooper and Wiedenmayer, 1994) http://www.environment.gov.au/ biodiversity/abrs/online-resources/fauna/afd/taxa/PORIFERA/ and several guides for online identification like "Sponges of Britain and Ireland" http:// www.habitas.org.uk/marinelife/sponge_guide/ (Picton *et al.*, 2011), "South

Florida Sponges: an online guide to identification" (Messing *et al.*, 2010), http://www.nova.edu/ncri/sofia_sponge_guide/index.html, or "The Sponge Guide: A web-based interactive photographic guide to identify Caribbean sponges" (Zea *et al.*, 2009) http://www.spongeguide.org/. With the increasing numbers of divers, numerous websites dedicated to marine fauna and with a section for sponges are flourishing (for example in France: http://doris.ffessm.fr or http://www.souslesmers.fr; in Norway: the Marine Fauna and Flora of Norway http://www.seawater.no/fauna/porifera/). It is quite surprising to meet so many "amateurs" interested in sponge taxonomy and biology.

The revival of sponge taxonomy and expertise nowadays is unquestionably linked to these web tools. If we had a dream it would be to merge most of these tools to make available a huge collaborative database giving the taxonomic keys at the species level, distribution maps, relevant illustrations and bibliographic references.

2.2. Exploration of new geographical areas, or new biotopes (seamount, reef sponge bank, bathyal and abyssal zone) with new tools and techniques

Sponge biodiversity surveys in several poorly sampled areas have increased: Indonesia (de Voogd and Cleary, 2008), South America (Hajdu and Desqueyroux-Faúndez, 2008; Goodwin *et al.*, 2011a; Muricy *et al.*, 2011), Pacific coast of North America (Carballo *et al.*, 2004; Austin *et al.*, 2007), Central Pacific (van Soest *et al.*, 2011a) and South Africa (Samaai and Gibbons, 2005). At the same time, exploration of well-known areas (e.g. Western Mediterranean sea, NE Atlantic or Antarctic Ocean) is frequently also bringing new species. The high number of faunistic surveys in the beginning of the twenty-first century reminds us of the turn of the nineteenth century when so many pioneer expeditions where undertaken such as the Challenger, or the Prince Albert I de Monaco campaigns for instance.

To illustrate this increase in biodiversity surveys, the case of Brazil is interesting and worth mentioning. Thanks to a governmental programme, the Project REVIZEE 2000–2009, which aimed at listing and mapping the biological resources of the Brazilian Exclusive Economic Zone (EEZ; e.g. Amaral *et al.*, 2003; Amaral and Rossi-Wongtschovski, 2004; Lavrado and Ignacio, 2006) and the project of development of Brazilian marine sponge taxonomy (Petrobras-ANP SAP 4600177470), an extraordinary joint effort has been made by the Brazilian universities to succeed in publishing a large number of new records of sponges for Brazil (Hajdu *et al.*, 1999, 2003, 2004; Mothes *et al.*, 2004a; Muricy *et al.*, 2006, 2007; Cedro *et al.*, 2011; Fernandez *et al.*, 2011; Lopes *et al.*, 2011) as well as chapters or books about sponge biodiversity (Mothes *et al.*, 2004b, 2006; Moraes *et al.*, 2006; Muricy and Hajdu, 2006; Muricy *et al.*, 2008). It is therefore no surprise when news from the Zootaxa Web site (14/03/2011) announced that "1,244 is the number of authors from Brazil, exceeding that of the USA (1243), which held the record of 1st place for ten years. More authors from the top biodiversity-rich country is great news for taxonomy". The sponge fauna of Brazil reaches nowadays 443 species (Muricy *et al.*, 2011), with, respectively, 380 species for Demospongiae (including 53 species of freshwater sponges), 47 for Calcispongiae and 16 for Hexactinellida. About 90 new species were described during the past 20 years. Quite naturally, Brazilian spongologists ended up organizing the Seventh International Sponge Symposium, held in Armação dos Búzios (Rio de Janeiro, Brazil) in May 2006. Whereas only one Brazilian spongologist attended the First International Sponge Symposium in London (1968), Brazil was the third most important delegation during the Eighth International Sponge Symposium in Girona (2010) with 19 participants, against 47 from the USA and 21 from Germany (Fig. 2.1). One challenge for the next few years is to promote the training of young systematicians in countries where there is still a significant hidden biodiversity. It includes countries close to hot spots of biodiversity, but also parts of the world or ecosystems that remain to be explored.

Deep-water sponge grounds are a key component of deep-sea ecosystems. They are quite diverse and occur worldwide. A huge effort has been made in the past few years to evaluate their global distribution and to understand their ecological role (Hogg *et al.*, 2010). The deep-water sponge grounds were discovered during the nineteenth century by the historic expeditions such as the HMS Porcupine, the Challenger and the Norwegian Cruises of Michael Sars. Sponge dwellers of particular habitats in the deep-sea are being continually investigated with traditional techniques. Boxcores, dredges and trawls are still used to study deep-sea coral reefs (Longo *et al.*, 2005; van Soest *et al.*, 2007a) or fjords (Cárdenas *et al.*, 2010), and scuba diving was used to study the top of seamounts (Xavier and van Soest, 2007). The availability of new tools such as deep-sea submersibles, beam trawls equipped with cameras and especially ROVs enables to explore and collect more efficiently in habitats difficult to reach and sample (i.e. in the deep sea). After the pioneer works with autonomous underwater vehicles in the middle of the twentieth century (Laborel *et al.*, 1961; Vacelet, 1969), the availability of ROVs at the end of the twentieth century clearly promoted studies of these deep ecosystems. Observing sponges in their natural habitats and carefully collecting specific individuals with such tools bring much more biological information than dredging sponges that arrive on deck in fairly bad shape, mixed with a lot of sediments and other organisms. They were used to obtain a better knowledge of the sponge fauna living in some remote ecosystems such as hydrothermal vents (Schander *et al.*, 2010), deep-sea coral reefs (Mastrototaro *et al.*, 2010; Lopes *et al.*, 2011), bathyal sponge reefs (Leys *et al.*, 2004; Conway *et al.*, 2005; Krautter *et al.*, 2006), seamounts (Aguilar *et al.*, 2011), canyons

(Schlacher *et al.*, 2007) and fjords (H. T. Rapp and P. Cárdenas, personal communication) (Fig. 2.3; Rützler, 1996). On the continental shelf of British Columbia, ROVs allowed to demonstrate that sponge reefs are one of the most remarkable areas in terms of sponge abundance and diversity. Hexactinellid species (*Heterochone calyx* Schulze, 1886,

Figure 2.3 Underwater photographs taken by ROVs (Remotely Operated Vehicles) on seamounts and canyons in the NW Mediterranean Sea. (A) ROV "Achille" from COMEX. (B) Undescribed Hexactinellida species from Valinco Canyon 41°41.227′N 008°47.446′E, 187 m deep. (C) Encrusting *Hamacantha falcula* from Galeria Canyon 42°28.974′N 008°34.402′E, 447 m deep. (D) Orange *Poecillastra compressa* and white *Geodia* sp. from Esquine Bank 43°03.105′N/005°32.973′E, 100–120 m deep. (E) Encrusting sponge species on an overhanging wall from Valinco Canyon 41°41.226′N 008°47.457′E, 178 m deep. (F) *Hamacantha falcula* with digitations from Castelsardo Canyon 41°17.300′N 008°44.489′E, 128 m deep. A (photo courtesy of the Comex); B–F (photos courtesy French Agency "Aires marines protégées"). (For interpretation of the references to colour in this figure legend, the reader is referred to the Web version of this chapter.)

Aphrocallistes vastus Schulze, 1886 and *Farrea occa* Bowerbank, 1862) are the main frame builders (Leys *et al.*, 2004; Conway *et al.*, 2005; Krautter *et al.*, 2006 (http://gsc.nrcan.gc.ca/marine/sponge/fauna_e.php); Chu and Leys, 2010). Mass occurrences of hexactinellids were also observed on the Meteor Seamount (Xavier *et al.*, 2010b: *Poliopogon amadou* Thomson, 1878); in the Rockall Bank (van Soest *et al.*, 2007a: *Rossella nodastrella* Topsent, 1915) and throughout the northeast Atlantic/west Mediterranean region (Reiswig and Champagne, 1995: *Pheronema carpenteri* Thomson, 1869). *Pheronema carpenteri*, also called the "bird's nest sponge" (Hogg *et al.*, 2010), is widely distributed in dense populations from the southern flank of the Iceland-Faroes Ridge to the Azores, the Canary Islands and off Morocco. In the northeast Atlantic, sponge grounds are dominated by *Astrophorida*[P] species (Klitgaard and Tendal, 2004; Klitgaard *et al.*, 1997). On the deep Antarctic shelf considered as a "sponge kingdom", about 300 species have been recorded with a high biomass density (Koltun, 1968; Barthel, 1992; Barthel and Gutt, 1992) and the number of species is increasing continuously in this region thanks to the numerous expeditions undertaken by the international research community.

2.3. Interest in molecular phylogeny by the whole biologist community, and new tools increase the amount of data

The first molecular phylogeny work on sponges was published in 1991 (Kelly-Borges *et al.*, 1991). It used partial 18S rDNA sequences to formulate phylogenetic hypotheses for sponges of the order Hadromerida. After a very slow start, we are faced nowadays with an increasing number of molecular phylogenetic works (see Erpenbeck and Wörheide, 2007; Chapter 1). According to Web of Science, there are about six papers/year investigating sponge molecular phylogenies.

Scientists at the Eighth Sponge World Conference of Girona (Spain) in September 2010 reached an important consensus regarding the four poriferan clades: *Hexactinellida*[P], *Demospongiae*[P], *Homoscleromorpha*[P] and *Calcispongiae*[P] (Manuel and Boury-Esnault, in press). However, the relationships between these four clades are still not fully consensual. Most studies now agree that the *Silicea*[P] Gray 1867 (*Hexactinellida*[P] + *Demospongiae*[P]) are monophyletic (Adams *et al.*, 1999; Borchiellini *et al.*, 2001; Medina *et al.*, 2001; Dohrmann *et al.*, 2008; Philippe *et al.*, 2009; Sperling *et al.*, 2010). This clade is further supported by a morphological synapomorphy: the same process of secretion of the siliceous spicules within sclerocytes around an axial filament (Leys, 2003). Meanwhile, *Homoscleromorpha*[P] and *Calcispongiae*[P] are considered either as sister groups within the *Porifera*[P] (Dohrmann *et al.*, 2008; Philippe *et al.*, 2009; Pick *et al.*, 2010) or as a paraphyletic group and closer to the Eumetazoa (Sperling *et al.*, 2007, 2010). For the time being, no synapomorphy has been found to support the sister-group relationship between *Homoscleromorpha*[P] and *Calcispongiae*[P] frequently suggested over the years (see Chapter 1). Since *Demospongiae*[P] and *Hexactinellida*[P] do not possess cross-striated rootlets, these were suggested as a possible synapomorphy for the clade (*Homoscleromorpha*[P] + *Calcispongiae*[P]) (Gazave et al., 2012). However, if they are known in the flagellated cells of amphiblastula and coeloblastula larvae of *Calcispongiae*[P] (Amano and Hori, 1992, 2001; Gallissian and Vacelet, 1992) and in those of cinctoblastula larvae of *Homoscleromorpha*[P] (Boury-Esnault et al., 2003), they are also present in placozoans and in all eumetazoans (Nielsen, 2001) as well as in *Monosiga* (Choanoflagellata), *Naegleria* (Percolozoa), *Noctiluca* (Dinophyta) and *Tritrichomonas* (Parabasalia) (Dilton, 1981; Vickerman et al., 1991; Nielsen, 2001). If the cross-striated rootlet is homologous between metazoans and protistan lineages, then the character is plesiomorphic for *Metazoa*[P] and has been lost in a common ancestor of *Demospongiae*[P] and *Hexactinellida*[P] (Boury-Esnault et al., 2003). If this character is not homologous between protistan lineages and *Metazoa*[P], it is a synapomorphy for *Metazoa*[P] which has been lost in a common ancestor of *Demospongiae*[P] and *Hexactinellida*[P] and in any case it cannot be considered as a synapomorphy for the (*Calcispongiae*[P] + *Homoscleromorpha*[P]) clade.

While morphological and molecular data sets for *Hexactinellida*[P] are largely congruent (Dohrmann et al., 2008, 2009), for *Calcispongiae*[P] they are congruent only for the deep nodes of the tree, the molecular results confirming the hypothesis of Bidder (1898) of two subclasses Calcinea and Calcaronea (Manuel et al., 2003, 2004; Dohrmann et al., 2006). For *Demospongiae*[P], molecular results (McCormack et al., 2002; Borchiellini et al., 2004a; Erpenbeck et al., 2005, 2007a,b,c; Nichols, 2005; Redmond and McCormack, 2008, 2009; Redmond et al., 2007; Gazave et al., 2010a; Cárdenas et al., 2011; Morrow et al., 2012) propose a quite different hypothesis from those obtained through cladistic analysis of the morphological characters (de Weerdt, 1985, 1986, 2000; van Soest, 1990, 1991; Hooper, 1991, 1996; Maldonado, 1993; de Weerdt et al., 1999; Alvarez et al., 2000; Alvarez and Hooper, 2009). An important re-evaluation of the current classification using both data sets is necessary to understand and overcome this incongruence.

3. THEORETICAL FRAMEWORK

3.1. Cladistics and morphology

Willi Hennig introduced cladistics in the 1950s (Hennig, 1950). Cladistics started to grow in the 1970s with the first phylogenetic algorithms, but it was used only in a few animal groups (e.g. myriapods, insects, mammals, birds). It is only in the 1980s that it inspired researchers working with all kinds of living organisms. The Hennigian philosophy was introduced in the sponge world in the middle of the 1980s by van Soest (1984a,b, 1987, 1990, 1991), de Weerdt (1989) and Hooper (1990, 1991). They favoured the cladistic approach because the resulting phylogeny not only provided considerable information but also had predictive value (van Soest, 1990). In cladistics, one specifically selects homologous characters in order to identify shared derived characters (synapomorphies) supporting clades (i.e. monophyletic groups of taxa all sharing an exclusive common ancestor). This is important as it is often misunderstood: cladistic methods do not test a phylogeny, but they test the hypotheses of homology of characters upon which it is based (Nelson, 1994). The first attempts to apply the cladistics principles were based mostly on one data set, the skeleton characters. The cladistic approach led to a reappraisal of the classification of the phylum and especially the Demospongiae (Hooper, 1991; van Soest, 1991) and resulted in the classification followed in the *Systema Porifera*. This Soest–Hooper classification differs from the Lévi–Bergquist–Hartman classification (Lévi, 1973; Bergquist, 1978; Hartman, 1980) mainly by the bursting of Axinellida Lévi, 1953 and the reallocation of (i) Axinellidae Carter, 1875, Heteroxyidae Dendy, 1905 and Bubaridae Topsent, 1894 to Halichondrida Gray, 1867; (ii) Hemiasterellidae Lendenfeld, 1889 and Trachycladidae Hallmann, 1917 to Hadromerida Topsent, 1894; and finally (iii) Raspailiidae Nardo, 1833, Euryponidae Topsent, 1928 and Rhabderemiidae Topsent, 1928 to Poecilosclerida Topsent, 1928.

This past decade has seen few publications building sponge morphological phylogenies (Calcispongiae: Manuel *et al.*, 2003; Hexactinellida: Dohrmann *et al.*, 2008; axinellids: Alvarez *et al.*, 2000; Verongida: Erwin and Thacker, 2007; *Petromica*: List-Amitage and Hooper, 2002; Homoscleromorpha: Muricy, 1999; Polymastiidae: Plotkin *et al.*, 2012; Guitarridae: Uriz and Carballo, 2001; Clionaidae: Rosell and Uriz, 1997). All of these studies used maximum parsimony (MP) to reconstruct their trees, but new likelihood methods for morphological characters are now available (Lewis, 2001a,b) and should be used in the future. Because morphological phylogenies often give poor resolution, another option is to use the morphological matrix to reconstruct character states at ancestral nodes on a molecular phylogeny. In our opinion, mapping of characters should be systematically investigated, since it can really help to understand the morphological evolution of the group studied and to reassess the characters used in the current classification. To map these characters, MP reconstruction was used in Calcispongiae (Manuel *et al.*, 2003), Demospongiae (Borchiellini *et al.*, 2004a), Verongida (Erwin and Thacker, 2007) or Astrophorida (Cárdenas *et al.*, 2010), while more sophisticated likelihood methods are just beginning to be used (Cárdenas *et al.*, 2011).

If cladistic is nowadays accepted and applied in most of the recent works, spongologists are still faced with a huge problem, that is, the absence of knowledge about the homology of morphological characters and the ancestral state of a character (Bergquist and Fromont, 1988; Fromont and Bergquist, 1990). Consequently, hypotheses regarding primary homology are established on exceedingly weak grounds. In spicule nomenclature (Boury-Esnault and Rützler, 1997), one name corresponds to a general form and is only descriptive without being indicative of homology (Fromont and Bergquist, 1990). For example, there are no arguments to be sure that the oxea present in Haplo-sclerida are homologous to those in Halichondrida or Tetractinellida, and the asters found in *Timea* Gray, 1867 are homologous to those found in *Adreus* Gray, 1867 or *Stelletta* Schmidt, 1862. The palaeontological information is too scarce to help answer this fundamental question: Which is ancestral, the tetraxon spicule or the monaxon spicule? The most accepted hypothesis is that the tetraxon is the ancestral state (Schulze, 1880; Dendy, 1921), although Reid (1970) suggested that it could be the monaxon. This last hypothesis was corroborated by the presence of monaxon spicules of Demospongiae in Precambrian fields (Li *et al.*, 1998; Pisera, 2006) and by molecular data (Borchiellini *et al.*, 2004a). The problem linked to primary homology was not taken sufficiently into account in the first sponge cladistic works as well as the character losses which appear much more common than previously thought (see Section 4.1).

3.2. From the Linnaean classification to the *PhyloCode*

To share scientific knowledge, in the most accurate way, scientists need (1) international rules to name taxa (2) to follow these rules in order to give clear, stable, unambiguous names to describe our rich biological environ-ment. This is the basis of the "biological nomenclature". Carl Linnaeus (1707–1778) clearly helped with the first of these requirements when he introduced binominal names for plant species in his *Species Plantarum* (1753), which was generalized to all living organisms in his 10th edition of *Systema Naturae* (1758). The official rules and recommendations to name animals came only later with the first edition of the *International Code of Zoological Nomenclature* (*ICZN*) (1905) and have evolved till the fourth edition of the *ICZN* (1999).

Meanwhile, scientists are also looking for a way to order this biological knowledge in a more useful way, in order to understand the world they live in. Therefore, scientists want to give names that (1) reflect a classification which itself (2) reflects the "true" tree of life in the best possible way, in the sense of being the result of evolutionary process. This science of classifica-tion, representation and analysis of organism relationships is called "taxon-omy". Carl Linnaeus (1758) was one of the first botanists to classify organisms in groups assigned to different ranks (kingdom, class, order, genus, species) in a hierarchical manner. But the classification was somewhat artificial by choosing arbitrary features, and without trying to interpret the relationships obtained (since it was thought that organisms were not historically related). Through the nineteenth century and with the Darwinian revolution, taxonomists felt that classification should now reflect common ancestry and thereby a truly "natural" tree of life (i.e. phylogeny), recapitulated in Darwin's (1859) famous quote "all true classification is genealogical". Fortunately, the Linnaean's system was adaptable to the Darwinian theory because the hierarchical ranks can mirror a phylogenetic tree and its nested sets of taxa (Dominguez and Wheeler, 1997; Gould, 2002). Phylogenetic trees started to grow and were interpreted in terms of patterns (evolutionary relationships) and processes (evolutionary mechanisms). Phylogeny and classification were now tightly linked for better and for worse. After the Hennigian revolution, most taxonomists have considered clades to be the only "natural" taxa that rightfully belonged in the biological classification. But with the overwhelming success of cladistics, phylogeneticists started to neglect or simply ignore Linnaean classification, thus creating an ever-growing phylogeny/classification gap: "fewer and fewer phylogenetic results are translated into proper Linnaean names and definitions" (Franz, 2005). Meanwhile, the Linnaean classification was under fire of many taxonomists reproaching among other things (i) its instability, (ii) the diversity of its codes (different codes for animals, for plants and fungi, for bacteria and for viruses), (iii) the absence of rules dealing with ranks above the superfamily (in the *ICZN*) and above all (iv) the subjectivity and inconsistencies of its ranks. Indeed, taxa at any given rank are not equivalent or comparable under any biological criterion, they just inform us on the hierarchical structure of the taxonomy, which of course can be useful for the storage and retrieval of taxonomic information (Dubois, 2007, 2011) as well as for memory purposes. These shortcomings of the Linnaean classification, notably the subjective nature of ranks, have a direct impact on evolutionary biology, ecology, palaeontology and biodiversity studies (Bertrand *et al.*, 2006; Laurin, 2010; Avise and Liu, 2011).

Actually, most taxonomists are aware of the problems raised by the Linnaean classification, and some of them think that it could be improved instead of creating a new system and a new code (Nixon *et al.*, 2003; Dubois, 2011). To facilitate discussions and the development of the future fifth edition of the *ICZN*, *ICZNwiki* (http://iczn.org) has been launched. Another way to address these criticisms is the initiative of the *BioCode*, which appeared in the 1990s and with the objective to replace the existing nomenclatural codes with a single universal one. After the *Draft BioCode* (1997) (Greuter *et al.*, 1998) rejected by the taxonomist community, a less contentious revised *Draft BioCode* (2011) (Greuter *et al.*, 2011) is now proposed: "a framework overarching the practices of the current series of codes, but which also addresses ways in which some of the key issues of current concern in systematics could be handled by all codes, for example, the registration of new names and electronic publication" (Hawksworth, 2011).

Yet another school of thought emerged in the 1990s, unsatisfied that the concept of evolution, notably the principle of descent, did not have a central role in the Linnaean system and nomenclature: "In order to make the definition of names evolutionary, they must be rooted in the concept of common ancestry" (de Queiroz and Gauthier, 1990). To achieve this, de Queiroz and Gauthier (1990, 1992, 1994) suggested that taxa should not be defined by their characters (a view inherited from Aristotle) but by their clade (i.e. ancestor + descendants). This "phylogenetic taxonomy" is to be governed by "phylogenetic nomenclature" following the rules of the *Phy-loCode*: "a formal set of rules governing phylogenetic nomenclature [...] designed to name the parts of the tree of life by explicit reference to phylogeny" (http://www.ohio.edu/phylocode). Basically, phylogenetic names are defined with respect to a specific point on a phylogenetic tree and the name refers to all organisms descending from that point (Pleijel and Rouse, 2003). To specify this point, the *PhyloCode* (Article 9) proposes three main ways (i.e the three types of "definitions of names"): node-based definition, branch-based definition and apomorphy-based definition (Fig. 2.4). All three types of definition make use of "specifiers" which can be species, specimens or apomorphies (Article 11). Species used in the definitions should refer to deposited specimens, thus reinforcing the use of vouchers, whose importance has been repeatedly stressed in molecular phylogenetic studies (Funk *et al.*, 2005; Pleijel *et al.*, 2008). One should also keep in mind that when species names (and not specimens) are used as specifiers, their name-bearing type specimens are *de facto* specifiers (*Phylo-Code*, Note 13.2.2). Of course, the phylogenetic relationships of these

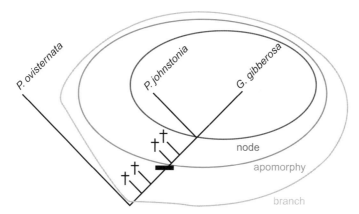

Figure 2.4 *PhyloCode* definitions. Modified from de Queiroz and Gauthier (1992; Fig. 4). The nested bubbles illustrate branch-, apomorphy- and node-based Geodiidae definitions. See text for wording of definitions. Crosses represent extinct and/or unknown species. (For colour version of this figure, the reader is referred to the Web version of this chapter.)

species or specimens used as specifiers should be well supported to reduce the risk of synonymy or homonymy in the future (Schander and Thollesson, 1995). For example, using species as specifiers, Cárdenas *et al.* (2010) gave a node-based definition to the sponge *Geodiidae*[p] clade (names established under the *PhyloCode* are always in italics and can be identified with the symbol "p" to avoid confusion when it comes to genera or species names of the Linnaean classification): the least inclusive clade containing *Geodia gibberosa* Lamarck, 1815 and *Pachymatisma johnstonia* (Bowerbank in Johnston, 1842) (Fig. 2.4). But they could have also decided to give an apomorphy-based definition—the clade originating with the first species to possess sterrasters as inherited by *P. johnstonia* (Bowerbank in Johnston, 1842)—or a branch-based definition—the most inclusive clade containing *Geodia gibberosa* Lamarck, 1815 and not *Pachastrella ovisternata* Lendenfeld, 1894 (Fig. 2.4). In this particular case, a node-based definition is preferable because the relationships within the *Geodiidae*[p] are better known and reliable than within its sister group (Cárdenas *et al.*, 2011). Not only one delineates taxa so that they are *explicitly* monophyletic (which is not the case in the Linnaean system), but also this phylogenetic definition will point to taxa, without any reference to ranks. Indeed, character-based definitions can be ambiguous, especially when new taxa (extant or fossil) are discovered with only parts of the characters expected or overlapping characters. With the *PhyloCode*, the placing of a new taxon does not depend on its characters, it rather depends on how the other taxa were defined (i.e. limited). Characters, instead of defining the taxa, become diagnostic so that, for example, secondary loss of synapomorphies is minor problems because only the occurrence of the character in the ancestor of the clade matters. As for the elimination of ranks, it should stimulate the use of more rigorous measures of biodiversity with direct reference to phylogenetic patterns and processes (Pleijel and Rouse 2003; Bertrand *et al.*, 2006) without the temptation to give ranks a biological meaning. The absence of ranks also clearly facilitates the naming of new clades, one at a time, without having to name all clades, use redundant taxa (e.g. monotypic genus and monogeneric families) or disturb the whole ranking system (a major source of name instability). At a time when sponge molecular phylogenies tend to overturn current or traditional classifications, rank disturbance is bound to happen with some families found nested within other families (e.g. Calthropellidae within Geodiidae (Cárdenas *et al.*, 2011); Fig. 2.5). Furthermore, as phylogenetic trees get bushier with ever-growing sampling, it seems that after a few bifurcations, all the common Linnaean ranks are exhausted (Pleijel and Rouse, 2003) calling thus for unconventional intermediate ranks which are not governed by the *ICZN* (e.g. supra-genera, infra-genera). This is typically what is bound to happen to particularly speciose sponge genera such as *Haliclona* (412 species, 6 subgenera), *Mycale* (232 species, 11 subgenera), *Callyspongia* (180 species, 5 subgenera), *Hyalonema* (107 species, 12

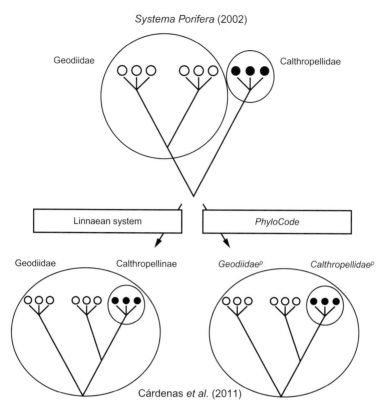

Figure 2.5 In the *Systema Porifera* (2002), Geodiidae and Calthropellidae are two separate families of Astrophorida (Demospongiae). But Cárdenas *et al.* (2011) suggest that the Calthropellidae may be part of the Geodiidae. In a Linnaean system, the Calthropellidae would be downgraded to a subfamily and see their name changed to Calthropellinae. With the *PhyloCode*, *Geodiidae*[P] and *Calthropellidae*[P] names would remain stable because the names are not linked to any rank. *Modified from de Queiroz and Gauthier (1994, Box 5).*

subgenera) or *Stelletta* (140 species, no subgenera) (van Soest *et al.*, 2011b). The recognition of many more ranks at low nomenclatural levels (i.e. just above genus, between genus and species and below species) in the *ICZN* is actually a request from some taxonomists (Dubois, 2011).

The *PhyloCode* deals with all the problems—seemingly governed by the *ICZN*—such as synonymy (Article 14), homonymy (Article 13) and priority of names (Article 12) (de Queiroz and Gauthier, 1990). For example, names are synonymous if they designate the same clade; priority is established by the first use of name to designate a particular clade (de Queiroz and Gauthier, 1994). Similar to the *BioCode*, the *PhyloCode* (i) will be applicable to all organisms, extant or extinct (vs. different codes today for different reigns) and (ii) will require the establishment of an online name registration database, to reduce the number of homonyms/synonyms. It has also been suggested that the *PhyloCode* (Article 16) copes better than the *ICZN* for the naming of hybrids (Laurin and Bryant, 2009), which in all likelihood exist in sponges. For species names, the *PhyloCode* (Article 21.5) recommends the use of binominal names, meaning "prenomen" (usually the genus name in the Linnaean system) and "species epithet" (Dayrat *et al.*, 2008). In case the genus has not been established as a clade name, the species epithet can be combined with the name of a more inclusive clade, which has a name under the *PhyloCode* (Note 21A.1) (e.g. *Demospongiae*[P] *panicea* for *Halichondria panicea* or *Geodiidae*[P] *johnstonia* for *P. johnstonia*).

The *PhyloCode* version 4c (http://www.ohio.edu/phylocode) is still a draft and has not been formally implemented yet. However, it is already widely applied in a variety of nomenclatural and phylogenetic contexts, mainly for vertebrates, Recent or fossils (Pleijel and Rouse, 2003, and references therein), but also for non-vertebrate taxa ranging from fungi (Hibbett and Donoghue, 1998), green algae (Nakada *et al.*, 2008), vascular plants (Cantino *et al.*, 2007), nemerteans (Härlin and Härlin, 2001), molluscs (Thollesson, 1999), annelids (Pleijel and Rouse, 2000) to Porifera (cf. following references). The *PhyloCode* will be published along with a "Companion Volume" including the first taxonomic names established under the code (Laurin and Bryant, 2009). Therefore, this "Companion Volume" will also officially present the first sponge *PhyloCode* names (Manuel and Boury-Esnault, in press). But in our opinion, this pre-*Phylocode* era is an opportunity not only to make the scientific community acquainted with the *PhyloCode* but also to discuss and polish the definitions we want to give for these clades, before their formal and official acceptance. Furthermore, the trial phylogenetic classifications and the *PhyloCode* definitions proposed should help focus future discussions of the *PhyloCode* on real definitions rather than simplified hypothetical ones (on which critics have been mainly focusing on) as well as explore ways to treat such taxon names (Pleijel and Rouse, 2003; Cantino *et al.*, 2007). As of today, there are 48 pre-*PhyloCode* Porifera names distributed in all groups of sponges except Hexactinellida (Manuel *et al.*, 2003; Borchiellini *et al.*, 2004a; Cárdenas *et al.*, 2010; Gazave *et al.*, 2010a, 2012; Cárdenas *et al.*, 2011; this review Section 10.1.2) (Table 2.2). Most of these names were taken from the Linnaean classification and converted to the *PhyloCode* (in order to take advantage of the previous 170 years of Porifera taxonomic studies, and to promote name stability), while seven new names define totally new clades which do not have corresponding names in the Linnaean classification: *Depressiogeodia*[P] (Cárdenas *et al.*, 2010), *Cymbaxinella*[P] (Gazave *et al.*, 2010a), *Geostelletta*[P] (Cárdenas *et al.*, 2011), *Plakostrella*[P] (Gazave *et al.*, 2012) and *Tetralophosa*[P] (Gazave *et al.*, 2012), *Haploscleromorpha*[P] and *Heteroscleromorpha*[P] (see Section 10.1.2) (Table 2.2). Of course, these names have no nomenclatural status (1) under the *PhyloCode*, since they were defined

Table 2.2 List of the *PhyloCode* names defined in the "Companion Book" and in the literature

Companion book	Definitions	References
Porifera	N	Manuel and Boury-Esnault (in press)
Calcispongiae	N	
Hexactinellida	B	
Homoscleromorpha	B	
Demospongiae	B	
From other sources		
Calcinea	B	Manuel *et al.* (2003)
Calcaronea	B	
Baeriida	B	
Lithonida	B	
Keratosa	B	Borchiellini *et al.* (2004a)
Myxospongiae	N	
Tetractinellida	A	
Acanthella	B	Gazave *et al.* (2010a)
Agelas	A	
Agelasida	N	
Axinellidae	B	
Axinella	B	
Cymbaxinella	B	
Scopalina	A	
Geodiidae	N	Cárdenas *et al.* (2010)
Geodinae	B	
Depressiogeodia	B	
cydonium	B	
Geodia	B	
Erylinae	B	
Pachymatisma	B	
Ancorinidae	B	Cárdenas *et al.* (2011)
Astrophorida	N	
Calthropella	B	
Dragmastra	B	
Erylus	B	
Geostelletta	B	
Pachastrella	B	
Penares	B	
Stelletta	B	
Stryphnus	B	
Synops	B	
Thenea	B	
Theonellidae	B	
Vulcanellidae	B	

(continued)

Table 2.2 *(continued)*

Companion book	Definitions	References
Oscarellidae	B	Gazave *et al.* (2012)
Plakinidae	B	
Corticium	A	
Plakostrella	A	
Tetralophosa	B	
Haploscleromorpha	B	This review
Heteroscleromorpha	B	
Spongillida	B	

In the first column, *PhyloCode* names, most of them are clade names converted from pre-existing names from the Linnaean classification, underlined names are new clade names. In the second column, a letter indicates the type of definition followed. A, apomorphy based; B, branch based; N, node based. In the third column, nominal authors of clade names (persons who established the name, including a phylogenetic definition for it under the *PhyloCode*).

before the *PhyloCode* was officially implemented (cf. Preamble of *Phylo-Code*), as well as (2) under the *ICZN*, since phylogenetic definitions do not obey the *ICZN* rules (notably the need of a rank and the designation of types) or (3) the *BioCode* (cf. Preamble). Consequently, these names will need to be made official once the *PhyloCode* is published. In comparison with other taxa, there are still few sponge phylogenies. As a consequence, informal unranked names are luckily still few. But who remembers the Demospongiae Clade C of Nichols (2005)? And how many sponge scientists remember what the G2 clade (Borchiellini *et al.*, 2004a) holds (not speaking of non-sponge scientists)? We need names and by providing formal phylogenetic definitions, the hope is to standardize the application of names for these important new clades (Cantino *et al.*, 2007), which are expected to multiply.

Critics of the *PhyloCode* have been numerous (for a review, see Pleijel and Rouse, 2003), including in the sponge scientific community (Hooper and van Soest, 2010) where it was considered as a potential "threat" to the *Systema Porifera*, for three main reasons: (1) "it doesn't fit" meaning that since most of the databases (e.g. EOL, GBIF, WoRMS (and therefore the WPD)) use a Linnaean system with ranks, the *PhyloCode* cannot be applied and (2) boundaries between "nomenclature and classification become blurred" (e.g. while a classification may change dramatically, informed by progress in constructing new phylogenies, the nomenclature to define life forms needs to remain unambiguous and constant) and (3) because the *PhyloCode* enables to name clades without any synapomorphies. The argument of not using the *PhyloCode* because "it wouldn't fit" is a false one: minds and databases evolve, they can adapt. The tentative inclusion of

Depressiogeodia[P] at the genus rank by R. van Soest in the WPD (accessed 20 September 2011) shows that solutions can be found. As for classification and nomenclature, they are obviously linked in the *PhyloCode* since name definitions are based on a phylogeny, but this does not mean that the names will change with new phylogenetic results. On the contrary, taxon names will remain stable because they unambiguously limit a clade, whose content will change eventually. de Queiroz (2006) showed that in fact both the *PhyloCode* and the Linnaean codes are not able to maintain a complete separation between nomenclature and classification. The Linnaean system also manages to name taxa without synapomorphies or without phylogenetic relationships: just think of all the sponge taxa without clear synapomorphies (e.g. Axinellidae, Halichondrida); this did not prevent researchers from communicating and using the Linnaean system. Actually, in the "Companion Volume", sponge *PhyloCode* names will all be associated with "diagnostic apomorphies" (Manuel and Boury-Esnault, in press). We recommend that whenever synapomorphies are present, they should be associated with the *PhyloCode* names but this will not always be possible. So should we refrain from naming the Haplosclerida clades although they were confirmed and well supported with many sets of independent molecular data? Synapomorphies could also be found after the clade has been named as we increase the sampling, or as we look for new characters. So we do not see the *PhyloCode* as a potential threat to the *Systema Porifera* and the Linnean classification, and we have a few suggestions on their respective uses (see Section 10.2).

3.3. Integrative taxonomy

α-Taxonomy is central to good systematics and phylogenetics. Sponge taxonomy is largely based on spicule/fibre morphology but in the past 20 years, sponge taxonomists have been slowly adopting new concepts and methods to gain access to additional sources of data. The use of multiple and complementary sources of data to evaluate the status of species is called integrative taxonomy (Dayrat, 2005). The integrative taxonomy approach combining all kinds of data (external morphology, spicules, embryology, geography, reproduction, genetic sequences, etc.) is now considered the most reliable and efficient way to evaluate the status of a species (Dayrat, 2005; DeSalle *et al.*, 2005; Padial and De La Riva, 2007; Padial *et al.*, 2009). Indeed, it is now understood that neither morphology nor genetics should be used as a single source of information (Page and Hughes, 2011) and that conflicts between morphologists and molecular biologists should be replaced by systematic integration of both fields. Obviously, with the spreading of molecular techniques, more and more taxonomists are already combining morphology and molecular data in their studies. Of the 17 PhDs defended since 2000 and involving sponge taxonomy/phylogeny, only 3 did not use molecular data. But further alternative independent character data sets are also an opportunity to reassess the validity of species and their morphological characters. Actually, Schlick-Steiner *et al.* (2009) recommend a minimum of three independent disciplines including morphology, genetics and a third data set. One way to illustrate how integrative taxonomy could work is "the taxonomic circle" (DeSalle *et al.*, 2005; Fig. 2.6). The lines that traverse the inner part of the circle indicate experimental routes that can be taken to corroborate taxonomic hypotheses. To reveal a new taxon, you need to confirm your species hypotheses with a different data set, once or more, depending on your strategy. This allows you to escape or "break out" of the taxonomic circle (represented by the solid arrow in Fig. 2.6) which is a metaphoric way to validate a new species hypothesis. Illustrations of how to read this taxonomic circle are given in Fig. 2.7, using two examples from the literature.

We reviewed the 26 integrative sponge systematics studies published since 2006 to understand how this approach was applied in the sponge field (Table 2.3). The number of independent data sets ranged from two to five per study. Molecular data now being more accessible and widespread, taxonomists most commonly combine morphology and molecular data in their studies: 23 of the 26 integrative taxonomy studies associate at least morphology and genetic data. Other sources of data (e.g. cytology, bacteria content, chemical fingerprint, reproduction) are more marginal. So a majority of these studies (17 of 26) only associate morphology and genetics (mitochondrial and/or nuclear data). Three studies associate four sources of data (Erpenbeck *et al.*, 2006b; Rützler *et al.*, 2007a,b; Ereskovsky *et al.*, 2011), one study

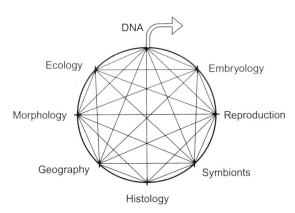

Figure 2.6 Taxonomic circle. Modified from DeSalle *et al.* (2005; Fig. 12). The lines that traverse the inner part of the circle indicate experimental routes that can be taken to corroborate taxonomic hypotheses. To reveal a new taxon, one needs to confirm your species hypotheses with a different data set, once or more, depending on your strategy. This allows you to escape or "break out" of the circle (solid arrow). (For colour version of this figure, the reader is referred to the Web version of this chapter.)

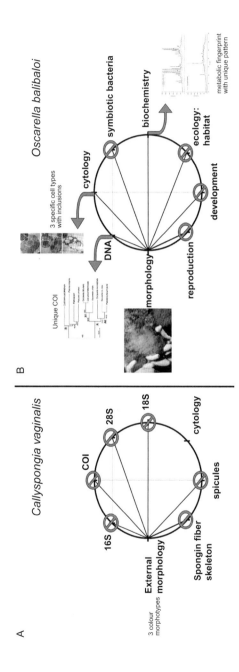

Figure 2.7 Illustration of the taxonomic circle (DeSalle *et al.*, 2005). The lines that traverse the inner part of the circle indicate experimental routes that can be taken to corroborate a taxonomic hypothesis. To reveal a new taxon, one needs to escape or "break out" of the circle (solid arrows) once or more, depending on your strategy. (A) Example of the Caribbean *Callyspongia vaginalis* (López-Legentil *et al.*, 2010). Three sympatric colour morphotypes (grey, red and orange) are hypothesized to be three different species. The taxonomists then used independent data sets to test the hypothesis. All of these data sets failed to detect a new taxon and the taxonomists cannot break out of the circle. Therefore, the initial hypothesis is rejected. (B) Example of *Oscarella balibaloi* (Ivanišević *et al.*, 2011a; Pérez *et al.*, 2011). A somewhat different *Oscarella* is noticed in the Mediterranean, quite singular in its colour; this sympatric morphotype is hypothesized to be a new species. Different data sets were investigated to test this hypothesis. Three of these data sets (biochemistry, cytology and mitochondrial DNA) confirm the species hypotheses so the taxonomists can escape or "break out" of the circle (solid arrows). (For interpretation of the references to colour in this figure legend, the reader is referred to the Web version of this chapter.)

Table 2.3 Data sets used in sponge integrative taxonomy studies since 2006 to this review: 26 studies

Disciplines	Total
Morphology	26 (all)
Mitochondrial DNA (CO1, Atp8)	14
Nuclear DNA (ITS, 16S, 18S, 28S, ATPS intron)	16
Whole mitochondrial genomes	1
Bacteria content	4
Cytology	6
Chemicals	3
Metabolic fingerprinting	3
Ecology	3
Reproduction	1

associates five sources of data (Reveillaud, 2011) and one study associates six sources of data: morphology, cytology, symbiotic bacteria, reproduction, ecology and metabolic fingerprint (Pérez *et al.*, 2011; Fig. 2.7B). Such an effort to bring so many alternative data sets together is generally stimulated by a lack of spicular characters in the species studied. Indeed, the studies aforementioned concern sponges with very poor spicule repertoires (Halichondrida, *Chondrilla* Schmidt, 1862) or no spicule at all (*Oscarella* Vosmaer, 1884; *Hexadella* Topsent, 1896; *Halisarca* Johnston, 1842).

This has notably led to the end of most cosmopolitan species (e.g. *Clathrina clathrus* (see Section 6.3), *Oscarella lobularis* (see Section 7.3), *Hexadella dedritifera* (Reveillaud, 2011), *Cliona celata* (Xavier *et al.*, 2010a), *Chondrilla nucula* and *Chondrosia reniformis* (see Section 8.2.1)); or species with bipolar distribution such as *Stylocordyla borealis* Lovén, 1868 (Uriz *et al.*, 2011).

In the best scenario, all sources of data agree on the delimitation of the species. For example, morphology, cytology, bacteria content and whole mitochondrial genomes clearly show that *Halisarca harmelini* is a new species, different from *Halisarca dujardini* Johnston, 1842 (Ereskovsky *et al.*, 2011). But often, especially if one deals with recently diverged populations, such as in polymorphic or cryptic species, sources of data may disagree. For example, the polymorphic morphology (spicule dimension, spongin fibres, external morphology and colour) disagrees with more conserved genetic data (16S, 28S and 18S rDNA, *cox1*) in *Callyspongia vaginalis* Lamarck, 1814 populations (López–Legentil *et al.*, 2010; Fig. 2.7A). On the other hand, over-conserved morphology disagrees with clear genetic differences (*cox1*, 28S rDNA, *Atp8*) in the species complexes *C. celata* Grant, 1826 (Xavier *et al.*, 2010a) or *C. nucula* Schmidt, 1862 (Klautau *et al.*, 1999; Cavalcanti *et al.*, 2007). As integrative taxonomy makes its way, sponge taxonomists are bound to encounter more and more of these ambiguous cases where discordance among lines of evidence does not automatically imply that a species hypothesis is invalid (Padial *et al.*, 2009) insofar as all characters may not change during speciation, and those that do not evolve at the same rate. Therefore, a major concern is "the degree of congruence that different characters must show to consider population or a group of populations as a separate species" (Padial *et al.*, 2010). Padial *et al.* (2010) see two distinct trends among taxonomists: (1) integration by congruence: a congruent combination of specific characters is necessary (e.g. congruence between molecular and morphological characters, congruence between molecular and ecological characters) and (2) integration by cumulation: any character data set, may it be alone (e.g. single mtDNA gene, additional spicule type), is sufficient to name a new species (congruence is desired but not necessary) (de Queiroz, 2007). "Integration by congruence" is a more conservative approach which tends to underestimate species numbers, while "integration by cumulation" is less conservative and consequently best suited to reveal recently diverged species, although it tends to overestimate species numbers (Padial *et al.*, 2010). This has led to the development of (i) work protocols for integrative taxonomy to rationalize the process of taxonomic decisions and species delimitation (Schlick-Steiner *et al.*, 2009; Padial *et al.*, 2010) and (ii) biodiversity informatics to integrate the full range of biological information (Sarkar, 2009). We recommend here the use of these protocols in sponge integrative taxonomy, in order to improve and clarify our taxonomic decisions, especially for complicated cases as in sibling species.

4. THE CHOICE OF CHARACTER DATA SETS IN TAXONOMY NOWADAYS

An often critical question is the choice of data sets for sponge systematics. As underlined by Jenner (2004a), no data set is the Holy Grail. Conflicts between the pros of "all molecules" and the pros of "all morphology" seem philosophical or political. The middle way is a much more promising approach, each data set bringing light on a different aspect. As underlined in Section 3.2, it is always necessary to have several data sets to propose a well-supported hypothesis. If there are no "ideal" data sets, a data set does not need to be rejected because the interpretations of results are difficult and are highly challenging. A correct use of skeleton characters, cytology, gene sequences or phylogenetic computer programs requires experience and training to be effective; however, each data set has its own set of problems (Table 2.4). Current powerful phylogenetic reconstruction softwares have their own set of problems and can induce themselves artefacts (e.g. long-branch attraction). These softwares, made always more complex with mixed models of reconstruction, may introduce additional artefactual results (Dohrmann *et al.*, 2009). Among other things, the complex relationships between model choice and the taxon sampling (in group and out

Table 2.4 Comparisons of problems linked to different techniques (i.e. morphology, anatomy, cytology, molecular phylogeny, secondary metabolites and software for phylogeny) used to acquire data sets

	Contamination	Artefact	Time consuming	Expensive	Experience
Spicules	Yes	No	No	No	Yes
Skeleton	No	No	Yes	No	Yes
Cytology	No	Yes	Yes	No	Yes
Molecular biology	Yes	Yes	Yes	Yes	Yes
Chemistry	Yes	Yes	Yes	Yes	Yes
Software for phylogeny	No	Yes	Yes	No	Yes

group) need to be carefully evaluated before discussing a phylogenetic tree (Dohrmann *et al.*, 2009; Pick *et al.*, 2010; Philippe *et al.*, 2011).

When different datasets give contradictory results, after checking possibilities of contamination, artefacts or misidentifications, it is absolutely necessary to question the admitted classification. Polyphyletic taxa can be found at all levels of the Linnaean hierarchy for sponges, from classes to genera. When taxonomists are confronted to such results, which would require some taxonomic actions and decisions, they often use the argument of the stability of names and/or the need of confirmation by additional results, thus leading to taxonomy over-conservatism. For example, it took a century to admit the hypothesis of Bidder (1898) for the classification of *Calcispongiae*[D]. The polyphyly of *Axinella* Schmidt, 1862; Axinellidae Carter, 1875; and Halichondrida Gray, 1867 suspected more than 15 years ago, even admitted nowadays (July 2011) by most of the taxonomists is not yet translated into the Linnaean classification of the WPD. As underlined by Jenner (2004b) "molecular systematics has at the very least provided a new set of hypotheses that encourage a detailed restudy of morphological characters". A back and forth investigation between the hypotheses provided by morphological, cytological, chemical and molecular data sets without preconceived ideas is the only way to propose the most robust hypothesis for sponge systematics (see a formalization of this method in Section 10.2).

4.1. Morphological characters: Skeleton, external features, anatomy, cytology

Systematics is based on specific characters. Most of the morphological characters used in taxonomy have been defined and illustrated in an essential "Thesaurus of Sponge Morphology" (Boury-Esnault and Rützler, 1997), while anatomical, cytological and reproductive characters were reviewed in an "Atlas of Sponge Morphology" (De Vos et al., 1991). Both books published by the Smithsonian Institution are precious tools to start working on sponges, to understand the diversity of known body plans and to acquire the fundamental basis of sponge taxonomy.

4.1.1. The skeleton
4.1.1.1. Diversity of the skeletons
The composition of the skeleton has been and remains the most widely used data set of morphological characters for sponges. These characters offer many advantages because they can be studied even if the specimen is not well preserved or even completely dried; a fragment of specimen can sometimes be enough for species identification. The skeleton characters are very useful when the number of spicule types is high and when the organization of the spicules in a precise framework is clear (Uriz, 2006; Fig. 2.8). In Hexactinellida[P] and Calcispongiae[P], spicules are well organized with specific spicule types occupying particular regions of the sponge (e.g. cortex, oscules, atrium, choanosome, etc.). In some Demospongiae[P] (e.g. Tetractinellida[P], Haplosclerida, Poecilosclerida, etc.), spicules are regularly arranged, while, in many other taxa with only one type of spicule, the skeleton organization can be very confused (see definition of Halichondriidae, Erpenbeck and van Soest, 2002). Furthermore, in some Demospongiae[P] and Homoscleromorpha[P], the skeleton can be (i) constituted by spongin only (all species of Keratosa[P] and some Myxospongiae[P]) or (ii) completely absent (Oscarellidae[P] and some Myxospongiae[P]) (Fig. 2.8).

These last 40 years, the increasing use of the scanning electron microscope (SEM) for spicules and skeletons allowed to better understand the three-dimensional organization of the sponge spicules and skeletons. For example, SEM is particularly essential for the study of skeletons in Hexactinellida[P] (see Uriz, 2006) or to compare Demospongiae[P] microscleres at a high magnification. SEM allows the microstructures of microscleres to be carefully observed. For example, the surface rosettes of Geodia[P] sterrasters bring additional characters to discriminate between species which have otherwise very similar spicules (Cárdenas et al., 2009). Recent works have also used thick sections of pieces of sponges included in epoxy resin and stained in toto before inclusion. This technique is relatively rapid and allows to observe the skeleton as well as the tissue organizations (Boury-Esnault et al., 2002; Plotkin and Janussen, 2007; Vacelet and Pérez, 2008; Cárdenas et al., 2009; Fig. 2.8). In a pioneer work, Heim and Nickel (2010) used X-ray microtomography to virtually reconstruct parts of the skeleton in 3D of a Tethya species. However, restricted access to synchrotron source beam lines makes it difficult to use in routine (Fig. 2.9).

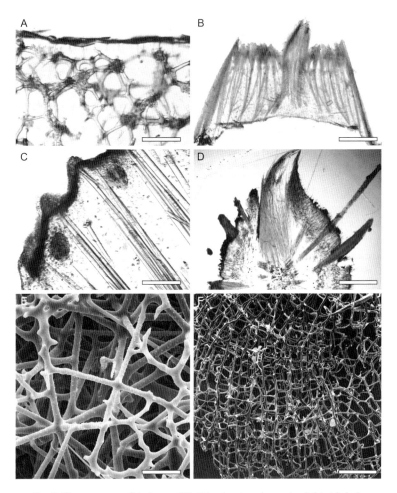

Figure 2.8 Different types of skeleton. (A) *Discodermia polymorpha*, lithistid skeleton made of ectosomal discotriaenes and articulating choanosomal megascleres called desmas, transverse section; scale bar = 400 μm. (B) *Tentorium semisuberites*, ectosomal skeleton made by a palisade of tylostyles and choanosomal skeleton constituted by bundles of principal tylostyles, longitudinal section; scale bar = 750 μm. (C) *Psammastra conulosa*, radial skeleton and cortex layer; scale bar = 1 mm. (D) *Chondrocladia* sp. radial skeleton of styles, transverse section; scale bar 2.7 mm. (E) *Oopsacas minuta*, hexactinellid skeleton constituted by anastomosed hexactines; scale bar 37 μm. (F) *Spongionella puchella* skeleton composed of fibres of spongin; scale bar 750 μm. A, C (photos P. Cárdenas); B, D (photos N. Boury-Esnault); E, F (photos courtesy J. Vacelet).

4.1.1.2. Homoplasy of skeleton characters

The term homoplasy refers to two major processes: convergent evolution and secondary loss (= reversal). Sponge taxonomists and phylogeneticists have always acknowledged morphological homoplasy (e.g. Dendy, 1921),

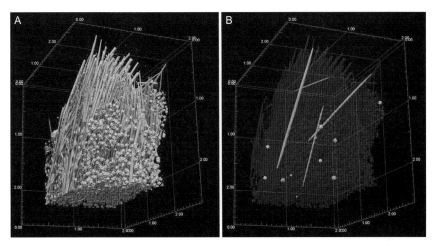

Figure 2.9 Virtual 3D isosurface rendering using VGStudio MAX of selected spicules within their skeletal context (A, B). 3D-reconstructed from synchrotron radiation-based X-ray micro-computed tomography images of the holotype of *Tethya leysae* Heim and Nickel, 2010. Virtual isolation of megasters (B). Micrasters are visualized as small dots, for example, in the peripheral region in (A) (photos courtesy M. Nickel, from Heim and Nickel, 2010).

but few studies have been able to show the extent of these evolutionary processes in sponges, notably because of the paucity of spicule types and other morphological characters. Secondary loss is particularly difficult to reveal in morphological studies and molecular studies of species with very few spicule types. Meanwhile, molecular phylogenetic studies revealing paraphyly and polyphyly of several sponge orders among *Demospongiae*[p] (e.g. Haplosclerida, Halichondrida) and *Calcispongiae*[p] (e.g. Clathrinida, Murrayonida) clearly suggest that the evolution of morphological characters (spicules especially) may be more intricate than currently thought.

An efficient way to reveal convergent evolution is to observe spicule formation (with SEM or TEM): spiculogenesis in sclerocytes (e.g. Rützler and Macintyre (1978) show that subglobular sterrasters and selenasters are not homologous). Another way is to consider the position and orientation of these spicules in the sponge architecture. For example, Cárdenas *et al.* (2011) consider that sterrasters (in *Geodia*) and sterrospherasters (in *Rhabdastrella*) are homologous: they have a similar morphology and they are both positioned in the ectocortex. One can also consider additional independent data sets (e.g. other spicule categories, embryology, biochemistry, histology, molecular phylogenetics). Finally, phylogenetic reconstructions methods can combine morphological and/or molecular data (mapping of morphological characters, cf. Section 3.1).

   If convergent characters are easier to consider with cladistics, the treatment of secondary loss has always been less obvious. Jenner (2002) emphasized that there was no sense to consider "absence" states as empirically empty as opposed to "presence" states, which furnish potential phylogenetic evidence. By doing so, we prevent these "absence" states to be optimized as plesiomorphies or apomorphies. An often ignored fundamental fact is that "simple can mean derived". In other words, an "absence" state can be a "gain", with the difference that this "gain" often leaves no trace of its past presence, and is therefore invisible. Identified secondary losses can therefore potentially represent synapomorphies and thus bring new characters with phylogenetic information. For example, one synapomorphy of *Geostelletta*[P] (a clade within the *Geodiidae*[P]) is the secondary loss of sterraster spicules (Cárdenas *et al.*, 2011; see Cárdenas, 2010 for a review on morphological homoplasy). The following examples are quite characteristic of secondary losses in sponge species or genera.

(a) *Crambe crambe* (Schmidt, 1862): losses of spicules among different specimens of a species. The most common Mediterranean species *C. crambe* has been described under a huge number of names (14 enumerated in the WPD) until the variability of its spicule content was clearly related to the silica content of the environment (Uriz and Maldonado, 1995; Maldonado *et al.*, 1999). In this genus, the full spicular content is composed of two size classes of tylostyles, isochelae and desmas (Fig. 2.10). The two last spicule types are most often absent in *C. crambe*. In their beautiful work, Maldonado *et al.* (1999) succeed in rearing newly settled *C. crambe* sponges at three concentrations of silicic acid Si(OH)$_4$ for 14 weeks and obtained the different spicule contents depending on the Si(OH)$_4$ concentration. This work demonstrates that specific Si(OH)$_4$ concentration thresholds induce the activation of different population of sclerocytes and thus the secretion of different spicules.

(b) *Merlia* Kirkpatrick, 1908: Losses of characters within species of a genus (Fig. 2.11). The genus *Merlia* has representatives with a circumtropical distribution (Caribbean Sea, Indian Ocean, NW and SW Pacific) and also occurs in warm temperate areas (subtropical Atlantic and Mediterranean Sea). The type-species *Merlia normani* Kirkpatrick, 1908 has a skeleton composed of tylostyles, raphides, comata and clavidiscs, together with a calcareous basal skeleton (Vacelet, 1980b; Gautret *et al.*, 1991; Hajdu and Soest, 2002; Vacelet *et al.*, 2010). Four species have been described in this genus, two of which lack the basal calcareous skeleton, *Merlia deficiens* Vacelet, 1980a,b, and *Merlia tenuis* Hoshino, 1990. Meanwhile, *Merlia lipoclavidisca* Vacelet and Uriz, 1991 possesses the basal calcareous skeleton and the tylostyles but lacks the microsclere complement including the diagnostic clavidisc.

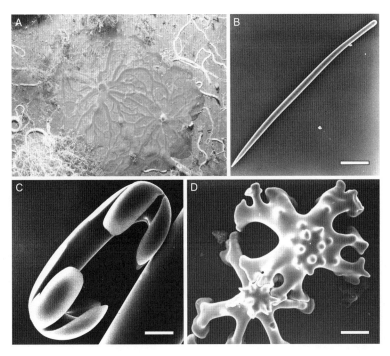

Figure 2.10 (A) *Crambe crambe in situ*. Spicules types present in *Crambe* species. (B) Styles from *Crambe acuata*; scale bar = 30 μm. (C) Isochelae from *C. acuata*; scale bar = 7 μm. (D) Desmas from *Crambe taillzei*; scale bar = 30 μm. A (photo T. Pérez); B–D (photos courtesy M.-J. Uriz). (For colour version of this figure, the reader is referred to the Web version of this chapter.)

van Soest (1984b) synonymized *M. normani* and *M. deficiens* by considering that the morphological variations quoted by Vacelet (1980b) were not relevant and therefore that these taxa represented a single cosmopolitan species. This was not followed by the subsequent authors (Hajdu and Soest, 2002). The apparent cosmopolitanism of *M. normani* and *deficiens* as well as the hypothetical loss of the basal calcareous skeleton or clavidiscs needs to be checked with a genetic approach.

(c) Homoplasy in the *Astrophorida*[P]. Chombard *et al.* (1998) was the first study to reveal secondary losses of spicule types in sponges thanks to molecular phylogenetics (within the Geodiidae: Astrophorida). Using a much larger sampling of *Astrophorida*[P] and additional molecular markers, Cárdenas *et al.* (2011) have not only confirmed this result but also shown how widespread spicule homoplasy was: convergences and secondary losses have happened many times and for all type of spicules, megascleres and microscleres. This high frequency of homoplasy in the Astrophorida is all the more impressive if we consider that these results

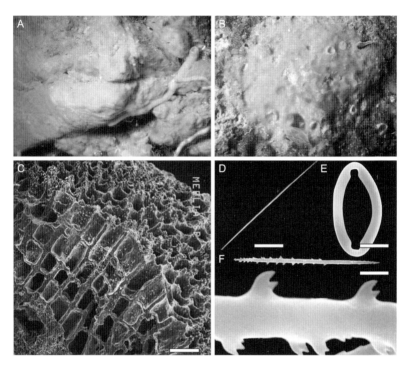

Figure 2.11   (A) *Merlia deficiens*, *in situ*, Pointe Fauconnière (NW Mediterranean). (B) *Merlia normani*, *in situ*, Lebanese coast (SE Mediterranean). (C) Transverse section through an hypercalcified skeleton of *Merlia lipoclavidisca*; scale bar = 400 μm. (D) Tylostyle from *Merlia deficiens*; scale bar = 35 μm. (E) Clavidisc *Merlia deficiens*; scale bar = 14 μm. (F) Commata and details of the spines *Merlia deficiens*; scale bar = 11 μm (photos courtesy J. Vacelet). (For colour version of this figure, the reader is referred to the Web version of this chapter.)

are certainly underestimated. Indeed, many cases of secondary losses are reported in other Astrophorida species, not sampled in Cárdenas *et al.* (2011). *Holoxea, Jaspis*, some *Stelletta*, some *Geodia*, some *Erylus*, some *Ecionemia, Lamellomorpha*, etc. are hypothesized to have also lost their triaenes. And there are no reasons to think that *Astrophorida*[P] have evolved differently than other sponge clades. The main consequence for the *Astrophorida*[P] (and *Porifera*[P]) taxonomy is that few spicule types (and secondary losses) are actually phylogenetically informative, at the order or family rank at least.

4.1.2. External features

The environment can influence external morphology and yet, specific characters can be singled out by the experienced field taxonomist. These characters were not properly considered due to the often bad preservation conditions of the specimen in the Museum collections. Nowadays, the specimens are often photographed underwater before collection which allows a good observation of the shape, characters of the surface, poral and oscular organization, texture and colour (Bergquist *et al.*, 1998; Pinheiro *et al.*, 2007). Again, there is no *a priori* good character, and this can be illustrated with the example of openings and colour.

The oscule/pore morphology has been shown to be homoplastic in some groups. Cribriporal and uniporal pores/oscules in *Geodia* spp. are not homologous (Cárdenas *et al.*, 2010). Meanwhile, pore sieves are homologous in the Hymedesmiidae Topsent, 1928: whereas it has been long affiliated to the Halichondriidae Gray, 1867, the genus *Hemimycale* Burton, 1934 has been definitively allocated to the Hymedesmiidae based on the possession of pore sieves shared with other genera of this family (van Soest, 2002; Fig. 2.12). This has been confirmed by molecular results (Goodwin *et al.*, 2010, p. 60). Since pore sieves have also been described within Crellidae Dendy, 1922 (Boury-Esnault, 1972), relationships between Crellidae and Hymedesmiidae should now be studied through a molecular phylogenetic work.

Colour can be a diagnostic character within a genus—all yellow *Clathrina* Gray, 1867 without tetractines constitute a monophyletic group (Rossi *et al.*, 2011)—or between species—*Phorbas tenacior* (Topsent, 1925) always has a pale blue colour, whereas *Phorbas topsenti* Vacelet and Pérez, 2008 has a bright red colour (Fig. 2.12). And the colour of the living specimens is often very specific, that is, a narrow range of colour. *Hexadella racovitzai* Topsent, 1896 has a pinkish colour, whereas *Hexadella pruvoti* has a yellow colour (Fig. 2.13). However, colour is not always a diagnostic character and can be influenced by the light exposition of the specimens. For example, populations of *Petrosia ficiformis* Poiret, 1789 and *C. reniformis* Nardo, 1847 living in caves are white, contrasting with the respectively reddish and grey-dark colour of populations living in luminous environments (Fig. 2.14). In the first case, this is due to the loss of cyanobacteria and in the second case due to the lack of expression of melanin. Seemingly, populations of *P. johnstonia* (Bowerbank in Johnston, 1842) living at low depth are purple coloured, while populations living in caves can be white. In other cases, the different colours of specimens may be linked to individual variations: *Poecillastra compressa* Bowerbank, 1866 has yellow and orange morphotypes (Fig. 2.3), and *Oscarella tuberculata* (Schmidt, 1868) can be green, yellow, blue, etc.

4.1.3. Cytology and choanosome anatomy

The use of cytological characters in sponge taxonomy has been neglected by most of the sponge systematicians. They have never been introduced systematically in sponge descriptions because cytological techniques are considered difficult to apply in routine analysis and because histological

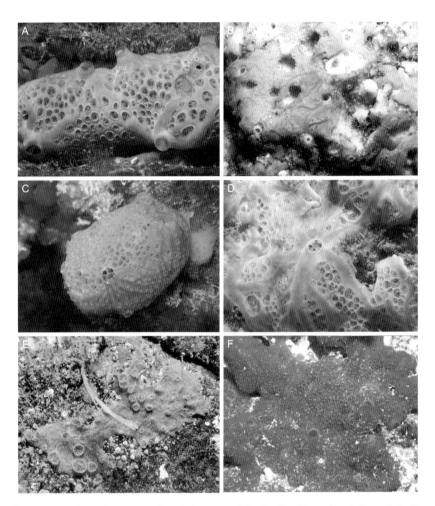

Figure 2.12 Pore sieves are a shared character of the families Hymedesmiidae and Crellidae. (A) *Hemimycale columella*, Frioul Island (NW Mediterranean), 30 m deep. (B) *Crella pulvinar*, Calanque coast (NW Mediterranean), 15 m deep. (C) *Phorbas fictitius*, Ceuta (SW Mediterranean), 30 m deep. (D) *Phorbas tenacior*, Monaco (NW Mediterranean), 25 m deep. (E) *Hymedesmia paupertas*, Irish Sea. (F) *Phorbas topsenti*, Calanque coast (NW Mediterranean), 7 m deep. A–D, F (photos T. Pérez); E (photo courtesy B. Picton). (For colour version of this figure, the reader is referred to the Web version of this chapter.)

slides and electron micrographs could be difficult to interpret (Erpenbeck, 2004). But the use of cytological features of the aquiferous system for sponge taxonomy was proposed by several authors who focused on the different aspects of choanocytes chambers, choanocytes and apopylar cells (Minchin, 1896; Bidder, 1898; Dendy and Row, 1913; Lévi, 1979;

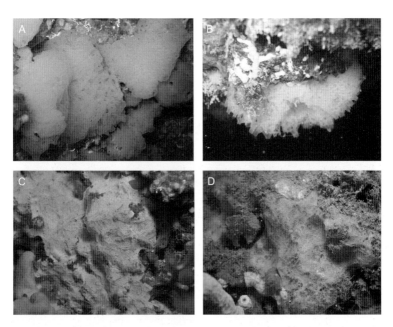

Figure 2.13   Yellow *Clathrina*. (A) *Clathrina clathrus*, Mediterranean Sea. (B) *Clathrina aurea*, SW Atlantic (Brazilian coast). Different colours of *Hexadella*: (C) *Hexadella racovitzai*, Mediterranean Sea. (D) *Hexadella pruvoti*, Mediterranean Sea. A, C, D (photos T. Pérez); B (photo courtesy E. Lanna). (For interpretation of the references to colour in this figure legend, the reader is referred to the Web version of this chapter.)

Bergquist, 1980, 1995; Vacelet *et al.*, 1989; Boury-Esnault *et al.*, 1990; De Vos *et al.*, 1990; Bergquist *et al.*, 1998; Bergquist and Cook, 2002a).

Cytological studies have been particularly developed for taxonomy of species without mineral skeleton or without skeleton at all, and the usefulness of these data has been thus demonstrated (Muricy, 1999; Ereskovsky *et al.*, 2011; Pérez *et al.*, 2011). These characters appeared particularly useful to describe demosponges without skeleton such as *Halisarca* (Vacelet and Donadey, 1987; Ereskovsky, 2007; Ereskovsky *et al.*, 2011), *Thymosiopsis* Vacelet and Pérez, 1998 and *Myceliospongia* Vacelet and Pérez, 1998 (Vacelet *et al.*, 2000), as well as *Homoscleromorpha*[P] (see Section 7).

The development of cytological and anatomical studies seems particularly important in spiculate orders or families for which we are faced with severe taxonomic problems such as the Haplosclerida where a body of knowledge remains unused in the recent phylogenetic hypothesis (Table 2.5). The use of the choanosome anatomy in taxonomy for Haplosclerida has been already suggested (Langenbruch, 1988, 1991; Langenbruch and Scalera-Liaci, 1990; Boury-Esnault, 2006) (Fig. 2.15A–F). Choanocyte chambers directly "hanging" in the inhalant space and covered by a pinacocyte layer into

Figure 2.14   Variability of the colours of sponges as a function of illumination. (A) *Petrosia ficiformis* on a horizontal surface at low depth. (B) *Petrosia ficiformis* on a wall of a cave. (C) *Chondrosia reniformis* on an illuminated surface. (D) *Chondrosia reniformis* on a wall of a cave (photos T. Pérez). (For interpretation of the references to colour in this figure legend, the reader is referred to the Web version of this chapter.)

which the prosopyles open have been described in several species belonging to the families Chalinidae Gray, 1867, Callyspongiidae de Laubenfels, 1936 and Petrosiidae van Soest, 1980 (Fig. 2.13 and Table 2.5). This type of choanocyte chambers is absent in Spongillidae Gray, 1867 (Haplosclerida Topsent, 1928) (Langenbruch and Weissenfels, 1987), in *Amphimedon compressa* Duchassaing and Michelotti, 1864, *Haliclona sarai* Pulitzer-Finali, 1969 and *Dendroxea lenis* Topsent, 1892 (Langenbruch, 1988, p. 20).

Among the different categories of cells present in the mesohyl (Simpson, 1984; Boury-Esnault, 2006), some are present in all species and therefore do not bring any phylogenetic signal. Other cells such as the cells with inclusions (spherulous cells, granular cells, glycocytes, etc.) can be useful for taxonomy (Topsent, 1900). Such cells contain large spherical or ovoid inclusions that occupy the main part of the cytoplasmic volume. They are involved in the elimination of metabolic wastes (Vacelet, 1967) and in the storage of bioactive molecules (Thompson *et al.*, 1983). Depending on the sponge species,

Table 2.5 Type of choanocyte chambers (CC) among Haplosclerida

	List of species	CC	References
Haplosclerina Topsent, 1928			
Chalinidae Gray, 1867	*Chalinula limbata* Montagu, 1818	H	Langenbruch (1991) (as *H. limbata*) and Langenbruch and Scalera–Liaci (1990)
	Chalinula saudiensis Vacelet, Al Sofyani, Al Lihaibi and Kornprobst, 2001	H	Vacelet et al. (2001; Fig. 2.12A)
	Haliclona (Reniera) cinerea Grant, 1826	H	Langenbruch and Scalera–Liaci (1986, 1990) and Langenbruch (1991) (as *H. elegans*)
	Haliclona (Reniera) mediterranea Griessinger, 1971	H	Langenbruch (1988, 1991), De Vos et al. (1991) and Boury-Esnault (2006)
	Haliclona (Rhizoniera) rosea Bowerbank, 1866	H	Langenbruch (1991) and Langenbruch and Scalera–Liaci (1990)
	Haliclona (R.) indistincta Bowerbank, 1886	E	Langenbruch (1991) and Langenbruch and Scalera–Liaci (1990)
	Haliclona (R.) sarai Pulitzer-Finali, 1969	E	Langenbruch (1988, 1991)
	Haliclona (Haliclona) simulans Johnston, 1842	H	Langenbruch (1991) (as *A. simulans*) and Langenbruch and Scalera–Liaci (1990)
	Haliclona (Haliclona) oculata Pallas, 1766	H	Langenbruch (1991), and Langenbruch and Scalera–Liaci (1990)
	Haliclona (Halichoclona) fulva Topsent, 1893	H	Langenbruch (1988, 1991) (as *R. fulva*)
	Haliclona (Halichoclona) fistulosa Bowerbank, 1866	H	Langenbruch (1991) and Langenbruch and Scalera–Liaci (1990)
	Haliclona (Soestella) mucosa Griessinger, 1971	H	Langenbruch (1988, 1991) (as *R. mucosa*) (Fig. 2.12B)

(continued)

Table 2.5 (continued)

	List of species	CC	References
	Haliclona (Gellius) rava Stephens, 1912	H	Langenbruch (1991) and Langenbruch and Scalera–Liaci (1990)
Callyspongiidae de Laubenfels, 1936	**Dendroxea lenis** Topsent, 1892	E	Langenbruch (1988, 1991; Fig. 2.12D)
	Callyspongia (Cladochalina) diffusa Ridley, 1884	H	Johnston and Hildemann (1982) and Smith and Hildemann (1986, 1990)
	Callyspongia (C.) vaginalis Lamarck, 1814	H	De Vos et al. (1991; Fig. 2.12C)
Niphatidae van Soest, 1980	**Niphates digitalis** Lamarck, 1814	H	Langenbruch (1988, 1991)
	Amphimedon compressa Duchassaing and Michelotti, 1864	E	Langenbruch (1988, 1991; Fig. 2.12E)
Petrosina Boury-Esnault and van Beveren, 1982			
Petrosiidae van Soest, 1980	**Petrosia (Petrosia) ficiformis** Poiret, 1789	H	Langenbruch (1983, 1991), Langenbruch et al. (1985) and Langenbruch and Scalera–Liaci (1990)
Calcifibrospongiidae Hartman, 1979	Calcifibrospongia actinostromarioides Hartman, 1979	H	Hartman and Willenz (1990)
Spongillina Manconi and Pronzato, 2002			
Spongillidae Gray, 1867	Ephydatia fluviatilis Linnaeus, 1759	E	Langenbruch and Weissenfels (1987), Langenbruch (1991) and De Vos et al. (1991; Fig. 2.12F)
	Ephydatia muelleri Lieberkühn, 1856	E	Langenbruch (1991)
	Eunapius fragilis Leidy, 1851	E	Langenbruch (1991)
	Spongilla lacustris Linnaeus, 1759	E	Langenbruch and Weissenfels (1987) and Langenbruch (1991)

Classification follows the World Porifera Database. Type species are in bold. In the column choanocyte chambers, H means hanging choanocyte chamber within incurrent canals, E means choanocyte chambers embedded in the mesohyl.

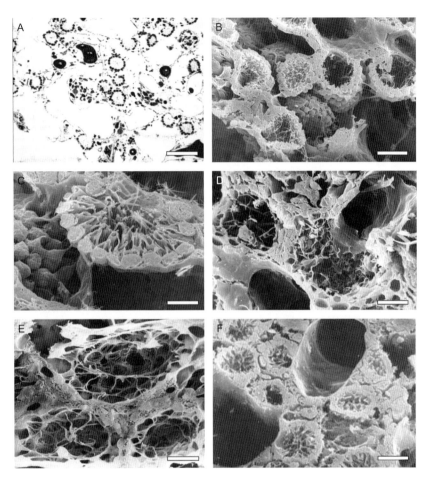

Figure 2.15   (A–C) SEM micrograph of hanging choanocyte chambers within an incurrent canal. (D–F) SEM micrograph of choanocyte chambers embedded in the mesohyl. (A) *Chalinula saudiensis*; scale bar = 52 μm. (B) *Haliclona mucosa*; scale bar = 17 μm. (C) *Callyspongia vaginalis*; 5.4 μm. (D) *Dendroxea lenis*; scale bar = 9.6 μm. (E) *Amphimedon compressa*; 8.4 μm. (F) *Ephydatia fluviatilis*; scale bar = 19 μm. A (photo courtesy J. Vacelet); B–F (photos courtesy L. De Vos; http://www.ulb.ac.be/sciences/biodic/homepage.html).

the spherules are either homogeneous or heterogeneous, often microgranular in size and often have a taxonomic value. For example, Verongida species have a very characteristic spherulous cell type (Vacelet, 1967). This type of cell has been found also in the aspiculate genus *Hexadella* Topsent, 1905, which led Bergquist and Cook (2002a) to reassign *Hexadella* species to the family Ianthellidae Hyatt, 1875 (Verongida Bergquist, 1978; Fig. 2.16). This result was confirmed later by molecular data (Borchiellini *et al.*, 2004a).

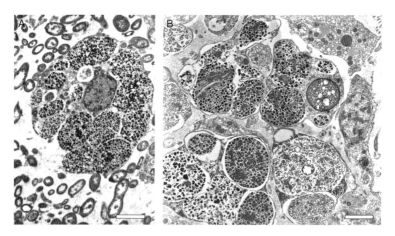

Figure 2.16   Spherulous cells with microgranular spherules. (A) *Aplysina aerophoba*; scale bar = 2 μm. (B) *Hexadella pruvoti*; scale bar = 1.8 μm (photos courtesy J. Vacelet).

These characters were also used to discriminate between spiculate species of *Polymastia* Bowerbank, 1864 (Boury-Esnault *et al.*, 1994) and *Spirastrella* Schmidt, 1868 (Boury-Esnault *et al.*, 1999b). Muricy (1999) used morphological, anatomical, skeletal and cytological data sets in a cladistic analysis of 10 species of *Homoscleromorpha*[P], and he showed that the topology of the tree obtained from the combined data sets gave a good resolution.

Among the component of the mesohyl, numerous sponge species display abundant communities of prokaryotic or/and eukaryotic microsymbionts. They are quite diverse and can be specific to different sponge ranks, from order to species. These specific symbionts can be transmitted vertically from one generation to another and may be even diagnostic to discriminate close species (Vacelet *et al.*, 1994; Ereskovsky *et al.*, 2005; Enticknap *et al.*, 2006; Vishnyakov and Ereskovsky, 2009; Schmitt *et al.*, 2012; Thacker and Freeman, 2012).

Fundamental cytological characters of such importance as the synapomorphy of the *Porifera*[P], the presence of choanocytes and of an aquiferous system, can also be secondarily lost: it has been completely lost in most carnivorous sponges of the family Cladorhizidae Dendy, 1922 (Vacelet and Boury-Esnault, 1995; Vacelet, 1999a, 2007).

4.2. Reproductive strategies

The mode of reproduction (oviparous vs. ovoviviparous) has been tentatively used to define two subclasses of Demospongiae: Tetractinomorpha/oviparous versus Ceractinomorpha/ovoviviparous by Lévi (1956). The mode of reproduction is actually homoplastic and cannot be used as a character shared by all taxa of these subclasses (van Soest, 1990, 1991). However, these two subclasses slightly emended had remained in the *Systema Porifera* (2002) because no alternative robust hypothesis had been proposed. Molecular studies with 28S partial length and 18S full-length rDNA as well as with the full length of mitochondrial DNA have proposed an alternative hypothesis (Borchiellini *et al.*, 2004a; Nichols, 2005; Lavrov *et al.*, 2008). Four supported clades have been recognized among Demospongiae *sensu stricto* named tentatively G1 (Keratosa = Dictyoceratida + Dendroceratida), G2 (Myxospongiae = Verongida + Chondrosida), G3 (marine Haplosclerida), G4 (all other remaining orders). When mapping the mode of reproduction on the molecular trees, ovoviviparity appears as the ancestral character, while oviparity has been acquired twice. Ovoviviparity has been reacquired (reversion to the ancestral character) on the branches leading to the Poecilosclerida and to the Halisarcidae, and this would mean that ovoviviparity in Poecilosclerida is not homologous to that of *Keratosa*[P] or of Halisarcidae (Fig. 2.17). *Tetractinellida*[P] are

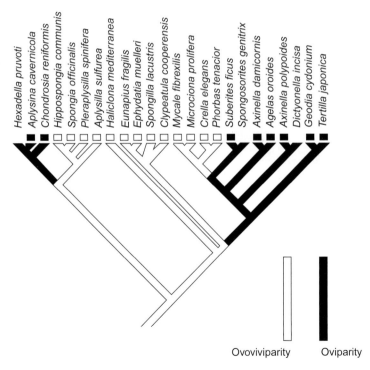

Figure 2.17 Evolution of character type of reproduction (oviparity in black, ovoviviparity in white), as optimized on the 18S rRNA tree by MacClade. The squares below taxon names give character state in the considered taxon *(modified from Borchiellini et al., 2004a).*

oviparous, but ovoviviparity has been reacquired on the branch leading to the Thoosidae Cockerel, 1925 (Vacelet, 1999a; Borchiellini *et al.*, 2004b).

In *Calcispongiae*[P], the *Calcinea*[P] share the presence of a coeloblastula (calciblastula), while *Calcaronea*[P] share an amphiblastula which presents a typical phenomenon of eversion during its morphogenesis (Manuel, 2006; for a review, see Ereskovsky, 2010) (Fig. 2.18).

Other larval types (see Ereskovsky, 2010 for a review) are cinctoblastula, a synapomorphy of *Homoscleromorpha*[P], coeloblastula shared by Chondrosida Boury-Esnault and Lopès, 1985 (Lévi and Lévi, 1976; Usher and Ereskovsky, 2005) and Verongida (Maldonado, 2009) and parenchymella for all other taxa of *Demospongiae*[P]. A direct development without larval stage is known among the Tetillidae Sollas, 1886 (Watanabe, 1978), and the Stylocordylidae (Sarà *et al.*, 2002). The larva of *Hexactinellida*[P] called the trichimella is known from few species: *Farrea sollasii* Schulze, 1886 (Okada, 1928) and *Oopsacas minuta* Topsent, 1927 (Boury-Esnault *et al.*, 1999a; Leys *et al.*, 2006; Fig. 2.19). It should therefore be emphasized that larval data are

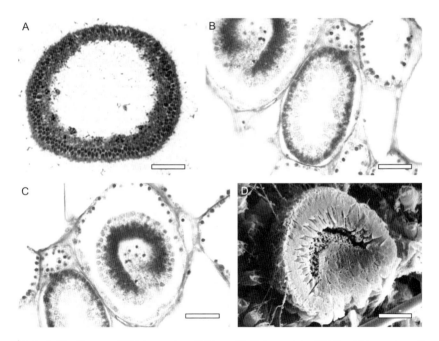

Figure 2.18   Larvae of Calcispongiae. *Calcinea, Clathrina contorta*. (A) Semithin section of a coeloblastula; scale bar = 40 µm. *Calcaronea, Leuconia* nivea. (B) Semithin section of a stomo-blastula with flagellated cells inside the cavity. (C) Amphiblastula after eversion of the stomoblastula; scale bar = 86 µm. *Sycon sycandra*. (D) Fractured amphiblastula in SEM; scale bar = 50 µm. A (photo courtesy A. Ereskovsky); B–D (photos N. Boury-Esnault). (For colour version of this figure, the reader is referred to the Web version of this chapter.)

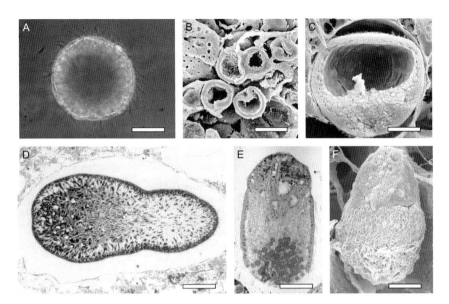

Figure 2.19 Different types of larva in Demospongiae and Hexactinellida. (A) External fertilization of an oocyte of *Chondrosia reniformis*; observe the spermatozoids surrounding the oocyte; scale bar = 35 μm. (B) SEM micrograph of a fracture of cinctoblastula larvae of *Oscarella lobularis* in the mesohyl; scale bar = 240 μm. (C) A meridian fracture of an *Ephydatia fluviatilis* parenchymella with a flotation cavity at the anterior pole; scale bar = 50 μm. (D) Semithin section of a parenchymella of *Vaceletia crypta*; scale bar = 60 μm. (E, F) TEM and SEM micrograph of a trichimella of *Oopsacas minuta*; scale bar = 16 μm. A, E, F (photo N. Boury-Esnault); B (from Boury-Esnault *et al.*, 2003); C (photo courtesy L. De Vos; http://www.ulb.ac.be/sciences/biodic/homepage.html); D *(courtesy J. Vacelet, from Vacelet, 1979)*.

missing for most species, especially for oviparous taxa (e.g. the free-living larva of *Tetractinellida*[p] is unknown to this day). The observation that spicules may appear and disappear during larval development (e.g. in *Alectona* Carter, 1879, discotriaenes disappear after the larvae have settled) suggests that these "larval spicules" may bring important phylogenetic characters, lost in the adults. In our opinion, a special effort should be made to find and describe these unknown larvae.

Ultrastructural comparison of the epithelial cells of the larvae of some Chondrosida (*C. reniformis*, *Chondrilla australiensis* Carter, 1873 and *H. dujardini*) and Verongida (*Aplysina aerophoba*) revealed that species from these two clades share: (1) a nonperpendicular orientation of the accessory centriole relative to the basal body; (2) a protruding nuclear apex; and (3) a Golgi apparatus encircling the nuclear apex and part of the organelles of the basal apparatus (Lévi and Lévi, 1976; Korotkhova and Ermolina, 1982; Usher and Ereskovsky, 2005; Gonobobleva, 2007; Maldonado, 2009).

These results are congruent with those obtained with mtDNA (Lavrov *et al.*, 2008) and 18S and 28S rDNA (Borchiellini *et al.*, 2004b). However, these potential synapomorphies have to be checked on more Verongida and Chondrosida species.

4.3. Molecular characters

4.3.1. Molecular markers for α-taxonomy

Molecular markers repeatedly reveal cryptic species and reject the hypothesis of cosmopolitanism for many sponge species. They are valuable for delineating species boundaries in *Porifera*[2] (Boury-Esnault *et al.*, 1992, 1999b; Solé-Cava *et al.*, 1992; Klautau *et al.*, 1994, 1999; Lazoski *et al.*, 1999, 2001; Miller *et al.*, 2001; Duran and Rützler, 2006; Wulff, 2006; Blanquer and Uriz, 2007; Cárdenas *et al.*, 2007; Wörheide *et al.*, 2008; Valderrama *et al.*, 2009; Ferrario *et al.*, 2010; Reveillaud *et al.*, 2010, 2011; Xavier *et al.*, 2010a). In the 1990s, allozyme electrophoresis became the method of choice for α-molecular systematics of marine organisms (reviewed in Thorpe and Solé-Cava, 1994; Solé-Cava and Boury-Esnault, 1999; Boury-Esnault and Solé-Cava, 2004). Although allozymes are good overall markers for population and species level systematics, they have the major drawback of requiring fresh or frozen samples. Alternative nuclear markers such as microsatellites (Duran *et al.*, 2002; Knowlton *et al.*, 2003; Blanquer *et al.*, 2009; Noyer *et al.*, 2009), internal transcribed spacers (López *et al.*, 2002; Wörheide *et al.*, 2002, 2004; Valderrama *et al.*, 2009), mitochondrial cytochrome *c* oxidase subunit 1 (*cox1*) gene and the D3–D5 region of the nuclear large ribosomal subunit (28S rDNA) have been used for α-taxonomy as well (Erpenbeck *et al.*, 2002; Duran and Rützler, 2006; Erpenbeck *et al.*, 2006b; Wulff, 2006; Blanquer and Uriz, 2007; Pöppe *et al.*, 2010; Xavier *et al.*, 2010a; Reveillaud *et al.*, 2011). The second intron of the nuclear ATP-synthetase β subunit gene has only recently been shown to provide a high resolution at the intraspecific level in sponge evolutionary studies (Bentlage and Wörheide, 2007; Wörheide *et al.*, 2008; Reveillaud *et al.*, 2010). Four new mitochondrial markers were compared to the Folmer *cox1* fragment in nine sponge species: *cox2*, partial sequence of ATP synthase 6, and two intragene spacers (SP1 and SP2) (Rua *et al.*, 2011). Although these results need to be confirmed with a wider intraspecific sampling and sibling species, the new markers appeared to be less restrained than *cox1*, for some species at least. It was also underlined in this work that the use of several molecular markers gives more robust results (see also Chapter 5).

4.3.2. Molecular markers for systematics and phylogeny

After the three classical molecular markers in sponge phylogenetics (28S rDNA, 18S rDNA and the Folmer *cox1* fragment) and the housekeeping genes (e.g. aldolase (ALD), catalase (CAT), elongation factor 1-alpha

(EF1α), heat-shock proteins (Hsp70)) (Borchiellini *et al.*, 1998; Erpenbeck *et al.*, 2005; Sperling *et al.*, 2009, 2010), more recent studies are now including large-scale multigene data: nuclear proteins (obtained from EST or full genomic sequencing) or complete mitochondrial genomes (Jiménez-Guri *et al.*, 2007; Dunn *et al.*, 2008; Lavrov *et al.*, 2008; Philippe *et al.*, 2009; Schierwater *et al.*, 2009; Gazave *et al.*, 2010b; Pick *et al.*, 2010; see Chapter 1). The comparison of the secondary structure for 18S and 28S rDNA also provides new molecular synapomorphies which may help resolve phylogeny of different problematic groups like Haplosclerida or *Axinella* or even between sister species (Erpenbeck *et al.*, 2004; Redmond and McCormack, 2008; Gazave *et al.*, 2010a).

DNA barcoding in sponges is still in its infancy and still somewhat controversial (Solé-Cava and Wörheide, 2007). It is now agreed that the Folmer *cox1* fragment is not the ideal sponge barcoding marker notably because (1) two different species may not be differentiated (Schröder *et al.*, 2003; Pöppe *et al.*, 2010), (2) there is substantial overlap between intra- and closest interspecific variation (i.e. no barcoding gap) (Huang *et al.*, 2008) and (3) CO1 is very difficult to sequence in *Calcispongiae*[P] and *Hexactinellida*[P] species (Dohrmann *et al.*, 2012; Rossi *et al.*, 2011). The Sponge Barcoding Project (SBP) (www.spongebarcoding.org) associating DNA barcodes with their voucher description can be viewed as an opportunity to develop integrative taxonomy. The SBP initiated at the Seventh International Sponge Symposium in 2006 (Wörheide *et al.*, 2007) is slowly building up with 1053 vouchers barcoded (July 2011), each with one, two or three barcoding markers. Unfortunately, most of the morphological descriptions associated with the barcodes are far from being complete, so "reference barcodes" are still too few. The more than 500 barcodes added to the SBP in August 2010 and sequenced from material from the Queensland Museum cannot be all trusted due to problems of contamination and/or identification; the morphological descriptions are incomplete for many of them and not always specific to the voucher. A few taxonomic publications are now trying to associate sponge descriptions with their DNA barcodes (Cárdenas *et al.*, 2009), and this is maybe a promising way to guarantee a full morphological description of the barcode vouchers. Such publications will not only ensure an unambiguous link between the voucher and its DNA sequences but also provide future sponge barcoding studies (or phylogenetic studies) with accurate data and testable species hypotheses.

Sponge-associated microorganisms are probably as old as the sponges themselves and, in many species, maintained through vertical transmission (Taylor *et al.*, 2007). Specificity of the sponge–microbe relationships suggests possible past and/or ongoing co-evolution as well as co-speciation events. Unfortunately, apart from a co-phylogeny pioneer study on halichondrids and their specific symbionts (Erpenbeck *et al.*, 2002), the use of symbiont phylogeny as an independent data set to infer sponge phylogenies is still poorly investigated. There is today a growing interest to understand the relationships between sponge microbial communities and their hosts and the field of sponge microbiology is rapidly expending with its First International Symposium held in Würzburg (Germany) in March 2011 (Taylor *et al.*, 2011; Thacker and Freeman, 2012). We strongly encourage sponge phylogeneticists/taxonomists and sponge microbiologists to keep close contacts in order to investigate new possibilities offered by co-phylogeny studies.

4.4. Chemical characters

In marine ecosystems, sponges are among the most important sources of secondary metabolites, with a great diversity of structures, biosynthetic pathways and biological activities, which evolved as products of natural selection. Secondary metabolites have an important role in many ecological processes that shape biodiversity, but their study has been mostly restricted to natural product chemistry and research of novel and potentially active compounds. So although the number of new compounds increases continuously (Kornprobst, 2005, 2010; Blunt, 2011), there is a poor knowledge of their biological or ecological functions in the wild.

4.4.1. Secondary metabolites as chemotaxonomical characters

Several attempts were undertaken to use these compounds in chemotaxonomy as an alternative or complementary tool to elucidate classification patterns and to propose potential synapomorphic chemical markers at different taxonomic ranks (see reviews by van Soest and Braekman, 1999; Erpenbeck and van Soest, 2007). In sponge systematics, additional chemotaxonomic markers can be particularly useful for taxa that do not possess characters essential for sponge systematics, the so-called sponges without skeleton (Boury-Esnault *et al.*, 1995; Bultel-Poncé *et al.*, 1999), or for sponges with highly polymorphic characters (some examples among the Halichondrida or Haplosclerida). This approach needs to verify the independence of the chemical characters (encoded by biosynthetic pathways or by some groups of dependent molecules, see for more details Genta-Jouve and Thomas, 2012), but this has been rarely done before relating the chemical diversity and the systematics classification of a given taxonomic group.

Since the pioneering works of Bergmann (1949, 1962) who searched for alternative data sets, taxonomists attempted to use the putative phylogenetic information carried by biochemical compounds which could provide some phylogenetic signal. However, the huge amount of data obtained from the chemical literature and indexed in the database MarinLit is hampered by the low level of confidence with respect to the identification of the specimens studied (van Soest and Braekman, 1999; Erpenbeck and van Soest, 2005, 2007).

As underlined by Bergquist and Wells (1983), chemotaxonomic studies need important requirements: (1) a precise characterization of the compounds, (2) an accurate identification of the specimen from which the compounds are extracted and (3) a sufficiently broad sample of species to allow a comprehensive characterization of genera, family, etc. Such studies require a close collaboration between chemists and biologists to obtain suitable results (Braekman *et al.*, 1992), and when efficient, such collaborations have raised several evaluations of the usefulness of chemotaxonomical markers (for an extensive review, see Erpenbeck and van Soest, 2007). For instance, the sterol and fatty acids composition of *Demospongiae*[P] allowed to re-evaluate the classification and to make hypotheses about the relationships of Verongida, and also to discuss the putative relationship between *Agelas* and some *Axinella* species (i.e. Bergquist, 1978; Bergquist *et al.*, 1980, 1984, 1986). Diterpene isocyanide in Halichondrida, furano/lactone terpenes in Dictyoceratida and Dendroceratida or polycyclic guanidine alkaloids in Crambeidae are among the most accepted chemotaxonomical markers (Berlinck *et al.*, 1992). Pyrrole-2-carboxylic derivatives have been long discussed as biochemical markers for the Agelasida and Axinellidae (Braekman *et al.*, 1992). Among them, pyrrole-2-aminoImidazoles (P2AI) were systematically expected in Axinellidae, Agelasidae and also in some representatives of the closely phylogenetically related family Dictyonellidae and Halichondriidae, about 200 distinct compounds of this family being described so far (Lejeune, 2010). However, the P2AI distribution among these sponge groups remained puzzling until a recent phylogeny performed by Gazave *et al.* (2010a) and a parallel study of the chemical diversity of Mediterranean representatives of these groups (Lejeune, 2010). These works confirmed that, as long suspected, Axinellidae and *Axinella* are polyphyletic assemblages (Gazave *et al.*, 2010a), with the studied *Axinella* species belonging to three distinct clades. One *Axinella* clade, named *Axinella*[P], contains the type species of the genus, *Axinella polypoides*, which actually does not contain P2AI. The chemical synapomorphy of this clade could be Verpacamides shared at least by *A. polypoides* and *Axinella vaceleti* (Vergne *et al.*, 2006). On the other hand, P2AI appeared to be shared between a new clade, *Cymbaxinella*[P], which includes the former "*Axinella damicornis*" and "*Axinella verrucosa*", and *Agelas*[P]. According to the most recent phylogeny (Gazave *et al.*, 2010a), P2AI would thus be a synapomorphy of *Agelasida*[P]. Finally, the Mediterranean species *Axinella cannabina* has been reallocated to a clade named *Acanthella*[P], this result being supported by the occurrence of terpene alkaloids in the representatives studied so far. These results obtained with a rather limited data set definitely indicate the need for an extensive revision of the worldwide distributed "*Axinella*" species, and for that purpose, chemical markers might be helpful in understanding unexpected phylogenetic results.

4.4.2. Metabolic fingerprinting

A strong limitation of the previous approach is that natural product chemists mainly focus on the description of original compounds, whereas only reports of similar ones in distinct organisms could provide phylogenetic information and useful synapomorphic chemical markers. Moreover, these approaches generally consider a low percentage of the whole metabolome. Similar to molecular phylogeny considering a higher number of genes, we might expect a better resolution of the interspecific relationships from methods analysing a broader portion of the metabolome. Such a more global metabolomics approach, called metabolic fingerprinting, can be used to screen the metabolic diversity of living systems (Fiehn, 2002; Weckwerth and Morgenthal, 2005; Nobeli and Thornton, 2006; Ellis *et al.*, 2007). The main objective of this metabolomics approach is to compare multiparametric patterns (or fingerprints) as dynamic metabolic phenotypes of a high number of samples (Wolfender *et al.*, 2009). This approach is widely applied in phytochemistry and microbiology for purposes such as classification of medicinal plants or prokaryotic strains (for a review, see Nielsen and Jewett, 2007), but rarely to discriminate between marine metazoan species. Metabolomics approaches proved useful to distinguish among individual signals and thus might serve in biomarker discovery, for the identification of new chemotaxonomic characters in systematics or biomarker of ecological processes (Fiehn, 2002; Wolfender *et al.*, 2009). Compared to an exhaustive metabolite analysis, metabolic fingerprinting is a rapid, untargeted and high-throughput method that can be used for a high number of samples, and last but not least it requires small amounts of biological material. A recent study using metabolic fingerprints as indicators of metabolomic diversity in order to assess interspecific relationships demonstrated that sponge chemical diversity may be useful for fundamental issues in systematics or evolutionary biology (Ivanišević *et al.*, 2011a) (Fig. 2.20). *Homoscleromorpha*[p] was particularly challenging because its chemistry was poorly studied and its phylogeny still debated. Aminosterols were sometimes considered as the chemical synapomorphy, although only recorded in *Plakina* Schulze, 1880 and *Corticium* Schmidt, 1862 species. A first validation of the metabolomics approach was to measure intraspecific variability, which was found significantly lower than the interspecific variability obtained between two *Oscarella* sister species (Ereskovsky *et al.*, 2009a; Ivanišević *et al.*, 2011a). This first analysis demonstrated that the divergence between these species was much more subtle than indicated by the natural products chemistry. Whereas Loukaci *et al.* (2004) only reported what could be considered as the major compounds of the two sister species, the metabolomic approach indicated that about 95% of both metabolomes were identical (Ivanišević, 2011; Ivanišević *et al.*, 2011a). Further research on their chemical composition actually showed that these species have exactly the same major compound

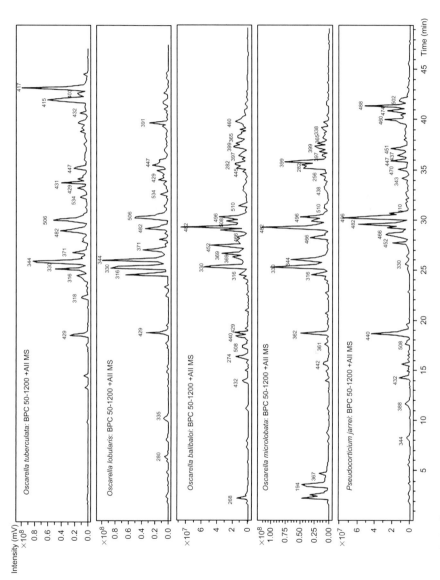

Figure 2.20 Metabolic fingerprints showing the interspecific variability among Homoscleromorpha without skeleton (Oscarellidae): HPLC-ESI (+)MS (BPC) with indications of the major m/z above peaks *(modified from Ivanišević, 2011)*.

(Ivanišević *et al.*, 2011b). Interspecific relationships among *Homoscleromorpha[P]* species were then inferred from the alignment of their metabolic fingerprints. The resulting classification appeared to be congruent with phylogenetic trees obtained for a DNA marker (*cox1* Folmer fragment) and demonstrates the existence of two distinct groups within *Homoscleromorpha[P]* (Ivanišević *et al.*, 2011a). Moreover, it appeared to be also congruent with a high-resolution molecular phylogeny (Gazave *et al.*, 2010a), thus proving the usefulness and potential of metabolic fingerprinting in sponge systematics. This case study called for a revision of *Homoscleromorpha[P]*, with two well-supported clades with (*Plakinidae[P]*) and without skeleton (*Oscarellidae[P]*), and to further identify additional chemical synapomorphic characters. For now, the aminosterol plakinamines are confirmed as a chemotaxonomical character of *Corticium[P]* and *Plakina*, but the *Plakinidae[P]* chemical synapomorphy remains to be found. The *Oscarellidae[P]* contains two well-supported clades: (i) one containing the two sister species *O. lobularis* and *O. tuberculata* and characterized by the recently described lysophosphatidylethanolamine (Ivanišević *et al.*, 2011b), and (ii) the other containing the new species *O. balibaloi* (Pérez *et al.*, 2011), *O. microlobata* and *Pseudocorticium jarrei*. Species belonging to this clade are characterized by mesohylar cells with paracristalline inclusions, while the description of the putative chemical synapomorphy is under progress (Ivanišević, 2011).

Today, metabolomics can revolutionize chemotaxonomy by allowing a fast analysis of a great number of samples and by considering a broader fraction of the metabolome in order to obtain classifications comparable to molecular ones. Metabolomics offer a rapid assessment method of an organism chemical diversity, thereby permitting the identification of bioactive compounds with pharmacological potential (in that case, it is an alternative to the biological screening approach) and highlighting rather easily ecological biomarkers and synapomorphic characters. We firmly believe that metabolomics can support some phylogenetic hypothesis and/or validate the current sponge classification. There is a number of recent works and forthcoming papers that will demonstrate its usefulness to discriminate between sister species or to describe cryptic species (Ivanišević, 2011; Pérez *et al.*, 2011; Reveillaud, 2011) among sponge groups lacking spicules (*Oscarellidae[P]* and Verongida). As it is the case for barcode sequences, we suggest to include a metabolic fingerprint for each new species described and to use this character to revisit a number of problematic sponge taxa (e.g. the former polyphyletic Axinellidae). However, chemotaxonomy and metabolomics defenders first have to overcome some fundamental challenges (i) to adopt consensual analytical processes, (ii) to find the proper method to exchange data files and (iii) to develop databases dedicated to chemotaxonomy.

5. CASE STUDIES: *HEXACTINELLIDA*[P.1] CONSENSUS BETWEEN MOLECULAR AND MORPHOLOGICAL DATA SETS

Today, Hexactinellida Schmidt, 1870 contains more than 580 extant species (7% of all described Recent *Porifera*[P]). Only five taxonomists are currently working on this group (Henry Reiswig, Martin Dohrmann, Dorte Janussen, Daniela Lopes and Konstantin Tabachnick). Hexactinellids have been collected from depths of 5 to 6770 m and are unknown in freshwater habitats. They constitute important members of deep-water marine communities which are undersampled. However, recent campaigns organized in the Antarctic Ocean as well as the exploration of deep-water habitat have led to the descriptions of numerous new taxa in many part of the world ocean (Janussen *et al.*, 2004; Lopes *et al.*, 2005, 2011; Menshenina *et al.*, 2007; Reiswig *et al.*, 2008; Tabachnick *et al.*, 2008; Janussen and Reiswig, 2009; Reiswig and Kelly, 2011). The recently discovered hexactinellid reefs of Canada's western continental shelf, analogues of long-extinct Jurassic sponge reefs have increased the attention of both palaeontologists and biologists to this clade (see Section 2.2). The *Hexactinellida*[P] clade has been recognized early (Thomson, 1868; Schmidt, 1870; see Reiswig, 2006 for a review) and never challenged. Synapomorphies of the *Hexactinellida*[P] include the possession of triaxonic spicules, an axial filament with a square section and a unique cellular and syncytial tissue organization (Leys *et al.*, 2007). The skeleton of *Hexactinellida*[P] is composed of "an amazing array of spicules of various shapes and sizes" (Leys *et al.*, 2007, p. 59) which remains loose during the whole life of the sponge and a rigid siliceous network formed by fusion of the main supporting spicules. Subclasses Amphidiscophora and Hexasterophora are discriminated by the respective possession of amphidiscs versus hexasters (Fig. 2.21). These two subclasses have been consistently upheld since their designation by Schulze (1887).

The skeleton is the only structure which has been well investigated and described in details (Schulze, 1887; Ijima, 1927). The recent studies on histology and reproduction have benefited from the occurrence of littoral hexactinellid populations on the Pacific Canadian coasts and in a Mediterranean cave. It is necessary to underline that in both cases, the collection sites are close to a marine station, respectively, Bamfield Marine Sciences Centre in Barkley Sound (British Columbia, Canada) and Endoume marine Station (Marseille, France) with diving facilities allowing the collection of well preserved specimens. The fragility of hexactinellid species does not

---

[1] Names established under the *PhyloCode* are in italics and can be identified with the symbol "p" to avoid confusion with names of the Linnaean classification.

Figure 2.21 Types of spicules present in Hexactinellida. (A) Hexactines characteristic of Hexactinellida; scale bar = 68 μm. (B) Amphidisc synapomorphy of the Amphidiscophora; scale bar = 53 μm. (C, D) Two types of discohexasters from *Oopsacas minuta*; hexasters are a synapomorphy of the Hexasterophora; scale bars = 28 and 13 μm, respectively (photos courtesy J. Vacelet).

ensure a good preservation of specimens when collected by traditional dredges, and it is only in the aquarium of Bamfield Center that they can be maintained in life during several weeks (Leys *et al.*, 2007). It is the main reason why histology and reproduction have been described only recently, with the exception of the work of Okada (1928). The larva is a trichimella (Boury-Esnault *et al.*, 1999a; Leys *et al.*, 2006; Fig. 2.19). The embryo is cellular and syncytial organization is acquired during embryogenesis (Leys *et al.*, 2006). The reproduction is known only from two species *Farrea sollasi* Schulze, 1886 (Okada, 1928) and *O. minuta* Topsent, 1927 (Boury-Esnault *et al.*, 1999a; Leys *et al.*, 2006), both Hexasterophora Schulze, 1886, but belonging, respectively, to Hexactinosida Schrammen, 1903 and

Lyssacinosida Zittel, 1877. The histology and cytology are known in several species of Hexasterophora (*Rhabdocalyptus dawsoni* Lambe, 1893: Mackie and Singla, 1983; *Farrea occa* Bowerbank, 1862: Reiswig and Mehl, 1991; *O. minuta*: Boury-Esnault and Vacelet, 1994; Pérez, 1996; *Caulophacus cyanae*: Boury-Esnault and De Vos, 1998; *A. vastus* Schulze, 1886: Leys, 1999), but for the time being nothing is known about histology, cytology or reproduction for Amphidiscophora Schulze, 1886.

The first molecular phylogeny studies included about 8% of all hexactinellid species and confirmed that *Amphidiscophora*[P] and *Hexasterophora*[P] are two well-supported clades (Dohrmann *et al.*, 2008, 2009, 2012). The molecular phylogeny of *Hexactinellida*[P] is fairly congruent with the current Linnaean classification. Among *Hexasterophora*[P], Lyssacinosida with 30 sampled species belonging to 3 families is monophyletic, these results being congruent with those obtained with morphological characters (Mehl, 1992). Hexactinosida represented by 13 species of Sceptrulophora Mehl, 1992 is also monophyletic; however, there is a need to improve the sampling in order to obtain a better representation of families and genera. Among *Hexasterophora*[P], taxa from Aulocalycoida Tabachnick and Reiswig, 2000, Lychniscosida Schrammen, 1903 and Fieldingida Tabachnick and Janussen, 2004 need to be added to better understand internal relationships. *Amphidiscophora*[P] is represented by only six species, distributed in two well-supported clades, three belonging to the Hyalonematidae Gray, 1857 and three to the Pheronematidae Gray, 1870.

Molecular phylogenetics of *Hexactinellida*[P] is just beginning, and we strongly believe that a significant effort in sampling will permit to test the first phylogenetic hypotheses, and to assess the congruence between molecular and morphological characters at lower taxonomical ranks. Hexactinellid taxonomists have a huge tendency to erect subspecies: 18 within *Farrea occa*, and 6 species and 7 subspecies in the genus *Asconema*. The taxonomic "tradition" has been extensively discussed by Lopes *et al.* (2011). The "cosmopolitanism" of sponge species rejected in most cases for *Demospongiae*[P] and *Calcispongiae*[P] (see Sections 3.2 and 4.3.1) has never been questioned for *Hexactinellida*[P]. Genetic studies are absolutely needed to resolve the question of cryptic species.

6. CASE STUDIES: *CALCISPONGIAE*[P]

Knowledge of the world fauna of *Calcispongiae*[P] is fragmentary. The total number of described species (ca. 675) represents about 9% of all described extant sponges. This is partially due to a bias in taxonomic effort and the common idea that calcareous sponges are difficult to identify. Fortunately, there has been a recent revival of taxonomic expertise (e.g. Michelle Klautau, Catarina Longo, Hans Tore Rapp, Oliver Voigt and Gert Wörheide) in several poorly studied biogeographical areas (e.g. Wörheide and Hooper, 1999; Klautau and Valentine, 2003; Rapp, 2006) and notably due to the interest of this clade as a model for evolutionary developmental studies (Adamska *et al.*, 2011). This trend can be illustrated by focusing on the 67 species of *Clathrina* described so far: 23 were described before 1900, 11 between 1900 and 1990, 11 between 1990 and 2000 and 22 since 2000 (Table 2.6).

Table 2.6 List of species of *Clathrina* present in the World Porifera Database on June 2011, classified by chronological order of descriptions (van Soest *et al.*, 2011b)

Scientific name accepted	Authority accepted	Geographical origin
Clathrina coriacea	Montagu, 1818	NE Atlantic
Clathrina reticulum	Schmidt, 1862	Mediterranean
⋆Clathrina clathrus	Schmidt, 1864	Mediterranean
Clathrina contorta	Bowerbank, 1866	Mediterranean
Clathrina canariensis	Miklucho-Maclay, 1868	NE Atlantic
Clathrina panis	Haeckel, 1870	Central Atlantic
Clathrina cerebrum	Haeckel, 1872	NE Atlantic
Clathrina cordata	Haeckel, 1872	SW Indian Ocean
Clathrina decipiens	Haeckel, 1872	Adriatic
Clathrina densa	Haeckel, 1872	SW Pacific
Clathrina dictyoides	Haeckel, 1872	SW Pacific
⋆Clathrina primordialis	Haeckel, 1872	Mediterranean
Clathrina sceptrum	Haeckel, 1872	NW Atlantic
Clathrina cancellata	Verrill, 1873	N Atlantic
Clathrina compacta	Schuffner, 1877	SW Indian Ocean
⋆Clathrina procumbens	von Lendenfeld, 1885	SW Pacific
⋆Clathrina laminoclathrata	Carter, 1886	SW Pacific
Clathrina dubia	Dendy, 1891	SW Pacific
Clathrina pelliculata	Dendy, 1891	SW Pacific
Clathrina laxa	Kirk, 1896	SW Pacific–N Zealand
Clathrina minoricensis	Lackschewitsch, 1896	Mediterranean
Clathrina nanseni	Breitfuss, 1896	Arctic Ocean
Clathrina multiformis	Breitfuss, 1898	White Sea
⋆Clathrina ceylonensis	Dendy, 1905	Central Indian Ocean (Ceylon)
Clathrina tenuipilosa	Dendy, 1905	Central Indian Ocean (Ceylon)
Clathrina atlantica	Thacker, 1908	NE Atlantic–Cape Verde Islands
Clathrina gardineri	Dendy, 1913	E Indian Ocean

(continued)

Table 2.6 *(continued)*

Scientific name accepted	Authority accepted	Geographical origin
Clathrina mutsu	Hozawa, 1928	NW Pacific
Clathrina sagamiana	Hozawa, 1929	NW Pacific
Clathrina soyo	Hozawa, 1933	NW Pacific
Clathrina izuensis	Tanita, 1942	NW Pacific
Clathrina rubra	Sarà, 1958	NE Atlantic–Mediterranean
Clathrina ascandroides	Borojevic, 1971	SW Atlantic
Clathrina biscayae	Borojevic and Boury-Esnault, 1987	NE Atlantic
*Clathrina aurea	Solé-Cava, Klautau, *et al.*, 1991	SW Atlantic
Clathrina brasiliensis	Solé-Cava, Klautau, *et al.*, 1991	SW Atlantic
Clathrina aspina	Klautau, Solé-Cava and Borojevic, 1994	SW Atlantic
**Clathrina cylindractina*	Klautau, Solé-Cava and Borojevic, 1994	SW Atlantic
Clathrina paracerebrum	Austin, 1996	NE Pacific
Clathrina adusta	Wörheide and Hooper, 1999	SW Pacific
Clathrina helveola	Wörheide and Hooper, 1999	SW Pacific
Clathrina heronensis	Wörheide and Hooper, 1999	SW Pacific
Clathrina luteoculcitella	Wörheide and Hooper, 1999	SW Pacific
Clathrina parva	Wörheide and Hooper, 1999	SW Pacific
Clathrina wistariensis	Wörheide and Hooper, 1999	SW Pacific
Clathrina chrysea	Borojevic and Klautau, 2000	SW Pacific
Clathrina conifera	Klautau and Borojevic, 2001	SW Atlantic
Clathrina quadriradiata	Klautau and Borojevic, 2001	SW Atlantic
Clathrina tetractina	Klautau and Borojevic, 2001	SW Atlantic
Clathrina cribrata	Rapp, Klautau and Valentine, 2001	NE Atlantic–Norway
Clathrina septentrionalis	Rapp, Klautau and Valentine, 2001	Arctic Ocean
Clathrina clara	Klautau and Valentine, 2003	E Indian Ocean
Clathrina africana	Klautau and Valentine, 2003	SW Indian Ocean
Clathrina sueziana	Klautau and Valentine, 2003	Red Sea
Clathrina hirsuta	Klautau and Valentine, 2003	SW Indian Ocean
Clathrina hispanica	Klautau and Valentine, 2003	Mediterranean
Clathrina hondurensis	Klautau and Valentine, 2003	W Central Atlantic
Clathrina rotunda	Klautau and Valentine, 2003	SW Indian Ocean
Clathrina sinusarabica	Klautau and Valentine, 2003	Red Sea
Clathrina sueziana	Klautau and Valentine, 2003	Red Sea
Clathrina tetrapodifera	Klautau and Valentine, 2003	SW Pacific–N Zealand

(continued)

Table 2.6   *(continued)*

Scientific name accepted	Authority accepted	Geographical origin
Clathrina corallicola	Rapp, 2006	Arctic and NE Atlantic
Clathrina jorunnae	Rapp, 2006	NE Atlantic–S Norway
Clathrina alcatraziensis	Lanna, Rossi, *et al.*, 2007	SW Atlantic
Clathrina angraensis	Azevedo and Klautau, 2007	SW Atlantic
Clathrina antofagastensis	Azevedo, Hajdu, Willenz and Klautau, 2009	SE Pacific
Clathrina fjordica	Azevedo, Hajdu, Willenz and Klautau, 2009	SE Pacific
Clathrina broendstedi	Rapp, Janussen and Tendal, 2011	Antarctic

6.1. Minchin–Bidder hypothesis confirmed by modern techniques. The precursors of an integrative taxonomy

The possession of a skeleton made of calcium carbonate spicules makes the *Calcispongiae*^{P} unique with respect to all other sponges. All calcisponge species have calcareous spicules, which can be associated with a massive calcareous skeleton in a small number of species (e.g. *Petrobiona massiliana* Lévi and Vacelet, 1958, *Murrayona phanolepis* Kirkpatrick, 1910). Over the past two centuries, the monophyletic origin of calcareous sponges has never been seriously doubted, and molecular phylogenies using the full 18S and partial 28S rDNA sequences confirm with high support the monophyly of *Calcispongiae*^{P} (Manuel *et al.*, 2003, 2004; Dohrmann *et al.*, 2006). The unique morphological synapomorphy is monocrystalline calcareous spicules.

   All along the twentieth century, two systems of classification were used prioritizing different characters. The first system based on the arrangement of the aquiferous system discriminates between homocoel (asconoid) and heterocoel (syconoid + leuconoid) grades of organization (Poléjaeff, 1883; Tuzet, 1973). The second system proposed by Bidder (1898) following observations of Minchin (1896) is based on the position of the nucleus within the choanocytes, the shape of the spicules, the type of larva and the first type of spicule to appear during ontogeny. Bidder's (1898) solid classification based on several independent data sets and recognized by several subsequent authors (Dendy and Row, 1913; Hartman, 1958; Borojevic, 1979) was only adopted at the end of the twentieth century and validated by the first molecular phylogenies on *Calcispongiae*^{P} at the beginning of the twenty-first century (Manuel *et al.*, 2003, 2004; Dohrmann *et al.*, 2006). The two clades recognized within *Calcispongiae*^{P} are the *Calcinea*^{P} and the *Calcaronea*^{P} defined by Bidder (1898). *Calcinea*^{P} has equiangular triactine spicules, a basal nucleus in the choanocytes, a flagellum arising independently from the nucleus, a coeloblastula larva and triactines as the first spicules to appear during ontogenesis. *Calcaronea^P* possesses inequiangular triactines, an apical nucleus in the choanocytes, a flagellum arising from the nucleus, a stomoblastula larva which after eversion becomes an amphiblastula and diactines as the first spicules to appear during ontogenesis. Although most of the diagnostic characters cannot be polarized, it is possible to assume that the embryological development of *Calcaronea^P* represents a synapomorphy of this clade: the internally flagellated blastula (stomoblastula) turns inside out to give the amphiblastula larva by a process called eversion (Duboscq and Tuzet, 1935; Manuel, 2006; Fig. 2.18). In larvae of both *Calcinea^P* and *Calcaronea^P*, the nucleus of the flagellated cells is apical and linked to the flagellum. During metamorphosis, the larval flagellated cells differentiate into choanocytes. In *Calcaronea^P*, the nucleus keeps the apical position, whereas in *Calcinea^P*, it becomes basal. Furthermore, in *Calcinea^P*, during the division of choanocytes, the nucleus transiently becomes apical (Hartman, 1958). We can assume from these observations that an apical nucleus linked to the flagellum is the plesiomorphic state in *Calcispongiae^P*, and consequently, a basal nucleus with no relation to the flagellum is a synapomorphy for *Calcinea^P* (Manuel, 2006).

6.2. Absence of consensus between current classification and molecular phylogeny inside the two clades *Calcinea^P* and *Calcaronea^P*

The congruence between the molecular results and the current morphology-based classification (see Manuel *et al.*, 2002) is not retrieved at the rank of orders, families or even genera (Manuel *et al.*, 2003, 2004; Dohrmann *et al.*, 2006; Manuel, 2006), probably because of a high level of homoplasy of morphological characters. The results obtained in these molecular phylogenies are still not translated in any classification either Linnaean or phylogenetic. For example, the genus *Sycon* Risso, 1826 has been suspected to be polyphyletic through a cladistic analysis of morphological characters (Manuel, 2001). All molecular phylogenies performed so far (Manuel *et al.*, 2003, 2004; Dohrmann *et al.*, 2006) suggest that *Sycon* and most of the families and genera of *Calcispongiae^P* are polyphyletic and thus in need of a thorough revision through an integrative approach.

6.3. Revision of taxa. Case study: *Clathrina* Gray, 1867

The genus *Clathrina* is defined almost exclusively by negative characters (Borojevic and Boury-Esnault, 1987; Klautau and Valentine, 2003). The aquiferous system is asconoid, all the cavities are lined by choanocytes and the skeleton comprises few spicule types. Numerous species have been long considered as cosmopolitan: *Clathrina cerebrum* Haeckel, 1872 (Fig. 2.22),

Figure 2.22   Clathrinidae *in situ* showing the different aspects of the cormus. (A) *Clathrina cerebrum*, Calanque coast (NW Mediterranean), 10 m deep. (B) *Clathrina coriacea*, Calanque coast (NW Mediterranean), 13 m deep. (C) *Clathrina reticulum*, Calanque coast (NW Mediterranean), 13 m deep. (D) *Guancha lacunosa* Calanque coast (NW Mediterranean), 12 m deep. A–C (photos N. Boury-Esnault); D (photo T. Pérez).

C. clathrus Schmidt, 1864 (Fig. 2.13) and *Clathrina coriacea* Montagu, 1818 (Fig. 2.22). In order to test the morphological variability and distribution of *Clathrina* species, sympatric and allopatric populations were studied using genetic and morphological approaches (Solé-Cava *et al.*, 1991; Klautau *et al.*, 1994; Borojevic and Klautau, 2000; Klautau and Borojevic, 2001). Actually, the species previously considered cosmopolitan have more restricted distributions. These genetic studies allowed the systematicians to adopt a less conservative attitude and to reassess small variations of the diagnostic characters. Consequently, the number of *Clathrina* species increased from 36 to 67 between 1987 and 2011 (Table 2.6), and since the 1990s, all the cosmopolitan *Clathrina* were shown to be species complexes. The bright yellow-coloured *C. clathrus* (type species of the genus) is nowadays restricted to the Mediterranean Sea and the nearby Atlantic. Several "yellow" *Clathrina* have been described worldwide and not automatically allocated to *C. clathrus*. *Clathrina aurea* Solé-Cava, Klautau, Boury-Esnault, Borojevic and Thorpe, 1991 has been described from the SW Atlantic (Fig. 2.13), *C. luteoculcitella* Wörheide and Hooper, 1999, from the SW Pacific (Great Barrier Reef) and *C. chrysea* Borojevic and Klautau, 2000 from the SW Pacific (New Caledonian lagoons and coral reefs).

According to the revision of the genus *Clathrina* (Klautau and Valentine, 2003), the most diagnostic characters are those from the organization of the cormus: the type of anastomosis between the tubes and the presence of one or several oscules. In the organization of the skeleton, the most diagnostic characters are the spicule types, the size of the spicules, the shape of the actine of triactines/tetractines, the presence of spines on the actines, the presence of diactines and possibly cytological features such as the density of choanocytes, patterns of choanocyte arrangement, presence/absence of granular cells, occurrence, size and shape of porocytes (Wörheide and Hooper, 1999). This very detailed revision has led to a complete reassessment of morphological and anatomical characters logically leading to a complete redescription of all *Clathrina* type specimens.

A recent molecular phylogenetic analysis based on 20 species of *Clathrina*, 3 species of *Guancha* (Clathrinidae Minchin, 1900), 1 species of *Ascandra* (Leucaltidae Dendy and Row, 1913) and 2 species of *Leucetta* (Leucettidae de Laubenfels, 1936) has actually shown that *Clathrina* was polyphyletic (Rossi *et al.*, 2011). A well-supported clade includes all *Clathrina* without tetractines, the type-species *C. clathrus* and two species of *Guancha*. Since then, the absence of tetractines is considered by the authors as a synapomorphy of this clade. They have allocated all the *Clathrina* without tetractines to the genus *Clathrina* and propose for all the other species a new generic allocation. Therefore, *Clathrina* would be restricted to the 24 species described without tetractines (Table 2.6). Important diagnostic characters such as the colour, the presence/absence of tetractines and the shape of the actines (undulated vs. cylindrical) support the different clades (Rossi *et al.*, 2011).

7. Case Studies: *Homoscleromorpha*[P]

7.1. Homoscleromorpha, the fourth class of Porifera

The monophyly of *Homoscleromorpha*[P] has been accepted for many years (Lévi, 1956, 1957, 1973; Bergquist, 1978; Muricy and Díaz, 2002). It was formally erected to the rank of a subclass of Demospongiae by Bergquist (1978). However, recent molecular phylogenies have shown that *Homoscleromorpha*[P] are not part of *Demospongiae*[P] (Borchiellini *et al.*, 2004a; Dohrmann *et al.*, 2008; Philippe *et al.*, 2009; Sperling *et al.*, 2009; Gazave *et al.*, 2010b; Pick *et al.*, 2010; Sperling *et al.*, 2010). Finally, Homoscleromorpha was formally proposed as the fourth class of Porifera (Gazave *et al.*, 2012). The interest for this small clade has arisen because of the presence of characters like basal lamina underlining the pinacoderm and choanoderm in the adult/larval epithelium (Boute *et al.*, 1996; Boury-Esnault *et al.*, 2003; de Caralt *et al.*, 2007; Ereskovsky *et al.*, 2007). Furthermore, *O. lobularis* Schmidt, 1862 (type species of the genus *Oscarella*) has been suggested as a promising model for evolutionary developmental biology (Gazave *et al.*, 2008; Ereskovsky *et al.*, 2009a; Lapébie *et al.*, 2009).

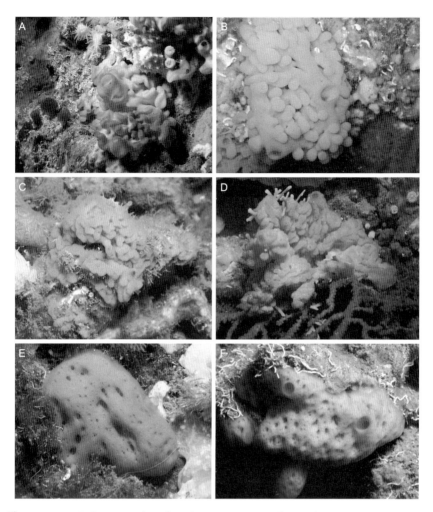

Figure 2.23 A few examples of Mediterranean *Homoscleromorpha* species. (A) *Oscarella lobularis*. (B) *Oscarella tuberculata*. (C) *Oscarella viridis*. (D) *Oscarella balibaloi*. (E) *Corticium candelabrum*. (F) *Pseudocorticium jarrei* (photos T. Pérez).

7.2. Lévi's classification recovered by modern techniques: Molecular and chemical

Homoscleromorpha is the smallest class of Porifera with 7 genera and 87 species described so far: 6 species of *Corticium* Schmidt, 1862, 3 of them described since 1990, 16 species of *Oscarella*, 10 of them described since 1990, 6 of *Placinolopha* Topsent, 1897, 28 species of *Plakina* Schulze, 1880,

13 of them described since 1990, 11 of *Plakinastrella* Schulze, 1880, 2 of them described since 1990, 19 species of *Plakortis* Schulze, 1880, 11 of them described since 1990 and *Pseudocorticium* Boury-Esnault, Muricy, Gallissian and Vacelet, 1995 which has been described after 1990. Altogether, 40 species have been described these past 20 years, representing an increase of 42% of the number of Homoscleromorpha species (Table 2.7 and Fig. 2.23). This clade has thus the highest rate of new species descriptions (Ereskovsky *et al.*, 2009c; Muricy, 2011; Pérez *et al.*, 2011). This high rate of description of new species is linked to the genetic studies of the 1990s which showed that morphological variability between sympatric populations correspond to low levels of genetic identity between them, indicating the absence of gene flow (Boury-Esnault *et al.*, 1992; Muricy *et al.*, 1996b).

All possible morphological data sets, from external features (colour, consistency, aspect of the surface), spicule shapes when present, anatomical and cytological characters, microsymbionts, are used as diagnostic characters to discriminate between species (Boury-Esnault *et al.*, 1995; Muricy *et al.*, 1998; Vishnyakov and Ereskovsky, 2009). More recently, molecular (rDNA, full mtDNA genome, partial sequence of *CO1*) and chemical characters (Gazave *et al.*, 2010b; Ivanišević *et al.*, 2011a; Pérez *et al.*, 2011; Fig. 2.7) arose as complementary diagnostic characters. The cytological data set of Homoscleromorpha allows to discriminate between cryptic aspiculate species of *Oscarella* (Boury-Esnault *et al.*, 1992, 1995; Muricy *et al.*, 1996a; Muricy and Pearse, 2004; Ereskovsky, 2006; Ereskovsky *et al.*, 2009b; Pérez *et al.*, 2011), as well as spiculate species of *Plakina* (Muricy *et al.*, 1998; Muricy, 1999). As underlined by Muricy (2011), the taxonomy of other spiculate homoscleromorphs such as *Plakortis*, *Plakinastrella*, *Placinolopha* and *Corticium* would greatly benefit from the inclusion of histological and cytological characters.

7.3. Revision of taxa. Case study: *Oscarella* Vosmaer, 1884 (Table 2.8)

In the 1990s, a genetic study on colour morphs of sympatric populations of "*O. lobularis*" led to a reappraisal of the characters which highlighted the existence of many cryptic species among *Oscarella*. Firstly, two species synonymized by Schulze (1877) were redescribed: *O. lobularis* (Schmidt, 1862) and *O. tuberculata* (Schmidt, 1868). These species can be discriminated by their consistency (soft vs. cartilaginous) as indicated in the descriptions of Schmidt (1868) and by their cytological contents (Boury-Esnault *et al.*, 1992). Three additional species were described from Mediterranean submarine caves (Muricy *et al.*, 1996a; Fig. 2.23) on cytological and genetics

Table 2.7 List of species of Homoscleromorpha present in the World Porifera Database on June 2011, classified by chronological order of descriptions (van Soest *et al.*, 2011b)

Scientific name accepted	Authority accepted	Geographical origin
*Oscarella lobularis	Schmidt, 1862	Mediterranean
*Oscarella tuberculata	Schmidt, 1868	Mediterranean
Oscarella rubra	Hanitsch, 1890	NE Atlantic
Oscarella cruenta	Carter, 1881	NE Atlantic
Oscarella membranacea	Hentschel, 1909	SE Indian Ocean
Oscarella tenuis	Hentschel, 1909	SE Indian Ocean
*Oscarella imperialis	Muricy, Boury-Esnault, Bézac and Vacelet, 1996a	Mediterranean
*Oscarella microlobata	Muricy, Boury-Esnault, Bézac and Vacelet, 1996a	Mediterranean
*Oscarella viridis	Muricy, Boury-Esnault, Bézac and Vacelet, 1996a	Mediterranean
*Oscarella carmela	Muricy and Pearse, 2004	NE Pacific
Oscarella ochreacea	Muricy and Pearse, 2004	W Indian Ocean
Oscarella nigraviolacea	Bergquist and Kelly, 2004	W Indian Ocean
Oscarella stillans	Bergquist and Kelly, 2004	W Pacific–Philippines
*Oscarella malakhovi	Ereskovsky, 2006	NW Pacific
*Oscarella kamchatkensis	Ereskovsky, Sanamyan and Vishnyakov, 2009b	NW Pacific
*Oscarella balibaloi	Pérez, Ivanišević, Dubois, Pedel, Thomas, *et al.*, 2011	Mediterranean
*Pseudocorticium jarrei	Boury-Esnault, Muricy, Gallissian and Vacelet, 1995	Mediterranean
Corticium candelabrum	Schmidt, 1862	Mediterranean
Corticium simplex	Lendenfeld, 1907	E Indian Ocean
Corticium quadripartitum	Topsent, 1923	W Central Atlantic
Corticium acanthastrum	Thomas, 1968	N Indian Ocean
Corticium bargibanti	Lévi and Lévi, 1983	SW Pacific
Corticium niger	Pulitzer-Finali, 1996	NW Pacific
Placinolopha bedoti	Topsent, 1897	N Indian Ocean
Placinolopha spinosa	Kirkpatrick, 1900	SW Indian Ocean
Placinolopha moncharmonti	Sarà, 1960	Mediterranean
Placinolopha acantholopha	Thomas, 1970	N Indian Ocean
Placinolopha europae	Vacelet and Vasseur, 1971	SW Indian Ocean
Placinolopha sarai	Lévi and Lévi, 1989	W Pacific–Philippines
Plakinastrella copiosa	Schulze, 1880	Mediterranean
Plakinastrella clathrata	Kirkpatrick, 1900	S Pacific
Plakinastrella oxeata	Topsent, 1904	NE Atlantic
Plakinastrella ceylonica	Dendy, 1905	Indian–Pacific

(*continued*)

Table 2.7 (*continued*)

Scientific name accepted	Authority accepted	Geographical origin
Plakinastrella mammillaris	Lendenfeld, 1907	SE Indian Ocean
Plakinastrella minor	Dendy, 1916	N Indian Ocean
Plakinastrella trunculifera	Topsent, 1927	NE Atlantic
Plakinastrella onkodes	Uliczka, 1929	W Central Atlantic
Plakinastrella polysclera	Lévi and Lévi, 1989	W Pacific–Philippines
Plakinastrella mixta	Maldonado, 1992	Mediterranean
Plakinastrella microspiculifera	Moraes and Muricy, 2003	SW Atlantic
Plakina australis	Gray, 1867	SW Indian Ocean
Plakina monolopha	Schulze, 1880	Mediterranean
Plakina dilopha	Schulze, 1880	Mediterranean
★Plakina trilopha	Schulze, 1880	Mediterranean
Plakina versatilis	Schmidt, 1880	W Central Atlantic
Plakina brachylopha	Topsent, 1927	NE Atlantic
Plakina elisa	de Laubenfels, 1936	W Central Atlantic
Plakina bowerbanki	Sarà, 1960	Mediterranean
Plakina tetralopha	Hechtel, 1965	W Central Atlantic
Plakina topsenti	Pouliquen, 1972	Mediterranean
Plakina corticioides	Vacelet, Vasseur and Lévi, 1976	SW Indian Ocean
Plakina corticolopha	Lévi and Lévi, 1983	SW Pacific
Plakina reducta	Pulitzer-Finali, 1983	Mediterranean
Plakina bioxea	Green and Bakus 1994	NE Pacific–California
Plakina fragilis	Desqueyroux-Faúndez and van Soest, 1997	SE Pacific
Plakina microlobata	Desqueyroux-Faúndez and van Soest, 1997	SE Pacific
Plakina pacifica	Desqueyroux-Faúndez and van Soest, 1997	SE Pacific
Plakina jamaicencis	Lehnert and van Soest, 1998	W Central Atlantic
★Plakina crypta	Muricy, Boury-Esnault, Bézac and Vacelet, 1998	Mediterranean
★Plakina endoumensis	Muricy, Boury-Esnault, Bézac and Vacelet, 1998	Mediterranean
★Plakina jani	Muricy, Boury-Esnault, Bézac and Vacelet, 1998	Mediterranean
Plakina tetralophoides	Muricy, Boury-Esnault, Bézac and Vacelet, 1998	NW Pacific
Plakina weinbergi	Muricy, Boury-Esnault, Bézac and Vacelet, 1998	Mediterranean
Plakina atka	Lehnert, Stone and Heimler, 2005	NE Pacific–Aleutian

(*continued*)

Table 2.7 (*continued*)

Scientific name accepted	Authority accepted	Geographical origin
Plakina tanaga	Lehnert, Stone and Heimler, 2005	NE Pacific–Aleutian
Plakortis simplex	Schulze, 1880	Mediterranean NE Atlantic
Plakortis angulospiculata	Carter, 1882	W Central Atlantic
Plakortis halichondrioides	Wilson, 1902	W Central Atlantic
Plakortis zyggompha	de Laubenfels, 1934	W Central Atlantic
Plakortis lita	de Laubenfels, 1954	Indo-West Pacific
Plakortis nigra	Lévi, 1953	Red Sea
Plakortis erythraena	Lévi, 1958	Red Sea
Plakortis japonica	Hoshino, 1977	NW Pacific
Plakortis kenyensis	Pulitzer-Finali, 1993	SW Indian Ocean
Plakortis copiosa	Pulitzer-Finali, 1993	SW Indian Ocean
Plakortis quasiamphiaster	Diaz and van Soest, 1994	SW Pacific
Plakortis galapagensis	Desqueyroux-Faúndez and van Soest, 1997	SE Pacific
Plakortis insularis	Moraes and Muricy, 2003	SW Atlantic
Plakortis microrhabdifera	Moraes and Muricy, 2003	SW Atlantic
Plakortis albicans	Cruz-Barraza and Carballo 2005	NE Pacific
Plakortis fromontae	Muricy, 2011	SE Indian Ocean
Plakortis hooperi	Muricy, 2011	SW Pacific
Plakortis bergquistae	Muricy, 2011	Indo-West Pacific
Plakortis communis	Muricy, 2011	Indo-West Pacific

An asterisk * indicates the species which have an available cytological data set. Type species of genera are in bold.

grounds. Three species were described more recently from the North Pacific (Muricy and Pearse, 2004; Ereskovsky, 2006; Ereskovsky *et al.*, 2009b) and three from the West Indian Ocean (Bergquist and Kelly, 2004; Muricy and Pearse, 2004), these descriptions being mostly based on cytological characters (Table 2.8). *O. balibaloi* Pérez, Ivanišević, Dubois, Pedel, Thomas, Tokina and Ereskovsky, 2011 is the last described *Oscarella*. This new Mediterranean species has become quite abundant in the past years. This description was based on several data sets: morphology, cytology, symbionts, reproduction traits and life cycle, ecology and metabolic fingerprint (Fig. 2.7B). Today, the complete cytological description is available only for 9 of the 16 species of *Oscarella* (Fig. 2.24). In particular, the cytology and the genetic relationships of the northeast Atlantic and Indo-Pacific species remain to be investigated.

Table 2.8 *Oscarella* species: main morphological, anatomical, cytological (cells with inclusions, bacteria) and ecological characters

Oscarella	*lobularis*	*tuberculata*	*viridis*	*microlobata*	*imperialis*	*balibaloi*	*carmela*	*malakhovi*	*kamchatkensis*	*stillans*	*nigraviolacea*
Locality	Mediterranean	Mediterranean	Mediterranean	Mediterranean	Mediterranean	Mediterranean	N–E Pacific	N–W Pacific	N–W Pacific	Indo-Pacific	Indo-Pacific
Habitat	Vertical walls	Vertical walls	Obscure caves	Obscure caves	Vertical walls	Semi-obscure caves	Boulders	Bivalve shells, stones	Boulders, rocks	Vertical walls	Underside of plate coral
Depth (m)	5–60	5–60	6–15	12–15	25–40	5–40	Intertidal	0.4–5	10–23	12 m	6 m
Colour	Pink-rose	Yellow-blue	Light-green	Light-brown	Yellowish-white	White and orange	Light-brown–rusty orange	Pinky-beige to yellow	Orange-yellow	Dark honey yellow	Dark violet, almost black
Consistency	Soft	Cartilaginous	Soft, very fragile	Soft, fragile	Soft	Soft, mucous, very slimy	Extremely soft, slimy	Soft, slimy	Soft, slimy	Collagenous	Soft, limb, mushy
Surface	Smooth	Wrinkled	Lumpy, microlobate	Lumpy, microlobate	Lumpy, microlobate	Lumpy, microlobate	Lumpy, microlobate	Lumpy, microlobate	Lumpy, microlobate	Very smooth	Inflate, bubbly
Shape	Thick lobate	Thick lobate	Thin lobate	Thin lobate	Thick lobate	Thin microlobate	Thin microlobate	Thin microlobate	Thin lobate	Fused thin tubes	Lobate, convoluted
Ectosome thickness (µm)	5–30	5–90	10–50	10–50	15–50	7–18	5–10	8–15	4–25	25–30 µm	20–30 µm
Chamber diameter (µm)	35–90	40–75	30–75	40–75	45–80	52–73	25–65	12–32	15–23	17–30 µm	30–40 µm
Vacuolar cells	Two types	One type	No	One type	Two types	One type	One type	One type	No	?	?
Spherulous cells	No	No	Two types	Three types	Two types	Two types	One type	One type	Three types	?	?
Types of bacteria	3	2	2	4	3	2	2	2	3	?	?
References	Schmidt (1862) and Boury-Esnault et al. (1992)	Schmidt (1868) and Boury-Esnault et al. (1992)	Muricy et al. (1996a)	Muricy et al. (1996a)	Muricy et al. (1996a)	Pérez et al. (2011)	Muricy and Pearse (2004)	Ereskovsky (2006)	Ereskovsky et al. (2009b)	Bergquist and Kelly (2004)	Bergquist and Kelly (2004)

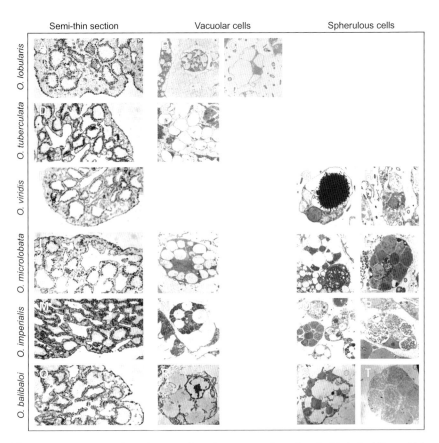

Figure 2.24 Cells with inclusions of Mediterranean *Oscarella*. *O. lobularis*: (A) semithin section showing the aquiferous system and the mesohylar cells. (B) Vacuolar cells type I. (C) Vacuolar cells type II. *O. tuberculata*: (D) semithin section showing the aquiferous system and the numerous vacuolar cells of the mesohyl. (E) Vacuolar cells. *O. viridis*: (F) semithin section showing the aquiferous system and the mesohylar cells. (G) Spherulous cell with a huge vacuole and fibrillar inclusions with a zigzag arrangement. (H) Spherulous cell with microgranular inclusions. *O. microlobata*: (I) semithin section showing the aquiferous system and the mesohylar cells. (J) Vacuolar cell. (K) Spherulous cell with microgranular inclusions (m) and spherulous cell with a single inclusion (s). (L) Spherulous cell with paracrystalline inclusion. *O. imperialis*: (M) semithin section showing the aquiferous system and the mesohylar cells. (N) Vacuolar cells with paracrystalline inclusion. (O) Vacuolar cells (v), spherulous cell with paracrystalline inclusion (p) and spherulous cell with a single spherule (s). (P) Detail of a spherulous cell with paracrystalline inclusion. *O. balibaloi*: (Q) semithin section showing the aquiferous system and the mesohylar cells. (R) Vacuolar cell. (S) Spherulous cell with granular paracrystalline inclusions. (T) Spherulous cell with paracrystalline inclusions. A, D, F, I, M, Q (photos courtesy A. Ereskovsky); B, C, E, G, H, J–L, N–P (photos N. Boury-Esnault), R–T (photos T. Pérez).

8. Case Studies: *Demospongiae*[P]: Molecular Phylogeny Proposes a New Classification. The Four New Clades Within *Demospongiae*[P]

Demospongiae[P] contains about 85% of all extant sponge species, with about 6900 living species described in the literature. This estimate is probably conservative, and new species are continuously described thanks to the investigation of new areas and to new techniques that allow to differentiate sibling species. The composition of Demospongiae has always been subject to debate contrary to that of the *Hexactinellida*[P] and the *Calcispongiae*[P] both clearly defined since the end of the nineteenth century. Basically, all taxa that are not hexactinellid or calcareous sponges are considered demosponges. This is reflected in the name itself of *Demospongiae*, from the Greek *dēmos* "*people*" and *spongiá* "*sponge*" which means "the common sponge" (Voultsiadou and Gkelis, 2005). Demospongiae, defined by either a combination of plesiomorphic characters or negative characters, had a fluctuating content over the years (Boury-Esnault, 2006; Erpenbeck and Wörheide, 2007). As previously said, Borchiellini *et al.* (2004a) have shown that *Homoscleromorpha*[P] does not group with *Demospongiae*[P], and this was confirmed by several subsequent studies (Dohrmann *et al.*, 2008; Philippe *et al.*, 2009; Sperling *et al.*, 2009; Gazave *et al.*, 2010b; Pick *et al.*, 2010; Sperling *et al.*, 2010; see also Section 7).

Thanks to molecular data, four inclusive clades have been recognized among *Demospongiae*[P], well supported by high boostrap value (>90%): *Keratosa*[P] (G1), *Myxospongiae*[P] (G2), *Haploscleromorpha*[P] new clade name (marine Haplosclerida, G3) and *Heteroscleromorpha*[P] new clade name (G4) (Borchiellini *et al.*, 2004a; Holmes and Blanch, 2007; Lavrov *et al.*, 2008; Sperling *et al.*, 2009) (Fig. 2.25; see Sections 10.1.1 and 10.1.2 for diagnosis and *PhyloCode* definition). The relationships between these four clades are not completely resolved and differ according to the data sets and the methods used in the phylogenetic analyses. However, a sister relationship between *Keratosa*[P] and *Myxospongiae*[P] (species without siliceous megascleres) and between *Haploscleromorpha*[P] and *Heteroscleromorpha*[P] (species with siliceous megascleres) have been obtained with partial 28S rDNA sequences (D4–D7 domains, 806 bp), mtDNA sequences and 18S rDNA sequences (Holmes and Blanch, 2007; Lavrov *et al.*, 2008; Fig. 2.25).

8.1. *Keratosa*[P] (Bowerbank, 1864) (Borchiellini *et al.*, 2004a).

This clade includes Dictyoceratida (491 species) and Dendroceratida (70 species) (Bergquist and Cook, 2002b; Cook and Bergquist, 2002a). A total of 561 species have been described: 36 of which between 1990 and 2000

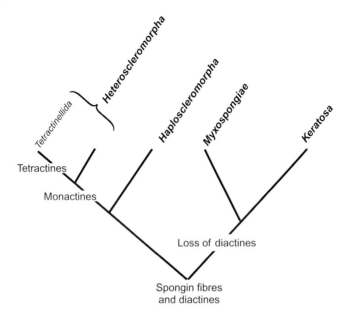

Figure 2.25 Current consensus tree for *Demospongiae*[P]. The ancestral characters for skeleton are supposed to be spongin fibres and monaxon-diactines. Diactines have been lost in *Myxospongiae*[P] and *Keratosa*[P]. Acquisition of monaxon-monactines occurred in the clade *Heteroscleromorpha*[P] and tetractines in *Tetractinellida*[P].

and 32 since 2000. *Keratosa*[P] include the commercial sponges belonging to the genera *Spongia* Linnaeus, 1759, *Hippospongia* Schulze, 1879, *Coscinoderma* Carter, 1883 and *Rhopaleoides* Thompson, Murphy, Bergquist and Evans, 1987. The skeleton of *Keratosa*[P] is exclusively constituted by spongin fibres sometimes reinforced by foreign material. "Spongin" initially served to designate sponge structures made of microfibrils of about 10 nm in diameter. However, we do not know if all spongin assemblages are homologous (Simpson, 1984; Garrone, 1985) and whether or not they are entirely made of sponge short-chain collagens (Exposito *et al.*, 1991; Aoucheria *et al.*, 2006) (see Section 10.1.1 for a diagnosis).

To build a solid framework for the classification, several data sets were used to investigate the relationships within this clade: morphological (skeleton and external features), soft-tissue organization (anatomy and cytology), reproduction and secondary-metabolite chemistry (Bergquist 1980, 1996; Bergquist *et al.*, 1990; Bergquist and Cook, 2002b; Cook and Bergquist, 2002a). Molecular phylogenies of this clade are not published yet; however, some could be published soon (Erpenbeck *et al.*, 2010; see Chapter 1).

Four families are recognized within the Dictyoceratida Minchin, 1900: Dysideidae Gray, 1867 (5 genera), Spongiidae Gray, 1867 (6 genera), Thorectidae Bergquist, 1978 (23 genera) and Irciniidae Gray, 1867 (4 genera) (see

Cook and Bergquist, 2002b for a detailed review). Cook (2007) has undertaken a clarification of dictyoceratid taxonomic characters with a detailed redescription of the characteristics of the skeleton, the hierarchy of the fibres, surface characters, choanocyte chambers, collagen, cortical armour as well as general form. A molecular study using the Folmer *cox1* fragment (Pöppe *et al.*, 2010) allowed to discriminate two well-supported clades: *Ircinia* Nardo, 1833 and *Psammocinia* Lendenfeld, 1889. However, the Folmer *cox1* fragment is not sufficient to resolve relationships within the irciniid species sampled.

Molecular investigations propose to allocate the Sphinctozoan *Vaceletia* Pickett, 1982 to *Keratosa*[P] and more specifically to Dictyoceratida (Lavrov *et al.*, 2008; Wörheide, 2008). And yet, *Vaceletia crypta* (Vacelet, 1977) has a chambered skeleton composed of an irregular arrangement of aragonite crystals, the larva is a parenchymella incubated in the maternal body (Vacelet, 1977, 1979, 2002). Thus, the relationships of *Vaceletia* within Dictyoceratida have had a lot of consequences on the palaeontological comprehension of Sphinctozoans (see Chapter 1).

Dendroceratida Minchin, 1900 is composed of two families, Darwinellidae Merejkowsky, 1879 (four genera) and Dictyodendrillidae Bergquist, 1980 (four genera). The content of Dendroceratida has changed over the past 50 years. The current classification is the results of a complete reassessment of the different characters (Bergquist *et al.*, 1990; Bergquist and Cook, 2002b).

8.2. *Myxospongiae*[P] Haeckel, 1866 (Borchiellini *et al.*, 2004a)

This clade is composed of the two orders Chondrosida and Verongida. The name "Myxospongiae" was chosen by Borchiellini *et al.* (2004a) because this clade includes most of the *Demospongiae*[P] without a skeleton: *Hexadella* within Verongida and *Chondrosia*, *Thymosiopsis* and *Halisarca* within Chondrosida (see Section 10.1.1 for a diagnosis). Today, this clade is represented by 137 species (83 Verongida and 54 Chondrosida), 15 of which were described between 1990 and 2000 and 23 since 2000. With the exception of the Halisarcidae, most of the species are oviparous. The two orders have been found as sister groups in all the molecular phylogenetic works so far (Borchiellini *et al.*, 2004a; Nichols, 2005; Lavrov *et al.*, 2008). Except for some ultrastructural traits shared by larvae (see Section 4.2), few synapomorphies have been recognized for this clade. de Laubenfels (1948) had allocated *Thymosia* Topsent, 1895 to Verongida because he found similarities between the fibres of *Thymosia* and *Aplysina* Nardo, 1834. Biochemical analyses have shown that chitin and silica–chitin–aragonite biocomposite are present in *A. aerophoba*, *A. cavernicola*, *A. cauliformis* Carter, 1882 and *Verongula gigantea* Hyatt, 1875 (Ehrlich *et al.*, 2007, 2010; Kurek *et al.*, 2010). It would be important to check if these compounds are present in the other families of Verongida as well as in the fibres of *Thymosia*.

8.2.1. Chondrosida Boury-Esnault and Lopès, 1985: A case study the genus *Chondrilla* Schmidt, 1862

The clade *Chondrosida*[P] includes two families: Chondrillidae Gray, 1872 with four genera *Chondrilla*, *Chondrosia*, *Thymosia* and *Thymosiopsis*, and Halisarcidae Schmidt, 1862 recently allocated to this order with only one genus *Halisarca* (Vacelet *et al.*, 2000; Boury-Esnault, 2002; Ereskovsky *et al.*, 2011). *Halisarca* has been allocated successively to the Myxospongiae with several other sponges without skeleton such as *Hexadella* (see Lévi, 1956), then to Dendroceratida within a family Halisarcidae in the absence of any clear alternative affiliation (de Laubenfels, 1936). This prompted Bergquist (1996) to create the order Halisarcida (Bergquist and Cook, 2002c). However, all the molecular phylogenies performed so far have indicated clear relationships between Halisarcidae and Chondrillidae (Borchiellini *et al.*, 2004a; Ereskovsky *et al.*, 2011). In this chapter, an emended diagnosis of the Chondrosida is proposed (see Section 10.1.1).

Chondrilla, *Chondrosia* and *Halisarca* were thought to include cosmopolitan species (e.g. *C. nucula*, *C. reniformis*, *H. dujardini*) until the first genetic studies of the 1990s (Klautau *et al.*, 1999; Lazoski *et al.*, 2001; Usher *et al.*, 2004; Duran and Rützler, 2006; Zilberberg *et al.*, 2006; Ereskovsky *et al.*, 2011). Since then, 17 species have been described in the genus *Chondrilla* (Table 2.9), 6 since 1997 and many more are still pending formal descriptions (Klautau *et al.*, 1999; Zilberberg *et al.*, 2006) and are for the time being phantom species (see Section 10.1.3). The main problems are encountered in the species complex of West tropical Atlantic *C.* aff. *nucula*. The comparison of numerous characters, such as diameter of spherasters, localization of the spicules as well as of spherulous cells, aspect of the surface, among specimens of *Chondrilla* from the same and different localities in the western Atlantic and in the Mediterranean (Bavestrello *et al.*, 1993; Cavalcanti *et al.*, 2007; Rützler *et al.*, 2007a) have failed to find discriminating characters within the *C. nucula* species complex. However, a possible taxonomic character could be the analysis of the sulphated polysaccharides by agarose gel electrophoresis which seems a promising biochemical technique to distinguish cryptic species even with preserved *Chondrilla* specimens. The sulphated polysaccharides have discriminated *C. nucula*, *C. australiensis* and three undescribed species of *Chondrilla* from the Bahamas and the Brazilian coast (Vilanova *et al.*, 2007). DNA sequences (D2 region in the 28S rDNA and ITS1–5.8S–ITS2 region) of *C. nucula* from the Mediterranean, *Chondrilla* from Bermudas and three Australian species clearly allow to distinguished five species (Usher *et al.*, 2004). The Australian species can be discriminated by mean size and presence or absence of two size classes of spherasters, as well as by presence or absence of oxyasters. Based on this conclusion and on a genetic study (Duran and Rützler, 2006), Rützler *et al.* (2007a) have described the Caribbean species under the name *Chondrilla caribensis* Rützler, Duran and Piantoni, 2007. For all of these species, it is important to undertake cytological studies using TEM.

Table 2.9 List of species of *Chondrilla* present in the World Porifera Database on June 2011, classified by chronological order of descriptions (van Soest *et al.*, 2011b)

Species names accepted	Authority accepted	Biogeographical area
Chondrilla nucula	Schmidt, 1862	Mediterranean, NE Atlantic
Chondrilla australiensis	Carter, 1873	SW Pacific (Australia)
Chondrilla mixta	Schulze, 1877	Indo-Pacific
Chondrilla sacciformis	Carter, 1879	W Indian Ocean
Chondrilla secunda	Lendenfeld, 1885	SW Pacific (Australia)
Chondrilla grandistellata	Thiele, 1900	NE Indian Ocean (Indonesia)
Chondrilla jinensis	Hentschel, 1912	NE Indian Ocean (Indonesia)
Chondrilla kilakaria	Kumar, 1925	N Indian Ocean (India)
Chondrilla euastra	de Laubenfels, 1949	W Pacific (Palau)
Chondrilla acanthastra	de Laubenfels, 1954	W Pacific (Palau)
Chondrilla oxyastera	Tanita and Hoshino, 1989	NW Pacific Ocean (Japan)
Chondrilla verrucosa	Desqueyroux-Faúndez and van Soest, 1997	SE Pacific Ocean
Chondrilla montanusa	Carballo, Gómez, Cruz-Barraza and Flores-Sánchez, 2003	NE Pacific Ocean
Chondrilla pacifica	Carballo, Gómez, Cruz-Barraza and Flores-Sánchez, 2003	NE Pacific Ocean
Chondrilla tenochca	Carballo, Gómez, Cruz-Barraza and Flores-Sánchez, 2003	NE Pacific Ocean
Chondrilla caribensis	Rützler, Duran and Piantoni, 2007a,b	W Tropical Atlantic
Chondrilla linnaei	Fromont, Usher, Sutton, Toze and Kuo, 2008	SW Pacific (Australia)

Note the high number of species described in the Indo-Pacific region (15 species) for only two described in the whole Atlantic.

8.2.2. Verongida Bergquist, 1978

The order Verongida was erected by Bergquist, 1978 based on the type of fibres, cytological characters as well as biochemical compounds, in particular, tyrosine-brominated compounds. The current composition of this order has been corroborated by few molecular phylogenies (Schmitt *et al.*, 2005; Erwin and Thacker, 2007). Erwin and Thacker (2007) nonetheless show that Aplysinidae could be monophyletic by including *Pseudoceratina* (the only genus of the Pseudoceratinidae) and leaving *Aiolochroia* with *Ianthella* and *Aplysinella*. The new Aplysinidae would then be sister group to a moderately supported (*Aiolochroia* + *Ianthella* + *Aplysinella*) clade. This phylogeny also demonstrates that characteristics of the spongin fibres (diameter, pith content, fibre branching pattern) are homoplastic between families. The taxonomy of many verongid species and genera has to be revisited, still using all the possible data sets, especially the external characters on fresh material (as Pinheiro *et al.*, 2007 for *Aplysina*; Reveillaud, 2011 for *Hexadella*), or metabolic fingerprinting (Reveillaud, 2011).

8.3. *Haploscleromorpha*[P] new clade name (G3, Haplosclerina Topsent, 1928 and Petrosina Boury-Esnault and van Beveren, 1982)

Haplosclerina include 792 species allocated in three families: Chalinidae (462 species), Niphatidae (120 species) and Callyspongiidae (210 species). A total of 424 species are included in the genus *Haliclona* (subdivided in six subgenera), 182 in the genus *Callyspongia* (subdivided in 5 subgenera) and 52 in *Amphimedon*. Petrosina include 248 species allocated in 3 families: Calcifibrospongiidae (1 species), Petrosiidae (125 species) and Phloedictyidae (122 species) (Fig. 2.26). Molecular phylogenies have shown a huge discrepancy between morphological and molecular data sets (see for molecular works Raleigh *et al.*, 2007; Redmond *et al.*, 2007; Redmond and McCormack, 2008, 2009; and for morphological works, see de Weerdt 1985, 1986, 1989, 2000, 2002). In most of the molecular phylogenies, the *Haploscleromorpha*[P] are a well-supported clade. However, internal relationships within this clade never matched the current phylogenetic hypotheses based on morphological cladistic analyses (see Section 3.1). Suborders, families, genera and even subgenera appeared polyphyletic in the molecular trees (Raleigh *et al.*, 2007; Redmond *et al.*, 2007, 2011; Redmond and McCormack, 2008). Haplosclerida is a typical case, perfectly illustrating what sponge taxonomists are often facing. Because haplosclerids have a skeleton, most early taxonomists considered it was enough to use those characters for classification; they did not try to use another data set such as cytology for example (contrary to what happened for groups without a skeleton). In our opinion, a re-evaluation of the morphological characters under the light of the molecular results is definitively needed. It appears that a bottom-up strategy should be the best alternative, studying first the type species of each haplosclerid genera. It would also seem reasonable to start with clades which are highly supported by molecular evidence and attempt to find morphological, cytological, reproduction or chemical synapomorphies for theses clades (see Section 4.1.3). See Sections 10.1.1 for an emended diagnosis and 10.1.2 for a *PhyloCode* definition.

Figure 2.26 Diversity of shape, size and colour of Haploscleromorpha species. (A) *Callyspongia vaginalis*, Martinique (Caribbean Sea), 15 m deep. (B) *Xestospongia testudinaria*, Martinique (Caribbean Sea), 40 m deep. (C) *Haliclona mucosa*, Calanques coast (NW Mediterranean), 15 m deep. (D) *Haliclona mediterranea*, Le Veron (NW Mediterranean), 30 m deep. (E) *Haliclona fulva*, Frioul island (NW Mediterranean), 30 m deep (photos T. Pérez). (For interpretation of the references to colour in this figure legend, the reader is referred to the Web version of this chapter.)

8.4. G4: *Heteroscleromorpha*[p] new clade name

The clade G4 (Borchiellini *et al.*, 2004a) contains about 5000 species and is the most important group of *Demospongiae*[p] in number of species. This clade has been named "Democlavia" by Sperling *et al.* (2009); however, this name is a *nomen nudum* because it has not been given a formal Linnaean (rank-based) definition. Because it is the *Demospongiae*[P] clade with the highest diversity of spicule types, we propose formally to name it: *Heteroscleromorpha*[P] (see Sections 10.1.1 and 10.1.2 for a diagnosis and a *PhyloCode* definition). The internal phylogeny of the *Heteroscleromorpha*[P] is poorly understood (Morrow *et al.*, 2012). At least two inclusive clades are well supported: Spongillina (Addis and Peterson, 2005; Meixner *et al.*, 2007; Redmond *et al.*, 2007; Itskovich *et al.*, 2007, 2008) and *Tetractinellida*[P] (Chombard *et al.*, 1998; Cárdenas *et al.*, 2011). On the contrary, Halichondrida, Hadromerida and Poecilosclerida orders, as defined by the *Systema Porifera*, are polyphyletic (see Erpenbeck and Wörheide, 2007).

8.4.1. Well-supported inclusive clades: *Spongillida*[P] and *Tetractinellida*[P]

8.4.1.1. *Spongillida*[P] new clade name

Freshwater sponges belong to the monophyletic suborder Spongillina (Manconi and Pronzato, 2002) which contains 243 species, 46 genera (19 of which are monospecific) and 8 families (2 of which are monogeneric and monospecific). The family Spongillidae has a worldwide distribution, whereas the other families have a more restricted distribution. Freshwater environments in Africa (Brien, 1970; Boury-Esnault, 1980; Manconi and Pronzato, 2009), South America (Manconi and Pronzato, 2005; Volkmer-Ribeiro, 2007) and Siberia (Efremova, 2001, 2004; Masuda, 2009) harbour a high number of freshwater sponge families, whereas the ones of North America (Frost *et al.*, 2001; Manconi and Pronzato, 2005) and Europe (Manconi and Pronzato, 2002) mostly contain Spongillidae species. Because of the absence of a recent general revision of European freshwater fauna, we tend to perhaps underestimate the species diversity of this area.

Even if the monophyly of Spongillina is well supported by all phylogenetic works performed so far (Addis and Peterson, 2005; Meixner *et al.*, 2007; Redmond *et al.*, 2007; Itskovich *et al.*, 1999, 2007, 2008; Morrow *et al.*, 2012), the internal relationships of the families and genera are not resolved, mostly because there are too few sequences available (about 30 sequences of 18S rDNA and Folmer fragment of the *CO1* in the sponge genetree server on 19 July 2011). However, Spongillidae and Lubomirskiidae as well as *Spongilla* and *Ephydatia* genera appeared not to be monophyletic with 18S rDNA and *CO1* sequences (Harcet *et al.*, 2010; Itskovich *et al.*, 2007; Meixner *et al.*, 2007; Redmond *et al.*, 2007; Erpenbeck *et al.*, 2011). With ITS1 and ITS2 the endemic Lubomirskiidae are a well-supported clade but not *Lubomirskia*, *Baikalispongia* and Spongillidae (Itskovich *et al.*, 2008). A taxonomic revision of Spongillina with several data sets seems to be necessary. For a *PhyloCode* definition of *Spongillida*[P], see Section 10.1.2.

8.4.1.2. *Tetractinellida*[P] Marshall, 1876 (Borchiellini *et al.*, 2004a)

The *Tetractinellida*[P] is a worldwide group comprising the sister-orders Spirophorida Bergquist and Hogg, 1969 and Astrophorida Sollas, 1888. The *Astrophorida*[P] Sollas, 1888 (Cárdenas *et al.*, 2011) represents about 820 species (Cárdenas *et al.*, 2011) and the Spirophorida 154 species (van Soest *et al.*, 2011b). *Tetractinellida*[P] species have a unique synapomorphy: four-rayed megascleres called *triaenes*. The monophyly of the *Tetractinellida*[P] along with the sister relationship of Spirophorida and *Astrophorida*[P] has been repeatedly shown with strong support by molecular data (Chombard *et al.*, 1998; Borchiellini *et al.*, 2004a; Nichols, 2005; Lavrov *et al.*, 2008; Voigt *et al.*, 2008). *Astrophorida*[P] species are characterized by the simultaneous presence of asters and triaenes but lack a synapomorphy. On the contrary, Spirophorida species have a clear synapomorphy: microscleres with a "S" or "C" shape called *sigmaspires*. Monophyly of the Astrophorida and Spirophorida has also been shown (Szitenberg *et al.*, 2010; Cárdenas *et al.*, 2011) although few species of Spirophorida were considered. The first comprehensive molecular phylogeny of the *Astrophorida*[P] including 153 specimens (9 families, 29 genera, 89 species) shows that many families are polyphyletic (Geodiidae, Pachastrellidae and Ancorinidae) as well as many genera (*Ecionemia, Erylus, Poecillastra, Penares, Rhabdastrella, Stelletta* and *Vulcanella*) and proposes a revised classification of this order following the Linnaean system and the *PhyloCode* (Cárdenas *et al.*, 2010, 2011). It has been also shown how widespread convergent evolution and secondary loss are in *Astrophorida*[P] spicule evolution: these processes have taken place many times, in all taxa, and concerned both megascleres and microscleres, even when these seem to be adaptive and under selective pressures (Cárdenas *et al.*, 2011). As of today, the Spirophorida phylogeny has not been fully investigated with molecular data, but preliminary molecular studies show that some genera belonging to the Tetillidae Sollas, 1886 (*Cinachyrella* Wilson, 1925, *Craniella* Schmidt, 1870, *Tetilla* Schmidt, 1868) are probably polyphyletic (Szitenberg *et al.*, 2010) and need to be revised.

8.4.2. Polyphyletic orders: Poecilosclerida Topsent, 1928, Halichondrida Gray, 1867 and Hadromerida Topsent, 1894

Phylogenetic studies always show that these three orders (as defined by the *Systema Porifera*) are polyphyletic. Poecilosclerida contains more than 2500 described species, but very few sequences are available. Halichondrida (740 species) and Hadromerida (660 species) have a relatively higher number of sequences available in GenBank (Erpenbeck *et al.*, 2004, 2005, 2007a,b). In many molecular phylogenetic trees, Suberitidae Schmidt, 1870 (belonging to the Hadromerida in the *Systema Porifera*) and Halichondriidae Gray, 1867 constitute together a well-supported clade (Erpenbeck *et al.*, 2005, 2007a; Morrow *et al.*, 2012) as suspected in earlier works (Chombard, 1998; Chombard and Boury-Esnault, 1999). For Poecilosclerida, the most important result revealed by the molecular phylogenetic studies is that Raspailiidae

Hentschel, 1923 does not belong to Poecilosclerida. This family seems more closely related to some axinellid species (Erpenbeck *et al.*, 2007b; Morrow *et al.*, 2012) as in the Lévi–Bergquist–Hartman classification (see Chapter 1 for more details). The Desmacellidae Ridley and Dendy, 1886 in the *Systema Porifera* are considered to be Poecilosclerida without chelae; however, molecular data suggest that some Desmacellidae are not related to the Poecilosclerida (Mitchell *et al.*, 2011; Morrow *et al.*, 2012).

8.4.2.1. A case study of polyphyly within a genus of Halichondrida: *Axinella* Schmidt, 1862

The genus *Axinella* Schmidt, 1862 currently includes a heterogeneous assemblage of species (Fig. 2.27), and it is quite difficult to give a satisfactory diagnosis of it as underlined by Ridley and Dendy (1887, p. 178). Several species included in *Axinella* lack some of the features which are generally recognized as typical of the genus (see the diagnosis in Alvarez and Hooper, 2002, p. 727). The polyphyly of *Axinella* was suspected due to (1) the high levels of intrageneric genetic differentiation found between *Axinella* species (Solé-Cava and Boury-Esnault, 1999) and (2) the results obtained with 28S rDNA (Alvarez *et al.*, 2000; Erpenbeck *et al.*, 2007b).

A phylogeny based on 18S and partial 28S rDNA resolves part of this problem even if the number of *Axinella* sequences (9/96) is still low (Gazave *et al.*, 2010a; Fig. 2.28). However within this data set, two main groups of *Axinella* are well supported. The first clade includes the type species *A. polypoides* Schmidt, 1862 with *A. vaceleti* Pansini, 1983, *Axinella infundibuliformis* (Linnaeus, 1759), *Axinella dissimilis* (Bowerbank, 1866) and *Axinella aruensis* Hentschel, 1912. The second clade is composed of *Axinella damicornis* Esper, 1794, *A. verrucosa* Esper, 1794 and *Axinella corrugata* George and Wilson, 1919. This clade includes also a *Cymbastela* and a *Ptilocaulis* species and has close affinity with *Agelas* species as suspected earlier (Lafay *et al.*, 1992). Both clades are clearly separated. The first clade, including the type species of *Axinella*, is named *Axinella^P^*. The second clade has to be renamed; they are not *Axinella* species. The name *Cymbaxinella^P^* has been proposed (Gazave *et al.*, 2010a). It is quite important to clearly split *Axinella* to avoid confusion for the end-users of taxonomy such as chemists working on natural products, as both groups of "*Axinella*" have a quite different chemical composition (Vergne *et al.*, 2006; Lejeune, 2010; see also Section 4.4). Synapomorphies of *Cymbaxinella^P^* are the presence of isocyanids (Braekman *et al.*, 1992) and of a homologous secondary structure of V4 of 18S rDNA (Gazave *et al.*, 2010a). A ninth species, *A. cannabina* (Esper, 1794), forms a clade with *Acanthella* and *Dictyonella* species. A proposed synapomorphy of this clade is flexuous spicules. Following this, *cannabina* species has been reallocated to *Acanthella*. Except for this last case where a clear morphological synapomorphy has been found, the main problem of this group is the lack of morphological synapomorphies (Gazave *et al.*, 2010b).

Figure 2.27 Diversity of shape, aspect of the surface of axinellid species. (A) *Axinella*[P] *polypoides*, Morgiou Calanque (NW Mediterranean), 35 m deep. (B) *Axinella*[P] *vaceleti*, "Grotte à Corail" (NW Mediterranean), 9 m deep. (C) *Axinella*[P] *dissimilis* from Rathlin Island, NE Atlantic, 35 m deep. (D) *Acanthella*[P] *cannabina*, Crete (NW Mediterranean), 35 m deep. (E) *Cymbaxinella*[P] *damicornis*, Calanques coast (NW Mediterranean), 15 m deep. (F) *Cymbaxinella*[P] *verrucosa*, Corsica (NW Mediterranean), 35 m deep (photos T. Pérez). (For colour version of this figure, the reader is referred to the Web version of this chapter.)

8.4.2.2. A case study of polyphyly within a family of Halichondrida: Axinellidae Carter, 1875

As assumed by Erpenbeck and Wörheide (2007), the family Axinellidae as traditionally defined is polyphyletic. Axinellidae polyphyly was first suspected through a chemical study that had revealed two groups of Axinellidae (Braekman *et al.*, 1992) and then confirmed by several molecular

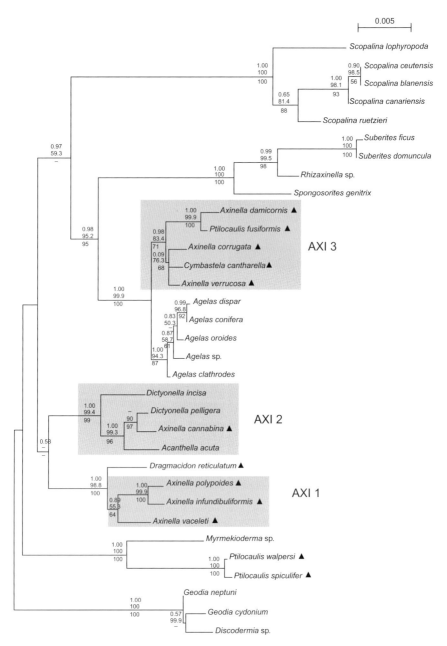

Figure 2.28 Tree resulting from the analyses of the 18S rRNA dataset of *Axinella* species. The tree is rooted on *Tetractinellida*[P] species. Three "*Axinella*" clades were found which confirms the polyphyly of *Axinella*. Black triangle indicates the species currently belonging to the Axinellidae family. *(modified from Gazave et al., 2010a).*

phylogenetic studies (Alvarez *et al.*, 2000; Erpenbeck *et al.*, 2005, 2007b; Gazave *et al.*, 2010a; Morrow *et al.*, 2012). The traditional family Axinellidae is composed of 12 genera: *Axinella* Schmidt, 1862, *Auletta* Schmidt, 1870, *Cymbastela* Hooper and Bergquist, 1992, *Dragmacidon* Hallmann, 1917, *Dragmaxia* Hallmann, 1916, *Ophiraphidites* Carter, 1876, *Pararhaphoxya* Burton, 1934, *Phakellia* Bowerbank, 1962, *Pipestela* Alvarez, Hooper and van Soest, 2008, *Ptilocaulis* Carter, 1885, *Phycopsis* Carter, 1883 and *Reniochalina* Lendenfeld, 1888. The clade *Axinellidae*^{P} recovered in the molecular trees so far (Erpenbeck *et al.*, 2007b; Gazave *et al.*, 2010a; Morrow *et al.*, 2012) is composed of *Axinella*^{P} (*Axinella*^{P} *polypoides*, *A. aruensis*, *A. dissimilis*, *A. infundibuliformis* + *A. vaceleti*), and *Dragmacidon*^{P} (*D. reticulatum* + *D.* aff. *mexicana*). The genus *Cymbastela*, except *C. cantharella* which is now included in the clade *Cymbaxinella*^{P}, seems to have close relationships with *Acanthella*^{P} clade (Erpenbeck *et al.*, 2007b) as well as *Phakellia ventilabrum*, type species of the genus *Phakellia* (Morrow *et al.*, 2012). The genus *Reniochalina* (type-species *R. stalagmites*) and the genus *Ptilocaulis* (type-species *P. gracilis*) are included in a clade of Raspailiidae (Erpenbeck *et al.*, 2007b). As far as we know, no sequences are available for *Dragmaxia*, *Ophiraphidites*, *Pipestela*, *Phycopsis*, *Auletta* and *Pararhaphoxya*. This is just an example of the huge reappraisal that has to be done in the so-called Halichondrida + Hadromerida + Poecilosclerida. A general framework of the phylogeny of this clade has to be built and checked through different data sets. It is also necessary to translate every bit of phylogenetic results, when it is consistent and well supported, into a system of classification.

9. SPONGE TAXONOMY NOWADAYS

9.1. Taxonomic changes accepted in the WPD since the publication of the *Systema Porifera*

The polyphyletic subclasses of Demospongiae, Ceractinomorpha Lévi, 1953 and Tetractinomorpha Lévi, 1953, suggested in the first cladistic morphological analysis (van Soest, 1990, 1991) have been definitively rejected after numerous congruent molecular results obtained by molecular phylogenies (Lafay *et al.*, 1992; Borchiellini *et al.*, 2004a; Nichols, 2005).

The monophyletic Homoscleromorpha is clearly not part of the Demospongiae (Borchiellini *et al.*, 2004a; Dohrmann *et al.*, 2008; Philippe *et al.*, 2009; Sperling *et al.*, 2009, 2010; Ereskovsky *et al.*, 2010; Gazave *et al.*, 2010b; Pick *et al.*, 2010). Consequently, a fourth class of Porifera has been defined following the Linnaean classification (Gazave *et al.*, 2012). In Homoscleromorpha, the two families Plakinidae and Oscarellidae have been resurrected on the basis of molecular, chemical and morphological data sets (Gazave *et al.*, 2010b; Ivanišević *et al.*, 2011a). The *PhyloCode* definition of *Homoscleromorpha*[P] is given in Manuel and Boury-Esnault (in press) and those of the more inclusive clades in Gazave *et al.* (2012).

Within Demospongiae, the order Halisarcida Bergquist, 1996 has been abandoned following the conclusions of Ereskovsky *et al.* (2011) based on mtDNA results and many other data sets like rDNA (Borchiellini *et al.*, 2004a) or cytology (Vacelet and Donadey, 1987; Bergquist *et al.*, 1998). The family Halisarcidae has been reallocated to the Chondrosida.

The family Thoosidae Cockerell, 1925 including the genera *Alectona* Carter, 1879, *Delectona* de Laubenfels, 1936, *Neamphius* de Laubenfels, 1953 and *Thoosa* Hancock, 1849 has been reallocated to the Astrophorida based on molecular, cytological and skeletal characters (Vacelet, 1999b; Borchiellini *et al.*, 2004b; Cárdenas *et al.*, 2011).

More importantly, since the publication of *Systema Porifera*, the polarity of morphological characters used in phylogenetic reconstruction is now better understood. In Demospongiae, contrary to what has been thought during many years, spongin fibres in the *Keratosa*[P] or the exclusive occurrence of diactines in the Haplosclerida are not derived characters, and tetractines are not the ancestral state for demosponge spicules (van Soest, 1987, 1991; de Weerdt, 1989; Fig. 2.25). This change of paradigm led us to re-evaluate most of the morphological phylogenetic hypotheses on which the current classification is based.

9.2. Problems to avoid

9.2.1. The loss of knowledge of old literature

The increasing amount of literature in all fields of sponge biology including sponge systematics and the tendency to read only what is available online slowly pushed us to forget the literature older than 20 years. A huge and most welcome effort has been made by the editors of WPD to digitize the oldest literature concerning sponge systematics, especially milestones like the Challenger expedition volumes or the Prince Albert I de Monaco Expeditions. In our opinion, it is of the responsibility of Natural History Museums which often possess historical libraries with a considerable patrimony to make their old literature available online. The whole spongologist community also has an important role to play to make all the knowledge on *Porifera*[P] available online for the next generations of taxonomists. Even if all the current scientific literature shares English as an international language, it is absolutely necessary in many cases to return to the primary description of a taxon whatever the language used (e.g. German, Russian, French, Spanish, English or even Latin) to avoid misinterpretations repeated from one work to another (Jenner, 2004c).

The following two examples show how publications have been overlooked. In 1996, a sponge species was found in South Brittany (France), in a well-studied area (the Ria of Etel). This species overgrowing numerous other organisms had never been recorded before, and it was suggested to be an invasive species. It became a dominant species in the gulf of Morbihan (France) (Pérez *et al.*, 2006) as well as the Dutch coastal waters (van Soest *et al.*, 2007b). This unknown sponge was finally described as a new species and genus, *Celtodoryx girardae* by Pérez *et al.* (2006). Later Henkel and Janussen (2011), in a survey of the Chinese Yellow Sea sponge fauna, found an abundant species with close morphological similarities to *C. girardae* and which looked very much like what Burton (1935) had described from Posiet Bay, Sea of Japan under the name *Cornulum ciocalyptoides*. The morphological characteristics of both species are quite similar, and Henkel and Janussen (2011) decided to put in synonymy both species under the name *Celtodoryx ciocalyptoides*. The origin of the invasive species could have occurred through the importation of the oyster *Crassostrea gigas* from the northeast Pacific to the northeast Atlantic European coasts. To conclude, in our times of economical globalization, we emphasize the necessity for taxonomists to review worldwide species of a genus when describing a new species. Another example of loss of knowledge concerns *Acanthostylotella* Burton and Rao, 1932, a genus overlooked and therefore missing in the *Systema Porifera*. Fortunately, de Voogd *et al.* (2010) have resurrected this genus and reallocated it to Agelasidae due to the possession of verticillated acanthostyles, and molecular/chemical data as previously suggested by Laubenfels (1936).

9.2.2. The errors or uncertainties in the databases

Many online open databases do not have a moderator or expert to check the diverse entries. These databases maintain a lot of errors which are transmitted to other databases and to publications. It is particularly true for GenBank (and all databases where sequences are deposited) where the validity of sequences and the identification of species are not checked. Furthermore, in most of these databases (e.g. GenBank, LifeDesk, MarinLit), the classification is not automatically updated (using the WPD, for example). Contaminations of sequences, misidentifications even when detected by further works are seldom corrected. In GenBank, only author(s) of the sequences can expressively ask them to be modified. Meanwhile, PorToL (www.portol.org) is making an effort to list those problematic sponge sequences so that phylogenetic studies stop propagating them. The same kind of problems occurs with MarinLit, database of the marine natural products literature, where the classification is not updated and the level of confidence for identification of species is not accurate (Erpenbeck and van Soest, 2007).

In phylogenetic or chemical works, "unidentified taxa" need to be avoided. For example, Halichondrida sp. or Axinellidae sp. will create phylogenetic noise in molecular or chemical databases and is absolutely useless for future studies. A minimum of the genus name is required for any sponge study, together with deposition of a voucher specimen in a recognized Museum.

9.2.3. The loss of type material

A species name should always be linked to a type material (e.g. specimen, slide and/or SEM preparation) deposited in a Museum and available for systematicists who want to compare it with new specimens. However, this type material has always been subject to many troubles such as problems of labels (mistake of describer, of subsequent taxonomists or of Museum personal), destruction of buildings during wars, or great fires (e.g. fire in the University of Lisbon (1978) with destruction of the library and the du Bocage collection), moving from a building to another, lack of staff to curate collection, retirement of a taxonomist whose collection is not taken care of or even thrown away. Consequently, many old specimens have been lost but can sometimes come back into view thanks to the perspicacity of a curator or the reorganization of a collection. The *Systema Porifera* project was in this sense a formidable opportunity to retrieve and list most of the type species for genera and families. It is also important to underline that the first sponge taxonomists did not have to follow the rules of the *ICZN* which were implemented only at the beginning of the twentieth century. Descriptions of these species were often very succinct, and it is often absolutely necessary to return to the type material to understand their morphology and anatomy. In our opinion, redescriptions of these old sponge species are therefore as important as description of new species (e.g. Rützler *et al.*, 2007b). However, it is not always easy to know where the type material was deposited (e.g. of Haeckel, Schmidt or Nardo). The worst being for the collection made during the eighteenth century, and it seems that the collection of Esper, Pallas, Olivi, Ellis and Solander, Fabricius has disappeared (Table 2.10). Some European taxonomists in the 1970–1980s realized the importance to know where the "old" collections are stored to be able to compare the type material to newly collected species. During these years, several catalogues on important sponge collections have been published co-authored by Shirley Stone: collections of Oscar Schmidt (Desqueyroux-Faúndez and Stone, 1992), Duchassaing and Michelotti (van Soest *et al.*, 1983), Dendy (Ayling *et al.*, 1982), Allan Hancock (Rützler and Stone, 1986). Equally important are specimen lists from Museums unfortunately rarely published and therefore not easily accessible, with some exceptions such as the Zoological Museum of Strasbourg (France) where a list is accessible   (http://www.musees.strasbourg.eu/index.php?page=mzoo-collections-invertebres) or the type catalogue of the Porifera collections of the Senckenberg Museum in Main, Germany (Barnich and Janussen, 2006). The Yale University Peabody Museum collections or the Zoological Museum of Amsterdam Porifera collections are accessible through the

Table 2.10 List of type material of the collections of Pallas, Esper, Olivi, Ellis and Solander, Fabricius, Risso, Nardo and their status[a]

Status following WPD	Names of Pallas, 1766	Status of the type
Amphimedon rubens Pallas, 1766	*Spongia rubens* Pallas, 1766	Unknown type
Aplysina fistularis Pallas, 1766	*Spongia fistularis* Pallas, 1766	Neotype
Aplysina fulva Pallas, 1766	*Spongia fulva* Pallas, 1766	Neotype
Callyspongia villosa Pallas, 1766	*Spongia villosa* Pallas, 1766	*species inquirenda*
Callyspongia tubulosa Pallas, 1766	*Spongia tubulosa* Pallas, 1766	Unknown type
Carteriospongia foliascens Pallas, 1766	*Spongia foliascens* Pallas, 1766	Neotype
Dendrilla membranosa Pallas, 1766	*Spongia membranosa* Pallas, 1766	Unknown type
Echinodictyum fibrillosum Pallas, 1766	*Spongia fibrillosa* Pallas, 1766	Unknown type
Halichondria papillaris Pallas, 1766	*Spongia papillaris* Pallas, 1766	*incertae sedis*
Halichondria fastigiata Pallas, 1766	*Spongia fastigiata* Pallas, 1766	Neotype
Halichondria panicea Pallas, 1766	*Spongia panicea* Pallas, 1766	*incertae sedis*
Haliclona cervicornis Pallas, 1766	*Spongia cervicornis* Pallas, 1766	Unknown type
Haliclona oculata Pallas, 1766	*Spongia oculata* Pallas, 1766	Holotype?
Hyattella cavernosa Pallas, 1766	*Spongia cavernosa* Pallas, 1766	Unknown type
Hyattella sinuosa Pallas, 1766	*Spongia sinuosa* Pallas, 1766	Unknown type
Ianthella basta Pallas, 1766	*Spongia basta* Pallas, 1766	Unknown type
Ianthella flabelliformis Pallas, 1766	*Spongia flabelliformis* Pallas, 1766	Unknown type
Isodictya frondosa Pallas, 1766	*Spongia frondosa* Pallas, 1766	Unknown type
Lubomirskia baikalensis Pallas, 1766	*Spongia baikalensis* Pallas, 1776	Unknown type
Phakellia crateriformis Pallas, 1766	*Spongia crateriformis* Pallas, 1766	Unknown type
Sarcotragus fasciculatus Pallas, 1766	*Spongia fasciculata* Pallas, 1766	Neotype
Spongia agaricina Pallas, 1766	*Spongia agaricina* Pallas, 1766	*species inquirenda*
Spongia floribunda Pallas, 1766	*Spongia floribunda* Pallas, 1766	*species inquirenda*
Spongia lichenoides Pallas, 1766	*Spongia lichenoides* Pallas, 1766	*incertae sedis*

(continued)

Table 2.10 (continued)

Status following WPD	Names of Pallas, 1766	Status of the type
Spongia tupha Pallas, 1766	*Spongia tupha* Pallas, 1766	**Neotype**
Tethya aurantium Pallas, 1766	*Alcyonium aurantium* Pallas, 1766	**Neotype**
Trikentrion muricatum Pallas, 1766	*Spongia muricata* Pallas, 1766	Unknown type
"All type material of Pallas was destroyed by fire in Küstrin (Poland) in 1758 during the Seven Years' War" (Wiedenmayer, 1977).		

Status following WPD	Names of Fabricius, 1780	Status of the type
Sycon ciliatum Fabricius, 1780	*Spongia ciliata* Fabricius, 1780	Unknown type
Grantia compressa Fabricius, 1780	*Spongia compressa* Fabricius, 1780	Unknown type
"If they still exist, at the Hunterian in Glasgow'"" (Stone, 1986).		

Status following WPD	Name of Ellis and Solander, 1786	Status of the type
Leucosolenia botryoides Ellis and Solander, 1786	*Spongia botryoides* Ellis and Solander, 1786	Unknown type
Isodictya palmata Ellis and Solander, 1786	*Spongia palmata* Ellis and Solander, 1786	Unknown type
Clathria prolifera Ellis and Solander, 1786	*Spongia prolifera* Ellis and Solander, 1786	Unknown type
Stelligera stuposa Ellis and Solander, 1786)	*Spongia stuposa* Ellis and Solander, 1786	Unknown type
Unlocated collection (Stone, 1986).		

Status following WPD	Name of Poiret, 1789	Status of the type
Petrosia ficiformis Poiret, 1789	*Spongia ficiformis* Poiret, 1789	Lectoype
Clathria coralloides Olivi, 1792	Names of Olivi, 1792 *Spongia coralloides* Olivi, 1792	Unknown type
Suberites domuncula Olivi, 1792	*Alcyonium domuncula* Olivi, 1792	**Neotype**

Status following WPD	Names of Esper, 1794; 1797	Status of the type
Amphilectus fucorum Esper, 1794	*Spongia fucorum* Esper, 1794	Unknown type
Auletta lyrata Esper, 1794	*Spongia lyrata* Esper, 1794	Unknown type
Axinella cannabina Esper, 1794	*Spongia cannabina* Esper, 1794	Unknown type

Current name	Original name	Status
Axinella damicornis Esper, 1794	*Spongia damicornis* Esper, 1794	Unknown type
Axinella damicornis Esper, 1794	*Spongia lactuca* Esper, 1794	Unknown type
Axinella verrucosa Esper, 1794	*Spongia verrucosa* Esper, 1794	Unknown type
Callyspongia foliacea Esper, 1797	*Spongia foliacea* Esper, 1797	Unknown type
Callyspongia tubulosa Esper, 1797	*Spongia tubulosa* Esper, 1797	Unknown type
Clathria rubicunda Esper, 1794	*Spongia rubicunda* Esper, 1794	*species inquirenda*
Clathria Thalysias cratitia Esper, 1797	*Spongia cratitia* Esper, 1797	Unknown type
Clathria surculosa Esper, 1794	*Spongia surculosa* Esper, 1794	Unknown type
Fasciospongia cellulosa Esper, 1794	*Spongia cellulosa* Esper, 1794	*species inquirenda*
Halichondria cartilaginea Esper, 1794	*Spongia cartilaginea* Esper, 1794	Unknown type
Halichondria suberosa Esper, 1794	*Spongia suberosa* Esper, 1794	Unknown type
Haliclona (Gellius) cymaeformis Esper, 1794	*Spongia cymaeformis* Esper, 1794	Unknown type
Hyattella pertusa Esper, 1794	*Spongia pertusa* Esper, 1794	Unknown type
Iotrochota membranosa Esper, 1794	*Spongia membranosa* Esper, 1794	Unknown type
Ircinia solida Esper, 1794	*Spongia solida* Esper, 1794	Unknown type
Isodictya compressa Esper, 1794	*Spongia compressa* Esper, 1794	Unknown type
Lendenfeldia plicata Esper, 1794	*Spongia plicata* Esper, 1794	Unknown type
Myrmekioderma granulatum Esper, 1794	*Alcyonium granulatum* Esper, 1794	Unknown type
Petrosia (Petrosia) clavata Esper, 1794	*Spongia clavata* Esper, 1794	Unknown type
Phyllospongia alcicornis Esper, 1794	*Spongia alcicornis* Esper, 1794	Unknown type
Phyllospongia lamellosa Esper, 1794	*Spongia lamellosa* Esper, 1794	Unknown type
Phyllospongia papyracea Esper, 1794	*Spongia papyracea* Esper, 1794	Unknown type
Spongia fruticosa Esper, 1794	*Spongia fruticosa* Esper, 1794	Unknown type
Spongia linteiformis Esper, 1797	*Spongia linteiformis* Esper, 1797	*species inquirenda*
Spongia polychotoma Esper, 1794	*Spongia polychotoma* Esper, 1794	Unknown type
Ulosa stuposa Esper, 1794	*Spongia stuposa* Esper, 1794	*species inquirenda*
Ulosa stuposa Esper, 1794	*Spongia crispata* Esper, 1794	Unknown type

(continued)

Table 2.10 (continued)

Status following WPD	Names of Esper, 1794; 1797	Status of the type
Verongula rigida Esper, 1794	*Spongia rigida* Esper, 1794	Unknown type
"No longer at the Zoologishes Museum, Erlingen. All the types were transferred after 1969 either to Frankfurt (NMS) or to the Bayerische Staatsammlung, Munchen". To be confirmed (Stone, 1986).		

Status following WPD	Name of Risso, 1826	Status of the type
Calyx nicaeensis Risso, 1826	*Spongia nicaeensis* Risso, 1826	Unknown type
Sycon humboldti Risso, 1826	*Sycon humboldti* Risso, 1826	Unknown type
	Names of Nardo, 1833, 1847	

Status following WPD		Status of the type
Aplysina aerophoba Nardo, 1833	***Aplysina aerophoba*** Nardo, 1833	Unknown type
Chondrosia reniformis Nardo, 1847	**Chondrosia reniformis** Nardo, 1847	**Neotype**
Axinella cinnamomea Nardo, 1833	*Grantia cinnamomea* Nardo, 1833	Unknown type
Haliclona flava Nardo, 1847	*Reniera flava* Nardo, 1847	Unknown type
Haliclona forcellata Nardo, 1847	*Reniera forcellata* Nardo, 1847	Unknown type
Haliclona typica Nardo, 1847	*Reniera typica* Nardo, 1847	*species inquirenda*
Suberites massa Nardo, 1847	*Suberites massa* Nardo, 1847	Unknown type
Stelletta pumex Nardo, 1847	*Tethia pumex* Nardo, 1847	Unknown type
"Types in the hands of daughter (in Venezia, Italy) in 1933; labels mixed and only a few identifiable" (Stone, 1986).		

[a] In the first column is the current accepted name; the name is in bold when it is the type species of a genus; the name is underlined if a holotype or a neotype has been designated. In the second column is the name under which the species was described, and in the third column the status of the type. Information compiled from *Systena Porifera* (Hooper and van Soest, 2002) and from the World Porifera database (van Soest *et al.*, 2011b).

Global Biodiversity Information Facility (GBIF) data portal (http://data. gbif.org; accessed 20 September 2011).

The European Union convinced by the richness of the collections of the Museums has developed since 2004 the SYNTHESYS project (http:// www.synthesys.info/) which aims to produce an accessible, integrated European resource for research users in natural sciences. SYNTHESYS will create a shared, high-quality approach to management and preservation, while giving access to leading European natural history collections. Collections will be perhaps better protected with this project because it stimulates researchers to visit museum collections. Accordingly, Museums see that their collections are used and put more effort and resources into curation. For example, the Calcispongiae collection at the National Museum of Natural History in Paris (France) was fully reorganized and renovated in 2011 because a collection manager realized its bad shape during a visit from a foreign researcher.

9.2.3.1. Lost specimen: The example of *Polymastia mamillaris* Müller, 1806

Polymastia was erected by Bowerbank (1864) with *Halichondria mamillaris* "Johnston, 1842" as the type species. In fact, the species "*mamillaris*" was not described by Johnston but by Müller (1806) under the name *Spongia mamillaris*. Johnston (1842) transferred "*mamillaris*" from the genus *Spongia* to the genus *Halichondria* and also synonymized *S. mamillaris* Müller, 1806 with *Spongia penicillus* Montagu, 1818. Bowerbank (1864) seems to have ignored the description of Müller, as there is no mention of this author in any text of Bowerbank concerning *P. mamillaris*. All subsequent spongologists have only taken into account Bowerbank's view and ignored Müller's description. According to Vosmaer (1935), in fact "nobody has been able to see Müller's original specimen". Fortunately, the type specimen of Müller was found in the Zoological Museum in Copenhagen (Denmark), while the type specimen of Montagu, 1818 was at the Natural History Museum, London (BMNH 30.7.3.26). The type specimen of Müller is quite different from that of Montagu, and they clearly belong to different species which means that Johnston (1842) was wrong when he put them in synonymy (Morrow and Boury-Esnault, 2000).

9.2.3.2. Description of a genus without type species and without type material: The tribulation of *Ircinia* type species

Ircinia was established by Nardo in 1833, with only a brief but clear description. Nardo did not describe any species but did include a list of five names all of which considered as *nomina nuda* (Cook and Bergquist, 2002b). The first species actually described as an *Ircinia* was the Mediterranean *Spongia fasciculata* Pallas, 1766, by Schmidt (1862). However, Pallas's Mediterranean types have been lost (Table 2.10). de Laubenfels (1948) designated *Ircinia fasciculata*, *sensu* Schmidt (1862) as the type species of the genus *Ircinia* and designated a neotype for *I. fasciculata*. Unfortunately, he used a specimen from the Dry Tortugas, Florida, and not from the type locality of the Mediterranean. This was in contradiction with the *ICZN* (Article 75.3.6), so the neotype is considered invalid (Cook and Bergquist, 2002b). Moreover, Schmidt's specimen of *I. fasciculata* has to be allocated to the genus *Sarcotragus* (Pronzato *et al.*, 2004).

Nardo had studied specimens collected exclusively in the canals and the lagoon of Venice (Adriatic Sea). In this lagoon, the only species of *Ircinia* is *Ircinia variabilis* (Schmidt, 1862). For this species, a lectotype from the Schmidt collection exists in the British Museum of Natural History (BMNH 1867:7:26:51) with Sebenico, Adriatic Sea as type locality (Pronzato *et al.*, 2004). *I. variabilis* was eventually designated as the type species of *Ircinia* by Pronzato *et al.* (2004).

9.2.3.3. Designation of neotypes for species without type material
We have lost the type material for numerous species, described mainly in the eighteenth and at the beginning of nineteenth centuries. In Table 2.10, we provide a list of type species described in the eighteenth century by Esper, Pallas, Nardo, Risso, etc. for which type material has disappeared. All the type material of Pallas, for example, was destroyed by a fire in Küstrin (Poland) in 1758 during the Seven Years' War, (Wiedenmayer, 1977). In few cases, a neotype has been designated, but it seems urgent to designate neotypes for all these species to avoid misinterpretations and misidentifications. This has to be done following the rules of the *ICZN*. The *ICZN* lists seven qualifying conditions for recognition of the validity of any neotype designation (Pinheiro *et al.*, 2007) (*ICZN* Articles 75.3.1–75.3.7). Among the most important conditions, (i) the neotype designation must be accompanied by data and a description sufficient to ensure recognition of the specimen designated, (ii) it must include evidence that the neotype came as nearly as practicable from the original type locality and (iii) it must include the author's reasons for believing the name-bearing type specimen(s) to be lost or destroyed.

9.2.4. The importance of a precise knowledge of the geographical area
The cosmopolitanism of many sponge species admitted until the end of the twentieth century is now most often doubted after revision of specimens from different parts of the distribution area. Among many examples two recent cases are described.

9.2.4.1. Indian Ocean and Mediterranean Elephant Ear sponges
The Elephant Ear sponge had been described by Pallas (1766) under the name *Spongia agaricina* without a precise type locality. In the Mediterranean Sea, this species is well known due to its vernacular name, well characterized

Figure 2.29 Elephant Ear sponge, *Spongia lamella*, "Grotte à Corail", Calanques coast (NW Mediterranean) (photo T. Pérez). (For colour version of this figure, the reader is referred to the Web version of this chapter.)

by its shape and was considered as one of the largest Mediterranean species (Fig. 2.29). A species described by Schulze (1879) from the Adriatic as *Spongia lamella* has been considered as a synonym of *S. agaricina* as well as of the Indo-Pacific *Spongia thienemanni* Arndt, 1943. Pallas left no specimen but, according to de Laubenfels (1948), the studied species originated from the Indian Ocean. Castritsi-Catharios *et al.* (2007) have compared the morphology and the tensile strength of two populations of Elephant Ear sponges, respectively, from Mediterranean and Indian Ocean (Philippines). The morphology, the skeleton and the tensile strength of both populations appeared quite different, and the authors concluded that they were faced with two species; however, they did not take a clear taxonomical decision. Finally, Pronzato and Manconi (2008) resolved the problem by designating two neotypes one for the Mediterranean Elephant Ear *S. lamella* Schulze, 1879 and the other for the Indian Ocean *S. agaricina* Pallas, 1766.

9.2.4.2. Mediterranean/NE Atlantic *Hymedesmia–Phorbas paupertas* (Bowerbank, 1866)

Hymedesmia paupertas is a species described by Bowerbank (1866) from the north European coasts. Topsent (1934) has allocated red, thick Mediterranean specimens collected in Tunisia and on Mediterranean French coasts to *Anchinoe* (= *Phorbas*) *paupertas* Bowerbank, 1866, and consequently *H. paupertas* and *A. paupertas* have been considered as synonyms. Recent revisions and underwater photograph have shown that *H. paupertas* is an encrusting thin blue species with a hymedesmioid skeleton (van Soest *et al.*, 2000; Picton *et al.*, 2011) quite different from the Mediterranean population with a columnar arrangement of the acanthostyles (Fig. 2.12). A new name

Phorbas topsenti Vacelet and Pérez, 2008 was recently proposed for the Mediterranean species (Vacelet and Pérez, 2008).

 ## 10. PERSPECTIVES

10.1. Propositions which could be adopted immediately

10.1.1. Propositions for the Linnaean classification

The four main demosponge clades recognized in all molecular phylogenies should be recognized as subclasses (Table 11). The name Keratosa Bowerbank, 1864 has been resurrected for the Dictyoceratida + Dendroceratida: Demospongiae with a skeleton made of spongin fibre only or with an hypercalcified skeleton (*Vaceletia* Pickett, 1982). Spongin fibres are either homogenous or pithed and strongly laminated with pith grading into bark. The family Verticillidae has to be reallocated to Dictyoceratida (Wörheide, 2008; Lavrov *et al.*, 2008).

Myxospongiae Haeckel, 1866 is proposed for the subclass which includes Verongida and Chondrosida (Maldonado, 2009). This name is proposed because all Demospongiae without skeleton belong to this subclass: *Halisarca*, *Chondrosia* and *Thymosiopsis* within Chondrosida and *Hexadella* within Verongida. Myxospongiae are Demospongiae without skeleton or with a skeleton made of siliceous asters (*Chondrilla*) or spongin fibres with a laminated bark and a finely fibrillar or granular pith (most of the Verongida and *Thymosia*).

An emended diagnosis of the order Chondrosida is proposed due to the recent allocation of Halisarcidae to this order: Myxospongiae, encrusting to massive, with a marked ectosome or cortex enriched by a highly organized fibrillar collagen. The skeleton is composed of nodular spongin fibres or aster microscleres or is absent. Collagen is always very abundant.

The third subclass corresponds to the marine Haplosclerida (Haplosclerina + Petrosina). We propose to name this subclass Haploscleromorpha: Demospongiae with an isodictyal anisotropic or isotropic choanosomal skeleton; spicules are diactinal megascleres (oxeas or strongyles); microscleres, if present, are sigmas and/or toxas, microxeas or microstrongyles.

The fourth subclass includes Spongillina, Tetractinellida, Poecilosclerida, Halichondrida, Agelasida and Hadromerida. It corresponds to the G4 clade of Borchiellini *et al.* (2004a). We propose for this subclass the name Heteroscleromorpha (see Section 8.4). Heteroscleromorpha are Demospongiae with a skeleton composed of siliceous spicules which can be monaxones and/or tetraxones and when they are present, microscleres are highly diversified. Within this subclass which contains most of the Demospongiae taxa (about 5000 species), the internal relationships are not resolved yet. However, the Spongillida and Tetractinellida are two well-supported monophyletic groups, whereas the other orders (Poecilosclerida, Halichondrida, Hadromerida) are polyphyletic.

The freshwater Spongillina is a well-supported clade, even if the intra-clade relationships are not resolved and need a re-evaluation with different data sets (e.g. Manconi and Pronzato, 2002; Itskovich *et al.*, 2007). We formally propose to upgrade Spongillina to the order rank and to name it Spongillida Manconi and Pronzato, 2002 with the same diagnosis.

The order Tetractinellida Marshall, 1876 has to be resurrected for Astrophorina Sollas, 1888 and Spirophorina Bergquist and Hogg, 1969 which would have the ranks of suborders (as in the Poecilosclerida Topsent, 1928, the suborders Latrunculina Kelly and Samaai, 2002, Mycalina Hajdu, van Soest and Hooper, 1994, Microcionina Hajdu, van Soest and Hooper, 1994 and Myxillina Hajdu, van Soest and Hooper, 1994). Tetractinellida is a clade strongly supported by molecular and morphological data. This order disappeared from the *Systema Porifera* without justification but has been used continuously by some authors (e.g. van Soest, 1991; Chombard *et al.*, 1998; Borchiellini *et al.*, 2004a; Cárdenas *et al.*, 2009; Morrow *et al.*, 2012). A new diagnosis is proposed: Heteroscleromorpha with a more or less radial skeleton associating tetractinal megascleres with asters or sigmaspires as microscleres.

The Astrophorida family Pachastrellidae Carter, 1875, as defined by the *Systema Porifera*, is polyphyletic and has been divided in three families by Cárdenas *et al.* (2011). A new diagnosis of Pachastrellidae Carter, 1875 is given: "Astrophorida with a majority of amphiasters as streptasters (never spirasters) in combination with large calthrops and/or short-shafted mesotriaenes or mesotrider desmas. A variety of monaxonic spicules can be present: microxeas, microrhabds, microstrongyles and microrhabdose streptasters" (Cárdenas *et al.*, 2011). The family Theneidae Carter, 1883 is resurrected with the following definition "Astrophorida with long-shafted triaenes (sometimes lost) in combination with diverse categories of streptasters: spirasters, metasters and plesiasters (sometimes with annulate actines)" (Cárdenas *et al.*, 2011). Finally, a new Astrophorida family called Vulcanellidae is erected for *Vulcanella* and *Poecillastra* with the following diagnosis "Astrophorida with calthrops, short-shafted triaenes or long-shafted triaenes. Aster microscleres include several categories of streptasters (spirasters, metasters, amphiasters and plesiasters). Monaxonic spicules consist of one to three categories of spiny microxeas" (Cárdenas *et al.*, 2011). Within the Geodiidae (Astrophorida), we propose to include in the WPD the resurrected subfamilies Erylinae Sollas, 1888 (*Caminus, Erylus, Pachymatisma*) and Geodinae Sollas, 1888 (*Geodia*) both being well-supported clades (Cárdenas *et al.*, 2010, 2011).

The Lithistida is an "order" which has always been considered polyphyletic but kept mainly for convenience, essentially because it is used by palaeontologists. There is no scientific justification for this, and we strongly believe that usage of this artificial group is preventing taxonomists, palaeontologists, chemists and ecologists from truly understanding the respective evolution of its different families and genera. Now that we have a bit more phylogenetic information concerning parts of this group, we believe it is time to start reallocating the different families of lithistids to their closest relatives in the Demospongiae as it was done for coralline sponges when the class Sclerospongiae was abandoned (Vacelet, 1985; Chombard *et al.*, 1997). We are fully aware that some of the following reallocations are based on very few data, but the eventual harm of these reallocations will be smaller than keeping all lithistids together. Based on molecular and/or morphological data, Cárdenas *et al.* (2011) already proposed to reallocate the following eight extant families to the Astrophorida: Corallistidae Sollas, 1888, Isoraphiniidae Schrammen, 1924, Macandrewiidae Schrammen, 1924, Neopeltidae Sollas, 1888, Phymaraphiniidae Schrammen, 1924, Phymatellidae Schrammen, 1910, Pleromidae Sollas, 1888 and Theonellidae Lendenfeld, 1903. Meanwhile, the Scleritodermidae Sollas, 1888 with rhizoclone desmas and sometimes sigmaspires should be reallocated to Spirophorida as suggested by morphological (Pisera and Lévi, 2002a) and molecular data (Kelly-Borges and Pomponi, 1994). The Azorecidae Sollas, 1888 are polyphyletic, but the type of rhizoclone desmas found in the two genera (*Leiodermatium* Schmidt, 1870 and *Jereicopsis* Lévi and Lévi, 1983) and molecular data suggests a relationship with Scleritodermidae (Kelly-Borges and Pomponi, 1994; Pisera and Lévi, 2002b). Thus, we propose to allocate them to the Spirophorida until molecular data can confirm this. The Desmanthidae Topsent, 1893 (including *Desmanthus* Topsent, 1894, *Paradesmanthus* Pisera and Lévi, 2002, *Sulcastrella* Schmidt, 1879 and *Petromica* Topsent, 1898) has uncertain phylogenetic relationships and is probably polyphyletic (Pisera and Lévi, 2002c); species of this family share monaxial complex branching desmas and presence of oxeas/styles. Molecular data (Kelly-Borges and Pomponi, 1994) suggest that *Petromica* does not belong to the Tetractinellida, while Muricy *et al.* (2001) consider that *Petromica* belongs to the Halichondriidae. Vacelet (1969) underlined the similarities between *Sulcastrella* and the Bubaridae. He suggested a possible homology between the desmas of *Desmanthus* and the basal contort spicules of *Bubaris*. A well-supported clade between *Bubaris carcisis*, *Acanthella acuta*, *Dictyonella obtusa* and *Desmanthus incrustans* is recovered by Morrow *et al.* (2012). To conclude, we propose to provisionally allocate desmanthid genera to the Dictyonellidae. The Siphonidiidae Lendenfeld, 1903 include *Gastrophanella* Schmidt, 1879, *Siphonidium* Schmidt, 1879 and *Lithobactrum* Kirkpatrick, 1903; they share rhizoclone choanosomal desmas with choanosomal exotylostyles and/or styles and no microscleres. Quite unexpectedly, preliminary molecular data suggested a relationship between *Siphonidium ramosum* from the Gulf of Mexico, *Scleritoderma* sp. (Scleritodermidae) and *Leiodermatium lynceus* (Azorecidae) (Kelly-Borges and Pomponi, 1994). Until further data are collected, we propose to consider the

Siphonidiidae as Demospongiae *incertae sedis*. Vetulinidae Lendenfeld, 1903 is represented by a single enigmatic species (*Vetulina stalactites* Schmidt, 1879) with no microscleres and no ectosomal spicules. We also propose to consider this family/genus/species as Demospongiae *incertae sedis*.

10.1.2. Propositions for new names of clades following the *PhyloCode*

As underlined in Section 3.3, clade names following the rules of the *Phylo-Code* v.4c (http://www.ohiou.edu/PhyloCode/) have been published for Calcispongiae, Demospongiae and Homoscleromorpha since 2003 (see Table 2.2). We proposed below the definitions for three more clade names: *Haploscleromorpha^P*, *Heteroscleromorpha^P* and *Spongillida^P* (Table 2.11).

Haploscleromorpha^P (*nomen cladi conversum*), the most inclusive clade containing *Haliclona mediterranea* Griessinger, 1971 but not *Spongia officinalis* Linnaeus 1759, *Aplysina cavernicola* Vacelet, 1959 and *Geodia cydonium* Jameson, 1811. The definition is branch based. *Etymology*: from the Greek "haplos" which means simple, *Demospongiae^P* with simple megascleres, diactines (Voultsiadou and Gkelis, 2005). *Reference Phylogeny*: Borchiellini *et al.* (2004a), Redmond *et al.* (2007) and Lavrov *et al.* (2008).

Heteroscleromorpha^P (*nomen cladi conversum*), the most inclusive clade containing *G. cydonium* Jameson, 1811 but not *H. mediterranea* Griessinger, 1971, *S. officinalis* Linnaeus 1759 and *A. cavernicola* Vacelet, 1959. The definition is branch based. *Etymology*: from the Greek "hetero" which means diverse; *Demospongiae^P* with different types of spicules, monactines, diactines and tetractines. *Reference Phylogeny*: Borchiellini *et al.* (2004a), Redmond *et al.* (2007) and Lavrov *et al.* (2008).

Spongillida^P Manconi and Pronzato, 2002 (*nomen cladi conversum*), the most inclusive clade containing *Ephydatia fluviatilis* (Linnaeus 1759), but not *H. mediterranea* Griessinger, 1971, *Scopalina lophyropoda* Schmidt, 1862, *G. cydonium* Jameson, 1811, *Biemna variantia* Bowerbank, 1858 and *P. ventilabrum* Linnaeus, 1767. The definition is branch based. *Etymology*: the word means "sponge" in Greek; *Demospongiae^P* living in freshwater environment. *Reference Phylogeny*: Borchiellini *et al.* (2004a), Redmond *et al.* (2007), Itskovich *et al.* (2007) and Morrow *et al.* (2012).

10.1.3. Description of phantom species: The concept of candidate species

With the increasing use of molecular data in sponge systematics, phylogenetic studies are revealing many new potential species (Klautau *et al.*, 1999; Lazoski *et al.*, 2001; Pöppe *et al.*, 2010; Reveillaud *et al.*, 2010; Xavier *et al.*, 2010a), faster than they can be formally described and named (or resurrected). And DNA barcoding as well as the coming use of large-scale

Table 2.11 Correspondence between *Systema Porifera*, the current classification following the Linnaean system and a classification following the *PhyloCode*

Systema Porifera classification (2002)	Linnean classification (2011)	*PhyloCode*
	Class Homoscleromorpha	*Homoscleromorpha*
Class Demospongiae	Order Homosclerophorida	
Subclass Homoscleromorpha	Family Plakinidae	*Plakinidae*
Order Homosclerophorida	Family Oscarellidae	*Oscarellidae*
Family Plakinidae	Class Demospongiae	*Demospongiae*
Sub Class Tetractinomorpha	**Subclass Heteroscleromorpha**	***Heteroscleromorpha***
Order Astrophorida	**Order Tetractinellida**	*Tetractinellida*
Order Spirophorida	Suborder Astrophorina	*Astrophorida*
Order Hadromerida	Suborder Spirophorina	
Subclass Ceractinomorpha	Order Hadromerida	
Order Agelasida	Order Agelasida	*Agelasida*
Order Halichondrida	Order Halichondrida	
Order Poecilosclerida	Order Poecilosclerida	
Suborder Mycalina	Suborder Mycalina	
Suborder Myxillina	Suborder Myxillina	
Suborder Microcionina	Suborder Microcionina	
Suborder Latrunculina	Suborder Latrunculina	
Order Haplosclerida	**Order Spongillida**	*Spongillida*
Suborder Spongillina	**Subclass Haploscleromorpha**	*Haploscleromorpha*
Suborder Haplosclerina	**Order Haplosclerida**	
Suborder Petrosina	**Subclass Keratosa**	*Keratosa*
Order Dictyoceratida	**Order Dictyoceratida + Verticillidae**	
Order Dendroceratida	Order Dendroceratida	
Order Verticillitida	**Subclass Myxospongiae**	*Myxospongiae*
Order Verongida	Order Verongida	
Order Chondrosida	**Order Chondrosida + Halisarcidae**	
Order Halisarcida		

New clade names defined in this chapter are in bold.

sequencing will dramatically increase the number of these Molecular Operational Taxonomic Units (MOTU; Floyd *et al.*, 2002) or "genospecies". What should the sponge community do with these phantom species without any names? Can sponge MOTUs be of any use for sponge systematics or biology? Padial *et al.* (2010) proposed to recognize three subcategories of candidate species, as more and more data are added to these MOTUs, the third and last category being "Confirmed Candidate Species" (CCS) awaiting for a taxonomist to formally describe and name it. Indeed, if not given some special status and visibility, we fear that these MOTUs may be useless and easily forgotten. Therefore, we propose to start listing these MOTUs in the WPD in order to remind their availability and existence to taxonomists or even ecologists. Padial *et al.* (2010) proposed to "designate these species with binominal species name of the most similar or closely related nominal species, followed by the abbreviation "Ca" (for candidate) with an attached numerical code referring to the particular candidate species (more than one candidate might be recognized under a valid species) and terminating with the author name and year of publication of the article in which the lineage was first discovered". But this designation may be a bit confusing for people unfamiliar with this system, so we suggest to clearly designate the species as "aff.", and to write "candidate species" instead of using the abbreviation "Ca". For example, *C.* aff. *celata* (candidate species 1 Xavier *et al.*, 2010a), *C.* aff. *celata* (candidate species 2 Xavier *et al.*, 2010a), and *C.* aff. *celata* (candidate species 3 Xavier *et al.*, 2010a) would be the three cryptic candidate species revealed by Xavier *et al.* (2010a). The numbers used after the candidate species are purely arbitrary. If no tentative assignment to a sister species can be made, one could only refer to the genus or family of the candidate species (Padial *et al.*, 2010). For example, *Psammocinia* sp. (candidate species 1 Pöppe *et al.*, 2010) and *Psammocinia* sp. (candidate species 2 Pöppe *et al.*, 2010) identify what has been called *Psammocinia* A and *Psammocinia* 394 by Pöppe *et al.* (2010). Due to the use of the traditional binominal name, these names would be easily integrated and searchable into the WPD.

10.2. Propositions for a general strategy for the *Systema Porifera* of the twenty-first century

Clearly, relationships between sponge species are far from being resolved. And the more phylogenetic trees we obtain, the more baffled we are by how sponges have evolved, different groups often retaining very similar solutions during skeleton evolution (i.e. convergent evolution), while other groups are loosing those derived characters later on during the evolution (i.e. secondary losses). Sponges have evolved probably since the Precambrian some 600 millions years ago, a lot of those processes have happened since and this phylogenetic noise needs to be reduced (and therefore understood) to find traces of the phylogenetic signal. We essentially see two main ways to succeed in this endeavour: the "top-down" strategy or the "bottom-up" strategy.

The top-down strategy is usually chosen by molecular biologists and phylogeneticists. These researchers are trying to obtain a robust phylogeny of sponges by sampling the main groups of the taxa studied. Species are not important *per se* since they are representative of a larger group. Consequently, morphology of the species is generally not reassessed with respect to the molecular results; morphology is seldom mapped on the molecular tree. Most of these studies concern relationships within the four classes (Demospongiae: Borchiellini *et al.*, 2004a; Nichols, 2005; Lavrov *et al.*, 2008; Calcispongiae: Dohrmann *et al.*, 2006; Hexactinellida: Dohrmann *et al.*, 2008, 2009, 2012; Homoscleromorpha: Gazave *et al.*, 2010a) or within orders (Halichondrida: Erpenbeck *et al.*, 2005, 2006a,b; Haplosclerida: Redmond *et al.*, 2007; Redmond and McCormack, 2008, 2011). In those studies, few phylogenetic results are translated into the classification and when they are, they mainly concern higher ranks (classes, orders or families).

The bottom-up strategy is preferred by many sponge taxonomists, each trying to resolve a small part of the tree separately (a genus, family or order). These investigators rely first and foremost on the species sampled, which are identified and carefully described before being sequenced. The molecular phylogeny hypothesis obtained is then confronted to the morphological data; synapomorphies of the clades are determined. Incongruence leads to reassessment of the morphology and eventually re-identification of the samples in order to understand discrepancies between the molecular tree and the current classification. Then, the molecular tree is reinterpreted in light of the new morphological data obtained. For example, this is the strategy followed to resolve relationships within the Verongida (Erwin and Thacker, 2007); the Astrophorida (Chombard *et al.*, 1998; Cárdenas *et al.*, 2011), the Geodiidae (Cárdenas *et al.*, 2010), the Sollasellidae (van Soest *et al.*, 2006; Erpenbeck *et al.*, 2007a,b), the Axinellidae (Alvarez *et al.*, 2000; Gazave *et al.*, 2010a), *Clathrina* (Rossi *et al.*, 2011) or *Scopalina* (Blanquer and Uriz, 2007, 2008). We strongly advocate this constant "yo–yo effect" or interplay between molecular and morphological data not only as a heuristic method to understand the trees obtained but also to be able to revise the taxonomy. We also strongly recommend future bottom-up studies to privilege the sequencing of type species of genera and families in order to take sound taxonomic decisions. Sequences from type species, type material or from specimens from the type locality should be explicitly identified on the resulting tree (Chakrabarty, 2010).

In short, a top-down strategy would like to resolve the tree by starting at the root toward the branches, while the bottom–up strategy starts from the smallest twigs towards the branches and the trunk. Hopefully, these two strategies will meet in order to cover the whole tree.

What strategy is the PorToL project (www.portol.org) following? Por-ToL is currently trying to obtain a robust phylogeny for sponges by sampling as many orders, families and genera from the four classes of sponges as possible. Since the idea here was to represent all the main groups and since species were not a priority in sampling, one could see the PorToL project as having a top-down strategy. But the taxonomy and examination of the species sampled were done during a workshop at Harbor Branch Oceano-graphic Institute in August 2011. Identification and sequencing is still ongoing, but we hope that morphological data of these species will later be confronted to the resulting phylogenies and that a bottom-up strategy will be applicable to well-sampled groups. In that case, the PorToL may manage to make both ends of the PorToL meet. The aim of PorToL should not only be to produce a tree but also to use morphological data to understand character distribution, provide explicit arguments to identify apomorphies and re-examine contradictions using the "yo–yo effect". In other words, it should also have a character-based approach, the latter being much too often taken over by statistical and tree-based phylogenetics (Mooi and Gill, 2010).

In poecilosclerids and haplosclerids, it is remarkable to notice an increase in the number of ranks: order, suborder, family, subfamily, genus, subgenus, species, subspecies or varieties or even "forma". Why such an increase and is it useful and based on solid phylogenetic assumptions? This increase seems to be linked more to "schools" of thought of systematicians than to solid systematics based on several data sets. In Haplosclerida, for example, after a cladistic analysis of morphological characters, de Weerdt (1986) merged most of the genera of the family Chalinidae in a large concept of the genus *Haliclona*. The consequence is a genus with a broad definition and including more than 400 species. Later she proposed six subgenera (de Weerdt, 2000, 2002). As under-lined in Section 3.1, the absence of knowledge on the ancestral state of a morphological character in the absence of fossil records leads to very weak hypothesis, and the classification proposed is not recovered with any other data set (Redmond *et al.*, 2007). The increase in number of subspecies is mostly due to hexactinellid taxonomists (Lopes *et al.*, 2011; see Section 5).

One thing is sure though: hypotheses of sponge phylogenetic relation-ships will continue to change with discovery and description of new species, collection of new data, use of new data sets and improvement of phyloge-netic reconstruction methods. So absolute nomenclatural stability is not only impossible but also undesirable. Name changes simply reflect the growth of phylogenetic knowledge (Dominguez and Wheeler, 1997). This is certainly bad news for many end-users of the sponge classification, but it is a reality that they have to accept and understand. The good news is that end-users and non-sponge specialists can now rely on a large choice of taxonomic databases, the WPD being the most complete and updated of them. Yes, the *Systema Porifera* published in 2002 is already out of date

(Hooper and van Soest, 2010), but it still represents a milestone for sponge researchers, due to the amount of taxonomic information it contains.

For the WPD to continue to reflect ongoing phylogenetic results, the phylogeny/classification link must be preserved. Phylogeneticists should therefore make a special effort to take taxonomic decisions to conclude their studies, either following a rank-based system and the *ICZN*, or the *PhyloCode*, or both. Indeed, the Linnaean nomenclature and the *PhyloCode* are based on different principles; as such they are not mutually exclusive and can be used in parallel manner. The *PhyloCode* is still poorly known or criticized in the sponge world, but we hope that this review will help sponge taxonomists appreciate the usefulness of this tool, tailored for today's world of phylogenies. One thing to keep in mind is that the use of the *PhyloCode* will not prevent bad taxonomy, just like in the Linnaean system, in both systems the taxonomist is still the one analysing the data and making the nomenclatural decisions (Schander and Thollesson, 1995). But would phylogenetic nomenclature really "cripple our ability to teach, learn and use taxonomic names in the field or in publications" as suggested by Nixon *et al.* (2003)? Or can taxonomists adapt to this new state of mind? Should we wait for more sponge phylogenies before giving phylogenetic definitions? Is phylogenetic nomenclature user-friendly enough to be adopted by non-taxonomists? These questions will be answered only when people will try the *PhyloCode* out. Already, Alessandro Minelli, former President of the International Commission on Zoological Nomenclature was writing in 1999, "In the future, Linnaean and not-Linnaean classification might exist side-by-side. [...] both parties are likely to go astray: Linnaean-style taxonomists on one side, patiently continuing to produce names that others are unwilling to use, and phylogeneticists, on the other, perhaps too ready to change the rules". Only one thing is sure, the Linnaean system of nomenclature must evolve in order to meet today's challenges, notably the increasing use of phylogenies or the implementation of more integrative taxonomic studies.

ACKNOWLEDGEMENTS

We would like to acknowledge all the participants of the Eighth World Sponge Conference (Girona, Spain, September 2010) as well as the PorToL workshop (Harbor Branch Oceanographic Institute, Florida, August 2011) for so many fruitful discussions on sponge systematics. A special thanks to Jean Vacelet and Michaël Manuel with whom we have shared many discussions on systematics and phylogeny all along these years. A part of this review has benefited by the framework of the ANR ECIMAR project. The COMEX (Fred Gauch and the Minibex crew), the French Marine Protected Areas Agency (Pierre Watremez), Jean Vacelet, Louis De Vos, Alexander Ereskovsky, Emilio Lanna, Julijana Ivanišević, Bernard Picton, Maria-Jesus Uriz, Xavier Turon and Mickel Nickel are acknowledged for providing illustrations.

REFERENCES

Adams, C. L., McInerney, J. O., and Kelly, M. (1999). Indications of relationships between Poriferan classes using full-length 18S rRNA gene sequences. *Memoirs of the Queensland Museum* **44**, 33–43.

Adamska, M., Degnan, B. M., Green, K., and Zwafink, C. (2011). What sponges can tell us about the evolution of developmental processes. *Zoology* **114**, 1–10.

Addis, J. S., and Peterson, K. J. (2005). Phylogenetic relationships of freshwater sponges (Porifera, Spongillina) inferred from analyses of 18S rDNA, COI mtDNA, and ITS2 rDNA sequences. *Zoologica Scripta* **34**, 549–557.

Aguilar, R., Correa, M. L., Calcinai, B., Pastor, X., de La Torriente, A., and Garcia, S. (2011). First records of *Asbestopluma hypogea* Vacelet and Boury-Esnault, 1996 (Porifera, Demospongiae Cladorhizidae) on seamounts and in bathyal settings of the Mediterranean Sea. *Zootaxa* **2925**, 33–40.

Alvarez, B., and Hooper, J. N. (2002). Family Axinellidae Carter, 1875. In "Systema Porifera: A Guide to the Classification of Sponges" (J. N. A. Hooper and R. W. M. van Soest, eds), pp. 742–747. Kluwer Academic/Plenum Publishers, New York.

Alvarez, B., and Hooper, J. N. A. (2009). Taxonomic revision of the order Halichondrida (Porifera: Demospongiae) from northern Australia. Family Axinellidae. *The Beagle. Records of the Museums and Art Galleries of the Northern Territory* **25**, 17–42.

Alvarez, B., and Hooper, J. N. A. (2010). Taxonomic revision of the order Halichondrida (Porifera: Demospongiae) of northern Australia. Family Dictyonellidae. *The Beagle, Records of the Museums and Art Galleries of the Northern Territory* **26**, 13–36.

Alvarez, B., Crisp, M. D., Driver, F., Hooper, J. N. A., and van Soest, R. W. M. (2000). Phylogenetic relationships of the family Axinellidae (Porifera: Demospongiae) using morphological and molecular data. *Zoologica Scripta* **29**, 169–198.

Amano, S., and Hori, I. (1992). Metamorphosis of calcareous sponges. I. Ultrastructure of free-swimming larvae. *Invertebrate Reproduction and Development* **21**, 81–90.

Amano, S., and Hori, I. (2001). Metamorphosis of coeloblastula performed by multipotential larval flagellated cells in the calcareous sponge *Leucosolenia laxa*. *The Biological Bulletin* **200**, 20–32.

Amaral, A. C. Z. and Rossi-Wongtschovski, C. L. B. (eds) (2004). Biodiversidade bentônica da região sul-sudeste do Brasil—plataforma externa e talude superiorIOUSP, São Paulo, Série Documentos REVIZEE—Score Sul.

Amaral, A. C. Z. Lana, P. C. Fernandes, F. C. and Coimbra, J. C. (eds) (2003). Biodiversidade bêntica da região sul-sudeste da costa brasileira. REVIZEE Score Sul—BentosMMA, Brasília117pp.

Aoucheria, A., Geourjon, C., Aghajari, N., Navratil, V., Deléage, G., Lethias, C., and Exposito, J.-Y. (2006). Insights into early extracellular matrix evolution: Spongin short chain collagen-related proteins are homologous to basement membrane type IV collagens and form a novel family widely distributed in Invertebrates. *Molecular Biology and Evolution* **23**, 2288–2302.

Austin, B., Ott, B., McDaniel, N., and Romagosa, P. (2007). Sponges of the cold temperate NE Pacific. http://www.mareco.org/KML/Projects/NEsponges_content.asp.

Avise, J. C., and Liu, J. X. (2011). On the temporal inconsistencies of Linnean taxonomic ranks. *Biological Journal of the Linnean Society* **102**, 707–714.

Ayling, A. L., Stone, S. M., and Smith, B. J. (1982). Catalogue of types of sponge species from Southern Australia described by Arthur Dendy. *Reports of the National Museum of Victoria* **1**, 97–109.

Barnich, R., and Janussen, D. (2006). Die Typen und Typoide des Naturmuseums Senckenberg, Nr. 86. Type catalogue of the Porifera in the collections of the Senckenberg Museum in Frankfurt am Main, Germany. *Seckenbergiana biologica* **86**, 127–144.

Barthel, D. (1992). Antarctic hexactinellids: A taxonomically difficult, but ecologically important benthic component. *Verhanlungen der Deutschen Zoologischen Gesellschaft* **85,** 271–276.

Barthel, D., and Gutt, J. (1992). Sponge associations in the eastern Weddell Sea. *Antarctic Science* **4,** 137–150.

Bavestrello, G., Bonito, M., and Sarà, M. (1993). Silica content and spicular size variation during an annual cycle in *Chondrilla nucula* Schmidt (Porifera, Demospongiae) in the Ligurian Sea. In "Recent Advances in Ecology and Systematics of Sponges" (M.-J. Uriz and K. Rützler, eds), Scientia Marina. **57,** pp. 421–425.

Bentlage, B., and Wörheide, G. (2007). Low genetic structuring among *Pericharax heteroraphis* (Porifera: Calcarea) populations from the Great Barrier Reef (Australia), revealed by analysis of nrDNA and nuclear intron sequences. *Coral Reefs* **26,** 807–816.

Bergmann, W. (1949). Comparative biochemical studies on the lipids of marine invertebrates with species reference to the sterols. *Journal of Marine Research* **8,** 137–176.

Bergmann, W. (1962). Sterol: Structure and distribution. In "Comparative Biochemistry" (M. Florkin and H. S. Mason, eds), pp. 122–157. Academic Press, New York.

Bergquist, P. R. (1978). Sponges. Hutchinson & Co., London.

Bergquist, P. R. (1980). A revision of the supraspecific classification of the orders Dictyoceratida, Dendroceratida, and Verongida (class Demospongiae). *New Zealand Journal of Zoology* **7,** 443–503.

Bergquist, P. R. (1995). Dictyoceratida, Dendroceratida and Verongida from the New Caledonia lagoon (Porifera: Demospongiae). *Memoirs of the Queensland Museum* **38,** 1–51.

Bergquist, P. R. (1996). The marine fauna of New Zealand: Porifera, Demospongiae, Part 5. Dendroceratida and Halisarcida. *New Zealand Oceanographic Institute Memoir* **107,** 1–53.

Bergquist, P. R., and Cook, S.de.C (2002a). Family Ianthellidae Hyatt, 1875. In "Systema Porifera: A Guide to the Classification of Sponges" (J. N. A. Hooper and R. W. M. van Soest, eds), pp. 1089–1093. Kluwer Academic/Plenum Publishers, New York.

Bergquist, P. R., and Cook, S. de C. (2002b). Order Dendroceratida Minchin, 1900. In "Systema Porifera: A Guide to the Classification of Sponges" (J. N. A. Hooper and R. W. M. van Soest, eds), p. 1067. Kluwer Academic/Plenum Publishers, New York.

Bergquist, P. R., and Cook, S. de C. (2002c). Order Halisarcida Bergquist, 1996. In "Systema Porifera: A Guide to the Classification of Sponges" (J. N. A. Hooper and R. W. M. van Soest, eds). Kluwer Academic/Plenum Publishers, New York, p. 1077.

Bergquist, P. R., and Fromont, P. J. (1988). The marine fauna of New Zealand: Porifera, Demospongiae, Part 4. Poecilosclerida. *New Zealand Oceanographic Institute Memoir* **96,** 5–138.

Bergquist, P. R., and Kelly, M. (2004). Taxonomy of some Halisarcida and Homosclerophorida (Porifera: Demospongiae) from the Indo-Pacific. *New Zealand Journal of Marine and Freshwater Research* **38,** 51–66.

Bergquist, P. R., and Wells, R. J. (1983). Chemotaxonomy of the Porifera: The development and current status of the field. In "Marine Natural Products: Chemical and Biological Perspectives" (P. J. Scheuer, ed.), pp. 1–50. Academic Press, London.

Bergquist, P. R., Hofheinz, W., and Oesterhelt, G. (1980). Sterol composition and the classification of the Demospongiae. *Biochemical Systematics and Ecology* **8,** 423–435.

Bergquist, P. R., Lawson, M. P., Lavis, A., and Cambie, R. C. (1984). Fatty acid composition and the classification of the Porifera. *Biochemical Systematics and Ecology* **12,** 63–84.

Bergquist, P. R., Lavis, A., and Cambie, R. C. (1986). Sterol composition and classification of the Porifera. *Biochemical Systematics and Ecology* **14,** 105–112.

Bergquist, P. R., Karuso, P., and Cambie, R. C. (1990). Taxonomic relationships within the Dendroceratida: A biological and chemotaxonomic appraisal. In "New Perspectives in Sponge Biology" (K. Rützler, ed.), pp. 72–78. Smithsonian Institution Press, Washington.

Bergquist, P. R., Walsh, D., and Gray, R. D. (1998). Relationships within and between the orders of Demospongiae which lack a mineral skeleton. In "Sponge Sciences—Multidisciplinary Perspectives" (Y. Watanabe and N. Fusetani, eds), pp. 31–40. Springer-Verlag, Tokyo.

Berlinck, R. G. S., Braekman, J.-C., Daloze, D., Bruno, I., Riccio, R., Rogeau, D., and Amade, P. (1992). Crambines C1 and C2: Two further ichthyotoxic guanidine alkaloids from the sponge *Crambe crambe*. *Journal of Natural Products (Lloydia)* **55**, 528–532.

Bertolino, M., Calcinai, B., and Pansini, M. (2009). Two new species of Poecilosclerida (Porifera: Demospongiae) from Terra Nova Bay (Antarctic Sea). *Journal of the Marine Biological Association of the United Kingdom* **89**, 1671–1677.

Bertrand, Y., Pleijel, F., and Rouse, G. W. (2006). Taxonomic surrogacy in biodiversity assessments, and the meaning of Linnaean ranks. *Systematics and Biodiversity* **4**, 149–159.

Bidder, G. P. (1898). The skeleton and classification of calcareous sponge. *Proceedings of the Royal Society* **64**, 61–76.

Blanquer, A., and Uriz, M. J. (2007). Cryptic speciation in marine sponges evidenced by mitochondrial and nuclear genes: A phylogenetic approach. *Molecular Phylogenetics and Evolution* **45**, 392–397.

Blanquer, A., and Uriz, M.-J. (2008). '*A posteriori*' searching for phenotypic characters to describe new cryptic species of sponges revealed by molecular markers (Dictyonellidae : Scopalina). *Invertebrate Systematics* **22**, 1–14.

Blanquer, A., Uriz, M. J., and Caujape-Castells, J. (2009). Small-scale spatial genetic structure in *Scopalina lophyropoda*, an encrusting sponge with philopatric larval dispersal and frequent fission and fusion events. *Marine Ecology Progress Series* **380**, 95–102.

Blunt, J. (2011). MarinLit: A comprehensive database of the literature for Marine Natural Products Maintained at the University of Canterbury, Auckland, New Zealand. http://www.chem.canterbury.ac.nz/marinlit/marinlit.shtml.

Borchiellini, C., Boury-Esnault, N., Vacelet, J., and Le Parco, Y. (1998). Phylogenetic analysis of the Hsp70 sequences reveals the monophyly of Metazoa and specific phylogenetic relationships between animals and fungi. *Molecular Biology and Evolution* **15**, 647–655.

Borchiellini, C., Manuel, M., Alivon, E., Boury-Esnault, N., Vacelet, J., and Le Parco, Y. (2001). Sponge paraphyly and the origin of Metazoa. *Journal of Evolutionary Biology* **14**, 171–179.

Borchiellini, C., Chombard, C., Manuel, M., Alivon, E., Vacelet, J., and Boury-Esnault, N. (2004a). Molecular phylogeny of Demospongiae: Implications for classification and scenarios of character evolution. *Molecular Phylogenetics and Evolution* **32**, 823–837.

Borchiellini, C., Alivon, E., and Vacelet, J. (2004b). The systematic position of *Alectona* (Porifera, Demospongiae): A tetractinellid sponge. *Bollettino dei Musei e degli Istituti Biologici della Università di Genova* **68**, 209–217.

Borojevic, R. (1979). Evolution des éponges Calcarea. In "Biologie des Spongiaires" (C. Lévi and N. Boury-Esnault, eds), pp. 527–530. Editions du C.N.R.S, Paris.

Borojevic, R., and Boury-Esnault, N. (1987). Calcareous sponges collected by N.O. Thalassa on the continental margin of the Bay of Biscay: I. Calcinea. In "Taxonomy of Porifera from the NE Atlantic and Mediterranean Sea" (J. Vacelet and N. Boury-Esnault, eds), NATO Asi Series, vol. G13, pp. 1–27. Springer-Verlag, Berlin, Heidelberg.

Borojevic, R., and Klautau, M. (2000). Calcareous sponges from New Caledonia. *Zoosystema* **22**, 187–201.

Boury-Esnault, N. (1972). Une structure inhalante remarquable des Spongiaires: le crible. Etude morphologique et cytologique. *Archives de Zoologie expérimentale & générale* **113**, 7–23.

Boury-Esnault, N. (1980). Spongiaires. In "Flore et Faune aquatiques de l'Afrique sahélo-soudanienne" (J. R. Durand and C. Lévêque, eds), ORSTOM. **44**, pp. 199–217.

Boury-Esnault, N. (2002). Order Chondrosida Boury-Esnault and Lopès, 1985. Family Chondrillidae Gray, 1872. In "Systema Porifera: A Guide to the Classification of Sponges" (J. N. A. Hooper and R. W. M. van Soest, eds), pp. 291–297. Kluwer Academic/Plenum Publishers, New York.

Boury-Esnault, N. (2006). Systematics and evolution of Demospongiae. *Canadian Journal of Zoology* **84**, 205–224.

Boury-Esnault, N., and De Vos, L. (1988). *Caulophacus cyanae*, n. sp., une éponge hexacti-nellide des sources hydrothermales. Biogéographie du genre *Caulophacus* Schulze, 1887. *Oceanologica Acta* **8**, 51–60.

Boury-Esnault, N., and Rützler, K. (1997). Thesaurus of Sponge Morphology. Smithsonian Press, Washington.

Boury-Esnault, N., and Solé-Cava, A. M. (2004). Recent contribution of genetics to the study of sponge systematics and biology. *Bollettino dei Musei e degli Istituti Biologici della Università di Genova* **68**, 3–18.

Boury-Esnault, N., and Vacelet, J. (1994). Preliminary studies on the organization and development of a hexactinellid sponge from a Mediterranean cave, *Oopsacas minuta*. In "Sponges in Time and Space. Proceedings of the Fourth International Porifera Congress" (R. W. M. van Soest, T. M. G. van Kempen and J. Braekman, eds), pp. 407–416. A.A. Balkema, Rotterdam.

Boury-Esnault, N., De Vos, L., Donadey, C., and Vacelet, J. (1990). Ultrastructure of choanosome and sponge classification. In "New Perspectives in Sponge Biology" (K. Rützler, ed.), pp. 237–244. Smithsonian Institution Press, Washington.

Boury-Esnault, N., Solé-Cava, A. M., and Thorpe, J. P. (1992). Genetic and cytological divergence between colour morphs of the Mediterranean sponge *Oscarella lobularis* Schmidt (Porifera, Demospongiae, Oscarellidae). *Journal of Natural History* **26**, 271–284.

Boury-Esnault, N., Hajdu, E., Klautau, M., Custódio, M., and Borojevic, R. (1994). The value of cytological criteria in distinguishing sponges at the species level: The example of the genus *Polymastia*. *Canadian Journal of Zoology* **72**, 795–804.

Boury-Esnault, N., Muricy, G., Gallissian, M.-F., and Vacelet, J. (1995). Sponges without skeleton: A new Mediterranean genus of Homoscleromorpha (Porifera, Demospongiae). *Ophelia* **43**, 25–43.

Boury-Esnault, N., Efremova, S., Bézac, C., and Vacelet, J. (1999a). Reproduction of a hexactinellid sponge: First description of gastrulation by cellular delamination in the Porifera. *Invertebrate Reproduction and Development* **35**, 187–201.

Boury-Esnault, N., Klautau, M., Bézac, C., Wulff, J., and Solé-Cava, A. M. (1999b). Comparative study of putative conspecific sponge populations from both sides of the Isthmus of Panama. *Journal of the Marine Biological Association of the United Kingdom* **79**, 39–59.

Boury-Esnault, N., Marschal, C., Kornprobst, J.-M., and Barnathan, G. (2002). A new species of *Axinyssa* Lendenfeld, 1897 (Porifera, Demospongiae, Halichondrida) from the Senegalese coast. *Zootaxa* **117**, 1–8.

Boury-Esnault, N., Ereskovsky, A. V., Bézac, C., and Tokina, D. B. (2003). Larval develop-ment in Homoscleromorpha (Porifera, Demospongiae). *Invertebrate Biology* **122**, 187–202.

Boute, N., Exposito, J. Y., Boury-Esnault, N., Vacelet, J., Noro, N., Miyazaki, K., Yoshigato, K., and Garrone, R. (1996). Type IV collagen in sponges, the missing link in basement membrane ubiquity. *Biology of the Cell* **88**, 37–44.

Bowerbank, J. S. (1864). A Monograph of the British Spongiadae, vol. 1. Ray Society, London.

Bowerbank, J. S. (1866). A Monograph of the British Spongiadae, vol. 2. Ray Society, London.

Braekman, J.-C., Daloze, D., Stoller, C., and van Soest, R. W. M. (1992). Chemotaxonomy of *Agelas* (Porifera: Demospongiae). *Biochemical Systematics and Ecology* **20**, 417–431.

Brien, P. (1970). Les Potamolépides africaines. Polyphylétisme des Eponges d'eau douce. *Archives de zoologie expérimentale et générale (Notes et Revues)* **110**, 527–562.

Bultel-Poncé, V., Brouard, J.-P., Vacelet, J., and Guyot, M. (1999). Thymosiosterol and delta24 thymosiosterol, new sterols from the sponge *Thymosiopsis*. *Tetrahedron Letters* **40**, 2955–2956.

Burton, M. (1935). Some sponges from the Okhotsk Sea and the Sea of Japan. *Exploration des Mers de l'URSS* **22**, 61–79.

Burton, M., and Rao, S. H. (1932). Report on the shallow-water marine sponges in the collection of the Indian Museum. Part I. *Records of the Indian Museum* **34**, 299–358.

Calcinai, B., and Pansini, M. (2000). Four new demosponges species from Terra Nova Bay (Ross Sea, Antarctica). *Zoosystema* **22**, 369–381.

Calcinai, B., Bavestrello, G., Cuttone, G., and Cerrano, C. (2011). Excavating sponges from the Adriatic Sea: Description of *Cliona adriatica* sp. nov. (Demospongiae: Clionaidae) and estimation of its boring activity. *Journal of the Marine Biological Association of the United Kingdom* **91**, 339–346.

Campos, M., Mothes, B., and Veitenheimer Mendes, I. L. (2007a). Antarctic sponges (Porifera, Demospongiae) of the South Shetland Islands and vicinity. Part I. Spirophorida, Astrophorida, Hadromerida, Halichondrida and Haplosclerida. *Revista Brasileira de Zoologia* **24**, 687–708.

Campos, M., Mothes, B., and Veitenheimer Mendes, I. L. (2007b). Antarctic sponges (Porifera, Demospongiae) of the South Shetland Islands and vicinity. Part II. Poecilosclerida. *Revista Brasileira de Zoologia* **24**, 742–770.

Cantino, P. D., Doyle, J. A., Graham, S. W., Judd, S. W., Olmstead, R. G., Soltis, D. E., Soltis, P. S., and Donaghue, M. J. (2007). Towards a phylogenetic nomenclature of Tracheophyta. *Taxon* **56**, 822–846.

Carballo, J. L., Gómez, P., Cruz-Barraza, J. A., and Flores-Sánchez, D. M. (2003). Sponges of the family Chondrillidae (Porifera: Demospongiae) from the Pacific coast of Mexico, with the description of three new species. *Proceedings of Biological Society of Washington* **116**, 515–527.

Carballo, J. L., Cruz-Barraza, J. A., and Gómez, P. (2004). Taxonomy and description of clionaid sponges (Hadromerida, Clionaidae) from the Pacific Ocean of Mexico. *Zoological Journal of the Linnean Society* **141**, 353–397.

Cárdenas, P. (2010). Phylogeny, Taxonomy and Evolution of the Astrophorida (Porifera, Demospongiae). Ph.D. University of Bergen, Bergen, 290pp.

Cárdenas, P., Xavier, J., Tendal, O. S., Schander, C., and Rapp, H. T. (2007). Redescription and resurrection of *Pachymatisma normani* (Demospongiae: Geodiidae), with remarks on the genus *Pachymatisma*. *Journal of the Marine Biological Association of the United Kingdom* **87**, 1511–1525.

Cárdenas, P., Menegola, C., Rapp, H. T., and Diaz, M. C. (2009). Morphological description and DNA barcodes of shallow-water *Tetractinellida* (Porifera: Demospongiae) from Bocas del Toro, Panama, with description of a new species. *Zootaxa* **2276**, 1–39.

Cárdenas, P., Rapp, H. T., Schander, C., and Tendal, O. S. (2010). Molecular taxonomy and phylogeny of the Geodiidae (Porifera, *Demospongiae*, Astrophorida)—Combining phylogenetic and Linnaean classification. *Zoologica Scripta* **39**, 89–106.

Cárdenas, P., Xavier, J. A., Reveillaud, J., Schander, C., and Rapp, H. T. (2011). Molecular phylogeny of the Astrophorida (Porifera, Demospongiae[P]) reveals an unexpected high level of spicule homoplasy. *PLoS One* **6**, e18318.

Castritsi-Catharios, J., Magli, M., and Vacelet, J. (2007). Evaluation of the quality of two commercial sponges by tensile strength measurement. *Journal of the Marine Biological Association of the United Kingdom* **87**, 1765–1771.

Cavalcanti, F. F., Zilberberg, C., and Klautau, M. (2007). Seasonal variation of morphological characters of *Chondrilla* aff. *nucula* (Porifera: Demospongiae) from the south-east coast of Brazil. *Journal of the Marine Biological Association of the United Kingdom* **87**, 1727–1732.

Cedro, V. R., Hajdu, E., and Correia, M. D. (2011). *Mycale alagoana* sp.nov. and two new formal records of Porifera (Demospongiae, Poecilosclerida) from the shallow-water reefs of Alagoas (Brazil). *Biota Neotropica* **11**, 161–171.

Chakrabarty, P. (2010). Genetypes: A concept to help integrate molecular phylogenetics and taxonomy. *Zootaxa* **2632**, 67–68.

Chombard, C. (1998). Les Demospongiae à asters: essai de phylogénie moléculaire. Homologie du caractère "aster". Ph.D. Thesis. Muséum national d'Histoire naturelle, Paris.

Chombard, C., and Boury-Esnault, N. (1999). Good congruence between morphology and molecular phylogeny of Hadromerida, or how to bother sponge taxonomists. *Memoirs of the Queensland Museum* **44**, 100.

Chombard, C., Tillier, A., Boury-Esnault, N., and Vacelet, J. (1997). Polyphyly of "sclerosponges" (Porifera, Demospongiae) supported by 28S ribosomal sequences. *Biological Bulletin (Woods Hole)* **193**, 359–367.

Chombard, C., Boury-Esnault, N., and Tillier, S. (1998). Reassesment of homology of morphological characters in tetractinellid sponges based on molecular data. *Systematic Biology* **47**, 351–366.

Chu, J. W. F., and Ley, S. P. (2010). High resolution mapping of community structure in three glass sponge reefs (Porifera, Hexactinellida). *Marine Ecology Progress Series* **417**, 97–113.

Coll, M., Pirodi, C., Steenbek, J., Kaschner, K., Ben Rais Lasram, F., Aguzzi, J., Ballesteros, E., Bianchi, C. N., Corbera, J., Dailianis, T., Danovoro, R., Estrada, M., Froglia, C., Galil, B. S., Gasol, J. P., Gertwagen, R., Gil, J., Guilhaumon, F., Kesner-Reyes, K., Kitsos, M.-S., Koukouras, A., Lampadariou, N., Laxamana, E., López-Fé de la Cuadra, C. M., Lotze, H. K., Martin, D., Mouillot, D., Oro, D., Raicevich, S., Rius-Barile, J., Saiz-Salinas, J. I., San Vicente, C., Somot, S., Templado, J., Turon, X., Vafidis, D., Villanueva, R., and Voultsiadou, E. (2010). The biodiversity of the Mediterranean Sea: Estimates, patterns, and threats. *PLoS One* **5**, e11842.

Conway, K. W., Barrie, J. V., and Krautter, M. (2005). Geomorphology of unique reefs on the western Canadian shelf: Sponge reefs mapped by multibeam bathymetry. *Geo-Marine Letters* **25**, 205–213.

Cook, S. de C. (2007). Clarification of dictyoceratid taxonomic characters, and the determination of genera. In "Porifera Research: Biodiversity, Innovation and Sustainability" (M. R. Custódio, G. Lôbo-Hajdu, E. Hajdu and G. Muricy, eds), pp. 265–274. Museu Nacional, Rio de Janeiro.

Cook, S. de C., and Bergquist, P. R. (2002a). Order Dictyoceratida Minchin, 1900. In "Systema Porifera: A Guide to the Classification of Sponges" (J. N. A. Hooper and R. W. M. van Soest, eds). Kluwer Academic/Plenum Publishers, New York, p. 1021.

Cook, S. de C., and Bergquist, P. R. (2002b). Family Irciniidae Gray, 1867. In "Systema Porifera: A Guide to the Classification of Sponges" (J. N. A. Hooper and R. W. M. van Soest, eds), pp. 1022–1027. Kluwer Academic/Plenum Publishers, New York.

Custódio, M. R. Lôbo-Hajdu, G. Hajdu, E. and Muricy, G. (eds) (2007). Porifera Research Biodiversity, Innovation and Sustainability. Museu Nacional, Rio de Janeiro.

Darwin, C. (1859). The Origin of Species by Means of Natural Selection or the Preservation of Favoured Races in the Struggle for Life. John Murray, London.

Dayrat, B. (2005). Towards integrative taxonomy. *Biological Journal of the Linnean Society* **85**, 407–415.

Dayrat, B., Cantino, P. D., Clarke, J. A., and de Queiroz, K. (2008). Species Names in the *PhyloCode:* The approach adopted by the International Society for Phylogenetic Nomenclature. *Systematic Biology* **57**, 507–514.

de Caralt, S., Uriz, M.-J., Ereskovsky, A. V., and Wijffels, R. H. (2007). Embryo development of *Corticium candelabrum* (Demospongiae: Homosclerophorida). *Invertebrate Biology* **126**, 211–219.

de Laubenfels, M. W. (1936). A discussion of the sponge fauna of the Dry Tortugas in particular and the West Indies in general, with material for a revision of the families and orders of the Porifera, Carnegie Institution, Washington, pp. 1–225, Tortugas Laboratory Paper 30.

de Laubenfels, M. W. (1948). The order Keratosa of the phylum Porifera—A monographic study. *Occasional Papers of the Allan Hancock Foundation* **3**, 1–217.

de Queiroz, K. (2006). The *PhyloCode* and the distinction between taxonomy and nomenclature. *Systematic Biology* **55**, 160–162.

de Queiroz, K. (2007). Species concepts and species delimitation. *Systematic Biology* **56**, 879–886.

de Queiroz, K., and Gauthier, J. (1990). Phylogeny as a central principle in taxonomy: Phylogenetic definitions of taxon names. *Systematic Biology* **39**, 307–322.

de Queiroz, K., and Gauthier, J. (1992). Phylogenetic taxonomy. *Annual Review of Ecology and Evolution* **23**, 449–480.

de Queiroz, K., and Gauthier, J. (1994). Toward a phylogenetic system of biological nomenclature. *Trends in Ecology & Evolution* **9**, 27–31.

de Voogd, N. J., and Cleary, D. F. R. (2008). An analysis of sponge diversity and distribution at three taxonomic levels in the Thousand Islands/Jakarta Bay reef complex, West-Java, Indonesia. *Marine Ecology* **29**, 205–215.

de Voogd, N. J., Erpenbeck, D., Hooper, J. N. A., and van Soest, R. W. M. (2010). *Acanthostylotella*, the forgotten genus. Affinites of the family Agelasidae (Porifera, Demospongiae, Agelasida): Chemical, molecular and morphological evidence. In "Ancient Animals, New Challenges." VIII World Sponge conference Girona, 20–24 September 2010 (M.-J. Uriz, M. Maldonado, M. Becerro and X. Turon, eds)., p. 172.

De Vos, L., Boury-Esnault, N., and Vacelet, J. (1990). The apopylar cell of sponges. In "New Perspectives in Sponge Biology" (K. Rützler, ed.), pp. 153–158. Smithsonian Institution Press, Washington.

De Vos, L., Rützler, K., Boury-Esnault, N., Donadey, C., and Vacelet, J. (1991). Atlas de morphologie des Eponges—Atlas of sponge morphology. Smithsonian Institution Press, Washington.

de Weerdt, W. H. (1985). A systematic revision of the North Eastern Atlantic shallow-water Haplosclerida (Porifera, Demospongiae). Part I: Introduction, Oceanapiidae and Petrosidae. *Beaufortia* **35**, 61–91.

de Weerdt, W. H. (1986). A sytematic revision of the North-Eastern Atlantic shallow-water Haplosclerida (Porifera, Demospongiae). Part 2: Chalinidae. *Beaufortia* **36**, 81–165.

de Weerdt, W. H. (1989). Phylogeny and vicariance biogeography of North Atlantic Chalinidae (Haplosclerida, Demospongiae). *Beaufortia* **39**, 55–90.

de Weerdt, W. H. (2000). A monograph of the shallow-water Chalinidae (Porifera, Haplosclerida) of the Caribbean. *Beaufortia* **50**, 1–67.

de Weerdt, W. H. (2002). Family Chalinidae Gray, 1867. In "Systema Porifera: A Guide to the Classification of Sponges" (J. N. A. Hooper and R. W. M. van Soest, eds), pp. 852–873. Kluwer Academic/Plenum Publishers, New York.

de Weerdt, W. H., de Kluijver, M. J., and Gomez, R. (1999). *Haliclona* (*Halichoclona*) *vansoesti* n.sp. a new chalinid sponge species (Porifera, Demospongiae, Haplosclerida) from the Caribbean. *Beaufortia* **49**, 47–54.

Dendy, A. (1921). The tetraxonid sponge-spicule: A study in evolution. *Acta Zoologica* **2**, 95–152.

Dendy, A., and Row, R. W. (1913). The classification and phylogeny of the Calcareous sponges, with a reference list of all the described species, systematically arranged. *Proceedings of the Zoological Society of London* **47**, 704–813.

DeSalle, R., Egan, M. G., and Siddall, M. (2005). The unholy trinity: Taxonomy, species delimitation and DNA barcoding. *Philosophical Transactions of the Royal Society B: Biological Sciences* **360**, 1905–1916.

Desqueyroux-Faúndez, R., and Stone, S. M. (1992). O. Schmidt Sponge catalogue. An illustrated guide to the Graz Museum collection, with notes on additional material. Muséum d'Histoire naturelle, Geneva.

Díaz, M. C., van Soest, R. M. W., Rützler, K., and Guzman, H. M. (2005). *Aplysina chiriquiensis*, a new pedunculate sponge from the Gulf of Chiriquí, Panamá, Eastern Pacific (Aplysinidae, Verongida). *Zootaxa* **1012,** 1–12.

Dilton, L. S. (1981). Ultrastructure, Macromolecules, and Evolution. Plenum Press, New York.

Dohrmann, M., Voigt, O., Erpenbeck, D., and Wörheide, G. (2006). Non-monophyly of most supraspecific taxa of calcareous sponges (Porifera, Calcarea) revealed by increased taxon sampling and partitioned Bayesian analysis of ribosomal DNA. *Molecular Phylogenetics and Evolution* **40,** 830–843.

Dohrmann, M., Janussen, D., Reitner, J., Collins, A. G., and Wörheide, G. (2008). Phylogeny and evolution of glass sponges (Porifera, Hexactinellida). *Systematic Biology* **57,** 388–405.

Dohrmann, M., Collins, A. G., and Wörheide, G. (2009). New insights into the phylogeny of glass sponges (Porifera, Hexactinellida): Monophyly of Lyssacinosida and Euplectellinae, and the phylogenetic position of Euretidae. *Molecular Phylogenetics and Evolution* **52,** 257–262.

Dohrmann, M., Haen, K. M., Lavrov, D., and Wörheide, G. (2012). Molecular phylogeny of glass sponges (Porifera, Hexactinellida): Increased taxon sampling and inclusion of the mitochondrial protein-coding gene, cytochrome oxidase subunit I. *Hydrobiologia* **687,** 11–20.

Dominguez, E., and Wheeler, Q. D. (1997). Taxonomic stability is ignorance. *Cladistics* **13,** 367–372.

Dubois, A. (2007). Phylogeny, taxonomy and nomenclature: The problem of taxonomic categories and of nomenclatural ranks. *Zootaxa* **1519,** 27–68.

Dubois, A. (2011). The *International Code of Zoological Nomenclature* must be drastically improved before it is too late. *Bionomina* **2,** 1–104.

Duboscq, O., and Tuzet, O. (1935). Un nouveau stade du développement des éponges calcaires. *Comptes Rendus Hebdomadaires des Séances de l'Académie des Sciences, Paris* **200,** 1788–1790.

Dunn, C. W., Hejnol, A., Matus, D. Q., Pang, K., Browne, W. E., Smith, S. A., Seaver, E., Rouse, G. W., Obst, M., Edgecombe, G. D., Sorensen, M. V., Haddock, S. H. D., Schmidt-Rhaesa, A., Okusu, A., Kristensen, R. M., Wheeler, W. C., Martindale, M. Q., and Giribet, G. (2008). Broad phylogenomic sampling improves resolution of the animal tree of life. *Nature* **452,** 745–749.

Duran, S., and Rützler, K. (2006). Ecological speciation in a Caribbean marine sponge. *Molecular Phylogenetics and Evolution* **40,** 292–297.

Duran, S., Estoup, P. A., and Turon, X. (2002). Polymorphic microsatellite in the sponge *Crambe crambe* (Porifera: Poecilosclerida) and their variation in two distant populations. *Molecular Ecology Notes* **2,** 478–480.

Efremova, S. M. (2001). Sponges (Porifera). In "Index of Animal Species Inhabiting Lake Baikal and Its Catchment Area" (O. A. Timoshkin, ed.), pp. 182–192. Nauka, Novosibirsk.

Efremova, S. M. (2004). New genus and new species of sponges from family Lubomirskiidae Rezvoj, 1936. In "Index of Animal Species Inhabiting Lake Baikal and Its Catchment Area" (O. A. Timoshkin, ed.), pp. 1261–1278. Nauka, Novosibirsk.

Ehrlich, H., Maldonado, M., Spindler, K.-D., Eckert, C., Hanke, T., Born, R., Goebel, C., Simon, P., Heinemann, S., and Worch, H. (2007). First evidence of chitin as a component of the skeletal fibers of marine sponges. Part I. Verongidae (Demospongia: Porifera). *The Journal of Experimental Zoology* **308B,** 1–10.

Ehrlich, H., Simon, P., Carrillo-Cabrera, W., Bazhenov, V. V., Botting, J. P., Ilan, M., Ereskovsky, A. V., Muricy, G., Worch, H., Mensch, A., Born, R., Springer, A.,

Kummer, K., Vyalikh, D. V., Molodtsov, S. L., Kurek, D., Kammer, M., Paasch, S., and Brunner, E. (2010). Insights into chemistry of biological materials: Newly discovered silica-aragonite-chitin biocomposites in Demosponges. *Chemistry of Materials* **22**, 1462–1471.

Ellis, J., and Solander, D. (1786). The Natural History of Many Curious and Uncommon Zoophytes, Collected from Various Parts of the Globe. Systematically Arranged and Described by the Late Daniel Solander. 4. Benjamin White & Son, London.

Ellis, D. I., Dunn, W. B., Griffin, J. L., Allwood, J. M., and Goodacre, R. (2007). Metabolic fingerprinting as a diagnostic tool. *Pharmacogenomics* **8**, 1243–1266.

Enticknap, J. J., Kelly, M., Peraud, O., and Hill, R. T. (2006). Characterization of a culturable Alphaproteobacterial symbiont common to many marine sponges and evidence for vertical transmission via sponge larvae. *Applied and Environmental Microbiology* **72**, 3724–3732.

Ereskovsky, A. V. (2006). A new species of *Oscarella* (Demospongiae: Plakinidae) from the Western Sea of Japan. *Zootaxa* **1376**, 37–51.

Ereskovsky, A. V. (2007). A new species of *Halisarca* (Demospongiae: Halisarcida) from the Sea of Okhotsk, North Pacific. *Zootaxa* **1432**, 57–66.

Ereskovsky, A. V. (2010). The Comparative Embryology of Sponges. Springer, Dordrecht.

Ereskovsky, A. V., Gonobobleva, E., and Vishnyakov, A. (2005). Morphological evidence for vertical transmission of symbiotic bacteria in the viviparous sponge *Halisarca dujardini* Johnston (Porifera, Demospongiae, Halisarcida). *Marine Biology* **146**, 869–875.

Ereskovsky, A. V., Tokina, D. B., Bézac, C., and Boury-Esnault, N. (2007). Metamorphosis of cinctoblastula larvae (Homoscleromorpha, Porifera). *Journal of Morphology* **268**, 518–528.

Ereskovsky, A. V., Borchiellini, C., Gazave, E., Ivaniševic, J., Lapébie, P., Pérez, T., Renard-Deniel, E., and Vacelet, J. (2009a). The homoscleromorph sponge *Oscarella lobularis* as model in evolutionary and developmental biology. *BioEssays* **31**, 89–97.

Ereskovsky, A. V., Sanamyan, K., and Vishnyakov, A. E. (2009b). *Oscarella kamchatkensis* sp. nov. from the North-West of Pacific. *Cahiers de Biologie Marine* **50**, 369–381.

Ereskovsky, A. V., Ivaniševic, J., and Pérez, T. (2009c). Overview on the Homoscleromorpha sponges diversity in the Mediterranean. In "Proceedings of the 1st Symposium on the Coralligenous and Other Bio-concretions of the Mediterranean Sea, Tabarka, 15–16 January 2009." pp. 89–95.

Ereskovsky, A. V., Konyukov, P. Y., and Tokina, D. B. (2010). Morphogenesis accompanying larval metamorphosis in *Plakina trilopha* (Porifera, Homoscleromorpha). *Zoomorphology* **129**, 21–31.

Ereskovsky, A. V., Lavrov, D. V., Boury-Esnault, N., and Vacelet, J. (2011). Molecular and morphological description of a new species of *Halisarca* (Demospongiae: Halisarcida) from Mediterranean Sea and a redescription of the type species *Halisarca dujardini*. *Zootaxa* **2768**, 5–31.

Erpenbeck, D. (2004). On the phylogeny of Halichondrid demosponges. Doctor Thesis, Universiteit van Amsterdam.

Erpenbeck, D., and van Soest, R. W. M. (2002). Family Halichondriidae Gray, 1867. In "A Guide to the Classification of Sponges" (J. N. A. Hooper and R. W. M. van Soest, eds), pp. 787–815. Kluwer Academic/Plenum Publishers, New York.

Erpenbeck, D., and van Soest, R. W. M. (2005). A survey for biochemical synapomorphies to reveal phylogenetic relationships of halichondrid demosponges (Metazoa: Porifera). *Biochemical Systematics and Ecology* **31**, 583–616.

Erpenbeck, D., and van Soest, R. W. M. (2007). Status and perspective of sponge chemosystematic. *Marine Biotechnology* **9**, 2–19.

Erpenbeck, D., and Wörheide, G. (2007). On the molecular phylogeny of sponges (Porifera). *Zootaxa* **1668**, 107–126.

Erpenbeck, D., Breeuwer, J. A. J., van der Velde, H. C., and van Soest, R. W. M. (2002). Unravelling host and symbiont phylogenies of halichondrid sponges (Demospongiae, Porifera) using a mitochondrial marker. *Marine Biology* **141,** 377–386.

Erpenbeck, D., McCormack, G. P., Breeuwer, J. A. J., and van Soest, R. W. M. (2004). Order level differences in the structure of partial LSU across demosponges (Porifera): New insights into an old taxon. *Molecular Phylogenetics and Evolution* **32,** 388–395.

Erpenbeck, D., Breeuwer, J. A. J., and van Soest, R. W. M. (2005). Implications from a 28S rRNA gene fragment for the phylogenetic relationships of halichondrid sponges (Porifera: Demospongiae). *Journal of Zoological Systematics and Evolutionary Research* **43,** 93–99.

Erpenbeck, D., Breeuwer, J. A. J., Parra-Velandia, F. J., and van Soest, R. W. M. (2006a). Speculation with spiculation? Three independent gene fragments and biochemical characters versus morphology in demosponge higher classification. *Molecular Phylogenetics and Evolution* **38,** 293–305.

Erpenbeck, D., Hooper, J. N. A., and Wörheide, G. (2006b). CO1 phylogenies in diploblasts and the "Barcoding of Life"—Are we sequencing a suboptimal partition? *Molecular Ecology Notes* **6,** 550–553.

Erpenbeck, D., Duran, S., Rützler, K., Paul, V., Wörheide, G., and Hooper, J. N. A. (2007a). Towards a DNA taxonomy of Caribbean demosponges: A gene tree reconstructed from partial mitochondrial CO1 gene sequences supports previous rDNA phylogenies and provides a new perspective on the systematics of Demospongiae. *Journal of the Marine Biological Association of the United Kingdom* **87,** 1563–1570.

Erpenbeck, D., List-Armitage, S. E., Alvarez, B., Degnan, B., Wörheide, G., and Hooper, J. N. A. (2007b). The systematics of Raspailiidae (Demospongiae: Poecilosclerida: Microcionina) re-analysed with a ribosomal marker. *Journal of the Marine Biological Association of the United Kingdom* **87,** 1571–1576.

Erpenbeck, D., Hooper, J. N. A., List-Armitage, S. E., Degnan, B. M., Wörheide, G., and van Soest, R. M. W. (2007c). Affinities of the family Sollasellidae (Porifera, Demospongiae). II. Molecular evidence. *Contributions to Zoology* **76,** 95–102.

Erpenbeck, D., Voigt, O., Gültas, M., and Wörheide, G. (2008). The sponge genetree server providing a phylogenetic backbone for poriferan evolutionary studies. *Zootaxa* **1939,** 58–60.

Erpenbeck, D., Sutcliffe, P., Cook, S. de C., Degnan, B. M., Hooper, J. N. A., and Wörheide, G. (2010). Horny sponges and their affairs: On the phylogenetic relationships of keratose sponges. In "Ancient Animals, New Challenges." VIII World Sponge Conference, Girona 20–24 September 2010 (M.-J. Uriz, M. Maldonado, M. Becerro and X. Turon, eds), p. 51.

Erpenbeck, D., Weier, T., de Voogd, N. J., Wörheide, G., Sutcliffe, P., Todd, J. A., and Michel, E. (2011). Insights into the evolution of freshwater sponges (Porifera: Demospongiae: Spongillina): Barcoding and phylogenetic data from Lake Tanganyika endemics indicate multiple invasions and unsettle existing taxonomy. *Molecular Phylogenetics and Evolution* **61,** 231–236.

Erwin, P. M., and Thacker, R. W. (2007). Phylogenetic analyses of marine sponges within the order Verongida: A comparison of morphological and molecular data. *Invertebrate Biology* **126,** 220–234.

Esper, E. C. J. (1794). Die Pflanzenthiere in Abbildungen nach der Natur mit Farben erleuchtet, nebst Beschreibungen. Zweyter TheilRaspe, Nürenberg.

Esper, E. J. C. (1797). Fortsetzungen der Pflanzenthiere in Abbildungennach der Natur mit Farben erleuchtet nebst Beschreibungen. Erster Theil, Nürenberg.

Exposito, J.-Y., Le Guellec, D., Lu, Q., and Garrone, R. (1991). Short chain collagens in sponges are encoded by a family of closely related genes. *The Journal of Biological Chemistry* **266,** 21923–21928.

Fabricius, O. (1780). Fauna Groenlandica: Systematice sistens Animalia Groenlandiae Occidentalis Hactenus Indagata, Quod Nomen Specificium. Hafniae et Lipsiae, Copenhagen, Denmark.

Fernandez, J. C. C., Peixinho, S., Pinheiro, U. S., and Menegola, C. (2011). Three new species of *Tetilla* Schmidt, 1868 (Tetillidae, Spirophorida, Demospongiae) from Bahia, northeastern Brazil. *Zootaxa* **2978**, 51–67.

Ferrario, F., Calcinai, B., Erpenbeck, D., Galli, P., and Wörheide, G. (2010). Two *Pione* species (Hadromerida, Clionaidae) from the Red Sea: A taxonomical challenge. *Organisms, Diversity and Evolution* **10**, 1–11.

Fiehn, O. (2002). Metabolomics—The link between genotypes and phenotypes. *Plant Molecular Biology* **48**, 155–171.

Floyd, R., Abebe, E., Papert, A., and Blaxter, M. (2002). Molecular barcodes for soil nematode identification. *Molecular Ecology* **11**, 839–850.

Franz, N. M. (2005). On the lack of good scientific reasons for the growing phylogeny/ classification gap. *Cladistics* **21**, 495–500.

Fromont, J. P., and Bergquist, P. R. (1990). Structural characters and their use in sponge taxonomy; when is a sigma not a sigma? In "New Perspectives in Sponge Biology" (K. Rützler, ed.), pp. 273–278. Smithsonian Institution Press, Washington, DC.

Fromont, J., Usher, K. L., Sutton, D. C., Toze, S., and Kuo, J. (2008). Species of the sponge genus *Chondrilla* (Demospongiae: Chondrosida: Chondrillidae) in Australia. *Records of the Western Australian Museum* **24**, 469–486.

Frost, T. M., Reiswig, H. M., and Ricciardi, A. (2001). Porifera. In "Ecology and Classification of NorthAmerican Freshwater Invertebrates" (J. H. Thorp and A. P. Covish, eds), pp. 97–133. Academic Press, New York.

Funk, V. A., Hoch, P. C., Prather, L. A., and Wagner, W. L. (2005). The importance of vouchers. *Taxon* **54**, 127–129.

Gallissian, M.-F., and Vacelet, J. (1992). Ultrastructure of the oocyte and embryo of the calcified sponge *Petrobiona massiliona* (Porifera, Calcarea). *Zoomorphology* **112**, 133–141.

Garrone, R. (1985). The collagen of Porifera. In "Biology of Invertebrate and Lower Vertebrate Collagens" (A. Bairati and R. Garrone, eds), pp. 157–175. Plenum Press, London.

Gautret, P., Vacelet, J., and Cuif, J. P. (1991). Caractéristiques des spicules et du squelette carbonaté des espèces actuelles du genre *Merlia* (Démosponges, Merliida), et comparaison avec des Chaetétides fossiles. *Bulletin du Muséum National d'Histoire Naturelle, Paris* **13**, 289–307.

Gazave, E., Lapébie, P., Renard, E., Bézac, C., Boury-Esnault, N., Vacelet, J., Pérez, T., Manuel, M., and Borchiellini, C. (2008). NK homeobox genes with choanocyte-specific expressionin homoscleromorph sponges. *Development Genes and Evolution* **218**, 479–489.

Gazave, E., Carteron, S., Chenuil, A., Richelle-Maurer, E., Boury-Esnault, N., and Borchiellini, C. (2010a). Polyphyly of the genus *Axinella* and of the family Axinellidae (Porifera: Demospongiae[P]). *Molecular Phylogenetics and Evolution* **57**, 35–47.

Gazave, E., Lapébie, P., Renard, E., Vacelet, J., Rocher, C., Lavrov, D. V., and Borchiellini, C. (2010b). Molecular phylogeny restores the supra-generic subdivision of Homoscleromorph sponges (Porifera, Homoscleromorpha). *PLoS One* **5**, e14290.

Gazave, E., Lapébie, P., Ereskovsky, A. V., Vacelet, J., Renard, E., Cárdenas, P., and Borchiellini, C. (2012). No longer Demospongiae: Homoscleromorpha formal nomination as a fourth class of Porifera. *Hydrobiologia* **687**, 3–10.

Genta-Jouve, G., and Thomas, O. P. (2012). Sponge chemical diversity: from biosynthetic pathways to ecological roles. *Advances in Marine Biology* **62**.

Gonobobleva, E. L. (2007). Basal apparatus formation in external flagellated cells of *Halisarca dujardini* larvae (Demospongiae: Halisarcida) in the course of embryonic development. In "Porifera Research: Biodiversity, Innovation and Sustainability" (M. R. Custódio,

G. Lôbo-Hajdu, E. Hajdu and G. Muricy, eds), pp. 345–351. Museu Nacional, Rio de Janeiro.

Goodwin, C., Morrow, C., and Picton, B. (2010). Unravelling the Hymedesmiidae: Aligning molecular and morphological evidence. In "Ancient Animals, New Challenges." VIII World Sponge conference Girona, 20–24 September 2010 (M.-J. Uriz, M. Maldonado, M. Becerro and X. Turon, eds)., p. 60.

Goodwin, C., Jones, J., Neely, K., and Brickle, P. (2011a). Sponge biodiversity of the Jason Islands and Stanley, Falkland Islands with descriptions of twelve new species. *Journal of the Marine Biological Association of the United Kingdom* **91**, 275–301.

Goodwin, C. E., Picton, B. E., and van Soest, R. W. M. (2011b). *Hymedesmia* (Porifera: Demospongiae: Poecilosclerida) from Irish and Scottish cold-water coral reefs, with a description of five new species. *Journal of the Marine Biological Association of the United Kingdom* **91**, 979–997.

Gould, S. J. (2002). I have landed. Three Rivers Press, New York.

Greuter, W., Hawksworth, D. L., McNeill, J., Mayo, M. A., Minelli, A., Sneath, P. H. A., Tindall, B. J., Trehane, P., and Tubbs, P. (1998). Draft BioCode (1997): The prospective international rules for the scientific names of organisms. *Taxon* **47**, 127–150.

Greuter, W., Garrity, G., Hawksworth, D. L., Jahn, R., Kirk, P. M., Knapp, S., McNeill, J., Michel, E., Patterson, D. J., Pyle, R., and Tindall, B. J. (2011). Draft BioCode (2011). Principles and Rules regulating the naming of organisms. New draft, revised in November 2010. *Bionomina* **3**, 26–44.

Hajdu, E., and Desqueyroux-Faúndez, R. (2008). A reassessment of the phylogeny and biogeography of *Rhabderemia* Topsent, 1890 (Rhabderemiidae, Poecilosclerida, Demospongiae). *Revue Suisse de Zoologie* **115**, 377–395.

Hajdu, E., and Soest, R. W. M. van (2002). Family Merliidae Kirkpatrick, 1908. In "Systema Porifera: A Guide to the Classification of Sponges" (J. N. A. Hooper and R. W. M. van Soest, eds), pp. 691–693. Kluwer Academic/Plenum Publishers, New York.

Hajdu, E., Berlinck, R. G. S., and Freitas, J. C. (1999). Porifera. In "Biodiversidade do Estado de São Paulo: Síntese do Conhecimento ao Final do Século XX. Volume 3. Invertebrados Marinhos" (A. E. Migotto and C. G. Tiago, eds), pp. 20–30. Fapesp, São Paulo.

Hajdu, E., Santos, C. P., Lopes, D. A., Oliveira, M. V., Moreira, M. C. F., Carvalho, M. S., and Klautau, M. (2003). Porifera. In "Biodiversidade bêntica da região sul-sudeste da costa brasileira" (A. C. Z. Amaral, P. C. Lana, F. C. Fernandes and J. C. Coimbra, eds), REVIZEE Score Sul—Bentos. pp. 28–33. MMA, Brasília.

Hajdu, E., Santos, C. P., Lopes, D. A., Oliveira, M. V., Moreira, M. C. F., Carvalho, M. S., and Klautau, M. (2004). Filo Porifera. In "Biodiversidade bentônica da região sul-sudeste do Brasil—plataforma externa e talude superior" (A. C. Z. Amaral and C. L. B. Rossi-Wongtschovski, eds), Série Documentos REVIZEE—Score Sul. pp. 49–56. IOUSP, São Paulo.

Harcet, M., Bilandžija, H., Bruvo-Mađarić, B., and Ćetković, H. (2010). Taxonomic position of *Eunapius subterraneus* (Porifera, Spongillidae) inferred from molecular data—A revised classification needed? *Molecular Phylogenetics and Evolution* **54**, 1021–1027.

Härlin, M., and Härlin, C. (2001). Phylogeny of the eureptantic nemerteans revisited. *Zoologica Scripta* **30**, 49–58.

Hartman, W. D. (1958). Re-examination of Bidder's classification of the Calcarea. *Systematic Zoology* **7**, 55–110.

Hartman, W. D. (1980). Systematics of the Porifera. In "Living and Fossil Sponges. (Notes for a Short Course)" (R. N. Ginsburg and P. Reid, eds), pp. 24–51. University of Miami, Miami.

Hartman, W. D., and Willenz, P. (1990). Organisation of the choanosome of three Caribbean Sclerosponges. In "New Perspectives in Sponge Biology" (K. Rützler, ed.), pp. 227–326. Smithsonian Institution Press, Washington.

Hawksworth, D. L. (2011). Introducing the *Draft BioCode* (2011). *Taxon* **60,** 199–200.

Heim, I., and Nickel, M. (2010). Description and molecular phylogeny of *Tethya leysae* sp. nov. (Porifera, Demospongiae, Hadromerida) from the Canadian Northeast Pacific with remarks on the use of microtomography in sponge taxonomy. *Zootaxa* **2422,** 1–21.

Henkel, D., and Janussen, D. (2011). Redescription and new records of *Celtodoryx ciocalyptoides* (Demospongiae: Poecilosclerida)—A sponge invader in the north east Atlantic Ocean of Asian origin? *Journal of the Marine Biological Association of the United Kingdom* **91,** 347–355.

Hennig, W. (1950). Grundzüge einer Theorie der phylogenetischen Systematik. Deutscher Zentralverlag, Berlin.

Hibbett, D. S., and Donoghue, M. J. (1998). Integrating phylogenetic analysis and classification in fungi. *Mycologia* **90,** 347–356.

Hogg, M. M., Tendal, O. S., Conway, K. W., Pomponi, S. A., van Soest, R. W. M., Gutt, J., Krautter, M., and Roberts, J. M. (2010). Deep-sea sponge grounds: Reservoirs of biodiversity. *UNEP-WCMC Biodiversity Series* **32,** 1–86.

Holmes, B., and Blanch, H. (2007). Genus-specific associations of marine sponges with group I crenarchaeotes. *Marine Biology* **150,** 759–772.

Hooper, J. N. A. (1990). Character stability, systematics and affinities between Microcionidae (Poecilosclerida) and Axinellida. In "New Perspectives in Sponge Biology" (K. Rützler, ed.), pp. 284–294. Smithsonian Institution Press, Washington, DC.

Hooper, J. N. A. (1991). Revision of the family Raspailiidae (Porifera: Demospongiae), with description of Australian species. *Invertebrate Taxonomy* **5,** 1179–1418.

Hooper, J. N. A. (1996). Revision of Microcionidae (Porifera: Poecilosclerida: Demospongiae), with description of Australian species. *Memoirs of the Queensland Museum* **40,** 1–626.

Hooper, J. N. A., and van Soest, R. W. M. (2002). Systema Porifera: A Guide to the Classification of Sponges Kluwer Academic/Plenum Publishers, New York.

Hooper, J. N. A., and van Soest, R. W. M. (2010). Threats to the system? Beyond the "Systema Porifera". In "Ancient Animals, New Challenges." VIII World Sponge Conference Girona, 20–24 September 2010 (M.-J. Uriz, M. Maldonado, M. Becerro and X. Turon, eds), p. 68.

Hooper, J. N. A., and Wiedenmayer, F. (1994). Porifera. (A. Wells, ed.), Zoological Catalogue of Australia. **12,** pp. 1–624. CSIRO, Melbournehttp://www.environment.gov.au/biodiversity/abrs/online-resources/fauna/afd/taxa/PORIFERA/checklist#selected.

Hooper, J. N. A., Kennedy, J. A., and van Soest, R. W. M. (2000). Annotated checklist of sponges (Porifera) of the South China Sea region. *Raffles Bulletin of Zoology* **8,** 125–207.

Hooper, J. N. A., Sutcliffe, P., and Schlacher-Hoelinger, M. (2008). New species of Raspailiidae (Porifera: Demospongiae: Poecilosclerida) from southeast Queensland. *Memoirs of the Queensland Museum* **54,** 1–22.

Hoshino, T. (1990). *Merlia tenuis* n. sp. Encrusting Shell Surfaces of Gastropods, *Chicoreus*, from Japan. In "New Perspectives in Sponge Biology" (K. Rützler, ed.), pp. 295–301. Smithsonian Institution Press, Washington.

Huang, D. W., Meier, R., Todd, P. A., and Chou, L. M. (2008). Slow mitochondrial COI sequence evolution at the base of the metazoan tree and its implications for DNA barcoding. *Journal of Molecular Evolution* **66,** 167–174.

Ijima, I. (1927). The Hexactinellida of the Siboga Expedition. *Siboga Expedition Reports* **6,** 1–383.

Itskovich, V. B., Belikov, S. I., Efremova, S. M., and Masuda, Y. (1999). Phylogenetic relationships between Lubomirskiidae, Spongillidae, and some marine sponges according partial sequences of 18S rDNA. *Memoirs of the Queensland Museum* **44,** 275–280.

Itskovich, V., Belikov, S., Efremova, S., Masuda, Y., Pérez, T., Alivon, E., Borchiellini, C., and Boury-Esnault, N. (2007). Phylogenetic relationships between freshwater and marine Haplosclerida (Porifera, Demospongiae) based on the full length 18S rRNA and partial COXI gene sequences. In "Porifera research: Biodiversity, Innovation and Sustainability" (M. R. Custódio, G. Lôbo-Hajdu, E. Hajdu and G. Muricy, eds), pp. 383–391. Museu Nacional, Rio de Janeiro.

Itskovich, V., Gontcharov, A., Masuda, Y., Nohno, T., Belikov, S., Efremova, S., Meixner, M., and Janussen, D. (2008). Ribosomal ITS sequences allow resolution of freshwater sponge phylogeny with alignments guided by secondary structure prediction. *Journal of Molecular Evolution* **67**, 608–620.

Ivanišević, J. (2011). Métabolisme secondaire des éponges Homoscleromorpha. Diversité et fluctuation de son expression en fonction des facteurs biotiques et abiotiques. Ph.D. Thesis, Université de la Méditerranée, Marseille, 205pp.

Ivanišević, J., Thomas, O. P., Lejeusne, C., Chevaldonné, P., and Pérez, T. (2011a). Metabolic fingerprinting as an indicator of biodiversity: Towards understanding interspecific relationships among Homoscleromorpha sponges. *Metabolomics* **7**, 289–304.

Ivanišević, J., Pérez, T., Ereskovsky, A. V., Barnathan, G., and Thomas, O. P. (2011b). Lysophospholipids in the Mediterranean sponge *Oscarella tuberculata:* Seasonal variability and putative biological role. *Journal of Chemical Ecology* **37**, 537–545.

Janussen, D., and Reiswig, H. M. (2009). Hexactinellida (Porifera) from the ANDEEP III Expedition to the Weddell Sea, Antarctica. *Zootaxa* **2136**, 1–20.

Janussen, D., Tabachnick, K. R., and Tendal, O. S. (2004). Deep-sea Hexactinellida (Porifera) of the Weddell Sea. *Deep-Sea Research Part II* **51**, 1857–1992.

Jenner, R. A. (2002). Boolean logic and character state identity: Pitfalls of character coding in metazoan cladistics. *Contributions to Zoology* **71**, 67–91.

Jenner, R. A. (2004a). When molecules and morphology clash: Reconciling conflicting phylogenies of the Metazoa by considering secondary character loss. *Evolution and Development* **6**, 372–378.

Jenner, R. A. (2004b). Towards a phylogeny of the Metazoa: Evaluating alternative phylogenetic positions of Platyhelminthes, Nemertea, and Gnathostomulida, with a critical reappraisal of cladistic characters. *Contributions to Zoology* **73**, 3–163.

Jenner, R. A. (2004c). Libbie Henrietta Hyman (1888–1969): From developmental mechanics to the evolution of animal body plans. *The Journal of Experimental Zoology* **2004**, 413–423.

Jeon, Y. J., and Sim, C. J. (2008). A new species of the genus *Biemna* (Demospongiae: Poecilosclerida: Desmacellidae) from Korea. *Animal Cells and Systems* **12**, 241–243.

Jiménez-Guri, E., Philippe, H., Okamura, B., and Holland, P. W. H. (2007). *Buddenbrockia* is a Cnidarian worm. *Science* **317**, 116–118.

Johnston, G. (1842). A History of British Sponges and Lithophytes. W.H. Lizars, Edinburgh.

Johnston, I. S., and Hildemann, W. H. (1982). Cellular organization in the marine demosponge *Callyspongia diffusa*. *Marine Biology* **67**, 1–7.

Kelly, M., Edwards, A. R., Wilkinson, M. R., Alvarez, B., Cook, S. de C., Bergquist, P. R., Buckeridge, J. S., Campbell, H. J., Reiswig, H. M., Valentine, C., and Vacelet, J. (2009). Phylum Porifera (sponges). In "New Zealand Inventory of Biodiversity" (P. D. Gordon, ed.), pp. 23–46. Canterbury University Press, Christchurch.

Kelly-Borges, M., Bergquist, P. R., and Bergquist, P. L. (1991). Phylogenetic relationships within the order Hadromerida (Porifera, Demospongiae, Tetractinomorpha) as indicated by ribosomal RNA sequence comparisons. *Biochemical Systematics and Ecology* **19**, 117–125.

Kelly-Borges, M., and Pomponi, S. (1994). Phylogeny and classification of lithistid sponges (*Porifera:Demospongiae*): a preliminary assessment using ribosomal DNA sequence comparisons. *Molecular Marine Biology and Biotechnology* **3**, 87–103.

Klautau, M., and Borojevic, R. (2001). Calcareous sponges from Arraial do Cabo—Brazil (I: the genus *Clathrina*). *Zoosystema* **23**, 395–410.

Klautau, M., and Valentine, C. (2003). Revision of the genus *Clathrina* (Porifera, Calcarea). *Zoological Journal of the Linnean Society* **139**, 1–62.

Klautau, M., Solé-Cava, A. M., and Borojevic, R. (1994). Biochemical systematics of sibling sympatric species of *Clathrina* (Porifera: Calcarea). *Biochemical Systematics and Ecology* **22**, 367–375.

Klautau, M., Russo, C., Lazoski, C., Boury-Esnault, N., Thorpe, J. P., and Solé-Cava, A. M. (1999). Does cosmopolitanism in morphologically simple species result from overconservative systematics? A case study using the marine sponge *Chondrilla nucula*. *Evolution* **53**, 1414–1422.

Klitgaard, A. B., and Tendal, O. S. (2004). Distribution and species composition of mass occurrences of large-sized sponges in the northeast Atlantic. *Progress in Oceanography* **61**, 57–98.

Klitgaard, A. B., Tendal, O. S., and Westerberg, H. (1997). Mass occurrences of large sponges (Porifera) Faroe Island (NE Atlantic) shelf and slope areas: Characteristics, distribution and possible causes. In "The Responses of Marine Organisms in Their Environments" (L. E. Hawkins and S. Hutchinson, eds), pp. 129–142. University of Southampton, Southampton.

Knowlton, A. L., Pierson, B. J., Talbot, S. L., and Highsmith, R. C. (2003). Isolation and characterization of microsatellite loci in the intertidal sponge *Halichondria panicea*. *Molecular Ecology Notes* **3**, 560–562.

Koltun, V. M. (1968). Spicules of sponges as an element of the bottom sediments of the Antarctic. Ocean Floor. In "Ocean Floor; Symposium on Antarctic Oceanogarphy," pp. 121–123. Scott Polar Research Institute, Cambridge. Santiago Chile, 13–16 September 1966.

Kornprobst, J.-M. (2005). Substances naturelles d'origine marine: chimiodiversité, pharmacodiversité, biotechnologie. Lavoisier, Paris.

Kornprobst, J.-M. (2010). Encyclopedia of Marine Natural Products. Wiley-Blackwell, Weinheim.

Korotkhova, G. P., and Ermolina, N. O. (1982). The larval development of *Halisarca dujardini* (Demospongiae). *Zoological Journal* **61**, 1472–1480.

Krautter, M., Conway, K. W., and Barrie, J. V. (2006). Recent hexactinosidan sponge reefs (silicate mounds) off British Columbia, Canada: Frame-building processes. *Journal of Paleontology* **80**, 38–48.

Kurek, D., Assal, Y., Born, R., Kammer, M., Ehlrich, A., Kljajic, Z., and Ehlrich, H. (2010). Comparative investigations of nature and origin of skeletal organic matrices of *Aplysina aerophoba* (Verongida) and *Dysidea avara* (Dendroceratida). In "Ancient Animals, New Challenges." VIII World Sponge Conference, Girona 20–24 September 2010 (M.-J. Uriz, M. Maldonado, M. Becerro and X. Turon, eds)., p. 238.

Laborel, J., Pérès, J. M., Picard, J., and Vacelet, J. (1961). Etude directe des fonds des parages de Marseille de 30 à 300 m avec la soucoupe plongeante COUSTEAU. *Bulletin de l'Institut océanographique de Monaco* **1206**, 1–16.

Lafay, B., Boury-Esnault, N., Vacelet, J., and Christen, R. (1992). An analysis of partial 28S ribosomal RNA sequences suggests early radiations of sponges. *Bio Systems* **28**, 139–151.

Langenbruch, P.-F. (1983). Body structure of marine sponges. 1. Arrangement of the flagellated chambers in the canal system of *Reniera* sp.. *Marine Biology* **75**, 319–325.

Langenbruch, P.-F. (1988). Body structure of marine sponges. V. Structure of choanocyte chambers in some Mediterranean and Caribbean haplosclerid sponges (Porifera). *Zoomorphology* **108**, 13–21.

Langenbruch, P. F. (1991). Histological indications of the phylogenesis of the Haplosclerida (Demospongiae, Porifera). In "Fossil and Recent Sponges" (J. Reitner and H. Keupp, eds), pp. 289–298. Springer-Verlag, Berlin.

Langenbruch, P. F., and Scalera-Liaci, L. (1986). Body structure of marine sponges. IV. Aquiferous system and choanocyte chambers in *Haliclona elegans* (Porifera, Demospongiae). *Zoomorphology* **106,** 205–211.

Langenbruch, P. F., and Scalera-Liaci, L. (1990). Structure of choanocyte chambers in Haplosclerid sponges. In "New Perspectives in Sponge Biology" (K. Rützler, V. Macintyre and K. P. Smith, eds), pp. 245–251. Smithsonian Institution Press, Washington.

Langenbruch, P.-F., and Weissenfels, N. (1987). Canal systems and choanocyte chambers in freshwater sponges (Porifera, Spongillidae). *Zoomorphology* **107,** 11–16.

Langenbruch, P. F., Simpson, T. L., and Scalera-Liaci, L. (1985). Body structure of marine sponges. III. The structure of choanocyte chambers in *Petrosia ficiformis* (Porifera, Demospongiae). *Zoomorphology* **105,** 383–387.

Lapébie, P., Gazave, E., Ereskovsky, A., Derelle, R., Bézac, C., Renard, E., Houliston, E., and Borchiellini, C. (2009). WNT/β-catenin signalling and epithelial patterning in the homoscleromorph sponge *Oscarella*. *PLoS One* **4,** e5823.

Laurin, M. (2010). The subjective nature of Linnaean categories and its impact in evolutionary biology and biodiversity studies. *Contributions to Zoology* **79,** 131–146.

Laurin, M., and Bryant, H. N. (2009). Third meeting of the International Society for Phylogenetic Nomenclature: A report. *Zoologica Scripta* **38,** 333–337.

Lavrado, H. P. and Ignacio, B. L. (eds) (2006). Biodiversidade bentônica da região central da Zona Econômica Exclusiva brasileira, **18,** Série Livros Museu Nacional, Rio de Janeiro, p. 389.

Lavrov, D., Wang, X., and Kelly, M. (2008). Reconstructing ordinal relationships in the Demospongiae using mitochondrial genomic data. *Molecular Phylogenetics and Evolution* **49,** 111–124.

Lazoski, C., Peixinho, S., Russo, C. A. M., and Solé-Cava, A. M. (1999). Genetic confirmation of the specific status of two sponges of the genus *Cinachyrella* (Porifera: Demospongiae: Spirophorida) in the Southwest Atlantic. *Memoirs of the Queensland Museum* **44,** 299–305.

Lazoski, C., Solé-Cava, A. M., Boury-Esnault, N., Klautau, M., and Russo, C. A. M. (2001). Cryptic speciation in a high gene flow scenario in the oviparous marine sponge *Chondrosia reniformis* Nardo, 1847. *Marine Biology* **139,** 421–429.

Lejeune, C. (2010). Métabolites d'éponges marines de la famille de la Palau'amine: Isolement et synthèse biomimétique. Ph.D. thesis, Université de Paris XI Sud, Orsay, 318pp.

Lévi, C. (1956). Etude des *Halisarca* de Roscoff. Embryologie et systématique des démosponges. *Archives de Zoologie expérimentale et générale* **93,** 1–184.

Lévi, C. (1957). Ontogeny and systematics in sponges. *Systematic Zoology* **6,** 174–183.

Lévi, C. (1973). Systématique de la classe des Demospongiaria (Démosponges). In "Spongiaires" (P. P. Grassé, ed.), pp. 577–632. Masson & Co., Paris.

Lévi, C. (1979). Remarques sur la taxonomie des Demospongea. In "Biologie des Spongiaires" (C. Lévi and N. Boury-Esnault, eds), pp. 497–502. Editions du C.N.R.S, Paris.

Lévi, C. (1993). Porifera Demospongiae: Spongiaires bathyaux de Nouvelle-Calédonie, récoltés par le "Jean Charcot" campagne Biocal, 1985. (A. Crosnier, ed.), Résultats des Campagnes Musorstom, vol. 11, pp. 9–87. MNHN, Paris.

Lévi, C., and Lévi, P. (1976). Embryogenèse de *Chondrosia reniformis* (Nardo), démosponge ovipare, et transmission des bactéries symbiotiques. *Annales des Sciences Naturelles Zoologie et Biologie Animale* **18,** 367–380.

Lewis, P. O. (2001a). A likelihood approach to estimating phylogeny from discrete morphological character data. *Systematic Biology* **50,** 913–925.

Lewis, P. O. (2001b). Phylogenetic systematics turns over a new leaf. *Trends in Ecology & Evolution* **16**, 30–37.

Leys, S. P. (1999). The choanosome of hexactinellid sponges. *Invertebrate Biology* **118**, 221–235.

Leys, S. P. (2003). Comparative study of spiculogenesis in Demosponge and Hexactinellid larvae. *Microscopy Research and Technique* **62**, 300–311.

Leys, S., Wilson, K., Holeton, C., Reiswig, H. M., Austin, W. C., and Tunnicliffe, V. (2004). Patterns of glass sponge (Porifera, Hexactinellida) distribution in coastal waters of British Columbia, Canada. *Marine Ecology Progress Series* **283**, 133–149.

Leys, S., Cheung, E., and Boury-Esnault, N. (2006). Embryogenesis in the glass sponge *Oopsacas minuta*: Formation of syncytia by fusion of blastomeres. *Integrative and Comparative Biology* **46**, 104–117.

Leys, S., Mackie, G. O., and Reiswig, H. M. (2007). The biology of glass sponges. *Advances in Marine Biology* **52**, 1–145.

Li, C.-W., Chen, J.-Y., and Hua, T.-E. (1998). Precambrian sponges with cellular structures. *Science* **279**, 879–882.

Linnaeus, C. (1753). Species Plantarum, exhibentes plantas rite cognitas, ad genera relatas, cum differentiis specificis, nominibus trivialibus, synonymis selectis, locis natalibus, secundum systema sexuale digestas. Salvii, Holmiae.

Linnaeus, C. (1758). Systema naturae per regna tria naturae, secundum classes, ordines, genera, species, cum characteribus, differentes, synonymis. locis Laurentii Salvii, Holmiae.

Linnaeus, C. (1759). Systema naturae per regna tria naturae, secundum classes, ordines, genera, species, cum characteribus, differentiis, synonymis, locis. Holmiae Salvii.

List-Amitage, S. E., and Hooper, J. N. A. (2002). Discovery of *Petromica* Topsent in the Pacific Ocean: A revision of the genus and a new subgenus (*Chaladesma*, subgen. nov.) and a new species (*P. (C.) pacifica*, sp. nov.) (Porifera: Demospongiae: Halichondrida: Halichondriidae). *Invertebrate Systematics* **16**, 813–835.

Longo, C., Mastrototaro, F., and Corriero, G. (2005). Sponge fauna associated with a Mediterranean deep-sea coral bank. *Journal of the Marine Biological Association of the United Kingdom* **85**, 1341–1352.

Lopes, D. A., Hajdu, E., and Reiswig, H. M. (2005). Redescription of two Hexactinosida (Porifera, Hexactinellida) from the southwestern Atlantic, collected by Programme REVIZEE. *Zootaxa* **1066**, 43–56.

Lopes, D. A., Hajdu, E., and Reiswig, H. M. (2011). Taxonomy of Farrea (Porifera, Hexactinellida, Hexactinosida) from the southwestern Atlantic, with description of a new species and a discussion on the recognition of subspecies in Porifera. *Canadian Journal of Zoology* **89**, 169–189.

López, J. V., Peterson, C. L., Willoughby, R., Wright, A. E., Enright, E., Zoladz, S., Reed, J. K., and Pomponi, S. A. (2002). Characterization of genetic markers for in vitro cell line identification of the marine sponge *Axinella corrugata*. *The Journal of Heredity* **93**, 27–36.

López-Legentil, S., Erwin, P. M., Henkel, T. P., Loh, T.-L., and Pawlik, J. R. (2010). Phenotypic plasticity in the Caribbean sponge *Callyspongia vaginalis* (Porifera: Haplosclerida). *Scientia Marina* **74**, 445–453.

Loukaci, A., Muricy, G., Brouard, J.-P., Guyot, M., Vacelet, J., and Boury-Esnault, N. (2004). Chemical divergence between two sibling species of *Oscarella* (Porifera) from the Mediterranean Sea. *Biochemical Systematics and Ecology* **32**, 893–899.

Mackie, G. O. (2006). Progress in sponge biology. *Canadian Journal of Zoology* **84**, 143–145.

Mackie, G. O., and Singla, C. L. (1983). Studies on hexactinellid sponges. I. Histology of *Rhabdocalyptus dawsoni* (Lambe, 1873). *Philosophical Transactions of the Royal Society of London Series B* **301**, 365–400.

Maldonado, M. (1993). The taxonomic significance of the short-shafted mesotriaene reviewed by parsimony analysis - Validation of *Pachastrella ovisternata* Von Lendenfeld (Demospongiae, Astrophorida). *Bijdragen tot de Dierkunde* **63**, 129–148.

Maldonado, M. (2009). Embryonic development of verongid demosponges supports the independent acquisition of spongin skeletons as an alternative to the siliceous skeletonof sponges. *Biological Journal of the Linnean Society* **97**, 427–447.

Maldonado, M., Carmona, C., Uriz, M.-J., and Cruzado, A. (1999). Decline in Mesozoic reef-building by silicon limitation. *Nature* **401**, 785–788.

Maldonado, M., Turon, X., Becerro, M., and Uriz, M.-J. (eds) (2012). Ancient animals new challenges: developments in sponge research. *Hydrobiologia*.

Manconi, R., and Pronzato, R. (2002). Suborder Spongillina subord. nov.: Freshwater sponges. In "Systema Porifera: A Guide to the Classification of Sponges" (J. N. A. Hooper and R. W. M. van Soest, eds), pp. 921–1019. Kluwer Academic/Plenum Publishers, New York.

Manconi, R., and Pronzato, R. (2005). Freshwater sponges of the West Indies: Discovery of Spongillidae (Haplosclerida, Spongillina) from Cuba with biogeographic notes and a checklist for the Caribbean area. *Journal of Natural History* **39**, 3235–3253.

Manconi, R., and Pronzato, R. (2009). Atlas of African freshwater sponges. Studies in Afrotropical Zoology, vol. 295. Royal Museum for Central Africa, Tervuren.

Manuel, M. (2001). Origine et évolution des mécanismes moléculaires contrôlant la morphogenèse chez les Métazoaires: un nouveau modèle spongiaire, *Sycon raphanus* (Calcispongia, Calcaronea). Ph.D. Thesis, Université de Paris XI Orsay, Paris, France.

Manuel, M. (2006). Phylogeny and evolution of calcareous sponges. *Canadian Journal of Zoology* **84**, 225–241.

Manuel, M. and Boury-Esnault, N. (in press). *Porifera*. In *Phylonyms Companion Volume* (K. de Queiroz, J. Gauthier and P. Cantino, eds.). The University of California Press, Berkeley.

Manuel, M., Borojevic, R., Boury-Esnault, N., and Vacelet, J. (2002). Class Calcarea Bowerbank, 1864. In "Systema Porifera: A Guide to the Classification of Sponges" (J. N. A. Hooper and R. W. M. van Soest, eds), pp. 1103–1110. Kluwer Academic/Plenum Publishers, New York.

Manuel, M., Borchiellini, C., Alivon, E., Le Parco, Y., Vacelet, J., and Boury-Esnault, N. (2003). Phylogeny and evolution of calcareous sponges: Monophyly of Calcinea and Calcaronea, high level of morphological homoplasy, and the primitive nature of axial symmetry. *Systematic Biology* **52**, 311–333.

Manuel, M., Borchiellini, C., Alivon, E., and Boury-Esnault, N. (2004). Molecular phylogeny of calcareous sponges using 18S rRNA and 28S rRNA sequences. *Bollettino dei Musei e degli Istituti Biologici della Università di Genova* **68**, 449–461.

Mastrototaro, F., D'Onghia, G., Corriero, G., Matarrese, A., Maiorano, P., Panetta, P., Gherardi, M., Longo, C., Rosso, A., Sciuto, F., Sanfilippo, R., Gravili, C., Boero, F., Taviani, M., and Tursi, A. (2010). Biodiversity of the white coral bank off Cape Santa Maria di Leuca (Mediterranean Sea): An update. *Deep Sea Research Part II: Topical Studies in Oceanography* **57**, 412–430.

Masuda, Y. (2009). Studies on the taxonomy and distribution of freshwater sponges in Lake Baikal. In "Biosilica in Evolution, Morphogenesis, and Nanobiotechnology" (W. E. G. Müller and M. A. Grachev, eds), Progress in Molecular and Subcellular Biology, Marine Molecular Biotechnology, vol. 47, pp. 81–110.

McCormack, G. P., Erpenbeck, D., and van Soest, R. W. M. (2002). Major discrepancy between phylogenetic hypotheses based on molecular and morphological criteria within the Order Haplosclerida (Phylum Porifera: Class Demospongiae). *Journal of Zoology Systematics and Evolution Research* **40**, 237–240.

Medina, M., Collins, A. G., Silberman, J. D., and Sogin, M. L. (2001). Evaluating hypotheses of basal animal phylogeny using complete sequences of large and small subunit rRNA. *Proceedings of the National Academy of Sciences of the United States of America* **98**, 9707–9712.

Mehl, D. (1992). Die Entwicklung der Hexactinellida seit dem Mesozoikum. Paläobiologie, Phylogenie und Evolutionsökologie. *Berliner geowissenschaftliche Abhandlungen* **E2**, 1–164.

Meixner, M. J., Lüter, C., Eckert, C., Itskovich, V., Janussen, D., von Rintelen, T., Bohne, A. V., Meixner, J. M., and Hess, W. R. (2007). Phylogenetic analysis of freshwater sponges provide evidence for endemism and radiation in ancient lakes. *Molecular Phylogenetics and Evolution* **45**, 875–886.

Menshenina, L. L., Tabachnick, K. R., and Janussen, D. (2007). Revision of the subgenus *Neopsacas* (Hexactinellida, Rossellidae, *Crateromorpha*) with the description of new species and subspecies. *Zootaxa* **1463**, 55–68.

Messing, C. G., Diaz, M.-C., Kohler, K. E., Reed, J. K., Rützler, K., van Soest, R. W. M., Wulff, J., and Zea, S. (2010). South Florida sponges: An online guide to identification. In "Ancient Animals, New Challenges." VIII World Sponge conference Girona, 20–24 September 2010 (M.-J. Uriz, M. Maldonado, M. Becerro and X. Turon, eds)., p. 268.

Miller, K., Alvarez, B., Battershill, C. N., Northcote, P. T., and Parthasarathy, H. (2001). Genetic, morphological, and chemical divergence in the sponge genus *Latrunculia* (Porifera: Demospongiae) from New Zealand. *Marine Biology* **139**, 235–250.

Minchin, E. A. (1896). Suggestions for a natural classification of the Asconidae. *Annals and Magazine of Natural History* **18**, 349–362.

Minelli, A. (1999). The names of Animals. *Trends in Ecology & Evolution* **14**, 462–463.

Mitchell, K. D., Hall, K. A., and Hooper, J. N. A. (2011). A new species of *Sigmaxinella* Dendy, 1897 (Demospongiae, Poecilosclerida, Desmacellidae) from the Tasman Sea. *Zootaxa* **2901**, 19–34.

Montagu, G. (1818). An essay on sponges, with descriptions of all the species that have been discovered on the Coast of Great Britain. *Memoirs of the Wernerian Natural History Society* **2**, 67–121.

Mooi, R. D., and Gill, A. C. (2010). Phylogenies without synapomorphies—A crisis in fish systematics: Time to show some character. *Zootaxa* **2450**, 26–40.

Moraes, F. C., Ventura, M., Klautau, M., Hajdu, E., and Muricy, G. (2006). Biodiversidade de esponjas das ilhas oceânicas brasileiras. In "Ilhas Oceânicas Brasileiras—da pesquisa ao manejo" (R. J. V. Alves and J. W. A. Castro, eds), pp. 147–178. Ministério do Meio Ambiente, Secretaria de Biodiversidade e Florestas, Brasília.

Morrow, C., and Boury-Esnault, N. (2000). Redescription of the type species of the genus *Polymastia* Bowerbank, 1864 (Porifera, Demospongiae, Hadromerida). *Zoosystema* **22**, 327–335.

Morrow, C. C., Picton, B. M., Erpenbeck, D., Boury-Esnault, N., Magss, C., and Allock, L. (2012). Congruence between nuclear and mitochondrial genes in Demospongiae: A new hypothesis for internal relationships within the G4 clade (Porifera: Demospongiae). *Molecular Phylogenetics and Evolution* **62**, 174–190.

Mothes, B., Hajdu, E., Lerner, C., and van Soest, R. W. M. (2004a). Species of *Ulosa* and *Biemna* (Porifera, Demospongiae, Poecilosclerida) from the N-NE Brazilian continental shelf. *Bollettino dei Musei e degli Istituti Biologici della Università di Genova* **68**, 477–482.

Mothes, B., Capítoli, R. R., Lerner, C., and Campos, M. A. (2004b). Filo Porifera—Região Sul. In "Biodiversidade bentônica da região sudeste-sul do Brasil—plataforma externa e talude superior" (A. C. Z. Amaral and C. L. D. B. Rossi-Wongtschowki, eds), Série Documentos REVIZEE—Score Sul. pp. 57–63. IOUSP, São Paulo.

Mothes, B., Lerner, C. B., and Silva, C. M. M. (2006). Illustrated Guide of the Marine Sponges from the Southern Coast of Brazil/Guia Ilustrado de Esponjas Marinhas da Costa Sul-Brasileira. (2nd edn.) USEB, Pelotas. 119pp (revised and extended).

Müller, O. F. (1806). Zoologia danica seu animalium Daniae et Norvegiae rariorum ac minus notorum. Descriptiones et Historia. N. Christensen, Hauniae.

Muricy, G. (1999). An evaluation of morphological and cytological data sets for the phylogeny of Homosclerophorida (Porifera: Demospongiae). *Memoirs of the Queensland Museum* **44,** 399–409.

Muricy, G. (2011). Diversity of Indo-Australian *Plakortis* (Demospongiae: Plakinidae), with description of four new species. *Journal of the Marine Biological Association of the United Kingdom* **91,** 303–319.

Muricy, G., and Díaz, M. C. (2002). Order Homosclerophorida Dendy, 1905, family Plakinidae Schulze, 1880. In "Systema Porifera: A Guide to the Classification of Sponges" (J. N. A. Hooper and R. W. M. van Soest, eds), pp. 71–82. Kluwer Academic/Plenum Publishers, New York.

Muricy, G., and Hajdu, E. (2006). Porifera Brasilis. Guia de identificação das esponjas mais comuns do Sudeste do Brasil. Eclesiarte, Rio de Janeiro.

Muricy, G., and Pearse, J. S. (2004). A new species of *Oscarella* (Demospongiae: Plakinidae) from California. *Proceedings of the California Academy of Sciences* **55,** 598–612.

Muricy, G., Boury-Esnault, N., Bézac, C., and Vacelet, J. (1996a). Cytological evidence for cryptic speciation in Mediterranean *Oscarella* species (Porifera, Homoscleromorpha). *Canadian Journal of Zoology* **74,** 881–896.

Muricy, G., Solé-Cava, A. M., Thorpe, J. P., and Boury-Esnault, N. (1996b). Genetic evidence for extensive cryptic speciation in the subtidal sponge *Plakina trilopha* (Porifera: Demospongiae: Homoscleromorpha) from the Western Mediterranean. *Marine Ecology Progress Series* **138,** 181–187.

Muricy, G., Boury-Esnault, N., Bézac, C., and Vacelet, J. (1998). Taxonomic revision of the Mediterranean *Plakina* Schulze (Porifera, Demospongiae, Homoscleromorpha). *Zoological Journal of the Linnean Society* **124,** 169–203.

Muricy, G., Hajdu, E., Minervino, J. V., Madeira, A. V., and Peixinho, S. (2001). Systematic revision of the genus *Petromica* (Demospongiae, Halichondrida), with a new species from the southwestern Atlantic. *Hydrobiologia* **443,** 103–128.

Muricy, G., Santos, C. P., Batista, D., Lopes, D. A., Pagnoncelli, D., Monteiro, L. C., Oliveira, M. V., Moreira, M. C. F., Carvalho, M. de S., Melão, M., Klautau, M., Rodriguez, P. R. D., Costa, R. N., Silvano, R. G., Schwientek, S., Ribeiro, S. M., Pinheiro, U. S., and Hajdu, E. (2006). Filo Porifera. In "Biodiversidade bentônica da região central da Zona Econômica Exclusiva brasileira" (H. P. Lavrado and B. L. Ignacio, eds), Série Livros, vol. 18, pp. 109–145. Museu Nacional, Rio de Janeiro.

Muricy, G., Hajdu, E., Oliveira, M. V., Heim, A. S., Costa, R. N., Lopes, D. A., Melão, M., Rodriguez, P. R. D., Silvano, R. G., Monteiro, L. C., and Santos, C. (2007). Filo Porifera. In "Atlas de invertebrados marinhos da região central da Zona Econômica Exclusiva brasileira. Parte 1" (H. P. Lavrado and M. S. Viana, eds), Série Livros. **25,** pp. 25–57. Museu Nacional, Rio de Janeiro.

Muricy, G., Esteves, E. L., Moraes, F., Santos, J. P., Silva, S. M., Almeida, E. V. R., Klautau, M., and Lanna, E. (2008). Biodiversidade Marinha da Bacia Potiguar: Porifera. Série Livros, vol. 29. Museu Nacional, Rio de Janeiro.

Muricy, G., Lopes, D., Hajdu, E., Carvalho, M. S., Moraes, F., Klautau, M., Silva, C. M. M., and Pinheiro, U. (2011). Catalogue of Brazilian Porifera. Serie Livros **47.** Museu Nacional, Rio de Janeiro.

Nakada, T., Misawa, K., and Nozaki, H. (2008). Molecular systematics of Volvocales (Chlorophyceae, Chlorophyta) based on exhaustive 18S rRNA phylogenetic analyses. *Molecular Phylogenetics and Evolution* **48,** 281–291.

Nardo, G. D. (1833). Auszug aue einem neuen System der Spongiarien, wonach bereits die. In "Isis, oder Encyclopädische Zeitung Coll," pp. 519–523. Aufstellung in der Universitäts-Sammlung zu Padua gemacht ist.

Nardo, G. D. (1834). De Spongiis. *Isis (Oken), Coll.* 714–716.

Nardo, G. D. (1847). Prospetto della fauna marine volgare del Veneto estuario con cenni sulle principali specie commestibili dell'Adriatico, ecc. G. Antonelli, Venice, pp. 1–45.

Nelson, G. (1994). Homology and systematics. In "Homology: The Hierarchical Basis of Comparative Biology" (B. K. Hall, ed.), pp. 101–149. Academic Press, San Diego.

Nichols, S. A. (2005). An evaluation of support for order-level monophyly and interrelationships within the class Demospongiae using partial data from the large subunit rDNA and cytochrome oxidase subunit I. *Molecular Phylogenetics and Evolution* **34**, 81–96.

Nielsen, C. (2001). Animal Evolution. Interrelationships of the Living Phyla. (2nd edn.) Oxford University Press, Oxford.

Nielsen, J., and Jewett, M. C. (2007). The role of metabolomics in systems biology. In "Metabolomics: A Powerful Tool in Systems Biology" (J. Nielsen and M. C. Jewett, eds), pp. 1–10. Springer-Verlag, Berlin Heidelberg.

Nixon, K. C., Carpenter, J. M., and Stevenson, D. W. (2003). The *PhyloCode* is fatally flawed, and the "Linnaean" system can easily be fixed. *The Botanical Review* **69**, 111–120.

Nobeli, I., and Thornton, J. M. (2006). A bioinformatician's view of the metabolome. *BioEssays* **28**, 534–545.

Noyer, C., Agell, G., Pascual, M., and Becerro, M. A. (2009). Isolation and characterization of microsatellite loci from the endangered Mediterranean sponge *Spongia agaricina* (Demospongiae: Dictyoceratida). *Conservation Genetics* **10**, 1895–1898.

Okada, Y. (1928). On the development of a hexactinellid sponge, *Farrea sollasii. Journal of the Faculty of Science Imperial University of Tokyo* **4**, 1–29.

Olivi, G. (1792). Zoologia Adriatica, ossia Catalogo ragionato degli Animali del Golfo e delle Lagune di Venezia: preceduto da una dissertazione sulla storia fisica e naturale del Golfo. e accompagnada da memorie, ed osservazioni di fisica storia naturale ed economia, Bassano.

Ott, J., Pansini, M., and Manconi, R. (eds.) (2008). Advances in sponge research. A tribute to Klaus Rützler, *Marine Ecology*, **29**, 134–320.

Padial, J. M., and De la Riva, I. (2007). Integrative taxonomists should use and produce DNA barcodes. *Zootaxa* **1586**, 67–68.

Padial, J. M., Castroviejo-Fisher, S., Kohler, J., Vila, C., Chaparro, J. C., and De La Riva, I. (2009). Deciphering the products of evolution at the species level: The need for an integrative taxonomy. *Zoologica Scripta* **38**, 431–447.

Padial, J., Miralles, A., De La Riva, I., and Vences, M. (2010). The integrative future of taxonomy. *Frontiers in Zoology* **7**, 16.

Page, T. J., and Hughes, J. M. (2011). Neither molecular nor morphological data have all the answers; with an example from *Macrobrachium* (Decapoda: Palaemonidae) from Australia. *Zootaxa* **2874**, 65–68.

Pallas, P. S. (1766). Elenchus Zoophytorum sistens generum adumbrations generaliores et specierum cognitarum succinctas descriptiones cum selectis auctorum synonymis. P. van Cleef, The Hague.

Pansini, M. (1982). Notes on some Mediterranean *Axinella* with description of two new species. *Bollettino dei Musei e degli Istituti Biologici della Università di Genova* **50–51**, 79–98.

Pansini, M., and Longo, C. (2003). A review of the Mediterranean Sea sponge biogeography with, in appendix, a list of the demosponges hitherto recorded from this sea. *Biogeographia* **24**, 59–90.

Pansini, M., and Longo, C. (2008). Porifera. *Biologia Marina Mediterranea* **15**, 44–70.

Pansini, M., Pronzato, R., Bavestrello, G., and Manconi, R. (eds.) (2004). Sponge Science in the New Millenium. *Bolletino dei Musei e degli Istituti Biologici dell'Università di Genova,* **68**.

Pérez, T. (1996). La rétention de particules par une éponge hexactinellide, *Oopsacas minuta* (Leucopsacasidae): le rôle du réticulum. *Comptes Rendus de l'Académie des Sciences Paris, Sciences de la Vie* **319**, 385–391.

Pérez, T., Perrin, B., Carteron, S., Vacelet, J. And, and Boury-Esnault, N. (2006). *Celtodoryx girardae* gen. nov. sp. nov., a new sponge species (Poecilosclerida: Demospongiae) invading the Gulf of Morbihan (North East Atlantic, France). *Cahiers de Biologie Marine* **4**, 205–214.

Pérez, T., Ivanišević, J., Dubois, M., Thomas, O. P., Tokina, D., and Ereskovsky, A. V. (2011). *Oscarella balibaloi*, a new sponge species (Homoscleromorpha: Plakinidae) from the Western Mediterranean Sea: Cytological description, reproductive cycle and ecology. *Marine Ecology* **32**, 174–187.

Philippe, H., Derelle, R., Lopez, P., Pick, K., Borchiellini, C., Boury-Esnault, N., Vacelet, J., Renard, E., Houliston, E., Quéinnec, E., Da Silva, C., Wincker, P., Le Guyader, H., Leys, S., Jackson, D. J., Schreiber, F., Erpenbeck, D., Morgenstern, B., Wörheide, G., and Manuel, M. (2009). Phylogenomics revives traditional views on deep animal relationships. *Current Biology* **19**, 1–7.

Philippe, H., Brinkmann, H., Lavrov, D. V., Littlewood, T. J., Manuel, M., Wörheide, G., and Baurain, D. (2011). Resolving difficult phylogenetic questions: Why more sequences are not enough. *PLoS Biology* **9**, e1000602.

Pick, K. S., Philippe, H., Schreiber, F., Erpenbeck, D., Jackson, D. J., Wrede, P., Wiens, M., Alie, A., Morgenstern, B., Manuel, M., and Wörheide, G. (2010). Improved phylogenomic taxon sampling noticeably affects nonbilaterian relationships. *Molecular Biology and Evolution* **27**, 1983–1987.

Picton, B. E., and Goodwin, C. E. (2007). Sponge biodiversity of Rathlin Island, Northern Ireland. *Journal of the Marine Biological Association of the United Kingdom* **87**, 1441–1458.

Picton, B. E., Morrow, C., and van Soest, R. W. M. (2011). Sponges of Britain and Ireland. http://www.habitas.org.uk/marinelife/sponge_guide/.

Pinheiro, U. S., Hajdu, E., and Custódio, M. R. (2007). *Aplysina* Nardo (Porifera, Verongida, Aplysinidae) from the Brazilian coast with description of eight new species. *Zootaxa* **1609**, 1–51.

Pisera, A. (2006). Palaeontology of sponges—A review. *Canadian Journal of Zoology* **84**, 242–261.

Pisera, A., and Lévi, C. (2002a). Family Scleritodermidae Sollas, 1888. In "Systema Porifera: A Guide to the Classification of Sponges" (J. N. A. Hooper and R. W. M. van Soest, eds), pp. 302–311. Kluwer Academic/Plenum Publishers, New York.

Pisera, A., and Lévi, C. (2002b). Family Azoricidae Sollas, 1888. In "Systema Porifera: A Guide to the Classification of Sponges" (J. N. A. Hooper and R. W. M. van Soest, eds), pp. 352–355. Kluwer Academic/Plenum Publishers, New York.

Pisera, A., and Lévi, C. (2002c). Family Desmanthidae, Topsent, 1893. In "Systema Porifera: A Guide to the Classification of Sponges" (J. N. A. Hooper and R. W. M. van Soest, eds), pp. 356–362. Kluwer Academic/Plenum Publishers, New York.

Pleijel, F., and Rouse, G. W. (2000). A new taxon, *capricornia* (Hesionidae, Polychaeta), illustrating the LITU ('Least-Inclusive Taxonomic Unit') concept. *Zoologica Scripta* **29**, 157–168.

Pleijel, F., and Rouse, G. W. (2003). "Ceci n'est pas une pipe": Names, clades and phylogenetic nomenclature. *Journal of Zoological Systematics and Evolutionary Research* **41**, 162–174.

Pleijel, F., Jondelius, U., Norlinder, E., Nygren, A., Oxelman, B., Schander, C., Sundberg, P., and Thollesson, M. (2008). Phylogenies without roots? A plea for the use of vouchers in molecular phylogenetic studies. *Molecular Phylogenetics and Evolution* **48**, 369–371.

Plotkin, A. S., and Janussen, D. (2007). New genus and species of Polymastiidae (Demospongiae: Hadromerida) from the Antarctic deep sea. *Journal of the Marine Biological Association of the United Kingdom* **87,** 1395–1401.

Plotkin, A. S., and Janussen, D. (2008). Polymastiidae and Suberitidae (Porifera: Demospongiae: Hadromerida) of the deep Weddell Sea, Antarctic. *Zootaxa* **1866,** 95–135.

Plotkin, A. S., Gerasimova, E., and Rapp, H. T. (2012). Phylogenetic reconstruction of Polymastiidae (Demospongiae: Hadromerida) based on morphology. *Hydrobiologia* **687,** 21–41.

Poiret, J. L. M. (1789). Voyage en Barbarie, ou Lettres Ecrites de l'Ancienne Numidie pendant les Années 1785 et 1786, avec un Essai sur l'Histoire naturelle de ce Pays. Deuxième Partie. 1. (Paris).

Poléjaeff, N. (1883). Report on the Calcarea dredged by H.M.S. 'Challenger', during the years 1873–1876. Report on the Scientific Results of the Voyage of H.M.S. 'Challenger', 1873–1876. *Zoology* **8,** 1–76.

Pöppe, J., Sutcliffe, P., Hooper, J. N. A., Wörheide, G., and Erpenbeck, D. (2010). COI barcoding reveals new clades and radiation patterns of Indo-Pacifisponges of the family Irciniidae (Demospongiae: Dictyoceratida). *PLoS One* **5,** e9950.

Pronzato, R., and Manconi, R. (2008). Mediterranean commercial sponges: Over 5000 years of natural history and cultural heritage. *Marine Ecology* **29,** 146–166.

Pronzato, R., Malva, R., and Manconi, R. (2004). The taxonomic status of *Ircinia fasciulata, Ircinia felix,* and *Ircinia varaibilis* (Dictyoceratida, Ircinidae). *Bollettino dei Musei e degli Istituti Biologici della Università di Genova* **68,** 553–563.

Pulitzer-Finali, G. (1993). A collection of marine sponges from east Africa. *Annali del Museo civico di Storia naturale Giacomo Doria* **89,** 247–350.

Raleigh, J., Redmond, N. E., Delahan, E., Torpey, S., van Soest, R. W. M., Kelly, M., and McCormack, G. P. (2007). Mitochondrial cytochrome oxidase 1 phylogeny supports alternative taxonomic scheme for the marine Haplosclerida. *Journal of the Marine Biological Association of the United Kingdom* **87,** 1577–1584.

Rapp, H. T. (2006). Calcareous sponges of the genera *Clathrina* and *Guancha* (Calcinea, Calcarea, Porifera) of Norway (north-east Atlantic) with the description of five new species. *Zoological Journal of the Linnean Society* **147,** 331–365.

Rapp, H. T., Janussen, D., and Tendal, O. S. (2011). Calcareous sponges from abyssal and bathyal depths in the Weddell Sea, Antarctica. *Deep-Sea Research II* **58,** 58–67.

Redmond, N. E., and McCormack, G. P. (2008). Large expansion segments in 18S rDNA support a new sponge clade (Class Demospongiae, Order Haplosclerida). *Molecular Phylogenetics and Evolution* **47,** 1090–1099.

Redmond, N. E., and McCormack, G. P. (2009). Ribosomal internal transcribed spacer regions are not suitable for intra- or inter-specific phylogeny reconstruction in haplosclerid sponges (Porifera: Demospongiae). *Journal of the Marine Biological Association of the United Kingdom* **89,** 1251–1256.

Redmond, N. E., van Soest, R. W. M., Kelly, M., Raleigh, J., Travers, S. A. A., and McCormack, G. P. (2007). Reassessment of the classification of the Order Haplosclerida (Class Demospongiae, Phylum Porifera) using 18S rRNA gene sequence data. *Molecular Phylogenetics and Evolution* **43,** 344–352.

Redmond, N. E., Raleigh, J., van Soest, R. W. M., Kelly, M., Travers, S. A. A., Bradshaw, B., Vartia, S., Stephens, K. M., and McCormack, G. P. (2011). Phylogenetic relationships of the marine Haplosclerida (Phylum Porifera) employing ribosomal (28S rRNA) and mitochondrial (cox1, nad1) gene sequence data. *PLoS One* **6,** e24344.

Reid, R. E. H. (1970). Tetraxons and demosponge phylogeny. In "The Biology of the Porifera" (W. G. Fry, ed.), pp. 63–90. Academic Press, London.

Reiswig, H. M. (2006). Classification and phylogeny of Hexactinellida (Porifera). *Canadian Journal of Zoology* **84,** 195–204.

Reiswig, H. M., and Champagne, P. (1995). The NE Atlantic glass sponges *Pheronema carpenteri* (Thompson) and *P. grayi* Kent (Porifera: Hexactinellida) are synonyms. *Zoological Journal of the Linnean Society* **115**, 373–384.

Reiswig, H. M., and Kelly, M. (2011). The Marine Fauna of New Zealand: Hexasterophoran Glass Sponges of New Zealand (Porifera: Hexactinellida: Hexasterophora): Orders Hexactinosida, Aulocalycoida and Lychiniscosida. *NIWA Biodiversity Memoir* **124**, 1–17.

Reiswig, H. M., and Mehl, D. (1991). Tissue organization of *Farrea occa* (Porifera, Hexactinellida). *Zoomorphology* **110**, 301–311.

Reiswig, H. M., Dohrmann, M., Pomponi, S., and Wörheide, G. (2008). Two new tretodictyids (Hexactinellida: Hexactinosida: Tretodictyidae) from the coasts of North America. *Zootaxa* **1721**, 53–64.

Reveillaud, J. (2011). Distribution and evolution of atlanto-mediterranean sponges from shallow-water and deep-sea coral ecosystems: A molecular, morphological and biochemical approach. Ph.D. Ghent University, Ghent, 261pp.

Reveillaud, J., Remerie, T., van Soest, R. W. M., Erpenbeck, D., Cárdenas, P., Derycke, S., Xavier, J. R., Rigaux, A., and Vanreusel, A. (2010). Species boundaries and phylogenetic relationships between Atlanto-Mediterranean shallow-water and deep-sea coral associated *Hexadella* species (Porifera, Ianthellidae). *Molecular Phylogenetics and Evolution* **56**, 104–114.

Reveillaud, J., van Soest, R. M. W., Derycke, S., Picton, B. E., Rigaux, A., and Vanreusel, A. (2011). Phylogenetic relationships among NE Atlantic *Plocamionida* Topsent (1927) (Porifera, Poecilosclerida): Under-estimated diversity in reef ecosystems. *PLoS One* **6**, 1–10.

Ridley, S. O., and Dendy, A. (1887). Report on the Monaxonida collected by H.M.S. 'Challenger' during the years 1873–1876. Report on the Scientific Results of the Voyage of H.M.S. 'Challenger', 1873–1876. *Zoology* **20**, 1–275.

Ríos, P., Cristobo, F. J., and Urgorri, V. (2004). Poecilosclerida (Porifera, Demospongiae) collected by the Spanish antarctic expedition BENTART-94. *Cahiers de Biologie Marine* **45**, 97–119.

Risso, A. (1826). Histoire naturelle des principales productions de l'Europe Méridionale et particulièrement de celles des environs de Nice et des Alpes Maritimes. Levrault, Paris.

Rosell, D., and Uriz, M.-J. (1997). Phylogenetic relationships within the excavating Hadromerida (Porifera), with a systematic revision. *Cladistics* **13**, 349–366.

Rossi, A. L., Russo, C. A. M., Solé-Cava, A. M., Rapp, H. T., and Klautau, M. (2011). Phylogenetic signal in the evolution of body colour and spicule skeleton in calcareous sponges. *Zoological Journal of the Linnean Society* **163**, 1024–1032.

Rua, C. P. J., Zilberberg, C., and Solé-Cava, A. M. (2011). New polymorphic mitochondrial markers for sponge phylogeography. *Journal of the Marine Biological Association of the United Kingdom* **91**, 1–8.

Rützler, K. (1996). The role of psammobiontic sponges in the reef community. In "8th International Coral Reef Symposium, June 24–29, 1996, Panama, Abstract." (H. A. Lessios, ed.), p. 171. ICRS.

Rützler, K., and Macintyre, I. G. (1978). Siliceous sponge spicules in coral reef sediments. *Marine Biology* **49**, 147–159.

Rützler, K., and Stone, S. M. (1986). Discovery and significance of Albany Hancock's microscope preparations of excavating sponges (Porifera: Hadromerida: Clionidae). *Proceedings of Biological Society of Washington* **99**, 658–675.

Rützler, K., Duran, S., and Piantoni, C. (2007a). Adaptation of reef and mangrove sponges to stress: Evidence for ecological speciation exemplified by *Chondrilla caribensis* new species (Demospongiae, Chondrosida). *Marine Ecology* **28**, 95–111.

Rützler, K., Piantoni, C., and Diaz, M. C. (2007b). *Lissodendoryx*: Rediscovered type and new tropical western Atlantic species (Porifera: Demospongiae: Poecilosclerida:

Coelosphaeridae). *Journal of the Marine Biological Association of the United Kingdom* **87**, 1491–1510.

Rützler, K., Soest, R. W. M. van, and Piantoni, C. (2009). Sponges (Porifera) of the Gulf of Mexico. In "Gulf of Mexico Origin, Waters, and Biota Volume 1, Biodiversity" (D. L. Felder and D. K. Camp, eds), pp. 285–313. A&M University Press, Texas.

Saleuddin, A. S. M. and Fenton, M. B. (eds) (2006). Biology of neglected groups: Porifera (sponges) *Canadian Journal of Zoology* **84**.

Samaai, T., and Gibbons, M. J. (2005). Demospongiae taxonomy and biodiversity of the Benguela region on the west coast of South Africa. *African Natural History* **1**, 1–96.

Sarà, A., Cerrano, C., and Sarà, M. (2002). Viviparous development in the Antarctic sponge *Stylocordyla borealis* Lovén, 1868. *Polar Biology* **25**, 425–431.

Sarkar, I. (2009). Biodiversity Informatics: The emergence of a field. *BMC Bioinformatics* **10**, S1.

Schander, C., and Thollesson, M. (1995). Phylogenetic taxonomy-some comments. *Zoologica Scripta* **24**, 263–268.

Schander, C., Rapp, H. T., Kongsrud, J. A., Bakken, T., Berge, J., Cochrane, S., Oug, E., Byrkjedal, I., Todt, C., Cedhagen, T., Fosshagen, A., Gebruk, A., Larsen, K., Levin, L., Obst, M., Pleijel, F., Stöhr, S., Warén, A., Mikkelsen, N. T., Hadler-Jacobsen, S., Keuning, R., Petersen, K. H., Thorseth, I. H., and Pedersen, R. B. (2010). The fauna of hydrothermal vents on the Mohn Ridge (North Atlantic). *Marine Biology Research* **6**, 155–171.

Schierwater, B., Eitel, M., Jakob, W., Osigus, H.-J., Hadrys, H., Dellaporta, S. L., Kolokotronis, S.-O., and DeSalle, R. (2009). Concatenated analysis sheds light on early Metazoan evolution and fuels a modern "Urmetazoon" hypothesis. *PLoS Biology* **7**, e20.

Schlacher, T. A., Schlacher-Hoenlinger, M. A., Williams, A., Althaus, F., Hooper, J. N. A., and Kloser, R. (2007). Richness and distribution of sponge megabenthos in continental margin canyons off southeastern Australia. *Marine Ecology Progress Series* **340**, 73–88.

Schlick-Steiner, B. C., Steiner, F. M., Seifert, B., Stauffer, C., Christian, E., and Crozier, R. H. (2009). Integrative Taxonomy: A multisource approach to exploring biodiversity. *Annual Review of Entomology* **55**, 421–438.

Schmidt, O. (1862). Die Spongien des adriatischen Meeres. W. Engelman, Leipzig.

Schmidt, O. (1868). Die Spongien der Küste von Algier mit Nachträgen zu den Spongien des Adriatischen Meeres (drittes Supplement). Verlag von Wilhelm Engelmann, Leipzig.

Schmidt, O. (1870). Grundzüge einer Spongien-fauna des Atlantischen Gebietes. Engelmann, Leipzig.

Schmitt, S., Hentschel, U., Zea, S., Dandekar, T., and Wolf, M. (2005). ITS-2 and 18S rRNA gene phylogeny of Aplysinidae (Verongida, Demospongiae). *Journal of Molecular Evolution* **60**, 327–336.

Schmitt, S., Tsai, P., Bell, J., Fromont, J., Ilan, M., Lindquist, N., Pérez, T., Rodrigo, A., Schupp, P., Vacelet, J., Webster, N., Hentschel, U., and Taylor, M. W. (2012). Assessing the complex sponge microbiota-core, variable and species-specific bacterial communities in marine sponges. *International Society for Microbial Ecology Journal* **6**, 564–576.

Schröder, H. C., Efremova, S. M., Itskovich, V. B., Belikov, S. I., Masuda, Y., Krasko, A., Müller, I. M., and Müller, W. E. G. (2003). Molecular phylogeny of the freshwater sponges in Lake Baikal. *Journal of Zoology, Systematics, and Evolution Research* **41**, 81–86.

Schulze, F. E. (1877). Untersuchungen über den Bau und die Entwicklung der Spongien. Die Familie der Chondrosidae. *Zeitschrift für wissenschaftliche Zoologie* **29**, 87–122.

Schulze, F. E. (1879). Untersuchungen über den Bau und die Entwicklung der Spongien. Siebente Mittheilung. Die Familie der Spongidae. *Zeitschrift für wissenschaftliche Zoologie* **32**, 593–660.

Schulze, F. E. (1880). Untersuchungen über den Bau und die Entwicklung der Spongien. Die Plakiniden. *Zeitschrift für wissenschaftliche Zoologie* **34**, 407–451.

Schulze, F. E. (1887). The Hexactinellida. In "Report on the Scientific Results of the Voyage of H.M.S. *Challenger* During the Years 1873–76" (T. H. Tizard, H. M. Moseley, J. Y. Buchanan and J. Murray, eds), pp. 437–451. Her Majesty's Government, London.

Sim, C. J., and Shin, E. J. (2006). A taxonomic study on marine sponges from Chujado Islands, Korea. *Korean Journal of Systematic Zoology* **22,** 153–168.

Sim, C. J., Kim, H. Y., and Byeon, H. S. (1990). A systematic study on the marine sponges in Korea. 8. Tetractinomorpha. *The Korean Journal of Systematic Zoology* **6,** 123–144.

Simpson, T. L. (1984). The Cell Biology of Sponges. Springer Verlag, New York.

Smith, L. C., and Hildeman, W. H. (1990). Cellular morphology of *Callyspongia diffusa*. In "New Perspectives in Sponge Biology" (K. Rützler, ed.), pp. 135–143. Smithsonian Institution Press, Washington, DC.

Smith, L. C., and Hildemann, W. H. (1986). Allogeneic cell interactions during graft rejection in *Callyspongia diffusa* (Porifera, Demospongia); a study with monoclonal antibodies. *Proceedings of the Royal Society of London* **266,** 465–477.

Solé-Cava, A. M., and Boury-Esnault, N. (1999). Patterns of intra and interspecific divergence in marine sponges. *Memoirs of the Queensland Museum* **44,** 591–602.

Solé-Cava, A. M., and Wörheide, G. (2007). The perils and merits (or the Good, the Badand the Ugly) of DNA barcoding of sponges—A controversial discussion. In "Porifera Research: Biodiversity, Innovation and Sustainability" (M. R. Custódio, G. Lôbo-Hajdu, E. Hajdu and G. Muricy, eds), pp. 603–612. Museu Nacional, Rio de Janeiro.

Solé-Cava, A. M., Klautau, M., Boury-Esnault, N., Borojevic, R., and Thorpe, J. P. (1991). Genetic evidence for cryptic speciation in allopatric populations of two cosmopolitan species of the calcareous sponge genus *Clathrina*. *Marine Biology* **111,** 381–386.

Solé-Cava, A. M., Boury-Esnault, N., Vacelet, J., and Thorpe, J. P. (1992). Biochemical genetic divergence and systematics in sponges of the genera *Corticium* and *Oscarella* (Demospongiae: Homoscleromorpha) in the Mediterranean Sea. *Marine Biology* **113,** 299–304.

Sperling, E. A., Pisani, D., and Peterson, K. J. (2007). Poriferan paraphyly and its implications for Precambrian paleobiology. In "The Rise and Fall of the Ediacara Biota" (P. Vickers-Rich and P. Komarower, eds), pp. 355–368. Geological Society, Special Publications, London.

Sperling, E. A., Peterson, K. J., and Pisani, D. (2009). Phylogenetic-signal dissection of nuclear housekeeping genes supports the paraphyly of sponges and the monophyly of Eumetazoa. *Molecular Biology and Evolution* **26,** 2261–2274.

Sperling, E. A., Robinson, J. M., Pisani, D., and Peterson, K. J. (2010). Where's the glass? Biomarkers, molecular clocks, and microRNAs suggest a 200-Myr missing Precambrian fossil record of siliceous sponge spicules. *Geobiology* **8,** 24–36.

Stone, S. M. (1986). Index to Important Sponge Collections. Natural History Museum, London.

Szitenberg, A., Rot, C., Ilan, M., and Huchon, D. (2010). Diversity of sponge mitochondrial introns revealed by cox 1 sequences of Tetillidae. *BMC Evolutionary Biology* **10,** 288.

Tabachnik, K. R., and Collins, A. G. (2008). Glass sponges (Porifera, Hexactinellida) of the northern Mid-Atlantic Ridge. *Marine Biology Research* **4,** 25–47.

Tabachnik, K. R., Janussen, D., and Menschenina, L. L. (2008). New Australian Hexactinellida (Porifera) with a revision of *Euplectella aspergillum*. *Zootaxa* **1866,** 7–68.

Taylor, M. N., Radax, R., Steger, D., and Wagner, M. (2007). Sponge-associated microorganisms: evolution, ecology and biotechnology potential. *Microbiology and Molecular Biology Reviews* **71,** 295–347.

Taylor, M. W., Hill, R. T., and Hentschel, U. (2011). Meeting Report: 1st International Symposium on Sponge Microbiology. *Marine Biotechnology* **13,** 1057–1061.

Thacker, R. W., and Freeman, C. J. (2012). Sponge-microbe symbioses: recent advances and new directions. *Advances in Marine Biology* 62.

Thollesson, M. (1999). Phylogenetic analysis of Euthyneura (Gastropoda) by means of the 16S rRNA gene: Use of a 'fast' gene for 'higher-level' phylogenies. *Proceedings of the Royal Society of London, Series B: Biological Sciences* **266**, 75–83.

Thompson, J. E., Barrow, K. D., and Faulkner, D. J. (1983). Localization of two brominated metabolites, aerothionin and homoaerothionin, in spherulous cells of the marine sponge *Aplysina fistularis* (= *Verongia thiona*). *Acta Zoologica (Stockholm)* **64**, 199–210.

Thomson, C. W. (1868). On the "vitreous" sponges. *Annals and Magazine of Natural History* **1**, 114–132.

Thorpe, J. P., and Solé-Cava, A. M. (1994). The use of allozyme electrophoresis in invertebrate systematics. *Zoologica Scripta* **23**, 3–18.

Topsent, E. (1900). Etude monographique des Spongiaires de France. III. Monaxonida (Hadromerina). *Archives de Zoologie Expérimentale et Générale* **8**, 1–131.

Topsent, E. (1934). Etudes d'éponges littorales du Golfe de Gabès. *Bulletin de la Station d'Aquiculture et de Pêche de Castiglione* **1932**, 71–102.

Tuzet, O. (1973). Eponges Calcaires. *Calcarea* Bowerbank (*Calcarosa* Haeckel, *Calcispongia* Nardo). In "Spongiaires" (P. P. Grassé, ed.), pp. 27–132. Masson & Co., Paris.

Uriz, M.-J. (ed.), (2003). Biology of silica deposition. *Microscopy Research and Technique* **62**, 277–278.

Uriz, M.-J. (2006). Mineral skeletogenesis in sponges. *Canadian Journal of Zoology/Revue Canadienne de Zoologie* **84**, 322–356.

Uriz, M. J., and Carballo, J. L. (2001). Phylogenetic relationships of sponges with placochelae or related spicules (Poecilosclerida, Guitarridae) with a systematic revision. *Zoological Journal of the Linnean Society* **132**, 411–428.

Uriz, M. J., and Maldonado, M. (1995). A reconsideration of the relationship between polyaxonid and monaxonid spicules in Demospongiae: New data from the genera *Crambe* and *Discorhabdella* (Porifera). *Biological Journal of the Linnean Society* **55**, 1–15.

Uriz, M. J., Maldonado, M., Becerro, M., and Turon, X. (2010). Preface. In "Ancient Animals, New Challenges." VIII World Sponge conference Girona, 20–24 September 2010 (M.-J. Uriz, M. Maldonado, M. Becerro and X. Turon, eds)., p. 5.

Uriz, M. J., Gili, J.-M., Orejas, A., and Perez-Porro, A.-R. (2011). Do bipolar distributions exist in marine sponges? *Stylocordyla chupachups* sp. nov. (Porifera: Hadromerida) from the Weddell Sea. (Antarctic), previously reported as *S. borealis* (Lovén, 1868). *Polar Biology* **34**, 243–255.

Usher, K. M., and Ereskovsky, A. V. (2005). Larval development, ultrastructure and metamorphosis in *Chondrilla australiensis* Carter, 1873 (Demospongiae, Chondrosida, Chondrillidae). *Invertebrate Reproduction and Development* **47**, 51–62.

Usher, K. M., Sutton, D. C., Toze, S., Kuo, J., and Fromont, J. (2004). Biogeography and phylogeny of *Chondrilla* species (Demospongiae) in Australia. *Marine Ecology Progress Series* **270**, 117–127.

Vacelet, J. (1967). Les cellules à inclusions de l'éponge cornée *Verongia cavernicola* Vacelet. *Journal de Microscopie (Paris)* **6**, 237–240.

Vacelet, J. (1969). Eponges de la Roche du Large et de l'étage bathyal de Méditerranée (récoltes de la Soucoupe plongeante Cousteau et dragages). *Mémoires du Muséum national d'Histoire naturelle* **59**, 145–219.

Vacelet, J. (1977). Une nouvelle relique du Secondaire.: un représentant actuel des éponges fossiles Sphinctozoaires. *Compte-Rendu hebdomadaire des Séances de l'Académie des Sciences* **285**, 509–511.

Vacelet, J. (1979). Quelques stades de la reproduction sexuée d'une éponge Sphinctozoaire actuelle. In "Biologie des Spongiaires" (C. Lévi and N. Boury-Esnault, eds), Colloques internationaux du Centre national de la Recherche scientifique, vol. 291, pp. 1–533.

Vacelet, J. (1980a). Les affinités du peuplement de Spongiaires de la Méditerranée. Journées d'Etudes Systématique Biogéographie Méditerranéenne CIESM, Cagliari, pp. 29–30.

Vacelet, J. (1980b). Squelette facultatif et corps de régénération dans le genre *Merlia*, Eponges apparentées aux Chaetétidés fossiles. *Comptes-Rendus de l'Académie des Sciences de Paris* **290**, 227–230.

Vacelet, J. (1985). Coralline sponges and the evolution of the Porifera. In "The Origins and Relationships of Lower Invertebrates" (S. Conway Morris, J. D. George, R. Gibson and H. M. Platt, eds), pp. 1–13. Clarendon Press, Oxford.

Vacelet, J. (1999a). Outlook to the future of sponges. *Memoirs of the Queensland Museum* **44**, 27–32.

Vacelet, J. (1999b). Planktonic armoured propagules of the excavating sponge *Alectona* (Porifera: Demospongiae) are larvae: Evidence from *Alectona wallichli* and *A. mesatlantica* sp. nov. *Memoirs of the Queensland Museum* **44**, 625–642.

Vacelet, J. (ed.) (2000). Special volume dedicated to Professor Lévi. *Zoosystema* **22**.

Vacelet, J. (2002). Recent "Sphinctozoa' Order Verticillitida, Family Verticillidae Steinmann, 1882. In "Systema Porifera: A Guide to the Classification of Sponges" (J. N. A. Hooper and R. W. M. van Soest, eds), pp. 1097–1098. Kluwer Academic/Plenum Publishers, New York.

Vacelet, J. (2007). Diversity and evolution of deep-sea carnivorous sponges. In "Porifera Research: Biodiversity, Innovation and Sustainability" (M. R. Custódio, G. Lôbo-Hajdu, E. Hajdu and G. Muricy, eds), pp. 107–115. Museu Nacional, Rio de Janeiro.

Vacelet, J., and Boury-Esnault, N. (1995). Carnivorous sponges. *Nature* **373**, 333–335.

Vacelet, J., and Donadey, C. (1987). A new species of *Halisarca* (Porifera, Demospongiae) from the Caribbean, with remarks on the cytology and affinities of the genus. In "European Contributions to the Taxonomy of Sponges" (W. C. Jones, ed.), pp. 5–12. Litho Press Co., Midleton.

Vacelet, J., and Pérez, T. (1998). Two new genera and species of sponges (Porifera, Demospongiae) without skeleton from a Mediterranean cave. *Zoosystema* **20**, 5–22.

Vacelet, J., and Pérez, T. (2008). *Phorbas topsenti* and *Phorbas taillezi* (Demospongiae, Poecilosclerida), new names for the Mediterranean "*Phorbas paupertas*" and "*Phorbas coriaceus*". *Zootaxa* **1873**, 26–38.

Vacelet, J., and Uriz, M.-J. (1991). Deficient spiculation in a new species of *Merlia* (Merliida, Demospongiae) from the Balearic Islands. In "Fossil and Recent Sponges" (J. Reitner and H. Keupp, eds), pp. 170–178. Springer-Verlag, Berlin.

Vacelet, J., Boury-Esnault, N., De Vos, L., and Donadey, C. (1989). Comparative study of the choanosome of Porifera: II. The Keratose Sponges. *Journal of Morphology* **201**, 119–129.

Vacelet, J., Vacelet, E., Gaino, E., and Gallissian, M.-F. (1994). Bacterial attack of spongin skeleton during the 1986-1990 Mediterranean sponge disease. In "Sponges in Time and Space. Proceedings of the Fourth International Porifera Congress" (R. W. M. van Soest, T. M. G. van Kempen and J. Braekman, eds), pp. 355–362. A.A. Balkema, Rotterdam.

Vacelet, J., Borchiellini, C., Pérez, T., Bultel-Poncé, V., Brouard, J.-P., and Guyot, M. (2000). Morphological, chemical, and biochemical characterization of a new species of sponges without skeleton (Porifera, Demospongiae) from the Mediterranean Sea. *Zoosystema* **22**, 313–326.

Vacelet, J., Al Sofyani, A., Al Lihaibi, S., and Kornprobst, J.-M. (2001). A new haplosclerid sponge species from the Red Sea. *Journal of the Marine Biological Association of the United Kingdom* **81**, 943–948.

Vacelet, J., Willenz, P., and Hartman, W. P. D. (2010). Part E, Revised, Volume 4, Chapter 1: Living hypercalcified sponges. In "Treatise of Invertebrate Paleontology Online," pp. 1–16. The University of Kansas, Paleontological Institute, Lawrence KS.

Valderrama, D., Rossi, A. L., Solé-Cava, A. M., Rapp, H. T., and Klautau, M. (2009). Revalidation of *Leucetta floridana* (Haeckel, 1872) (Porifera, Calcarea): A widespread species in the tropical western Atlantic. *Zoological Journal of the Linnean Society* **157**, 1–16.

van Soest, R. W. M. (1984a). Marine sponges from Curaçao and other Caribbean localities. Part III. Poecilosclerida. In "Studies on the Fauna of Curaçao and other Caribbean Islands" (P. Wagenaar Hummelinck and L. J. van der Steen, eds), pp. 1–167. Foundation for Scientific Research in Surinam and the Netherlands Antilles, Amsterdam.

van Soest, R. W. M. (1984b). Deficient *Merlia normani* Kirkpatrick, 1908, from the Curaçao reefs, with a discussion on the phylogenetic interpretation of sclerosponges. *Bijdragen tot de Dierkunde* **54**, 211–219.

van Soest, R. W. M. (1987). Phylogenetic exercises with monophyletic groups of sponges. In "Taxonomy of Porifera from the N.E. Atlantic and Mediterranean Sea" (J. Vacelet and N. Boury-Esnault, eds), pp. 227–242. Springer-Verlag, Berlin.

van Soest, R. W. M. (1990). Toward a phylogenetic classification of sponges. In "New Perspectives in Sponge Biology" (K. Rützler, ed.), pp. 344–350. Smithsonian Institution Press, Washington, DC.

van Soest, R. W. M. (1991). Demosponge higher taxa classification re-examined. In "Fossil and Recent Sponges" (J. Reitner and H. Keupp, eds), pp. 54–71. Springer-Verlag, Berlin.

van Soest, R. W. M. (2001). Porifera. In "European Register of Marine Species: A Checklist of the Marine Species in Europe and a Bibliography of Guides to Their Iidentification" (M. J. Costello *et al.*, eds.), Collection Patrimoines Naturels. **50**, pp. 85–103. SPN/IEGB/MNHN, Paris.

van Soest, R. W. M. (2002). Family Hymedesmiidae Topsent, 1928. In "Systema Porifera: A Guide to the Classification of Sponges" (J. N. A. Hooper and R. W. M. van Soest, eds), pp. 575–593. Kluwer Academic/Plenum Publishers, New York.

van Soest, R. W. M. (ed.), (2007). Sponge biodiversity. *Journal of the Marine Biological Association of the United Kingdom* **87**, (6), 1345–1348.

van Soest, R. W. M. (2009). New sciophilous sponges from the Caribbean (Porifera: Demospongiae). *Zootaxa* **2107**, 1–40.

van Soest, R. W. M., and Braekman, J.-C. (1999). Chemosystematics of Porifera: A review. *Memoirs of the Queensland Museum* **44**, 569–589.

van Soest, R. M. W., Stone, S. M., Boury-Esnault, N., and Rützler, K. (1983). Catalogue of extant and lost specimens of Duchassaing and Michelotti collection of West Indian Sponges. *Bulletin Zoologishes Museum Amsterdam* **9**, 189–205.

van Soest, R. W. M., Picton, B. E., and Morrow, C. C. (2000). Sponges of the North East Atlantic. World Biodiversity Database, Biodiversity Center of ETI, CD-ROM Series Springer-Verlag, Heidelberg.

van Soest, R. M. W., Hooper, J. N. A., Beglinger, E., and Erpenbeck, D. (2006). Affinities of the family Sollasellidae (Porifera, Demospongiae). I. Morphological evidence. *Contributions to Zoology* **75**, 133–144.

van Soest, R. W. M., van Duyl, F. C., Maier, C., Lavaleye, M. S. S., Beglinger, E. J., and Tabachnick, K. R. (2007a). Mass occurrence of *Rossella nodastrella* Topsent on bathyal coral reefs of Rockall Bank, W of Ireland (Lyssacinosida, Hexactinellida). In "Porifera Research: Biodiversity, Innovation and Sustainability" (M. R. Custódio, G. Lôbo-Hajdu, E. Hajdu and G. Muricy, eds), pp. 645–652. Museu Nacional, Rio de Janeiro.

van Soest, R. W. M., de Kluijver, M. J., van Bragt, P. H., Faasse, M., Nijland, R., Beglinger, E. J., de Weerdt, W. H., and de Voogd, N. J. (2007b). Sponge invaders in Dutch coastal waters. *Journal of the Marine Biological Association of the United Kingdom* **87**, 1733–1748.

van Soest, R. W. M., Kaiser, K., and van Syoc, R. (2011a). Sponges from Clipperton Island, East Pacific. *Zootaxa* **2839**, 1–46.

van Soest, R. W. M., Boury-Esnault, N., Hooper, J. N. A., Rützler, K., de Voogd, N. J., Alvarezde de Glasby, B., Hajdu, E., Pisera, A. B., Manconi, R., Schoenberg, C., Janussen, D., Tabachnick, K. R., Klautau, M., Picton, B., and Kelly, M. (2011b). World Porifera database. http://www.marinespecies.org/porifera.

Vergne, C., Boury-Esnault, N., Pérez, T., Martin, M.-T., Adeline, M.-T., Tran Huu Dau, E., and Al-Mourabit, A. (2006). Verpacamides A-D, a sequence of $C_{11}N_5$ diketopiperazines relating cyclo(Pro-Pro) to cyclo(Pro-Arg), from the marine sponge *Axinella vaceleti*: Possible biogenetic precursors of pyrrole-2-aminoimidazole alkaloids. *Organic Letters* **8**, 2421–2424.

Vickerman, K., Brugerolle, G., and Mignot, J.-P. (1991). Mastigophora. In "Microscopic Anatomy of Invertebrates, vol. I, Protozoa" (F. W. Harrison and J. O. Corliss, eds), pp. 13–159. Wiley-Liss, New York.

Vilanova, E., Zilberberg, C., Kochem, M., Custódio, M. R., and Mourão, P. A. S. (2007). A novel biochemical method to distinguish cryptic species of *Chondrilla* (Chondrosida, Demospongiae) based on its sulfated polysaccharides. In "Porifera Research: Biodiversity, Innovation and Sustainability" (M. R. Custódio, G. Lôbo-Hajdu, E. Hajdu and G. Muricy, eds), pp. 653–659. Museu Nacional, Rio de Janeiro.

Vishnyakov, A. E., and Ereskovsky, A. V. (2009). Bacterial symbionts as an additional cytological marker for identification of sponges without a skeleton. *Marine Biology* **156**, 1625–1632.

Voigt, O., Erpenbeck, D., and Wörheide, G. (2008). Molecular evolution of rDNA in early diverging Metazoa: First comparative analysis and phylogenetic application of complete SSU rRNA secondary structures in Porifera. *BMC Evolutionary Biology* **2008**, 8–69.

Volkmer-Ribeiro, C. (2007). South American continental sponges: State of the art of the research. In "Porifera Research: Biodiversity, Innovation and Sustainability" (M. R. Custódio, G. Lôbo-Hajdu, E. Hajdu and G. Muricy, eds), pp. 117–121. Museu Nacional, Rio de Janeiro.

Vosmaer, R. (1935). The Sponges of the Bay of Naples: Porifera Incalcaria with Analyses of Genera and Studies in the Variations of Species. Martinus Nijhoff, The Hague.

Voultsiadou, E., and Gkelis, S. (2005). Greek and the phylum Porifera: A living language for living organisms. *Journal of the Zoological Society of London* **267**, 143–157.

Watanabe, Y. (1978). The development of two species of *Tetilla* (Demosponge). *Natural Science Report of the Ochanomizu University* **29**, 71–106.

Weckwerth, W., and Morgenthal, K. (2005). Metabolomics: From pattern recognition to biological interpretation. *Drug Discovery Today* **10**, 1551–1558.

Wiedenmayer, F. (1977). Shallow-water sponges of the western Bahamas. *Experientia Supplementum* **28**, 1–287.

Wilson, E. O. (2003). The encyclopedia of life. *Trends in Ecology & Evolution* **18**, 77–80.

Wolfender, J. L., Glauser, G., Boccard, J., and Rudaz, S. (2009). MS-based plant metabolomic approaches for biomarker discovery. *Natural Product Communications* **4**, 1417–1430.

Wörheide, G. (2008). A hypercalcified sponge with soft relatives: *Vaceletia* is a keratose demosponge. *Molecular Phylogenetics and Evolution* **47**, 433–438.

Wörheide, G., and Hooper, J. N. A. (1999). Calcarea from the Great Barrier Reef. I: Cryptic Calcinea from Heron Island and Wistari Reef (Capricorn-Bunker group). *Memoirs of the Queensland Museum* **43**, 859–891.

Wörheide, G., Hooper, J. N. A., and Degnan, B. M. (2002). Phylogeography of western Pacific *Leucetta* "*chagoensis*" (Porifera: Calcarea) from ribosomal DNA sequences: Implications for population history and conservation of the Great Barrier Reef World Heritage Area (Australia). *Molecular Ecology* **11**, 1753–1768.

Wörheide, G., Nichols, N., and Goldberg, J. (2004). Intragenomic variation of the rDNA internal transcribed spacersin sponges (Phylum Porifera): Implications for phylogenetic studies. *Molecular Phylogenetics and Evolution* **33**, 816–830.

Wörheide, G., Erpenbeck, D., and Menke, C. (2007). The Sponge Barcoding Project: Aiding in the identification and description of poriferan taxa. In "Porifera Research: Biodiversity, Innovation and Sustainability" (M. R. Custódio, G. Lôbo-Hajdu, E. Hajdu and G. Muricy, eds), Série Livros. **28**, pp. 123–128. Museu Nacional, Rio de Janeiro.

Wörheide, G., Epp, L., and Macis, L. (2008). Deep genetic divergences among Indo-Pacific populations of the coral reef sponge *Leucetta chagosensis* (Leucettidae): Founder effects, vicariance, or both? *BMC Evolutionary Biology* **8**, 24.

Wulff, J. L. (2006). Sponge systematics by starfish: Predators distinguish cryptic sympatric species of Caribbean fire sponges, *Tedania ignis* and *Tedania klausi* n. sp. (Demospongiae, Poecilosclerida). *The Biological Bulletin* **211**, 83–94.

Xavier, J., and van Soest, R. W. M. (2007). Demosponge fauna of Ormonde and Gettysburg Seamounts (Gorringe Bank, north-east Atlantic): Diversity and zoogeographical affinities. *Journal of the Marine Biological Association of the United Kingdom* **87**, 1643–1653.

Xavier, J. R., Rachello-Dolmen, P. G., Parra-Velandia, F., Schönberg, C. H. L., Breeuwer, J. A. J., and van Soest, R. W. M. (2010a). Molecular evidence of cryptic speciation in the "cosmopolitan" excavating sponge *Cliona celata* (Porifera, Clionaidae). *Molecular Phylogenetics and Evolution* **56**, 13–20.

Xavier, J. R., Tojeira, I., and van Soest, R. W. M. (2010b). On a Hexactinellid sponge ground at the Great Meteor seamount (Northeast Atlantic). In "Ancient Animals, New Challenges." VIII World Sponge conference Girona, 20–24 September 2010 (M.-J. Uriz, M. Maldonado, M. Becerro and X. Turon, eds)., p. 370.

Zea, S. (2001). Patterns of sponge (Porifera: Demospongiae) distribution in remote, oceanic reef complexes of the Southwestern Caribbean. *Revista de la Academia de Colombia de Ciencias* **25**, 579–592.

Zea, S., Henkel, T. P., and Pawlik, J. R. (2009). The Sponge Guide: A Picture Guide to Caribbean Sponges. www.spongeguide.org.

Zilberberg, C., Solé-Cava, A. M., and Klautau, M. (2006). The extent of asexual reproduction in sponges of the genus *Chondrilla* (Demospongiae: Chondrosida) from the Caribbean and the Brazilian coasts. *Journal of Experimental Marine Biology and Ecology* **336**, 211–220.

CHAPTER THREE

THE ROLE OF SPONGES IN THE MESOAMERICAN BARRIER-REEF ECOSYSTEM, BELIZE

Klaus Rützler[1]

Contents

Abstract

Over the past four decades, sponge research has advanced by leaps and bounds through endeavours such as the Caribbean Coral Reef Ecosystems (CCRE) programme at the U.S. National Museum of Natural History in Washington, D.C.

Department of Invertebrate Zoology, National Museum of Natural History, Smithsonian Institution, Washington D.C., USA
[1]Corresponding author: Email: ruetzler@si.edu

Advances in Marine Biology, Volume 61

ISSN 0065-2881, DOI: 10.1016/B978-0-12-387787-1.00002-7

Since its founding in the early 1970s, the programme has been dedicated to a detailed multidisciplinary study of a section of the Mesoamerican Barrier Reef, the Atlantic's largest reef complex, and has generated data far beyond the capability of lone investigators and brief expeditions. This reef complex extends 250 km southward from Yucatan, Mexico, into the Gulf of Honduras, most of it lying 20–40 km off the coast of Belize. A relatively unspoiled ecosystem, it features a great variety of habitats in close proximity, ranging from mangrove islands, seagrass meadows, and patch reefs in its lagoon to the barrier reef along the margin of the continental shelf. Among its varied macrobenthos, sponges stand out for their ubiquity, range of colours, rich species and biomass, and ecological importance; they populate rocky substrates, some sandy bottoms, and the subtidal stilt roots and peat banks of mangroves.

Working from a field station established in 1972 on Carrie Bow Cay, a sand islet atop the reef off southern Belize, experts in numerous disciplines from both the Museum and academic institutions throughout the world have explored the area's biodiversity in the broadest sense and community development over time. At last count, 113 researchers (88 working on site) have focused on the biological and geological role of Porifera in Carrie Bow's reef communities, with the results reported in 125 scientific papers to date. The majority of these sponge studies have centred on systematics and faunistics, including quantitative distribution among the various habitats. Taxonomic approaches have ranged from basic morphology to fine structure, DNA barcoding, and ecological manipulations and culminated in a mini-workshop involving several experts on Caribbean Porifera. Ecological work has covered a broad spectrum as well: bioerosion, silica and nutrient cycling, symbiosis, mutualism, space competition, predation, disease, and the effects on sponge individuals and populations of environmental factors such as light, temperature, salinity, desiccation, substrate, and sedimentation. Many projects were enhanced by scientific illustration, laboratory studies of larvae settlement preferences and development, and investigations of microbial and invertebrate sponge associates, notably symbiotic cyanobacteria, parazoanthid epizoans, and crustacean and ophiuroid endobionts. Of the striking discoveries, the work on alpheid shrimps colonizing sponges off Carrie Bow Cay has yielded the first report of eusociality in marine organisms.

Key Words: systematics; ecology; reproduction; species interaction; disease; bioerosion; silicon cycle; coral reef; mangrove; seagrass meadow

1. Introduction

Belize is a small country in Central America with an area of less than 23,000 km^2 and a population of just over 300,000. Formerly British Honduras, it has been an independent state since 1981. In the north, west, and south, it is bordered by Mexico, Guatemala, and Honduras, respectively, and in the east by the Caribbean Sea. It shares with its neighbours the Mesoamerican Barrier Reef, the largest such structure in the Western Hemisphere, which parallels Belize's coastline for some 220 km, nearly the entire length of the country. The shallow lagoon between the mainland and the reef ranges in width from 14 to 40 km and serves as a protected shipping channel. A bit further out, 40–70 km east from the mainland, three large (30–50 km long), oval atolls—Turneffe Islands, Lighthouse Reef, Glovers Reef—rise to the sea surface, adding considerably to the region's coral reef diversity.

Despite Belize's many natural treasures—pristine reefs, mangroves, rivers, lagoons, and rainforest—most of its habitats, biology, and geological history remained by and large unexplored until the 1970s. In the absence of the necessary infrastructure, tourism, too, was slow to develop, attracting mainly visitors on boats or bird watchers and scuba divers who did not mind roughing it a bit. Its remarkable marine environment soon gained the attention of eight marine scientists, including myself, at the Smithsonian Institution's National Museum of Natural History in Washington, D.C. Though diverse in our interests—from marine botany and invertebrate zoology to palaeobiology and sedimentology—we all saw a great need for information on the development and biology of coastal marine ecosystems, particularly coral reefs, and could find no better place for studies of this nature than the reef systems off Belize, nearly untouched by pollution and only slightly affected by local fisheries. Their habitats also fitted perfectly my own central concern: the biology of reef sponges (Porifera), a prominent member of the reef system.

After surveying many sites and weighing financial and logistical considerations, the group decided to work mainly on the central barrier reef east of the town of Dangriga, including the Twin Cays and Blueground Range mangroves. There the reef lies about 25 km from the mainland, separated by a 15-m deep lagoon, and features a great variety of reef types alongside mangroves and seagrass meadows, with the nutrient-rich lagoon and oligotrophic ocean close by. In 1972 we established a field station on Carrie Bow Cay, a tiny island on the barrier reef made up of coral rubble (16°48.1′ N, 88°04.9′ W). At the time, we could not imagine how long-lasting and scientifically significant this "temporary" laboratory would become (Rützler and Macintyre, 1982a; Rützler, 2009). In later years, starting about 1995, we expanded our research area to the Pelican Cays to the south. Part of the region our team helped to define (Figs. 3.1 and 3.2) has been designated the South Water Cay Marine Protected Area by the Belize government.

Since its founding, the CCRE programme has focused on elucidating species and the composition of communities and their development over time; ecological factors and their ecophysiological impacts on individuals,

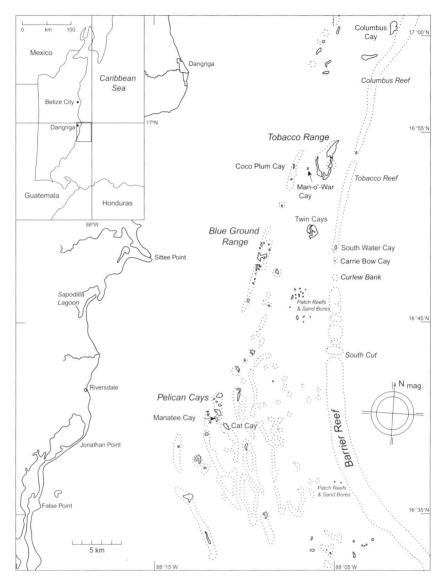

Figure 3.1 Map of barrier-reef platform near Carrie Bow Cay, the principal research area within easy access from the Smithsonian Carrie Bow Marine Field Station. (Drawing by Molly Kelly Ryan).

species, and populations; ecosystem processes, boundaries, and interactions (reef–mangrove–seagrass and land–sea linkages); and the consequences of natural or anthropogenic environmental change in these ecosystems.

Figure 3.2 Aerial photographs of key CCRE research areas on the Belize barrier reef. (A) Carrie Bow Cay with the buildings of the Smithsonian Carrie Bow Marine Field Station in the foreground; South Water Cay and the continuous barrier reef in the back. (B) Twin Cays looking southeast towards Carrie Bow and the barrier reef. (C) Pelican Cays, principal isles studied for their sponge diversity (Fishermans Cay, left; Manatee Cay, foregrounds; Cat Cay, right background). (Photographs: A, B, Ilka Feller from LightHawk; C, Tony Rath/tonyrath.com) (For colour version of this figure, the reader is referred to the online version of this chapter.)

As was apparent from the outset, sponges are the principal components of most habitats around Carrie Bow Cay, particularly the forereef (below a depth of 5 m, where it is less exposed to waves), patch reefs, mangroves, and seagrass meadows (Fig. 3.3). Indeed, sponges are vastly richer in species, abundance, and biomass than even the reef-building corals. Through microbial symbionts, they also facilitate or contribute directly to significant processes, such as primary production and nitrification, water filtration and

Figure 3.3 Representative habitats in the Carrie Bow research area. (A) Diver photograph-ing sponges on forereef, 8 m deep. (B) Shaded overhang on forereef slope showing high diversity and biomass of sponges (species of *Mycale*, *Agelas*, *Xestospongia*, *Monanchora*, *Spiras-trella*) and gorgonians, 15 m. (C) Tidal channel at Twin Cays with red-mangrove prop roots covered by sponges, 0.5 m. (D) Turtle-grass flat outside south Twin Cays, showing minia-ture patch reefs composed of sponges (*Tedania*, *Clathria*), calcified algae (*Halimeda*), and corals (*Porites*). (E) Slope of Sand Bore patch reef, with large specimen of *Neofibularia nolitangere* in the foreground, surrounded by species of *Callyspongia*, *Amphimedon*, *Aplysina*, and *Iotrochota*, 5 m. (F) Diverse sponge cluster on mangrove prop root at Manatee Cay in the Pelican group (*Spongia tubulifera*, dark grey, covered by species of *Plakortis*, *Scopalina*, *Amphimedon*, and *Haliclona*, and ascidians and sabellid polychaetes), 1 m. (Photo credits: (A–C) by Chip Clark; others by the author.) (For interpretation of the references to colour in this figure legend, the reader is referred to the online version of this chapter.)

bacterial retention, and carbonate framework cementation and bioerosion. Moreover, the siliceous skeleton present in most species aids in the cycling of silica in carbonate environments. In addition, being sessile and unprotected by solid structures, many forms produce and release natural products of ecological importance. These considerable contributions, not to mention the dearth of information on them, sparked our efforts in many venues—the literature, field courses, workshops, international symposia— to encourage colleagues and students to collaborate in our endeavours or to follow a similar path (Rützler and Feller, 1988, 1996; Feller and Sitnik, 1996; Macintyre and Aronson, 1997; Diaz and Rützler, 2001; Rützler, 2004, 2009).

2. METHODS

2.1. Data gathering

The following review covers primarily the published contributions of CCRE team members, supplemented here and there by information on other research for background or explanatory purposes. In a few appropriate instances, unpublished studies or works in progress are mentioned as well. The validity of species names was confirmed by consulting the World Porifera Database (Van Soest *et al.*, 2011).

2.2. Observational and monitoring techniques

Most work at the Carrie Bow Marine Field Station takes place under water, on the beach, or in simple laboratories supplied by solar-voltaic power and equipped with an open-circuit seawater aquarium and good-quality micro-scopes. These facilities make it possible to determine the taxonomy of many organisms before fixation, set up experiments for developmental biological studies, and examine live material by stereo and compound interference and fluorescence microscopy. An additional aid to research, the station's envir-onmental monitoring program provides semiannual records of benthos composition and production in selected reef, mangrove, and seagrass sites (following the Caribbean Coastal Marine Productivity—CARICOMP—protocol) and automatic and continuous readings of oceanographic condi-tions (Koltes *et al.*, 1998; Opishinski *et al.*, 2001). Such monitoring is particularly important to assess the impact of the region's frequent hurri-canes and associated periods of calm, water heating, UV penetration, and unusual tidal regime. Work at Carrie Bow has also generated many inno-vative techniques for qualitative and quantitative sampling and for *in situ* recording of microclimate, such as current flow, temperature, salinity, and hydrogen sulphide (e.g. Rützler, 1978, 1996; Rützler *et al.*, 1980, 2004, 2007a).

2.3. Scientific illustration

Scientific illustrators have been of invaluable assistance in documenting sponge species and communities off Belize. Photography and drawings have revealed in print and other media live sponges in their habitat, along with their myriad colour variations and growth forms. To show entire communities and prevailing processes, organisms are studied *in situ* and then taken to the laboratory for detailed analysis of their skeletal elements (spicules), cells, microbial symbionts, and other components under light and electron microscopes. Once sponges, substrate, epibionts, and endobionts are separated, photographed, and their roles in the community understood, they can be reassembled by the illustrator and rendered in pen and ink or colour paint for publication or teaching purposes. This technique was used to demonstrate details of *Svenzea* and *Scopalina* specimens and their embryos and larvae (Rützler *et al.*, 2003) and is being applied in several ongoing projects.

3. SYSTEMATICS

3.1. New species and taxonomic revisions

Before our first surveys off Belize in 1971 (Rützler and Macintyre, 1982b), information on the region's sponges was sparse, except for reports on eight species found off Belize City and at Turneffe Islands atoll, and references to a commercial sponge industry, also at Turneffe. The earliest record was that of *Polymastia biclavata* Priest (now accepted as *Coelosphaera biclavata* (Priest)), dredged by a collector from 25 to 31 m off Belize City and sent to B. W. Priest in England, who described it before a meeting of the Quekett Microscopical Club in London (Priest, 1881) (Fig. 3.4). The other seven species were collected by the British Rosaura Expedition (1937–1938) but were not listed or described until much later (Burton, 1954). They consisted of *Thenea fenestrata* Schmidt, *Radiella sol* Schmidt, and *Dragmatyle topsenti* Burton (now *Eurypon topsenti* (Burton)) collected from 900 m north of Turneffe Islands; *Haliclona spiculosa* Dendy (now *Amphimedon spiculosa* (Dendy)), *H. tenerrima* Burton, and *Tedania anhelans* Lieberkuehn (presumably *T. ignis* (Duchassaing and Michelotti)) from shallow water (2–3.5 m) at Indian Cay, Turneffe; and *Hymeniacidon glabrata* Burton from Belize City harbour (6 m). The bath-sponge varieties reported from many shallow-water locations along the coast of Belize included sheepswool (*Hippospongia lachne* (de Laubenfels), velvet (*Hippospongia gossypina* (Duchassaing and Michelotti)), and grass sponge (*Spongia graminea* Hyatt)—but none were of high commercial quality (Moore, 1910; Stuart, 1948).

Our own investigators were struck by the teeming poriferan life everywhere—not only just on Belize's reefs but also in the Twin Cays

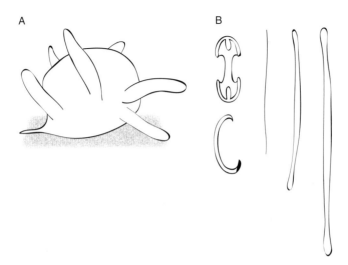

Figure 3.4 Illustration of *Coelosphaera biclavata*, first sponge recorded from Belize. (A) The body (without fistules) is reported as 7 × 13 mm in diameters, the spicules (B) measure (left to right, top to bottom): isocheles, 13–17 μm; sigmas, 50 μm; rhaphids, 250 μm; tylotes, 300–500 μm. (Adapted from Priest, 1881; plate XXIII, partial.)

mangrove, where a rich sponge fauna covers the red-mangrove prop roots and peat banks lining the ponds and tidal channels. The astonishing diversity and biomass in these predominantly tidal habitats is largely due to the Caribbean's small tidal range, which allows mangrove islands to develop under near-oceanic conditions (with low terrestrial input) on the outer reef platform (Rützler and Feller, 1988, 1996). In the mid-1990s, local fishermen guided us to the nearby Pelican Cays, a group of mangrove islands with deep, clear, craterlike ponds. Here, the red-mangrove trees are anchored not in mud but on a lush coral reef, abounding in species richness and live cover unparalleled in the Caribbean (Macintyre and Rützler, 2000).

Our study of this array of fresh material made clear that numerous Caribbean families and genera needed to be revised and revealed more than 30 new species, with many more still under investigation (Figs. 3.5 and 3.6). This work has been of great benefit to several chapters in the Systema Porifera monograph (Hooper and Van Soest, 2002). In reviewing the Tetillidae, a difficult family because of its great diversity and the variability of spicules, we coupled field observations with close study of a large specimen series obtained by the Mineral Management Service (U.S. Department of Interior) during continental shelf surveys along the southeastern U.S. and Gulf of Mexico coasts in the early 1980s (Rützler, 1987; Rützler and Smith, 1992). Although genera could be well defined by anatomical characteristics and the presence or absence of unusual accessory

Figure 3.5 Examples of sponge species described as new from the Carrie Bow region or representing a Caribbean-wide revised group. (A) *Svenzea zeai* on the reef, 7 m; the diver is Sven Zea. (B) Artist's rendering of detail (8 cm width) of *Svenzea zeai* and its larva, the largest known from sponges (>6 mm long). (C) *Haliclona magnifica*, Twin Cays, incorporating mangrove peat, 1 m (image width: 10 cm). (D) *Mycale citrina* encrusting mangrove peat, Twin Cays, 0.5 m (image width: 8 cm). (E) *Chondrilla caribensis* forma *hermatypica* overgrowing *Diploria* coral, Carrie Bow backreef, 1.5 m (picture width: 12 cm). (F) *Agelas clathrodes*, part of a revision of this phylogenetically interesting genus; Carrie Bow forereef, ca. 23 m (picture width: 50 cm). (G) *Tedania klausi* overgrowing *Lissodendoryx columbiensis*, turtlegrass flat at the south entrance of the Twin Cays main channel, 1 m (picture width: 20 cm). (Illustration credits: A, Mateo López-Victoria; B, Molly Kelly Ryan; F, Fernando Parra; G, Janie Wulff; others by the author.) (For colour version of this figure, the reader is referred to the online version of this chapter.)

spicules, some individuals lacked spicule types or combinations used to identify species—or these were too rare to show up in the tissue volume used for standard spicule preparations. Thus it is obviously advisable to study populations whenever possible, or at least take multiple microscopy samples from each specimen.

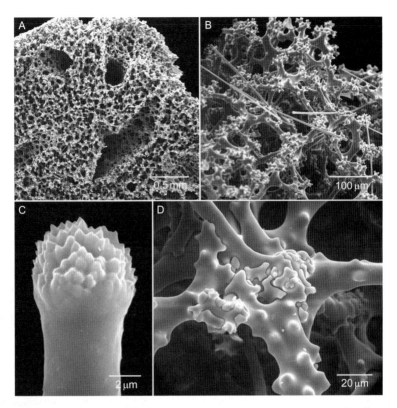

Figure 3.6 SEM micrographs of new shallow-water lithistid sponge, *Gastrophanella caverni-cola*, from Columbus Cay cave, 30 m. (A) Perpendicular section showing skeleton structure. (B) Detail, reticulation of fused desmas and associated tylostrongyles. (C) Microspined tylostyle head. (D) Articulation of desmas.

Field experience from numerous West Indian locations—including Bermuda, the Bahamas, Puerto Rico, and Belize—combined with high-resolution (scanning electron) microscopy and fatty acid chemistry (Vicente *et al.*, 1991) enabled us to clarify the taxonomic status of the common loggerhead sponge, *Spheciospongia vesparium* Lamarck, and its closest relatives. After studying live specimens in their environment on shallow sandy bottoms near Carrie Bow Cay, we placed the related species, the staghorn sponge *S. cuspidifera* Lamarck, into a separate, newly established genus, *Cervicornia* (Rützler and Hooper, 2000). Using fluorescent dyes, we found that *Cervicornia cuspidifera* possesses not only a specialized inhalant, ectosomal tissue complex—the staghorn-like structures protruding from the substrate surface and the only parts of the sponge obtained and described by previous collectors—but also an amorphous choanosomal base buried in sand with which it consolidates and excavates sand and rubble fragments, as well as discharges exhalant water (Rützler, 1997).

Multidisciplinary work made it possible to treat a number of sponge families comprehensively, particularly those from mangroves of the outer shelf. In our extensive collections of Chalinidae from the channels and ponds of Twin Cays, for example, we found and described in detail eight species in the genus *Haliclona*, three of them new (de Weerdt *et al.*, 1991). Another species earlier considered new, *H. pseudomolitba* de Weerdt, Rützler, and Smith, turned out to be an unusual form of *Chalinula molitba* de Laubenfels (de Weerdt, 2000). The composition of this chalinid fauna differs markedly from that in many other Caribbean mangrove habitats and is closer to the fauna of Florida Keys mangroves, perhaps because of the absence of nearby landmasses (and their runoff). Hence, it was concluded that differences in climatic conditions may be of less importance. (Subsequent studies, however, particularly on the Chalinidae of Panama and Venezuela, suggest that apparent differences may have been the result of superficial collecting rather than of ecological causes, as shown in distribution tables by M. C. Diaz and colleagues, so far unpublished but accessible under "sources" at http://www.marinespecies.org/porifera.)

Intensive collecting in these lagoon habitats revealed two new species in the genus *Terpios* (Suberitidae), defined after we revisited taxonomic characters to clarify their usefulness for this and other genera in the family (Rützler and Smith, 1993). In a comparable survey of the Mycalidae in the Twin Cays mangrove and its surrounding waters, we discovered eight species, two of them new (Hajdu and Rützler, 1998), bringing the total number of species known in the region to 17, all characterized in a dichotomous key.

An even larger and less investigated group in the area—despite its many conspicuous reef species—was the Axinellidae. Working with abundant material from Belize and the large collection obtained from the U.S. continental shelf survey along the southeastern and Gulf of Mexico coasts of the early 1980s, we revised and described 21 species under this family, five of them new (Alvarez *et al.*, 1998). Following subsequent redefinition of related families, five of the species were allocated to the family Dictyonellidae (order Halichondrida). A new genus had to be created for one species, named *Pseudaxinella* (?) *zeai* Alvarez, van Soest, and Rützler to honour our Colombian friend and colleague Sven Zea (Alvarez *et al.*, 1998). Since in our field notes it was already named *Svenzea*, Sven ended up with his name attached to both genus and species, *Svenzea zeai* Alvarez, van Soest, and Rützler (Alvarez *et al.*, 2002), probably a first in nomenclature but well deserved. Along with students and other collaborators, Sven was an active programme participant. Their important contribution to systematic studies included a revision of the *Cliona caribbaea* Carter complex, a much-discussed group of excavating sponges characterized by simple spiculation of tylostyles, delicate spirasters, and brown to blackish colour (due to symbiotic zooxanthellae). The group was found to consist of three species—*C. aprica* Pang, *C. caribbaea*, and *C. tenuis* Zea and Weil—the latter described as new, which can be distinguished in the field by their shades of the basic brown colour, growth form (papillate, thinly or thickly encrusting), and spicule details (Zea and Weil, 2003). Another study focused on the genus *Agelas*, which includes many conspicuous and colourful reef sponges and, using classical taxonomic characteristics, defined 13 Caribbean-wide species (Parra-Velandia *et al.*, 2012).

Some taxonomic studies arose by chance, triggered by some unforeseen event. During an excursion to Wee-Wee Cay, about 8 km to the Southwest of Carrie Bow, for example, we were able to map a few shallow (ca. 4 m deep) lagoon patch reefs visible from the boat because of dead-calm sea conditions. At one location, we encountered numerous specimens of an unusual stringy, deep greyish green sponge with encrusting base, which suggested naming its habitat spaghetti reef. Additional surveys revealed a sibling species, also encrusting but rather shaggy, with sand embedded in the surface, and more brownish-bluish green in colour. Both turned out to be undescribed species of the Halichondrida, *Dictyonella funicularis* Rützler and *D. arenosa* Rützler (originally described under the genus *Ulosa*), and distinguished by a high concentration of large cyanobacteria (*Aphanocapsa raspaigellae* (Hauck) Frémy) as symbionts (Rützler, 1981).

When our Brazilian colleague Guilherme Muricy of the Museu Nacional, Rio de Janeiro, studied some of our collections at the National Museum of Natural History in Washington, he found a lithistid sponge species, genus *Gastrophanella* (Siphonidiidae) collected in Belize, to be remarkably similar to one he discovered in a cave in Fernando de Naronha, northeast Brazil. Our material came from caves as well, from a flooded Karst cavern near Columbus Cay (25 km north of Carrie Bow Cay), and from the Blue Hole inside Lighthouse Reef atoll made famous by a Jacques Cousteau television documentary. The species was subsequently described as *Gastrophanella cavernicola* (Muricy and Minervino, 2000). A particularly surprising discovery was that the starfish *Oreaster reticulatus* Linnaeus could distinguish between different sponge species where we experts had failed. This common Caribbean starfish is very abundant in and near the mangroves of Twin Cays and was often seen consuming sponges attached to red-mangrove stilt roots or growing among turtlegrass. One of its prey, observed by collaborator Janie Wulff of Florida State University, was the Caribbean fire sponge *T. ignis* (Poecilosclerida), which is ubiquitous in these habitats, although not all individuals were being eaten. Those not consumed actually belong to a second sympatric species in the same genus—named *Tedania klausi* Wulff—as established from morphological and ecological characteristics as well as molecular traits (Wulff, 2006a). Another new species named after the present author—hereby gratefully acknowledged—is *Aka ruetzleri* Calcinai, Cerrano, and Bavestrello (Phloeodictyidae, Haplosclerida), an excavating sponge found during a visit of an Italian team who also clarified differences in micro- and macrostructure of limestone erosion patterns caused by members of this interesting genus (Calcinai *et al.*, 2007). Collaborators from Spain discovered an unusual growth form of the common *Iotrochota birotulata* Higgin (Poecilosclerida), which prompted further collecting and reexamination of similar material from the U.S. continental shelf survey mentioned earlier. Our revision confirmed that we had discovered an undescribed species, *I. arenosa* Rützler, Maldonado, Piantoni, and Riesgo, and clarified the taxonomic status of three other *Iotrochota*, bringing the total for the region to four species (Rützler *et al.*, 2007b).

One of the most common sponges in Belize's mangroves belongs to the genus *Lissodendoryx* (Coelosphaeridae, Poecilosclerida). Its many shapes and colour variants—from purple to pale blue, turquoise, green, and yellow to almost white—can create some confusion, as does the presence of two sizes of microscleres in some specimens and only one size in others. The common species in this environment had always been thought to be *L. isodictyalis* Carter, but its type material, collected from mangroves in Venezuela (now filled in and paved over) and stored in Britain, appeared to have been lost during World War II, making it difficult to ascertain what species we were dealing with. Through a lucky coincidence in the course of our inquiry, Carter's types from that period turned up at the National Museum Liverpool and were generously made available by curator Ian Wallace. Now able to review all species in the region, including unidentified material from the U.S. continental shelf survey, we redefined some and established and described three new ones (Rützler *et al.*, 2007c).

In the late 1990s, sponge systematists worldwide—under the leadership of John Hooper and Rob Van Soest—began revising all known genera for the above-mentioned Systema Porifera, a Guide to the Classification of Sponges. Although most of the work was done in museums and other depositories of preserved types and other material, it benefited from field experience with live sponges such as the clionaids, Spirastrellidae, acanthochaetetids, and tetillids of Carrie Bow Cay and elsewhere (e.g. Rützler, 1987, 2002a,b; Rützler and Vacelet, 2002; Van Soest and Rützler, 2002).

It is gratifying to observe that focused systematic research on sponges in the Carrie Bow region generated collaborations and comparative distributional studies elsewhere in the region and helped identify Belize as one of the Caribbean marine biodiversity hotspots that should be conserved at all cost (Miloslavich *et al.*, 2010; Diaz and Rützler, 2011).

3.2. Phylogeny and molecular genetics

Before 2004, sponge work in Belize concentrated on general morphology, skeletal structure determined in part by scanning electron microscopy (SEM), and some histology and cytology studied with light and transmission electron microscopy (TEM). Subsequently, in collaboration with colleagues from the coast of Catalonia, Spain, we made a few pioneering attempts to solve taxonomic and phylogenetic problems with molecular techniques (Erpenbeck *et al.*, 2007). One of the sponges selected for this work was *Chondrilla* cf. *nucula* Schmidt—a Mediterranean species lacking clear distinguishing features and having its name applied to specimens from the tropical and subtropical western Atlantic. We decided to study this species from the mangroves and reefs in Florida, Belize, and Panama because of its apparent morphological modifications in different habitats. As indicated by mtDNA sequences of the cytochrome *c* oxidase subunit I gene, haplotypes in specimens from mangroves and reefs differed enough to suggest a high level of reproductive isolation between the two populations (Duran and Rützler, 2006). This finding prompted careful morphological reexamination and ecological experimentation, which revealed previously overlooked characteristics, justified the description of a new species (*Chondrilla caribensis* Rützler, Duran, and Piantoni) and two forms (formae *caribensis* and *hermatypica*), and explained mechanisms for the speciation suggested by molecular data (Rützler *et al.*, 2007a). Although habitat quality obviously has considerable effect on the distribution and morphology of sponges, genetic study confirms that, even within a small region, populations of a species can become ecologically separated long enough to develop genetic differences.

3.3. Invasive species

Studies of recruitment, community development, and invasive species on settlement plates are still under way in temperate (Connecticut, Virginia) to subtropical and tropical (Florida, Belize, including the Carrie Bow reef) biogeographic zones (Freestone *et al.*, 2009). Thus far it appears that recruitment and community development rates of the various sessile invertebrates monitored, including sponges, decrease along the gradient from temperate to tropical habitats and are lowest in Belize. No evidence has yet been found that longer availability of space could facilitate invasion of non-native species, possibly because of the high diversity and competitive abilities of sessile fauna and resident predators. On the other hand, it could be argued that invasive species are difficult to identify in a region where many native taxa are still undescribed or little known. At least for sponges, it remains difficult to distinguish between new discoveries and new arrivals.

4. REPRODUCTION AND EARLY DEVELOPMENT

Over the years, researchers have observed a number of sponge spawning events around Carrie Bow, both on the reef and in the mangroves, but few have devoted much time to embryological developments.

Occasionally, tourist divers send the Museum photographs of "smoking" (sperm-releasing) *Aplysina* and *Agelas* tubes, or vase-shaped *Xestospongia*, and similar sponges covered by thick, gelatinous spaghetti- or grape-like masses (oocyte- and embryo-containing strands), which they assume to be symptoms of disease, a notion quickly put to rest by previous and recent research (Reiswig, 1976; Ritson-Williams *et al.*, 2005). Some of our programme participants looked for diversity in sponge larvae in routine plankton tows but found none. Snorkelling through mangrove channels during calm periods, we have noted conspicuous red larvae (later identified to be from *Tedania*) swimming in a spiral motion while remaining very close to the mangrove stilt roots and other substrates, such as peat banks. Because conventional plankton tows had to take place at a safe distance from the coral framework or mangrove roots, we tried pushing or pulling nets by hand while snorkelling or diving, keeping as close as possible to the substrate contour. Although we caught a variety of sponge, coral, and gorgonian larvae (the principal benthos of the reef), it took considerable time and effort to catch enough material with this technique. We then designed a stationary sampler made of an acrylic cylinder containing an electric outboard motor with a propeller and a standard plankton net and flow-metre attached (Rützler *et al.*, 1980). Mounted in a rigid frame and weighed down by a battery pack, it could be positioned anywhere on the sea bottom and could propel water through the net at a given speed and volume, usually overnight. The resulting catch allowed us to study larval behaviour, metamorphosis, fine-structure morphology, histochemistry, and generational transmission of symbiotic microbes (Fig. 3.7).

In cooperation with graduate student Simon Weyrer and his adviser Reinhard Rieger from the University of Innsbruck, Austria, we studied the cytology of larvae and early developmental stages of the fire sponge, *T. ignis*. This species reproduces almost year-round, and larvae can be obtained in large numbers by incubating a sponge with mature embryos in its tissue in the laboratory's seawater system. Applying histochemical methods, we found serotonin-like immunoreactivity, particularly in archaeocytes of the parenchymella larvae and settled juvenile sponges (Weyrer *et al.*, 1999). The discovery of a neuroactive substance in sponges, the only free-living invertebrates lacking a demonstrable nervous system, may bring new insights into the origin and evolution of nerve cells in metazoans. The results of another project mentioned earlier (Rützler *et al.*, 2003), which compared cytology, microbial symbionts, ciliary patterns, and larval shape and behaviour between *Svenzea* and *Scopalina*, supported our decision from morphology alone to place Svenzea in the family Dictyonellidae. Fine-structure (TEM) study also revealed that the microbes in *Svenzea*, which hosts a diverse population of symbiotic cyanobacteria and bacteria, become transferred to the next generation via the eggs and larvae. This process was confirmed by an independent genetic study (Lee *et al.*, 2009) and also by our finding from TEM morphological

Figure 3.7 Reproductive Biology. (A) Diver servicing automated benthic plankton sampler after overnight deployment. (B) Larvae of *Dysidea etheria* obtained during overnight sampling (picture width, 6 cm). (C) Young sponge, *D. etheria*, 3 days after settlement in laboratory (picture width, 1.5 cm). (D) *Tedania ignis*, two larvae and one newly settled juvenile sponge from Twin Cays (picture width, 8 mm). (E) SEM image of *Scopalina ruetzleri* larva collected at Twin Cays (picture width: 0.9 mm). (F) *Xestospongia muta* with egg mass on Carrie Bow reef (picture width, ca. 30 cm). (G) *X. muta* releasing a cloud of sperm (picture width, ca. 30 cm). (Photo credit: F, G, Raphael Ritson-Williams; others by author.) (For colour version of this figure, the reader is referred to the online version of this chapter.)

comparison that the symbionts are very different and therefore not derived from microbe populations of the surrounding water column.

By contrast, no such transmission was found in mass-spawning corals at Carrie Bow studied by Koty Sharp, one of our postdoctoral collaborators and colleagues (Sharp *et al.*, 2010). Sharp investigated gametes, planula larvae, and newly developing polyps in representatives of the genera *Montastraea*, *Acropora*, and *Diploria* using fluorescence *in situ* hybridization (FISH) to detect endobiotic bacteria that are present in great diversity in mature colonies. Rather than being transmitted through the planulae, in this case bacteria are acquired by the coral polyps during development. When Sharp applied similar molecular techniques (16S rRNA, FISH) to a tropical brooding sponge, *Corticium* sp., in Micronesia, however, microbial vertical transmission was confirmed (Sharp *et al.*, 2007).

5. ECOLOGY

5.1. Distribution and factor regime

5.1.1. Principles of horizontal distribution

From the outset of our programme, we made numerous general faunistic–floristic surveys as a basis for more specialized studies. We found sponges to be a prominent feature of the environment throughout our reconnaissance of the Carrie Bow reefs, a large submarine cave near Columbus Cay to the North, the mangroves of Twin Cays, Blueground Range, and the Pelican Cays, and in comparative studies of habitats on Australia's Great Barrier Reef (GBR).

A primary aim in those early years was to define reef zones and other principal communities in the immediate vicinity of Carrie Bow (Rützler and Macintyre, 1982b). To this end, we first established a transect starting in the lagoon just northwest of the island and running eastward and perpendicularly across the barrier reef. The lagoon bottom here consists of sand, rubble, some seagrass (turtlegrass, *Thalassia*), and patch reefs here and there composed primarily of massive coral heads (*Montastraea*, *Diploria*). Only excavating and small rock- and rubble-encrusting sponges are able to survive in this shallow, agitated zone. The same is true in the shallow backreef, reef crest, and inner forereef in depths of up to at least 3 m. As the obvious and fully exposed organisms captured our initial attention, it was not until much later that we found a diverse population of very thin, colourful sponge crusts and cushions on the undersides of heavy, platy coral rubble and boulders resembling material produced by the elkhorn coral, *Acropora palmata* Lamarck. Because they are not easily overturned and have large areas of the underside not buried in sediment, these platy slabs provide a shaded habitat exposed to strong water flow but protected from fish grazing. We found another interesting habitat inside conch shells (*Strombus gigas* L.) discarded by fishermen after they extracted the animal by chopping off the shell's apex with a machete to gain access to the adductor muscle. This creates a second opening in the shell, opposite the aperture, and thus provides sponges with adequate water flow.

On the wave-exposed section of our transect, massive sponges do not reach significant numbers on the low-relief coral spurs of the inner forereef until depths of 10 m or more. In similar depths on the outer forereef, they occur in considerable diversity and abundance, exceeding the ubiquitous corals and octocorals in biomass. The transect ends at a depth of 30 m on the forereef slope, but bounce dives to 60 m, with visibility to 80 m or more, revealed that large, massive sponges continue to dominate the community.

Patch reefs in the deeper, well-protected lagoon 1–2 km southwest of Carrie Bow are populated by numerous species of fairly large sponges that grow protected within the coral framework in depths of about 4 m; some also occur in turtlegrass on and in the sandy lagoon floor at a depth of 6 m or so. A bit further southwest, towards Wee-Wee Cay, similar patch reefs rise steeply from even deeper (12 m) lagoon bottoms to the surface. Exposed at low tide, their glistening sand and rubble tops are visible from a great distance and are locally known as sand bores. These patch reefs, particularly those in deeper water along steep flanks, are home to several unusually large sponge species, notably in the genera *Ircinia, Xestospongia, Iotrochota, Amphimedon,* and *Callyspongia,* also the toxic *Neofibularia nolitangere* Duchassaing and Michelotti and the highly competitive *Desmapsamma anchorata* Carter.

Twin Cays, the mangrove closest to Carrie Bow and about 3 km to the northwest, are so named because a relatively deep (2 m) tidal channel divides the island into two. The shallow (1–2 m) bottom just outside the entrance to the larger (southern) channel is sandy and covered by turtlegrass, but swift currents keep the area clear of fine-grained sediments, allowing a mini-reef to flourish here, with small patches of coral (*Manicina, Millepora*), calcified green algae (*Penicillus, Halimeda*), and sponges (*Tedania, Hyrtios, Lissodendoryx, Aplysina, Haliclona*). Small communities of large loggerhead sponges (*S. vesparium*) are also present. This scenery changes dramatically in the channel, which has a muddy bottom with patches of turtlegrass and banks lined with red mangrove (*Rhizophora mangle* L.). Some of the stilt roots are anchored, and some hang free as they have not yet reached the bottom. The roots and peat banks behind them have a thick cover of algae and invertebrates, the latter consisting predominately of sponges and ascidians (see also Rützler and Feller, 1988, 1996; Rützler *et al.*, 2000, 2004).

5.1.2. Submarine caves

A suspected stalactite retrieved from an undersea cave and sent to us by a Belizean recreational scuba diver in 1977 drew our attention to the Columbus Cay "blue hole," a feature often mentioned by local fishermen (Macintyre *et al.*, 1982). The "blue hole" lies due northwest of Columbus Cay, which is located on the lagoon side of the barrier-reef platform, about 24 km north of Carrie Bow. It is a sinkhole leading into a drowned Karst cavern, comparable to the famous Blue Hole in the Lighthouse Reef atoll lagoon off Belize City. The cavern is partly filled with sand, and its entrance is a ragged crack in the seafloor some distance above the sand. The water near the entrance is quite murky, and only a few hardy stony corals and octocorals and sponges typical for lagoon seagrass bottoms occur there (particularly, *Aplysina fulva* (Pallas), *Amphimedon compressa* Duchassaing and Michelotti, *I. birotulata*, and *Ircinia strobilina* Lamarck). Inside the large domed cave, stalactite-like structures protrude from the entire ceiling. These club-like projections turned out to be cemented serpulid (polychaete) worm tubes, which we named pseudostalactites. We were only able to examine steps in the gently curved ceiling because any vertical walls reaching the floor were too deep and far from the entrance to be reached in single-tank scuba gear. The cave floor consists of a huge sediment cone, up to 30 m deep under the cave entrance and unsuitable for sessile benthic organisms such as sponges. Along the dome ceiling, sponges do not occur beyond 20 m from the entrance. With the lack of water exchange and food supply except during very strong storm surges, as well as the dark conditions, the sponge fauna differs significantly from that of the lagoon bottom outside but includes species common in shaded substrates elsewhere on the fore- and backreef, such as framework crevices and lower surfaces of coral boulders and rubble. Some examples are *Geodia gibberosa* Lamarck, *Placospongia carinata*, Bowerbank, *Spirastrella coccinea* Duchassaing and Michelotti, and *C. caribensis* (as *C. nucula)*, and two lithistids typical of deep water or dark caves, *Desmanthus incrustans* Topsent, and the aforementioned new species *G. cavernicola* (first identified as *G. implexa* Schmidt). The sponges and associates occupying a number of framework caves of the forereef slope have recently been sampled, and their diversity and distribution are currently under study (Rützler and Piantoni, in preparation).

5.1.3. The Pelican Cays

In the early 1990s, Paul and Mary Shave, operators of a teaching laboratory on nearby Wee-Wee Cay, told us of the species-rich benthic fauna and flora in the Pelican Cays, about 16 km southwest of Carrie Bow. At that point, only a few of our investigators were prepared to make the relatively long trip on one of our small boats, but as word spread of the spectacular but delicate biota flourishing in a striking environment of mixed mangrove cays and steep coral ridges surrounding numerous deep ponds and lagoons, all barely touched by humans, a multidisciplinary team came together to survey the area's principal habitats, communities, and geological features (Macintyre *et al.*, 2000). Although this archipelago is situated in the main lagoon inside the barrier reef, it is awash with oceanic water and contains an abundance of coral and rubble. The depth of the ponds (as much as 10 m in places) prevents resuspension of most fine sediments during storms. The abundance of nutrients and low level of turbidity seem to account for the high diversity and biomass of the benthic communities there, the most striking of which are the sponges, ascidians, and seaweeds. Within view of the unparalleled species richness and lush sponge growth along the rims of the ponds, we held a workshop for experts focused on collecting and identifying the species and comparing the fauna of three of the Pelicans Cays—Cat Cay, Fisherman's Cay, and Manatee Cay—with that of the muddy mangrove islands more common elsewhere in the lagoon, Blue Ground Range, and Twin Cays, 10–15 km to the north (Rützler et al., 2000). Workshop participants, all familiar with Caribbean sponge faunas, were Belinda Alvarez (then at the National Institute of Water and Atmospheric Research, New Zealand), Cristina Diaz (then at the University of California at Santa Cruz), Kate Smith (then at the Smithsonian Institution), Rob van Soest (University of Amsterdam), Janie Wulff, Sven Zea (National University of Colombia), and myself. The team collected specimens from depths of 0–2 m and occasionally 5 m in all locations by snorkelling and free-diving (2-h excursions at each location), first photographing them in situ, if possible, and then back at the Carrie Bow lab. Their abundance was ranked, by consensus, between very common and very rare. Most identifications were based on hand sections and spicule mounts made in the field, supplemented by a few follow-up studies upon return. The total count from these locations combined was a surprising 182 species, but even more astonishing was that for a well-studied region like the Caribbean Sea, only 100 (55%) could readily be identified. The remaining 82 taxa were either undescribed species or varieties (formae) too poorly known to be confidently identified. Many of these are still under study or may become part of future revisions.

We concluded that the Pelican Cays, owing to their special topography, constitute a transitional environment between mangrove and reef, with an ecological regime ideal for sponge faunas derived from both. As colleagues from other disciplines have also pointed out (see Macintyre and Rützler, 2000), however, the remarkable diversity and productivity in the Pelicans depend on a delicate ecological balance that would be easily upset by human impact should these islands be discovered by ecotourism or, worse, made available for real estate development. Sadly, although the Belize government accepted our recommendation that the Southwater Cay Marine Reserve be expanded to include the Pelican Cays, commercial development has outpaced our warnings with recent mangrove clear-cutting in the area (Macintyre et al., 2009). Only time will tell whether its delicate sessile photosynthetic and filter-feeding communities can recover from the loss of stabilizing trees, high turbidity related to sediment dredging, and other severe impacts.

5.1.4. Comparisons with Indo-Pacific reefs

In pursuit of a more global perspective, our colleagues Clive Wilkinson and Anthony Cheshire of the Australian Institute of Marine Science in Townsville, Queensland, examined the sponges of Belize alongside those of Australia's GBR (Wilkinson, 1987; Wilkinson and Cheshire, 1990). Their quantitative and productivity study of sponges from reefs near the mainland and from similar habitats in oceanic conditions further offshore revealed that the Caribbean barrier reef supports on average up to six times more sponge biomass per unit area than its Pacific counterpart and that up to 15 times more organic carbon per unit area is consumed through filter feeding. Differences between the two areas were less pronounced on the inshore reefs. On the oceanic reefs, half of GBR sponge biomass consists of phototrophic species, sponges with photosynthetic symbionts that provide most of the host's energy.

5.1.5. Physico-chemical determinants

What determines sponge distribution? Why, for example, are reef and mangrove higher taxa such as families and orders so different in their morphology, reproduction, and symbionts? An effort is currently under way to explain distributional spectra by examining ecological mechanisms and evolutionary histories (Diaz, in preparation). In the case of Caribbean mangrove-sponge faunas, their interesting distributional schemes appear to be linked to the physico-chemical environment and to biological interactions.

Two principal types of mangrove swamp can be distinguished in Belize and elsewhere in the Caribbean, both having the pioneering red-mangrove tree, *R. mangle*, as the main component. One is the coastal or river mangrove, generally found in estuarine environments with a salinity gradient ranging from near oceanic to freshwater. Only a few well-protected marine invertebrates such as oysters and sedentary (tube-forming) polychaetes survive being exposed to brackish water and heavy silting from land runoff; sponges are not among them. The other type is the island mangrove, which is surrounded and flushed by more or less oceanic–quality water, and ranges from small tree clusters to large swamp islands. In Belize, these mangrove islands, the larger ones locally known as ranges, occur primarily on the barrier-reef platform along the outer perimeter of the lagoon (Rützler and Feller, 1988, 1996). Species richness of the root-fouling community increases with distance from the mainland (Ellison and Farnsworth, 1992). Although salinity gradients can be strong in island mangroves too, tidal flushing with sea water of normal salinity is adequate to allow almost reef-like flora and fauna to thrive there. However, diversity tends to decrease in locations where temperature and salinity vary greatly.

Sponges abound on submerged stilt roots and peat banks, in some areas with amazing diversity and biomass. Even so, sponge distribution in mangroves is extremely patchy and not easily explained. In a survey of the major groups of prop-root epibionts on different islands—including sponges ranging from centimetre size on single roots to kilometres—current- or wave-exposed, well-illuminated roots were populated mainly by algae, while leeward and darker substrates were inhabited primarily by sponges and ascidians. The relative importance of species groups is unclear, however, and can change over time (Farnsworth and Ellison, 1996). Still, some consistency is evident, despite differences in non-biotic factors and geographical distance between sites. In a comparative study of two adjacent locations, Sponge Haven and Hidden Creek at Twin Cays, and a third site near Bocas del Toro, Panama, 1200 km distant, for example, identical quantitative survey methods applied over 3 years indicated that nine sponge species made up 73–89% of the total sponge volume on censused roots at those locations (Wulff, 2009). Community changes over time, reflected by shifts in number or volume of sponge individuals, may therefore be misunderstood without some consideration of a species' life history characteristics and responses to environmental properties.

To study recruitment and growth in response to environmental conditions at Twin Cays, a pair of mangrove islands typical for most of the lagoon, we observed roughened acrylic settlement plates (90 cm^2 each) over several annual cycles, all placed with three replicates at various ecologically distinctive sites (K. Rützler, unpublished) (Fig. 3.8). At the same time, environmental variables were measured and related to parameters recorded by our environmental monitoring station on Carrie Bow Cay (Opishinski et al., 2001). Divers sampled larvae of common species by pushing plankton nets along the roots or collecting mature (larvae-containing) adult sponges and extracting larvae from them. Larvae were then studied under controlled laboratory conditions.

To date, we have determined, first, that substrate has little bearing on the settlement of mangrove sponges, whatever its composition. We tried the shell of mangrove oysters, tile, glass, red-mangrove and pine wood, and PVC and acrylic plastic materials without detecting any significant differences in settlement rates and diversity. Mature larvae take 5–10 days to settle after a layer of (bacteria-generated) biofilm is established. Second, larvae of some common sponges, such as Sycon sp., T. ignis, Mycale magniraphidifera van Soest, Haliclona tubifera George and Wilson, and Spongia tubulifera Lamarck, are released during all seasons, but recruitment slows down during the coldest winter months (25 °C average, 16 °C low). Of the settling plates, the dark, sediment-free substrates (lower surfaces of black, horizontal plates) were 15 times more likely to be chosen for settlement than semi-shaded vertical surfaces. Fully illuminated surfaces, whether exposed to sediment or not, were not settled by sponges at all. In laboratory experiments, upon release and after a dark period (night), larvae first headed towards light but became negative phototactic after hours in light and

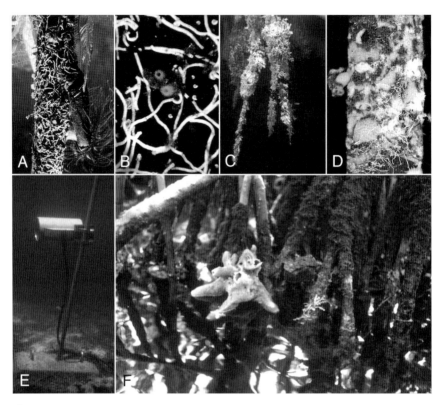

Figure 3.8 Sponge distribution at Twin Cays determined by abiotic factors. (A) Early settlers on new mangrove prop roots include serpulid worms, vermetid gastropods, and a few calcareous sponges (*Sycon* sp., *Clathrina* sp.) and demosponge crusts (picture width, 12 cm). (B) Similar assemblage on settling plate after 30-day exposure (picture width, 4 cm). (C) Advanced root community from a shaded mangrove channel dominated by sponges (picture width, 25 cm). (D) Comparable community on settling plate after 1.5-year exposure at the same location (picture width, 9 cm). (E) Flow metre with temperature probe deployed in mangrove channel; similar instrument packages recorded water level (tides) and salinity. (F) Sponge, *Lissodendoryx isodictyalis*, fully exposed to air during extremely low tide at noon (picture width, 60 cm). (Photo credit: C, Chip Clark; others by author.) (For colour version of this figure, the reader is referred to the online version of this chapter.)

before settlement. They were also more likely to settle near adult sponges than disperse to new areas despite good swimming ability. This behaviour is enhanced by prevailing current patterns.

The swim speed of tested larvae (of five species, representing Poecilosclerida, Halichondrida, Haplosclerida, and Dictyoceratida) ranges from 1.3–2.0 cm/s (for *Tedania*, *Dysidea*) to 2.0–5.0 cm/s (for *Haliclona*, *Scopalina*), which is near or faster than the mean speed of most water currents in the swamp, except for the faster periodic tidal flow in some narrow channels.

However, this swimming potential is not used for moving upstream (as determined in calibrated laboratory flow channels) and rarely (apart from one single incident revealed on test panels) for traversing a distance of about 10 m of open water (with a mud bottom and sparse turtlegrass).

Freshly exposed substrates are instantly fouled by biofilm-producing microbes. In illuminated habitats, the next to arrive are short-lived, quick-cycling algae, protozoans, and invertebrates, followed by more persistent, ubiquitous sessile invertebrates, particularly serpulid polychaetes and didemnid tunicates. Colonization on shaded substrates is slower, probably because there are fewer photosynthetic organisms there. In this case, the pioneering microbes are followed by coralline algae, foraminiferans, sponges, sessile polychaetes, bryozoans, and ascidians. Once established, encrusting sponges may spread rapidly (about 10 linear mm per month) and cover a test plate completely after 7 months. Growth may be temporarily delayed, halted, or reversed by competition for space with other sponges (and other organisms), by predators, or by invasion of endofauna such as polychaetes and crustaceans. If such activities are pronounced, it may take up to 28 months for a test plate to be entirely covered.

If substrate is in short supply but growth conditions are favourable, specialists may grow on top of suitable support sponges. S. tubulifera, for example, is often overgrown by other sponges and only recognized by its protruding oscular chimneys. If sponges were not part of the original settlement population—for instance, on plates deployed during winter—they may be excluded from this substrate indefinitely, or until events lead to a collapse of the established microcosm. New substrates, such as test panels, may also be readily invaded by adult sponges growing nearby, which may span distances of tens of centimetres by means of tissue bridges. Illuminated as well as sediment-exposed substrates not chosen by larvae may be colonized in this way. On the other hand, well-established sponge populations covering older stilt roots weakened by wood borers may suddenly be lost when the substrate breaks under the weight of the sponges. Clusters released in this manner may survive partly buried in mud for several weeks or months but will eventually die and disintegrate.

From these and similar experiments and from habitat surveys over many years (Diaz et al., 2004a; Rützler et al., 2004; Diaz and Rützler, 2009), we have found environmental conditions to be among the principal determinants of sponge distribution in the mangal. Furthermore, diversity and community structure of the sessile benthos, if carefully assessed, may be used as indicators of degrading environmental conditions. In the relatively pristine mangroves of Belize, the effect of above-average sediment load or change in temperature or salinity, in particular, is readily apparent in shifts in the sponge community. On a Caribbean-wide level, however, the composition of the mangrove-sponge fauna is quite unpredictable, so it is impossible to generalize local observations.

Water depth and tidal range control the availability of solid substrates, which, for the most part, consist of *Rhizophora* prop roots and peat banks lining the shores and tidal channels. The water depth in the study area ranges from 0 to 3 m, and the bottom consists of mud and detritus. A few sponge species here (e.g. *Haliclona magnifica* de Weerdt, Rützler, and Smith) have lost their substrate support and have adapted to living buried in mud, with only elongate oscular chimneys protruding from the surface. The upper limit of suitable substrate is controlled by the tide range, which is small on average (15 cm mean) but may ebb to 40 cm and more during spring tides and other oceanographic events. The tidal signature influences the zonation of sponges, depending on their resistance to desiccation, possibly through a mechanism of cellular osmoregulation (Rützler, 1995). In the study area, three species can tolerate more than 2 h of exposure to air, the hardiest being *Haliclona implexiformis* Hechtel, followed by *Lissodendoryx isodictyalis* and *Scopalina ruetzleri* (Wiedenmayer). These species were able to withstand tissue-water loss of 66%, 54%, and 38% after 6 h, as well as increases in the practical salinity of pore water ranging from the ambient 35 to 51 and 59. Even the hardiest members of the polychaete and crustacean endofauna had died or left their hosts at this point.

Below the tidal zone, vertical distribution is influenced by light. Most species grow well in the semi-shaded habitats of vertical roots and banks, protected from full sun by the mangrove tree canopy. Some, such as *Cinachyrella apion* (Uliczka), *Amorphinopsis* sp., *S. ruetzleri*, and the calcareous *Clathrina* sp., are most common on the ceilings of caves and overhangs of underwashed peat banks. Extended periods of extreme temperatures and salinity are limiting in some places, notably in some shallow ponds (and connecting creeks) exposed to particularly cold nights, hot days, heavy rain, or severe evaporation. Temperatures in such areas can range from 16 to 39 °C and practical salinity from 26 to 41. In most tidal channels, however, severe conditions last for only brief periods and are well tolerated by many species.

Sponge populations reach their highest densities in well-flushed, deep channels where peat banks, undercuts, and numerous mangrove prop roots reach below the water surface. The lowest sponge biomass values were recorded in shallow salt ponds without much firm, submerged substrate; only two sponge species are regularly found there, *Suberites aurantiacus* (Duchassaing and Michelotti) and *Halisarca* sp. It may be that abiotic pressures exclude certain species from some habitats, as demonstrated in a transplant experiment with five species common in Sponge Haven, a cove along the main channel awash with good-quality lagoon water (Wulff, 2004). The five species—*Mycale microsigmatosa* Arndt, *Calyx podatypa* (de Laubenfels), *Spirastrella mollis* Verrill, *Halichondria* cf. *poa* (de Laubenfels), and *T. klausi* Wulff (originally listed as aff. *T. ignis*)—were moved to an area where they did not occur, a narrow waterway known as Hidden Creek connecting the main channel with a shallow, saline lake, Hidden Lake. During each tidal change, the water in the creek undergoes a radical change in temperature, salinity, and suspended sediment load, often in the extreme, depending on the impact of weather conditions on the oceanography of the lake or open lagoon. Although the transplants showed no initial decline in health, 97% of the individuals had died after 1 year. Two of the *Calyx* survived and gained volume, but their subsequent fate is not known.

Although many situations described above are extreme cases, they illustrate that life in a mangrove swamp may be too challenging for reef species that are not physiologically adapted to rapid and pronounced changes in the physico-chemical environment. This is not to say that pure reef communities are devoid of environmental stress, particularly in shallow water. Water warming and enhanced ultraviolet radiation, as occur during a hurricane—along with increased light intensity and sedimentation—can have a devastating impact on shallow reef corals by inducing bleaching, stress, and eventually death (Shick *et al.*, 1996). In the short term, sponges seem less affected. Indeed, we were unable to induce bleaching in experimental setups using species with zooxanthellae (*Cliona varians* (Duchassaing and Michelotti)) as well as zoocyanellae (*C. caribensis* Rützler, Duran, and Piantoni). On the contrary, many sponges become successful competitors for space when corals are stressed. Also, modest increases in nutrients may harm hermatypic corals but promote the growth of bacteria that benefit sponges as food.

In both reef and mangroves environments, life-historical and morphological tactics and biological interactions play a central role in competitive success. As elegantly demonstrated by Janie Wulff, biotic factors such as competition for space and predation are instrumental in maintaining species diversity, an important measure in conservation efforts (Wulff, 2000, 2005). Although she acknowledged that many differences in abiotic conditions correlate well with differences in the two faunas of these environments, her experiments showed them to be of secondary importance (see section 5.2.1).

5.2. Species interactions

Sponges interact with other sponges and other organisms in a variety of ways. They compete for space with sessile invertebrates and plants; are involved with a large spectrum of microbes, plants, invertebrates, and fishes through epibiosis, endobiosis, and symbiosis; and are preyed upon by invertebrates, fishes, and seaturtles (for specific examples, see Wulff, 2006b). These interactions were the focus of many projects in the Carrie Bow region.

5.2.1. Competition and predation (Fig. 3.9)
Quite apart from the physico-chemical environment, post-recruitment processes play a large role in sponge distribution and community development in mangroves, as indicated by long-term monitoring of populations

Figure 3.9 Species interactions shaping distribution patterns. (A) Intricate association between reef corals (species of *Diploria*, *Montastraea*) and *Mycale laevis* protects the coral underside from attacks by bioeroders, the sponge from predation (picture width, 6 cm). (B) This orange–red *Clathria* sp., overgrowing *Lissodendoryx colombiensis*, seems inedible to the predatory starfish *Oreaster reticulatus*, thus protecting *L. colombiensis* from being consumed (picture width, 12 cm). (C) Unprotected *L. colombiensis* under attack by *Oreaster* (picture width, 9 cm). (D) *L. colombiensis*, appearance after removal of the feeding *Oreaster* (picture width, 10 cm). (E) Grey angel fish, another common reef-sponge predator (picture width, 80 cm). (F) *Desmapsamma anchorata*, a fast-growing competitor that overgrows and oftentimes smothers other sponges (here covering most of a *Geodia* species) and sessile invertebrates, such as gorgonians (picture width, 30 cm). (Photo credits: A, Carl Hansen; B–D, Janie Wulff; E, Chip Clark; F, by author.) (For interpretation of the references to colour in this figure legend, the reader is referred to the online version of this chapter.)

following settlement on artificial substrates off Belize (Wulff, 2004). When Wulff offered polyvinyl chloride pipes as mangrove root substitutes at Hidden Creek, Twin Cays, she found that even after 20 months and almost complete overgrowth by sessile organisms, mostly sponges, the species composition and relative abundance between pipes were still not the same as on roots. Ultimately, the fastest-growing species prevail and community-wide diversity is maintained only because substrate is not continuous but consists mainly of separate roots (Wulff, 2005). In comparing sponge diversity on mangrove roots between Twin Cays and the Pelican Cays, we found the latter to be about 2.5 times richer in species than the former (Rützler *et al.*, 2000; Wulff, 2000). The main difference between the two locations is that Twin Cays red-mangrove trees are anchored in mud and peat and support a typical Caribbean mangrove-sponge fauna, whereas the Pelican Cays trees are rooted on reef structures and populated by sponges commonly found on shallow coral reefs.

Again, Wulff used transplants to assess the impact of biotic interactions on species distribution in the different habitats (Wulff, 2005). She transferred genetically identical replicates (cut and healed pieces from one specimen) of six common and typical mangrove sponges from Twin Cays (including the prominent *T. ignis* and *L. isodictyalis*) to the Pelicans and six typical reef sponges taken from roots in the Pelicans (e.g. *I. birotulata*, *A. fulva*) to Twin Cays. She also provided caging to protect the samples from predation (with uncaged controls) and new PVC tubing to offer competitor-free space. The unprotected Twin Cays mangrove sponges at the Pelicans were almost immediately consumed by fishes, most notably the grey angelfish (*Pomacanthus arcuatus* L.), whereas caged specimens grew well as long as they were mechanically protected. Reefs are close to or part of the Pelicans and provide niches or hiding places for most spongivore fishes or invertebrates (such as the starfish *O. reticulata*), which determine species composition and diversity (Wulff 2000, 2005). Pelican Cays reef species did poorly at Twin Cays, some already deteriorating in transit. Although the remaining ones readily attached to roots and PVC pipes, most shrank in volume, as they did following death or partial death, and were covered by fast-growing neighbouring mangrove sponges, such as *T. ignis*, *Halichondria magniconulosa* Hechtel, and *Biemna caribea* Pulitzer-Finali.

Aside from predation, which is not necessarily terminal, sponges can be damaged by storms, disease, or other physical disturbances. Yet because they are cellular animals without highly specialized tissues, many recover well, by healing wounds and regenerating lost body parts. In fact, when we cut collected specimens into convenient pieces and reattached them to substrates of our choice, they healed within days unless traumatized by exposure to air or extreme temperatures. Janie Wulff, who frequently uses this method, applied it to study energetic costs of wound regeneration in an ecological context, focusing on mangrove roots with high population density and strong competition for space (Wulff, 2010). To this end, she wounded representatives of a number of species of different growth forms, some closely related, others not, and investigated how long it took for wounds to heal, how responses compared in species with close evolutionary history (congeners) or with similar growth forms or anatomy, and how regeneration affects processes such as growth, reproduction, and space competition. She found that wounds heal within a few days, but that the rate or mode of regeneration, of filling in missing tissue, could not be predicted from properties such as genetic similarity or rates of normal growth. Species with a rapid life cycle and known to be strong competitors (e.g. *B. caribea*, *Haliclona curacaoensis* van Soest, *H. manglaris* Alcolado) are the least likely to quickly cover substrate space made available to competitors in the course of wounding. By contrast, species with a low recruitment rate (e. g. *T. ignis*, *Spongia obscura* Hyatt, *H. implexiformis*) fill in an exposed substrate patch quickly and hold on to it well. In species targeted by one of the rare mangrove-sponge predators (e.g. *Halichondria magniconulosa* and *H. cura-caoensis*, preyed on by spotted trunk fish, *Lactophrys bicaudalis* (L.)), wounds from chunks bitten off are more rapidly filled in by tissue than in species not commonly attacked.

Some branching, fast-growing sponges are quite aggressive, particularly *D. anchorata*, and impact neighbouring sessile species, such as other sponges and gorgonians. Elisabeth McLean, in course of thesis work under the auspices of the University of Buffalo, New York, is conducting experiments to determine how recruitment of new aggressors takes place and what the victim's defences and chances of survival in a stand-off may be. Her results are not yet published but contain interesting data on growth rates, defence mechanisms, and conditions of coexistence, as well as a victim's possible recovery.

In seagrass meadows, sponge diversity and abundance have been shown to increase through collaboration between a reef species (*Lissodendoryx colombiensis* Zea and van Soest) and typical seagrass species (*Clathria schoenus* de Laubenfels, *C. caribensis*, *Amphimedon erina* de Laubenfels [=*A. viridis* Duchassaing and Michelotti ?], and *T. klausi*) (Wulff, 2008). These turtle-grass associates overgrow *L. colombiensis*, thus acquiring stable substrate in a sandy habitat and in turn protecting it from predatory feeding by the starfish *O. reticulatus*. As an added benefit, *Lissodendoryx* itself has efficient recruitment abilities and rapid regeneration and growth rates.

5.2.2. Symbioses with photosynthetic organisms (Fig. 3.10)

Many sponges in Belizean shallow-water habitats host inter- and intracellular photosynthetic symbionts, microbes, and algae, which have a large impact on the ecological balance of nutrient-deprived reefs and mangroves (Rützler, 1990). The majority of large reef sponges harbour the rather small (up to 2 µm) "*Aphanocapsa feldmanni*"-type cyanobacteria, some identified

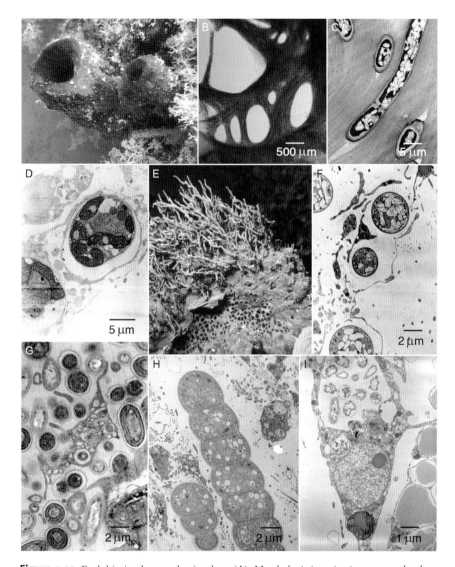

Figure 3.10 Endobiotic algae and microbes. (A) *Mycale laxissima, in situ* on peat bank at Twin Cays (picture width, 30 cm). (B) Cleaned fibre skeleton of *M. laxissima*, densely permeated by filamentous microalgae, the chlorophyte *Ostreobium* and the rhodophyte *Acrochaetium*. (C) Longitudinal section (TEM) of an *Acrochaetium* filament embedded among spongin fibrils. (D) *Gymnodinium* sp., the dinophycean symbiont of *Cliona varians*, differs by smaller size and in structural details of nucleus and chloroplast from the well-known *Gymnodinium microadriaticum* found in other clionaids and in hermatypic corals. (E) *Dictyonella funicularis in situ* on a patch reef, overgrowing a *Geodia*; its dull green colour is due to accumulations of "*Aphanocapsa raspaigellae*"-type cyanobacteria in its tissue (picture width, 20 cm). (F) The unicellular *D. funicularis* symbionts (TEM micrograph). (G) Unicellular "*Aphanocapsa feldmanni*"-type cyanobacteria in tissue of *Chondrilla caribensis*. (H) Filamentous "*Oscillatoria/Phormidium spongeliae*"-type cyanobacterium. (I) Bacteriocyte in larva of *Svenzea zeai*, with nucleolate nucleus, contains vesicle filled with bacteria, demonstrating vertical transmission of microbial symbionts. (For interpretation of the references to colour in this figure legend, the reader is referred to the online version of this chapter.)

through 16S rRNA sequences as *Synechococcus spongiarum* (Steindler *et al.*, 2005). These sponges include members of the genera *Geodia* (Astrophorida), *Chondrilla* (Chondrosida), *Neofibularia* (Poecilosclerida), *Xestospongia* (Haplosclerida), *Ircinia* (Dictyoceratida), *Aplysina* (Verongida), and more. Much rarer is the considerably larger "*A. raspaigellae*"-type cyanobacteria, found in two newly described species of mainly encrusting *Dictyonella* (Halichondrida) (Rützler, 1981; as *Ulosa*). A filamentous, "*Oscillatoria/Phormidium spongeliae*"-type cyanobacterium occurs in high density in *Hyrtios violaceus* Duchassaing and Michelotti (formerly known as *Oligoceras*), a common occupant of seagrass meadows in shallow lagoon waters. The same, or a very closely related symbiont, was recently discovered (first in Caribbean Panama) in two new species of the haplosclerid sponges *Haliclona walentinae* Diaz, Thacker, Rützler, and Piantoni and *Xestospongia bocatorensis* Diaz, Thacker, Rützler, and Piantoni (Diaz *et al.*, 2007; Thacker *et al.*, 2007), both described from shallow reef and mangrove habitats. We measured photosynthesis and respiration of the three compound species above and compared them with sponges symbiotic with high numbers of unicellular cyanobacteria (high content of chlorophyll *a*), namely, *C. nucula*, *Neopetrosia subtriangularis* Duchassaing, and *A. fulva* (Thacker *et al.*, 2007; K. Rützler, unpublished). Despite great variability, we calculated P:R ratios of 1.5 or more, showing that all qualify as phototrophic species. Nevertheless, even if a few more phototrophic sponges were added to the list, the population is much smaller than that in the oligotrophic zones of Australia's GBR, where 50% of the sponge biomass consists of small phototrophic species (Wilkinson, 1987; Wilkinson and Cheshire, 1990).

A few of our colleagues tested whether the production of dissolved organic nitrogen (DIN) through nitrification in some of the common symbiotic sponges might influence the nutrient cycle on the reef and in the mangrove. One study concentrated on three common species known to harbour cyanobacteria in high concentrations: *S. zeai* (as ?*Pseudaxinella*) from ca. 20 m depth on the forereef, *C. caribensis* (as *C. nucula*) from mangrove roots at Twin Cays, and *H. violaceus* (as *Oligoceras violacea*) from 2–3 m in turtlegrass in the lagoon; a fourth species devoid of cyanobacterial symbionts, the reef sponge *Plakortis halichondrioides* Wilson, served as a control (Diaz and Ward, 1997). In the DIN flux experiments, sponge-mediated nitrification proved to be an important process in these habitats, and species associated with cyanobacteria showed the highest production rates per weight of oxidized nitrogen reported from any benthic community. *P. halichondrioides* showed some nitrification activity as well, but to a much lower degree than in the species containing cyanobacteria; it is safe to assume that this nitrification is related to other types of endobiotic bacteria. Another study used molecular methods to detect ammonium-oxidizing bacteria of *Nitrosomonas* spp. lineage in common sponges from the Twin Cays mangrove (Diaz *et al.*, 2002, 2004b). Microbes of this nature were found in five of six widely distributed sponges: *H. implexiformis* (Haplosclerida), *L. isodictyalis*, *T. ignis* (Poecilosclerida), *Geodia papyracea* Hechtel (Astrophorida), and *C. caribensis* (as *C. nucula*) (Chondrosida). Apparently their occurrence is independent of symbiotic cyanobacteria (regularly present only in the latter two species). Only an undetermined species of *Spongia* (Dictyoceratida) gave negative results, possibly because of procedural problems in the experiments.

Because solid, subtidal substrates are in short supply in the mangal, most sponges grow on the prop roots of red mangrove (*R. mangle*) lining the ponds and tidal channels. Since it has already been established that the root-fouling community of sponges and ascidians protects mangroves from exposure to wood-boring isopods (Ellison and Farnsworth, 1990), some have questioned whether the trees might also receive a nutritional benefit from this association in an oligotrophic environment. Accordingly, specimens of the very common *T. ignis* (Poecilosclerida) and *H. implexiformis* (Haplosclerida) at Twin Cays were transplanted from populated to new, bare roots and nitrogen and carbon levels measured in these sponges (as well as in *S.* (as *Ulosa*) *ruetzleri* (Halichondrida) and in parts (rootlets, roots, twigs, leaves) of the new host tree (Ellison *et al.*, 1996). Wherever sponges were attached, prop-root growth increased significantly, and hair rootlets that permeate mangal epiphytes developed. The latter were comparable to those spreading into the substrate mud where new roots reach the bottom. Furthermore, when attached to the roots, the transplanted sponges grew faster than the controls fastened to plastic pipes. Stable isotope analyses indicated both flow of inorganic nitrogen from sponges to tree and tree-produced carbon into the fouling sponges, thus signalling a facultative mutualistic relationship between trees and sponges. This is a surprising outcome, further supported by controls with the similarly fouling red alga, *Acanthophora spicifera* Vahl, which did not seem to repay stilt roots for the use of their space as a substrate.

Unicellular eukaryotic symbionts (zooxanthellae) also occur in sponges, but almost exclusively in the hadromerid Clionaidae, genera *Cliona*, and *Cervicornia*. On the basis of morphological characteristics, established in part by transmission electron micrography, we distinguished two species of the dinophycean algae *Gymnodinium*. One, *G. microadriaticum* Freudenthal, occurs in the *C. caribbaea* complex, brown, papillate, or encrusting sponges, which, according to Zea and Weil (2003), actually include three species: *C. aprica*, *C. caribbaea*, and *C. tenuis*. *Gymnodinium* is also found in the inhalant fistules of the endopsammic *C. cuspidifera* (Vicente *et al.*, 1991, as *Spheciospongia*; Rützler and Hooper, 2000). This zooxanthella appears to be the same as that associated with protozoans, cnidarians (including the reef corals), and mollusks. The other alga, *Gymnodinium* sp., was found only in *C. varians* (formerly *Anthosigmella*). Its symbiotic stage differs from that of *G. microadriaticum* in size (being about half the diameter) and in some fine-structural details of the nucleus and chloroplast (Rützler, 1990). Identification will remain uncertain, however, until free-living and reproductive stages can be obtained (Anastazia Banaszak, personal communication). Although there is no reason as yet to doubt that symbioses with zooxanthellae are species-specific, recent molecular evidence suggests that infraspecific relations might be less rigid. Experiments conducted in the Florida Keys have shown that recolonization of partly bleached (aposymbiotic) *C. varians* by zooxanthellae does not necessarily involve the original algal strain left in undisturbed tissue areas of the same specimen (Hill and Wilcox, 1998). In this case, the repopulating algae were more related to a strain found in a sea anemone (*Aiptasia pallida*) that was attached near the experimental site than to that living in unbleached areas of the same sponge. Unfortunately, the study did not include fine-structure-morphological imaging that could be compared to our observations.

In 1998, an extended period of anomalously high seawater temperatures caused widespread coral bleaching on reefs in Belize and throughout the world (Aronson *et al.*, 2000). Under the temperature stress, host corals expelled their zooxanthellae and became bleached. To determine whether sponges might be affected in a similar way, we subjected specimens of *C. caribbaea* (which contains zooxanthellae) and *C. caribensis* (which contains zoocyanellae) to harsh temperatures. Placed tied down onto floating wooden frames with a stainless-steel mesh bottom, the sponges were kept fully exposed to sunlight but were cooled and fed by ambient water. Bleaching failed to occur, however, and the specimens died after 10–16 days of exposure. Elsewhere—both in the Florida Keys, as mentioned earlier (Hill and Wilcox, 1998), and in Puerto Rico—some bleaching has been reported, in the latter case involving the species *C. varians* (with zooxanthellae) and *Xestospongia muta* and *Petrosia pellasarca* de Laubenfels (Haplosclerida; with zoocyanellae) (Vicente, 1990). But unlike corals in the same habitats, sponges appear to experience rather localized bleaching, with only a few individuals of the population being affected. In Belize we saw or were informed about partly or fully "bleached" *X. muta*, but in each case the affected specimens were dead or dying, and the dead parts were held together only temporarily by their skeletons. The causes of these occasional die-offs are unknown but seem unrelated to recorded temperature highs.

Other sponge-endobiotic plants include algae that permeate or replace proper spongin skeletons. Two types of algae—the chlorophyte *Ostreobium* cf. *constrictum* and the rhodophyte *Acrochaetium spongicolum*—are found in the poecilosclerid sponge *Mycale laxissima*, a dark reddish to brown tubular species growing on shallow patch reefs and in well-flushed mangrove channels near Carrie Bow (Rützler, 1990). The sponge is supported by a rigid network of thick (0.3–3 mm) fibres that are cored by staggered bundles of large subtylostyloid spicules. Intertwined filaments of these algae are present along the spicules, boring and permeating the layered spongin of the fibres, and giving the otherwise clear fibres a strong greenish or reddish tint. No data are available on the possible benefit to either partner in this association, and it is unclear how the algae can obtain enough light for survival in such a dense, dark-coloured sponge. From ground and polished sections of samples embedded in epoxy resin, it appears that the strands of overlapping, glassy spicules could function much like a fibre-optics light conduit, as soon confirmed by the report of a similar phenomenon in an Antarctic sponge (Cattaneo-Vietti *et al.*, 1996). In some instances, relationships with higher algae are even closer, with the branching, calcified plants replacing the spongin skeleton of the sponges entirely. A common example in the tropical western Atlantic is the calcified rhodophyte alga *Jania adherens* Lamouroux which has become part of the dictyoceratid sponge *Dysidea janiae* Duchassaing and Michelotti and *J. capillacea* Harvey substituting for the skeleton of the poecilosclerid *Strongylacidon griseum* Schmidt. In both compound species, it is notable that the algae conform the shape of the sponges, occasionally protruding during a growth spurt, but then remaining confined within the exopinacoderm of the host. Similarly, the Indo-Pacific haplosclerid sponge *Haliclona cymaeformis* (Esper) has incorporated the red alga *Ceratodictyon spongiosum* Zanardini, which is thought to benefit directly from dissolved inorganic nitrogen (ammonia) excreted by the host sponge (Davy *et al.*, 2002).

5.2.3. Epizoism
Sponge-surface real estate can be quite valuable in some communities, serving an amazing variety of algae, fellow sponges, stoloniferan or embedded hydrozoans, anthozoans (*Parazoanthus*), crustaceans (barnacles), bryozoans (including Entoprocta), and ascidians. When Barry Spracklin, one of CCRE's earliest volunteer station managers and also a student of hydroid biology, investigated reef habitats off Carrie Bow, he found that of the 45 species collected only one, *Halecium bermudense* Congdon, which is dominant on the forereef and patch reefs, was regularly associated with sponges (Spracklin, 1982). An undescribed corynid hydroid, believed to belong to a new genus, also grew on the surface of some sponges: in the reef-crest zone, it was found on the excavating species *C. caribbaea* Carter, and in the spur-and-groove area of the forereef, it was on *Monanchora barbadensis* Hechtel (now accepted as *Monanchora arbuscula* (Duchassaing and Michelotti)). In a survey of hydroid distribution at Twin Cays and of their species composition, zonation, substrates, and quantitative abundance by our Canadian colleague Dale Calder (Calder, 1991a,b,c), 9 out of 49 species were found growing on sponges, including one described as new, *Turritopsoides brehmeri* Calder (1988), and one previously known only from the Seychelles islands in the Indian Ocean, which turned up in the Twin Cays main channel on the toxic *T. ignis*. Most other interactions between hydroids and sponges were found to be of a competitive nature, with

sponges, algae, and ascidians outcompeting the cnidarians on subtidal red-mangrove prop roots by overgrowing and smothering them.

Many reef sponges have a very close association with one of a half dozen species of the epizoic zoanthid *Parazoanthus*, some drab, others quite colourful, such as bright yellow or red and often a colour contrasting with that of the host (Fig. 3.11A). These zoanthids seem to be attracted to a select group of sponges, although not all individuals of a population may be affected. Sara Lewis, a graduate student at the time, studied the roles of the two partners of the association (Lewis, 1982), focusing on the sponges *Callyspongia vaginalis* (Lamarck) and *Niphates digitalis* (Lamarck) because they are common on the reef and a high percentage of specimens support the zoanthid *Parazoanthus parasiticus* (Duchassaing and Michelotti). Both sponges have tubular to vasiform bodies with single distal atrial openings, which makes it convenient to monitor the animals' metabolic activity levels with an unobtrusive thermistor-based flow-metre (Forstner and Rützler, 1969). Field observations and experiments showed that the frequency of zoanthid colonization is higher on larger, older sponges, and that, contrary to the original hypothesis, fish predation potential is not lowered, at least not by the fairly inconspicuous zoanthids studied. Also, flow measurements showed that *Parazoanthus*-populated specimens have a reduced variance of pumping rates, indicating an increased metabolic expenditure of the host. As the zoanthids clearly benefit from gaining competitor-free living space, in the particular species combinations investigated, the functional interactions tend to be of a more parasitic than mutualistic nature. However, the mutualism model may still apply to sponges associated with more brilliantly coloured zoanthids that may slow or prevent predation by fishes.

5.2.4. Endobioses (Fig. 3.11)

Much attention has been paid to sponges as habitats occupied by various specialists undeterred by repellent metabolites. These include a host of polychaetes, such as syllids—one even named for its habitat, *Haplosyllis spongicola* (Grube), another, *Syllis mayeri*, described as a new species from Carrie Bow where it was found in the sponge *I. strobilina* (Musco and Giangrande, 2005). Then there are various copepods, decapods (shrimps, crabs), amphipods, brittle stars, and a few fishes (blennies, eels). Many of these tenants do not hesitate to take a few bites out of the host tissue now and again. On the other hand, some turn out to be quite useful in fending off predators or clearing fine sediments off a surface containing incurrent pores.

In some instances, little is yet known about the tenant, such as the new species of parasitic copepod, *Asterocheres reginae* Boxshall and Huys (1994), described from the sponge *Agelas clathrodes* Schmidt by visiting researchers based at London's Natural History Museum. The sponge was collected from a reef off Carrie Bow, almost in view of the field station's kitchen and the crustacean named after one of our excellent cooks, Regina, who always

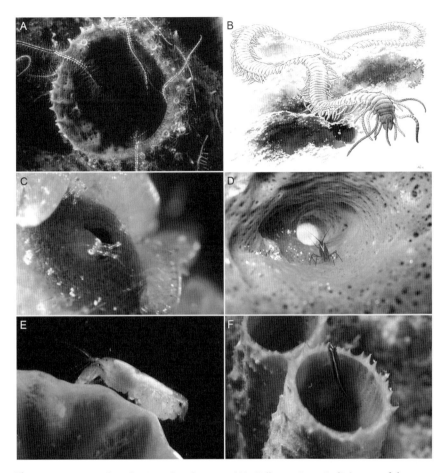

Figure 3.11 Examples of epi- and endozoans. (A) *Callyspongia vaginalis* is one of the most common tubular sponges on the reefs and many support epizoic *Parazoanthus parasiticus*; this specimen also offers shelter to brittlestars, such as *Ophiothrix lineata* (picture width, 10 cm). (B) Artist's rendering of *Syllis mayeri*, a newly described polychaete inhabiting *Ircinia strobilina* (picture width, ca. 5 cm). (C) *Leucothoe* sp., a new species of amphipod from *Aplysina fulva* (picture width, ca. 2.5 cm). (D) Shrimp, *Lysmata pederseni*, retreating into tube of *Callyspongia vaginalis* (picture width, ca. 10 cm). (E) Queen of eusocial shrimp *Synalpheus regalis* on host sponge, *Lissodendoryx colombiensis* (picture width, ca. 5 cm). (F) Sponge-endobiotic fish, *Gobiosoma* cf. *xanthiprora* Boehlke and Robins (picture width, 12 cm). (Illustration credits: A, Roger Hanlon; B, painting by Alberto Gennari, Museo dell'Ambiente, University of Salento; C, James Darwin Thomas; D, Antonio Baeza; E, Emmett Duffy; F, Chip Clark.) (For colour version of this figure, the reader is referred to the online version of this chapter.)

kept an eye on nearby divers while she prepared the next meal. The copepod's feeding habits and damage, if any, to the host remain unclear. However, studies of live specimens and electron micrographs have shed considerable light on other tenants, such as the amphipod (*Anamixis hanseni* Stebbing) found inside some calcareous sponge (as well as in ascidians). Long suspected of piercing and sucking host tissue, it was shown to be commensal in its feeding habit after close examination of the crustacean's modified and reduced mouth parts (Thomas and Taylor, 1981). While inside their hosts, the animals were found to trap small organic particles on a filter formed by tufts of setae on their gnathopods. Upon sampling many more sponges and other potential hosts for commensal amphipod species, our colleagues Jim Thomas and (then) graduate student Kristina Klebba described several new species from Belize, including *Leucothoe ashleyae* from *C. vaginalis*, *Niphates erecta* Duchassaing and Michelotti, *Aiolochroia crassa* (Hyatt) (listed as *Pseudoceratina*), and *Amphimedon compressa* (Duchassaing and Michelotti); *L. kensleyi*, named after Smithsonian carcinologist and longtime CCRE staff member Brian Kensley, from *C. vaginalis* (Thomas and Klebba, 2006); *L. barana* from *A, compressa, C. podatypa, I. birotulata, Leucetta imberbis* (Duchassaing and Michelotti), *N. erecta, S. zeai*, and *T. ignis*; *L. garifunae* (named for the Garifuna culture, which prevails on the mainland, in Dangriga) from *I. birotulata, A. crassa* (as Pseudoceratina), and *L. imberbis*; *L. saron* from *A. crassa* (as *Pseudoceratina*); and *L. ubouhu* (Garifuna is the term for "island") from *C. varians, Hyrtios* sp., *L. isodictyalis, M. laxissima*, and *Spongia obliqua* Duchassaing and Michelotti (as *S. officinalis* subsp. *obliqua*) (Thomas and Klebba, 2007). In all, 20 sponge species are listed as hosts for these amphipods.

A bit further up the evolutionary tree of crustacean sponge inquilines are the caridean shrimps. One of these, *Lysmata pederseni* Rhyne and Lin (Lysmatidae), is a common occupant of the tubular *C. vaginalis* where it often occurs in pairs. In studying the mating behaviour of these shrimps, Antonio Baeza, one of our postdoctoral fellows, found that they are monogamous hermaphrodites and that the pairs consist of either two hermaphrodites or one hermaphrodite and one male (Baeza, 2010). The hermaphrodites are biased towards the development of female reproductive structures, and the pairs seem to mate with each other exclusively for a long period of time. The monogamous mating behaviour appears to be highly beneficial on a reef, a structured habitat with many predators, where suitable shelters, such as this host sponge, are relatively rare.

In the late 1980s, (then) Smithsonian postdoctoral fellow Emmett Duffy joined our team to study the ecology, behaviour, and genetics of one of the most diverse groups of sponge "guests," the snapping shrimps (Alpheidae). Every diver is familiar with the loud clicking sound that these animals make with their large claw, which makes them look like a miniature version of the Maine lobster. Upon comparing species of the genus *Synalpheus* living in the sponges *S. vesparium* and *A. clathrodes* from Belize, Panama, and Florida, with and without planktonic larvae, Duffy noted that genetic diversity was much higher in the species with lower dispersal potential and developing directly without swimming larvae (Duffy, 1993). Subsequent research by Duffy and collaborators led to the discovery and description of a number of new alpheid taxa, among them *Synalpheus regalis* Duffy (1996a), so named because only a single female (queen) inhabited the colony. This species lives in sponges exclusively and was found in *Xestospongia* (now *Neopetrosia*) *subtriangularis* Duchassaing and *Hyatella intestinalis* Lamarck on the outer reef ridge off Carrie Bow Cay. Ten more new alpheid species were described by the team over the next decade (Duffy, 1998; Ríos and Duffy, 1999, 2007; Macdonald and Duffy, 2006). Eventually, Ríos and Duffy had accumulated enough material to recognize that a new genus may be required to separate most *Synalpheus* species because of two consistent morphological structures on their pereopods, including a brush of setae (Ríos and Duffy, 2007). The proposed new genus, *Zuzalpheus*, incorporated the Mayan word *zuz* (brush) in its name—most appropriate for our research area—but because of taxonomic technicalities, this new taxon was not generally accepted by specialists in the group (E. Duffy, personal communication). At least 16 sponge species from the Belize barrier reef off Carrie Bow Cay serve as hosts for *Synalpheus*, all in the genera *Spheciospongia*, *Lissodendoryx*, *Hymeniacidon*, *Agelas*, *Niphates*, *Calyx* (as *Pachychalina*), *Oceanapia*, *Xestospongia*, *Hyatella*, and *Aiolochroia* (as *Pseudoceratina*). One of the alpheids was the above-mentioned *S. regalis*, which became famous as the first marine species known to have eusocial organization, where reproduction is reserved for a few individuals in a colony, whereas most other members help raise and defend the offspring (Duffy, 1996b, 2007). This social organization had hitherto been known from only a few insects (ants, termites, honeybees) and from African mole-rats. Duffy and his group examined many more sponge-dwelling alpheid species, found eusociality to be widespread, and contributed a wealth of data to research on host specificity of the shrimps, their morphological differentiation, evolution, behaviour, and colony assembly, structure, and defence (Duffy and Macdonald, 1999; Duffy, 2002; Duffy *et al.*, 2000, 2002; Morrison *et al.*, 2004; Tóth and Duffy, 2005, 2008; Macdonald and Duffy, 2006; Macdonald *et al.*, 2006; Tóth and Bauer, 2007).

Brittle stars (Ophiuroidea) and certain fishes (apognids, blennioids) are often found inside or around sponges but do not seem as selective as alpheid shrimps. When mentioned in the literature, their species are not always listed (Greenfield and Johnson, 1981), although *C. vaginalis* is often named, probably because it is easily identified owing to its tubular shape, convenient for hiding (Hendler, 1984; Gilbert and Tyler, 1997). One of its associates is the ophiuroid *Ophiothrix lineata* Lyman, whose feeding behaviour has been studied by Gordon Hendler, formerly at the Smithsonian but now an echinoderm specialist at the Los Angeles County Museum (Hendler, 1984). Using experiments and cinematography, Hendler found that *Ophiothrix* not only lives in *C. vaginalis* but also sweeps up detritus particles stuck to the sponge surface and too large to be drawn through the inhalant pores. This sweeping removes silt that slows the sponge's pumping activity and is also of benefit to the brittle star in providing it shelter. Another sponge shown to be desirable as a hideout for brittle stars is *Mycale laevis* (Carter): a specimen extracted during a collecting trip to southern Belize contained nearly 250 specimens of an *Ophiactis* (Hendler and Pawson, 2000).

5.3. Disease and status of commercial sponge fishery

Although sponges seem to be much more resistant to environmental stress than corals, one must not ignore the agents that occasionally leave specimens dead or dying, both on the reef and in the mangrove (Fig. 3.12). A cover of whitish bacterial filaments (*Beggiatoa* sp.), a sure sign of decay, is particularly noticeable on some of the beautiful large reef sponges. At times, however, it may be impossible to distinguish between disease agents and tissue decay in a community, as in preserved samples of dead *X. muta* and *Callyspongia plicifera* (Lamarck) sent to us at the Smithsonian. A few cases of *Aplysina* red-band syndrome (ARBS), first reported in 2004 in the Bahamas (Olson *et al.*, 2006) and now considered widespread in the Caribbean, prompted extensive reef surveys off Carrie Bow Cay and monitoring of affected specimens (study in progress by D. G. Gochfeld, J. B. Olsen, C. Piantoni, K. Rützler, R. W. Thacker, and E. Villamizar, and M. C. Diaz). ARBS seems to be triggered by a filamentous cyanobacterium, still under study. The saprophagous disease community of bacteria, diatoms, ciliates, polychaetes, and others is dominated by the filamentous red alga *Polysiphonia*, which produces the characteristic red colour. Another syndrome caused by a filamentous cyanobacterium was observed in *N. digitalis*. Since cases of diseased sponges remain sporadic at this time, the study will be continued when more material becomes available.

Early in our CCRE programme, we found up to 90% of *G. papyracea* in the Twin Cays mangrove diseased or dying, without any obvious signs of a pathogen (Rützler, 1988). Under fine-structure examination, symbiotic coccoid cyanobacteria normally present looked healthy and appeared to be flourishing among the decaying sponge cells, while the sponge archaeocytes seemed unable to control symbiont numbers by phagocytosis and shedding. Inasmuch as the population replenished itself over the years and remained free of new infections, we assumed that stress due to an environmental event, not a disease agent, may have caused the epidemic.

In the late 1930s, an even more historic event occurred nearby Turneffe Islands when commercial sponges were devastated by mass mortality. The government of British Honduras (Belize) had established a sponge-planting industry in Turneffe lagoon in 1926, even though the two most promising species—sheepswool (*H. lachne*) and velvet (*H. gossypina*)—were considered to be of low or average quality, respectively (Stuart, 1948). The main

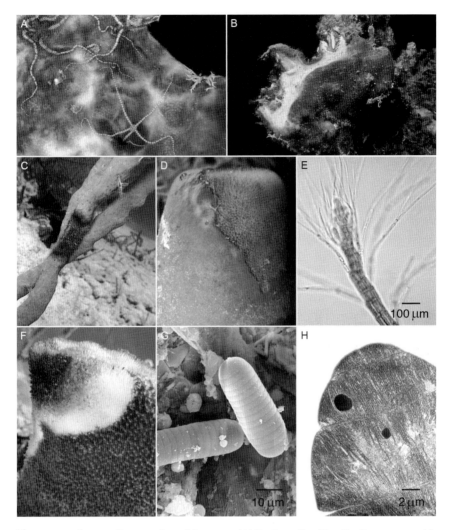

Figure 3.12 Sponge disease and possible agents. (A) Surface of healthy *Geodia papyracea* with associated ophiuroids, from Twin Cays mangrove (picture width, 12 cm). (B) Loss of endobionts and exposure of choanosome after histolysis caused by uncontrolled population explosion of symbiotic cyanobacteria; presumed cause was temperature stress affecting the sponge (picture width, 15 cm). (C) *Aplysina* red band syndrome (ARBS) affecting *A. cauliformis* on a patch reef (picture width, 8 cm). (D) Fresh lesion of undetermined cause on *Aplysina fistularis*, on the forereef (picture width, 8 cm). (E) Filamentous rhodophyte alga, *Polysiphonia* sp., associated with ARBS and lending characteristic red colour to the wound tissue. (F) Diseased *Niphates digitalis* on the forereef; the exposed, whitish skeleton area (to the left) is secondarily invaded by various filamentous algae and diatoms (picture width, 12 cm). (G) These filamentous cyanobacteria (SEM micrograph) are the only foreign microorganisms associated with the *Niphates* wound tissue and are under study as possible pathogen. (H) TEM micrograph of cyanobacterium shown in G. (Photo credits: C, Deborah Gochfeld; D, F, Carla Piantoni; others by author.) (For interpretation of the references to colour in this figure legend, the reader is referred to the online version of this chapter)

objective was to supply cuttings for concessions granted to private individuals. Despite considerable financial investments, the operation was not very successful, partly because of slow sponge growth and the government's inability to provide enough cuttings. By 1939, the established plantations of about 800,000 sponges (one-third wool, two-thirds velvet sponges) were destroyed by a disease attributed to a filamentous microbial organism, possibly a fungus that had spread from the Bahamas through the Florida Keys and Cuba to Belize (Smith, 1939, 1941; Galtsoff, 1942). Only about 250 kg of the sponges could be salvaged and sold in 1940. Subsequent test plantings of sheepswool cuttings showed a survival rate of only 2% after 5 years, indicating that the blight was still active in the area. The filamentous disease organism could not be positively identified but was thought to be different from bacteria or fungi that appear in sponges after death from other causes, such as extreme fluctuations in salinity. Although records of environmental conditions at the Turneffe Islands culture sites were scarce in those days, local observers had indeed reported a period of exposure to unusually high salinity, attributed to extremely low tides and rainfall and a spell of very hot weather (Smith, 1941).

In 1989, colleagues rediscovered and investigated the old Turneffe Islands lagoon sponge farms (Stevely and Sweat, 1994), focusing on the locally most valuable species, the velvet sponge (*H. gossypina*) (Fig. 3.13). Of the nine locations examined within the area of the old plantings (by divers with mask, snorkel, and fins towed for 30–90 min behind a boat), only one had a sizable population of velvet sponges in depths of 1–2 m. The finding was impressive enough, however, to suggest the potential for reviving a fishery and to call for further surveys to determine the magnitude of the resource. At one location, along the margin of one of the presumed old farms—the exact position and extent of which are not available from the literature—researchers found several of the characteristic concrete disks used by farmers to anchor cuttings from larger sponges until they regenerated and grew back to harvestable size.

Observations from the Bahamas and Turneffe sponge farms seem to agree with our assessment of possible impacts on noncommercial species, mainly that stress reduces sponge resistance to otherwise benign associated microorganisms. From a review of the literature on commercial species (of the genera *Spongia* and *Hippospongia*) throughout the West Indies, his own observations, and information from local fishermen, Vicente (1989) learned that up to the first half of the twentieth century, these sponges were ubiquitous in shallow water and commercially exploited to some degree throughout the area. After a series of widespread mortalities, most populations disappeared from all habitats in the southern parts of the region, particularly the Greater and Lesser Antilles (Hispaniola, Jamaica, Puerto Rico, Virgin Islands), and never returned. On the other hand, the equally affected populations in the northern parts of the region (Gulf of Mexico,

A

Figure 3.13 Turneffe Islands atoll, location of historical sponge farms during the 1930s. (A) Aerial view of southeast Turneffe, looking north across the reef towards the southern lagoon where some of the farms operated. (B) A fine specimen (24 cm diameter) of velvet sponge (*Hippospongia gossypina*) collected at Turneffe by John Stevely in 1989. (C) Two of the original concrete disks (9.5 cm, 11 cm diameter), also retrieved by Stevely, used to stabilize sponge cuttings at the farms. (Photo credit: A, Ilka Feller, taken during a LightHawk survey flight in 2008; others by author). (For colour version of this figure, the reader is referred to the online version of this chapter.)

Florida, Cuba, Bahamas) eventually recovered and could again be commercially harvested. In Vicente's view, the species in these two genera may have evolved under a somewhat cooler climate than exists at present and may have succumbed or become vulnerable to slight but long-lasting temperature elevation in the southern West Indies.

5.4. Impact of sediment

Calcareous sediment of all sizes, a ubiquitous component of the reef environment, has a strong impact on sessile organisms. Derived mostly from corals and calcified algae—to a lesser extent from foraminiferans and miscellaneous invertebrates such as calcareous sponges, mollusks, and echinoderms—it affects the development and ecology of many sponges, particularly in the sandy and muddy lagoon (Fig. 3.14). Although sponges contribute only a minor amount of skeletal material to modern seas, they play a substantial role in producing limestone mud through bioerosion (see Section 5.5). At the same time, sponges can benefit immensely from incorporating calcareous particles or adopting a morphological and physiological strategy for living buried in sand.

Figure 3.14 Psammobiontic and detritus–adapted sponges. (A) *Tectitethya crypta* inhabits shallow sandy lagoon bottoms and lives covered and permeated by sand (picture width, 30 cm). (B) *Spheciospongia vesparium* has similar appearance and distribution as *Tectitethya*; a perpendicular section through one of the oscula shows the distribution of sand grains in the body, an anatomical adaptation for survival and reattachment after being dislodged by storms (picture width, 12 cm). (C) Inhalant (ectosomal) fistules of *Cervicornia cuspidifera* protruding from sand near a lagoon patch reef; the diver is about to apply fluorescent dye to show water flow into the fistules and out through the surrounding sand bottom (picture width, 80 cm). (D) *Haliclona tubifera* anchored in and covered by mangrove detritus, with sea anemones, *Aiptasia*, protruding from its base (picture width, 30 cm). (For colour version of this figure, the reader is referred to the online version of this chapter.)

The Caribbean staghorn sponge, *C. cuspidifera* (Lamarck) (previously as *Spheciospongia*) (Hadromerida), is a good example. It is common in shallow reef sands but only the branching inhalant fistules are seen protruding from the bottom, whereas most of the sponge body is buried and was only discovered after extensive digging (Rützler, 1997; Rützler and Hooper, 2000). Although the presence of zooxanthellae in the exposed structures may help the uptake of nutrients, the sponge is a fully functional filter feeder and takes in water through pores in the fistules, filters it through the choanosomal parts of the buried body, and expels it through oscula deep below the sand surface. Sand is incorporated throughout the buried part of the body and stabilized around it so that the sponges serve as an early stage in the generation of hard-bottom benthos communities, such as patch reefs.

Two less conspicuous sponges in the same habitat are the mostly buried *Oceanapia peltata* (Schmidt) (Haplosclerida) and *Tectitethya crypta* (de Laubenfels) (Hadromerida). *Oceanapia* has the same general body structure as *Cervicornia*, although it belongs to a different order, and it too takes in water through fistules protruding from the sand, except that in this case the latter are small, delicate, and pagoda shaped. *Tectitethya* is built like a regular sponge, conical to cylindrical, massive, usually with just one large osculum and many ostia partly arranged in groups. This species is deeply embedded and anchored in sand and also regularly covered by sediment. When colleagues from the universities of Genova and Ancona, Italy, examined this and 12 other species found on sandy lagoon bottoms, they noticed that all incorporated sand in their bodies, 8 of them (including the very common *Lissodendoryx strongylata* van Soest, *Amphimedon viridis* Duchassaing and Michelotti, and *C. schoenus*) selecting for particles larger than 5 mm. Five (including *Haliclona caerulea* Hechtel and *T. crypta*) made no distinction and took up all size fractions occurring in the surrounding sand bottom (Cerrano *et al.*, 2004). Interestingly, *T. crypta*, which uses sand as structural support, is able to sort and organize the incorporated sediments in beneficial ways by cellular transport, depending on whether the sponge is in a rolling phase, is stabilized but resting on the substrate surface, or is buried.

5.5. Bioerosion

In the course of many of our surveys, we encountered and identified the principal limestone-excavating sponges, most of which were in the hadromerid family Clionaidae (primarily, *Cliona amplicavata* Rützler, *C. aprica* Pang, *C. caribbaea*, *C. delitrix* Pang, *C. schmidti* (Ridley), *C. varians*, and the haplosclerid Phloeodictyidae (*Aka coralliphaga* (Rützler), *A. xamaycaense* (Pulitzer-Finali)) (Fig. 3.15). We also investigated several ecological and palaeoecological factors, such as rates of excavation and production of fine sediments. We had already shown how sponge cells chemically carve out small but distinctive chips from the substrate (Rützler and Rieger, 1973) and thus contribute significantly

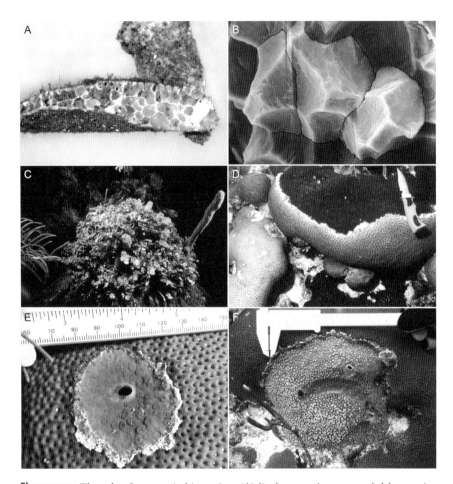

Figure 3.15 The role of sponges in bioerosion. (A) Broken-up *Acropora* coral slab exposing extensive bioerosion by the hadromerid *Cliona amplicavata* (picture width, 15 cm). (B) Carbonate chips edged by clionaids from coral and other calcareous substrata, leaving behind characteristically scalloped surface patterns; the chips make up much of the mud-size sediment on the reef (picture width, 70 μm). (C) The haplosclerid *Aka coralliphaga* is another highly corrosive reef sponge, commonly attacking live coral (picture width, 40 cm). (D) *C. caribbaea*, an encrusting species, can overpower heat-stressed corals, like this massive *Diploria*, by undermining the peripheral live polyps (picture width, 50 cm). (E) An early growth stage of *C. delitrix*, which too can overpower live corals. (F) More advanced growth of *C. delitrix*, which is commonly found associated with the zoanthid *Parazoanthus*. (Photo credits: C, Chip Clark; E, F, Andia Chaves-Fonnegra; others by author.) (For colour version of this figure, the reader is referred to the online version of this chapter.)

to the microgranular fraction (4–30 μm) of the carbonate sediment. By monitoring chip production and the loss of mass of previously weighed substrate, we obtained data on excavating activity and rates (Rützler, 1975).

A question of particular interest was how much faster the principal boring organisms (i.e. sponges, polychaetes, sipunculans, and bivalves) could penetrate corals with a highly porous texture than those with a very dense structure. Surprisingly, we found that penetration into dense substrate is quicker, and also that porous coral species are not necessarily the ones that grow faster (Highsmith, 1981). These traits may be of adaptive value because the endolithical lifestyle is geared to protection, and many predators, like triggerfish or wrasses, have powerful jaws and can easily consume or expose the borers. These conclusions were confirmed by measurements of growth rates, density, and bioerosion in three common massive corals near the Carrie Bow laboratory (Highsmith *et al.*, 1983). Sponges were found to be responsible for 85–94% of all excavations, and dense species were more bored than porous ones.

Many excavating sponges are fairly inconspicuous with most of the body hidden in the substrate and only oscular and ostial papillae showing. However, some rather corrosive species grow from a papillate (alpha) body into beta form, where the papillae merge and expand, ultimately covering the substrate with a more or less thin, continuous crust. One example is the brownish to blackish (owing to the presence of zooxanthellae) *C. caribbaea*, a common shallow-water species on reefs near Carrie Bow. Zea and Weil (2003) believe some members of this population belong to another species, *C. tenuis* Zea and Weil. Following a series of hurricanes preceded or accompanied by very calm seas and high temperatures, we noticed this sponge overgrowing many corals in shallow water, including very large, old, massive species of *Montastraea* and *Diploria*. An experimental study showed that the sponge is both undermining coral calices and growing over the live surfaces of corals that are temperature-stressed and unable to fend off the aggressor (Rützler, 2002c). Similar research was conducted by Andia Chaves-Fonnegra, for a dissertation with advisor Sven Zea of the National University of Colombia. The species she concentrates on is *C. delitrix* and preliminary results confirm the suspicion that it has become a threatening bioeroder of corals. These observations add strength to the idea that many kinds of coral stress (temperature, pollution) may lead to irreversible destruction of the reef framework.

6. Sponge Silica in a Carbonate Environment

Sponges contribute only modest amounts of minerals to modern seas, although excavating species convert solid limestone to mud by eroding their substrates. Most species produce siliceous spicules, which are released after death (Fig. 3.16). One organism, a newly described foraminifer, puts liberated spicules to good use by agglutinating them for its own skeleton

Figure 3.16 Silicon in a carbonate environment. (A) The large sessile foraminifer *Spiculidendron corallicolum*, first described from Carrie Bow reefs, is common in entrances of reef caves (picture width, ca. 7 cm). (B) *S. corallicolum* agglutinates exclusively sponge spicules for its test. (C) *S. corallicolum*, branch near the stem showing tight cementation of sponge spicules. (D) *Xestospongia muta* with diver on the forereef, demonstrating that substantial quantities of biogenic silica standing stocks in this carbonate environment, are contributed by sponges; on average, 58% of the dry weight of this sponge is made up of siliceous spicules (picture width, 2 m). (E) SEM micrograph of spheraster from freshly collected reef sponge, *Placospongia* (picture width, 100 μm). (F) Similar spherasters etched to varying degrees, as encountered in reef sediments after release from the sponge (picture width, 65 μm). (Photo credits: A, Paul Humann; D, Dustin Kemp; others by author.) (For colour version of this figure, the reader is referred to the online version of this chapter.)

(Rützler and Richardson, 1996). This unusually large (up to 50 mm) arborescent foraminifer, named *Spiculidendron corallicolum*, has been long known to divers as "spicule tree." Although it is normal for the group to agglutinate sediment particles to form a test, this organism selects and cements siliceous sponge spicules exclusively.

Generally, however, sponge spicules, once liberated, become etched and eventually dissolve in a calcium-carbonate-rich, silicon-undersaturated environment like a coral reef (Rützler and Macintyre, 1978). They are the main component of particulate silica in sediments on the outer reefs bordering the continental slope. The various stages of dissolution, revealed in SEM images, are particularly obvious on delicate surface ornaments of many microscleres and from the enlargement of axial canals in monaxon and tetraxon megascleres (oxeas, styles, triaenes, etc.) and asteroid microscleres (oxyasters, sterrasters, selenasters, etc.). As silicon is such a rare element on the reef, it is rapidly taken up and recycled by sponges growing in the neighbourhood.

Despite the obvious etching patterns on liberated sponge spicules, no reliable measurements of dissolution rates are yet available, although they are certainly slower than those of the much more delicate frustules of diatoms, the principal photosynthetic plankton organisms generally thought to be driving the silicon cycle in the sea. This may well be the case for the euphotic zone of the open ocean, but measurements of biogenic silica (bSi) in standing stocks contributed by planktonic diatoms and benthic sponges in the major habitats (reef, mangroves, turtlegrass beds) of the Belize outer reef platform suggest otherwise (Maldonado *et al.*, 2010). bSi in sponge communities alone was nearly 90% of the total there, with diatoms and ambient silicate accounting for the rest. Clearly, future silicon budgets for a coastal region should include the Si standing stock in sponges, which, in contrast to diatom bSi, cycles are independent of primary production.

7. NATURAL PRODUCT CHEMISTRY AND ALLELOPATHY

Sponges, like algae and representatives of several invertebrate phyla, are more or less attached for life, although a few are able to relocate over small distances or reattach when torn off their substrate. That sessile lifestyle and the lack of protective hard structures, such as shells, caused them to develop a vast arsenal of secondary metabolites to ward off predators. As many of these chemicals are thought to have strong pharmaceutical potential for humans, some collaborators have extracted and tested a number of them. Others have focused more on ecological effects, that is, allelopathic means of space competition and sessile invertebrate recruitment.

Chemists from the Scripps Institution of Oceanography and Stanford University, California, for example, launched a search for new sponge sterols. In specimens of the verongid *Aplysina cauliformis* Carter (listed as *Verongia*) collected on the forereef just off Carrie Bow Cay, Kokke *et al.* (1979) isolated not only two compounds previously discovered from this species but also five new sterols, bringing the total to seven. In comparing guanidine alkaloids from the poecilosclerid sponge *Monanchora arbuscula* collected in Brazil and Belize, another team, from Belgium and The Netherlands, came upon a new compound in this group, which is known for cytotoxic and antimicrobial components (Tavares *et al.*, 1995). This new knowledge of the sponge's chemistry led them to discover taxonomic inconsistencies in the myxillinid genera *Monanchora*, *Crambe*, and *Batzella*.

Two Scripps chemists focused on the ecological implications of secondary metabolites produced by the limestone-excavating haplosclerid sponge *Aka coralliphaga* (as *Siphonodictyon coralliphagum*) (Sullivan and Faulkner, 1990). Several species in this genus excavate live coral boulders by carving and etching out single cavities that can attain 10 cm or more in diameter. Their study showed that secondary metabolites (various forms of siphonodictyal, named A–H) produced a dead zone on the live coral surrounding the inhalant papillae and exhalant oscular chimneys, thus keeping the sponge from being overgrown by the host coral. This confirmed previous field observations (Rützler, 1971) suggesting that a mucuous secretion from the sponge kept the nearby coral tissue in check through the effect of metabolites in the mucus. The same compound may also aid sponge larvae in the initial penetration, which often appears to take place on the top of live corals. The chemical ecologists noted, too, that the metabolites appear to help transport calcium ions, a byproduct of the excavating process, from the sponge's interior to the surrounding sea water. Two decades later, both results—defence against coral overgrowth and removal of calcium ions produced by bioerosion—were independently confirmed by another team working with material from *S. coralliphagum* (= *Aka coralliphaga*) collected in the Bahamas (Bickmeyer *et al.*, 2010).

If sponge secondary metabolites aid aggression and serve as feeding deterrents, one might easily conclude that they may also inhibit the settlement of potential competitors in benthic habitats of limited space such as mangroves. To address this question, colleagues from Florida at the time conducted field experiments comparing Belizean and Florida mangrove and seagrass habitats (Bingham and Young, 1991). Contrary to the results of laboratory experiments presented in the literature, none of the sponges tested in the field repelled nearby settlement of invertebrate larvae. In fact, some ascidians, sea anemones, and oysters seemed to be somewhat attracted to established sponges. As the authors suggest, the laboratory trials may not have been effective in reproducing conditions in the undisturbed habitat. In studying chemical defences and space competition among sponges on reefs in Guam, Thacker *et al.* (1998) identified allelopathic terpenoid compounds that indeed were effective in space competition and deterred fish spongivory as well, showing that complex mechanisms are involved and thus require much more field research to delineate the compounds and their effects under natural conditions.

8. Conclusion and Outlook

The Mesoamerican Barrier Reef is a multifaceted ecosystem containing habitats ranging from several types of reefs and mangrove islands to seagrass meadows. Sponges, a prominent feature in most locations, play an important ecological role in the system, as established by a large body of research over the past 40 years, particularly in the Carrie Bow area. Of the 113 scientists who have worked there to date, 88 (78%) have conducted in-depth fieldwork, while the others have participated in analyses and co-authored resulting publications. Of the field researchers, 63 (72%) studied sponges directly, and the remaining 25 (28%) worked on sponge associates. Systematics, developmental biology, ecology, symbiosis, predators, and inquilines were the principal subjects. The results of this work—thus far published in 125 scientific papers, with many more still in progress—demonstrate the enormous value of a well-run field station located adjacent to the habitats under study and open to experts from many disciplines, not necessarily sponge-related ones. Collaboration or just the exchange of ideas and methods in such a setting broadens marine research immeasurably.

Needless to say, much remains to be discovered about Porifera. Many species—an estimated 35% of the population off Belize—are still unde-scribed or so poorly known that they cannot be identified with confidence. Although molecular methods are opening exciting opportunities and chan-ging many taxonomic and evolutionary concepts, it will take time to resolve methodological complications. Above all, we should not be tempted to discard more than two centuries of published morphological and natural historical information in order to arrive at quick assumptions about biodi-versity data. What is essential to learn from nature is not only how many species it has been able to accommodate, but also the organisms' morpho-logical diversity, life histories, behaviour, physiological adaptations and tolerances, and ecological roles in communities and ecosystems.

Sponge science has obviously benefited greatly from scuba diving, one of the striking innovations in research methods of the twentieth century, providing researchers a direct means to observe, experiment, photograph, film, or videorecord live communities (Rützler, 1996, 2004). Although mixed-gas equipment has extended the depth range of diving scientists considerably, safety concerns restrict most research to the uppermost 50 m. On the Mesoamerican Barrier Reef and elsewhere, submersibles and remote-controlled vehicles have shown that rich reef and sponge bottoms may only start at that depth, and hence that efforts to observe and sample these deep mesophotic reef communities in a safe manner should be encouraged, despite their high cost.

At the VIIIth World Sponge Conference in Girona, Spain, in 2010, a record number of sponge scientists and students demonstrated in over 350 contributions a strong investment in traditional and innovative research disciplines and strategies important to meeting many of the world's current environmental challenges. The following trends are particularly encouraging:

• Classical systematics based on morphology and ecological observations and experiments in the field is alive and well—and enhanced and complemented by ever more refined molecular techniques—which have become indispensable in recognizing phylogenetic and evolutionary relationships. Data bases posted on the Worldwide Web and interactive and richly illustrated (in colour) field guides will be essential tools in research and teaching.

- Studies of ecology and ecophysiology will shed more light on the role of sponges in marine communities—particularly in tropical shallow-water coral reefs—during times of climate change, ocean acidification, and increased pollution from growing human populations. They will also make clearer the kind of aquaculture needed to produce sustainable seafood and bioactive compounds. Many aspects of sponge feeding, species interactions, and space competition are weakly understood, as are their suitability as biomonitors or biological filters that may indicate or deal with certain pollutants, such as trace metals and oil seeps or spills.

- Technical advances and new allocations of resources will enable science to set new boundaries in the study of poorly known marine communities and ecosystems, such as deep reefs and deep-sea bottoms.

- New findings and study techniques in the field of microbial symbioses will improve the characterization of microbes, permit their possible culture, and clarify their role in sponge-metabolic processes and disease. There is also much to be learned about the kinds and importance of cyanobacteria and zooxanthellae in sponges, not to mention the fate of hosts and symbionts during temperature stress. Sponge diseases and their causes constitute another important topic in itself.

- Last but not least, sponge silica cycling in the sea merits close attention, with an emphasis on the relatively new field of sponge-generated bio-composits such as carbonate, silica, chitin, and spongin, and various combinations thereof.

ACKNOWLEDGEMENTS

I am grateful to the organizers of the VIIIth World Sponge Conference for inviting me to present a summary of our sponge research at the Carrie Bow Marine Field Station. Thanks are also due to all researchers, photographers, and artists who have joined us there and contributed to work related to the biology of Porifera.

Some friends and colleagues who have supported our research and logistics in many ways deserve special mention: Arnfried Antonius, Therese Bowman-Rath, Mike Carpenter, Chip Clark, Cristina Diaz, Ilka Feller, Joan Ferraris, Carl Hansen, Marty Joynt, Ernst Kirsteuer, Sarah Klontz, Michael Lang, Ian Macintyre, Vicky Macintyre, Dan Miller, Michelle Nestlerode, Tom Opishinski, Mary Parrish, Valerie Paul, Laurie Penland, Carla Piantoni, Hans Pulpan, Tony Rath, Raphael Ritson-Williams, Martha Robbart, Molly Ryan, Kjell Sandved, Marsha Sitnik, Kate Smith, Robyn Spittle, Yolanda Villacampa, and countless volunteer station managers who helped the Carrie Bow lab operation. Authors of illustrations are gratefully acknowledged; photos not credited are by myself.

Thanks also go to Belize Fisheries, the Bowman-Rath family, and the staff of Pelican Beach Resort, Dangriga, Belize (including captains, engineers, cooks, carpenters, and boatmen).

This is contribution number 905 of the Caribbean Coral Reef Ecosystems (CCRE) Programme, Smithsonian Institution, supported in part by the Hunterdon Oceanographic Research Fund.

REFERENCES

Alvarez, B., van Soest, R. W. M., and Rützler, K. (1998). A revision of Axinellidae (Porifera: Demospongiae) of the Central-West Atlantic region. *Smithsonian Contributions to Zoology* **598**, 1–47.

Alvarez, B., van Soest, R. W. M., and Rützler, K. (2002). *Svenzea*, a new genus of Dictyonellidae (Porifera: Demospongiae) from tropical reef environments, with description of two new species. *Contributions to Zoology* **71**, (4), 171–176.

Aronson, R. B., Precht, W. F., Macintyre, I. G., and Murdoch, T. J. T. (2000). Coral bleach-out in Belize. *Nature* **405**, 36.

Baeza, J. A. (2010). The symbiotic lifestyle and its evolutionary consequences: Social monogamy and sex allocation in the hermaphroditic shrimp *Lysmata pederseni*. *Naturwissenschaften* **97**, 729–741.

Bickmeyer, U., Grube, A., Klings, K. W., Pawlik, J. R., and Köck, M. (2010). Siphonodictyal B1from a marine sponge increases intracellular calcium levels comparable to the Ca2+-ATPase (SERCA) inhibitor thapsigargin. *Marine Biotechnology* **12**, 267–272.

Bingham, B. L., and Young, C. M. (1991). Influence of sponges on invertebrate recruitment: A field test of allelopathy. *Marine Biology* **109**, 19–26.

Boxshall, G. A., and Huys, R. (1994). *Asterocheres reginae*, a new species of parasitic copepod (Siphonostomatoida: Asterocheridae) from a sponge in Belize. *Systematic Parasitology* **27**, 19–33.

Burton, M. (1954). The Rosaura expedition. 5. Sponges. *Bulletin of the British Museum (Natural History) Zoology* **2**, 215–239.

Calcinai, B., Cerrano, Carlo, and Bavastrello, Giorgio (2007). Three new species and one re-description of *Aka* de Laubenfels, 1936. *Journal of the Marine Biological Association of the United Kingdom* **87**, 1355–1365.

Calder, D. R. (1988). *Turritopsoides brehmeri*, a new genus and species of athecate from Belize. (Hydrozoa, Clavidae). *Proceedings of the Biological Society of Washington* **101**, 229–233.

Calder, D. R. (1991a). Abundance and distribution of hydroids in a mangrove ecosystem at Twin Cays, Belize, Central America. *Hydrobiologia* **216/217**, 221–228.

Calder, D. R. (1991b). Associations between hydroid species assemblages and substrate types in the mangal Twin Cays, Belize. *Canadian Journal of Zoology* **69**, 2067–2075.

Calder, D. R. (1991c). Vertical zonation of the hydroid *Dynamena crisioides* (Hydrozoa, Sertulariidae) in a mangrove ecosystem at Twin Cays, Belize. *Canadian Journal of Zoology* **69**, 2993–2999.

Cattaneo-Vietti, R., Bavestrello, G., Cerrano, C., Sarà, M., Benatti, U., Glovine, M., and Gaino, E. (1996). Optical fibres in an Antarctic sponge. *Nature* **383**, 397–398.

Cerrano, C., Pansini, M., Valisano, L., Calcinai, B., Sarà, M., and Bavastrello, G. (2004). Lagoon sponges from Carrie Bow Cay (Belize): Ecological benefits of selective sediment incorporation. In "Sponge Science in the New Millennium" (M. Pansini, R. Pronzato, G. Bavestrello and R. Manconi, eds), Bolletino dei Musei e degli Istituti Biologici dell' Universitá di Genova, vol. 68, pp. 239–252.

Davy, S. K., Trautman, D. A., Borowitzka, M. A., and Hinde, R. (2002). Ammonium excretion by a symbiotic sponge supplies the nitrogen requirements of its rhodophyte partner. *The Journal of Experimental Biology* **205**, 3505–3511.

de Weerdt, W. H. (2000). A monograph of the shallow-water Chalinidae (Porifera, Haploslerida) of the Caribbean. *Beaufortia* **50**, (1), 1–67.

de Weerdt, W. H., Rützler, K., and Smith, K. P. (1991). The Chalinidae (Porifera) of Twin Cays, Belize, and adjacent waters. *Proceedings of the Biological Society of Washington* **104,** (1), 189–205.

Diaz, M. C., and Rützler, K. (2001). Sponges: An essential component of Caribbean coral reefs. *Bulletin of Marine Science* **69,** (2), 535–546.

Diaz, M. C., and Rützler, K. (2009). Biodiversity and abundance of sponges in Caribbean mangrove: Indicators of environmental quality. In "Proceedings of the Smithsonian Marine Science Symposium." (M. A. Lang, I. G. Macintyre and K. Rützler, eds), Smithsonian Contributions to the Marine Sciences, vol. 38, pp. 151–172.

Diaz, M. C., and Ward, B. B. (1997). Sponge-mediated nitrification in tropical benthic communities. *Marine Ecology Progress Series* **156,** 97–107.

Diaz, M. C., Rützler, K., Cary, C. S., and Feller, I. (2002). Distribution and phylogenetic affinities of nitrifying microbes associated to epibiont sponges of nutrient limited *Rhizophora mangle* stands. *Bolletino dei Musei e degli Istituti Biologici dell'Universitá di Genova* **66–67,** 56.

Diaz, M. C., Smith, K. P., and Rützler, K. (2004a). Sponge species richness and abundance as indicators of mangrove epibenthic community health. *Atoll Research Bulletin* **518,** 1–17.

Diaz, M. C., Akob, D., and Cary, C. S. (2004b). Denaturing gradient gel electrophoresis of nitrifying microbes associated with tropical sponges. In "Sponge Science in the New Millennium" (M. Pansini, R. Pronzato, G. Bavestrello and R. Manconi, eds), Bolletino dei Musei e degli Istituti Biologici del' Universitá di Genova, vol. 68, pp. 279–289.

Diaz, M. C., Thacker, R. W., Rützler, K., and Piantoni, C. (2007). Two new haplosclerid sponges from Caribbean Panama with symbiotic filamentous cyanobacteria, and an overview of sponge-cyanobacteria associations. In "Porifera Research: Biodiversity, Innovation, and Sustainability" (M. R. Custódio, G. Lôbo-Hajdu, E. Hajdu and G. Muricy, eds), pp. 31–39. Museu Nacional, Rio de Janeiro.

Diaz, M. C., and Rützler, K. (2011). Biodiversity of sponges: Belize and beyond, to the greater Caribbean. In "Too Precious to Drill: The Marine Biodiversity of Belize" (M. L. D. Palomares and D. Pauly, eds), Fisheries Centre Research Reports (University of British Columbia), vol. 19(6), pp. 57–65.

Duffy, J. E. (1993). Genetic population structure in two tropical sponge-dwelling shrimps that differ in dispersal potential. *Marine Biology* **116,** 459–470.

Duffy, J. E. (1996a). *Synalpheus regalis,* new species, a sponge-dwelling shrimp from the Belize Barrier Reef, with comments on host specificity in *Synalpheus. Journal of Crustacean Biology* **16,** 564–573.

Duffy, J. E. (1996b). Eusociality in a coral-reef shrimp. *Nature* **381,** 512–514.

Duffy, J. E. (1998). On the frequency of eusociality in snapping shrimps (Decapoda: Alpheidae), with description of a second eusocial species. *Bulletin of Marine Science* **62,** 387–400.

Duffy, J. E. (2002). The ecology and evolution of eusociality in sponge-dwelling shrimp. In "Genes, Behaviors, and Evolution of Social Insects" (T. Kikuchi, S. Higashi and N. Azuma, eds), pp. 217–254. Hokkaido University Press, Sapporo.

Duffy, J. E. (2007). Ecology and evolution of eusociality in sponge-dwelling shrimp. In "Evolutionary Ecology of Social and Sexual Systems: Crustaceans as Model Organisms" (J. E. Duffy and M. Thiel, eds), pp. 387–409. Oxford University Press, Oxford.

Duffy, J. E., and Macdonald, K. S. (1999). Colony structure of the social snapping shrimp *Synalpheus filidigitus* in Belize. *Journal of Crustacean Biology* **19,** (2), 283–292.

Duffy, J. E., Morrison, C. L., and Rios, R. (2000). Multiple origins of eusociality among sponge-dwelling shrimps (*Synalpheus*). *Evolution* **54,** (2), 503–516.

Duffy, J. E., Morrison, C. L., and Macdonald, K. S. (2002). Colony defense and behavioral differentiation in the eusocial shrimp *Synalpheus regalis. Behavioral Ecology and Sociobiology* **51,** 488–495.

Duran, S., and Rützler, K. (2006). Ecological speciation in a Caribbean marine sponge. *Molecular Phylogenetics and Evolution* **40**, 292–297.

Ellison, A. M., and Farnsworth, E. J. (1990). The ecology of Belizean mangrove-root fouling communities: I. Epibenthic fauna are barriers to isopod attack of red mangroves. *Journal of Experimental Marine Biology and Ecology* **142**, 91–104.

Ellison, A. M., and Farnsworth, E. J. (1992). The ecology of Belizean mangrove-root fouling communities: Patterns of epibionts distribution and abundance, and effects on root growth. *Hydrobiologia* **247**, 87–98.

Ellison, A. M., Farnsworth, E. J., and Twilley, R. R. (1996). Facultative mutualism between red mangroves and root-fouling sponges in Belizean mangal. *Ecology* **77**, (8), 2431–2444.

Erpenbeck, D., Duran, S., Rützler, K., Paul, V., Hooper, J. N. A., and Wörheide, G. (2007). Towards a DNA taxonomy of Caribbean demosponges: A gene tree reconstructed from partial mitochondrial CO1 gene sequences supports previous rDNA phylogenies and provides a new perspective on the systematics of Demospongiae. *Journal of the Marine Biological Association of the United Kingdom* **87**, 1563–1570.

Farnsworth, E. J., and Ellison, A. M. (1996). Scale-dependent spatial and temporal variability in biogeography of mangrove root epibiont communities. *Ecological Monographs* **66**, (1), 45–66.

Feller, I. C., and Sitnik, M. (1996). Mangrove Ecology: A Manual for a Field Course Focused on the Biocomplexity of Mangrove Ecosystems. Smithsonian Institution Press, Washington, DC.

Forstner, H., and Rützler, K. (1969). Two temperature-compensated thermistor current meters for use in marine ecology. *Journal of Marine Research* **27**, 263–271.

Freestone, A. L., Osman, R. W., and Whitlatch, R. B. (2009). Latitudinal gradients recruitment and community dynamics in marine epifaunal communities: Implications for invasion success. In "Proceedings of the Smithsonian Marine Science Symposium." (M. A. Lang, I. G. Macintyre and K. Rützler, eds), Smithsonian Contributions to the Marine Sciences, vol. 38, pp. 247–258.

Galtsoff, P. S. (1942). Wasting disease causing mortality of sponges in the West Indies and Gulf of Mexico. In "Proceedings of the Eighth American Scientific Congress, Washington, D.C., 1940." Biological Sciences, vol. 3, pp. 411–421.

Gilbert, C. R., and Tyler, J. C. (1997). *Apogon robbyi*, a new cardinalfish (Perciformes: Apogonidae) from the Caribbean Sea. *Bulletin of Marine Science* **60**, (3), 764–781.

Greenfield, D. W., and Johnson, R. K. (1981). The blennioid fishes of Belize and Honduras, Central America, with comments on their systematics, ecology and distribution (Blennidae, Chaenopsidae, Labrisomidae, Tripterygiidae). *Fieldiana Zoology* **8**, (1324), 1106.

Hajdu, E., and Rützler, K. (1998). Sponges, genus *Mycale* (Poecilosclerida: Demospongiae: Porifera), from a Caribbean mangrove and comments on subgeneric classification. *Proceeding of the Biological Society of Washington* **111**, (4), 737–773.

Hendler, G. (1984). The association of *Ophiothrix lineata* and *Callyspongia vaginalis*: A brittlestar-sponge cleaning symbiosis? *Marine Ecology* **5**, (1), 9–27.

Hendler, G., and Pawson, D. L. (2000). Echinoderms of the Rhomboidal Cays, Belize: Biodiversity, distribution, and ecology. *Atoll Research Bulletin* **479**, 275–299.

Highsmith, R. C. (1981). Coral bioerosion: Damage relative to skeletal density. *The American Naturalist* **117**, 193–198.

Highsmith, R. C., Lueptow, R. L., and Schonberg, S. C. (1983). Growth and bioerosion of three massive corals on the Belize Barrier Reef. *Marine Ecology Progress Series* **13**, 261–271.

Hill, M., and Wilcox, T. (1998). Unusual mode of symbiont repopulation after bleaching in *Anthosigmella varians*: Acquisition of different zooxanthellae strains. *Symbiosis* **25**, 279–289.

Hooper, J. N. A. and Van Soest, R. W. M. (eds) (2002). Systema Porifera, A Guide to the Classification of Sponges Kluwer Academic/Plenum Publishers, New York, vols. 1 and 2.

Kokke, W., Fenical, W., Pak, C., and Djerassi, C. (1979). XII. Occurrence of 24 $(R+S)$-isopropenylcholesterol, 24 $(R+S)$-methylcholesta-5, 25-dien3β-ol, and 24$(R+S)$-methylcholesta-7, 25-dien-3β-ol in the Caribbean sponge, *Verongia cauliformis*. *Helvetica Chimica Acta* **62**, 1310–1380.

Koltes, K. H., Tschirky, J. J., and Feller, I. C. (1998). Carrie Bow Cay, Belize. In "Caribbean Coastal Marine Productivity (CARICOMP): Coral Reef, Seagrass, and Mangrove Site Characteristics" (B. Kjerfve, ed.), pp. 79–94. UNESCO, Paris.

Lee, O. O., Chui, P. Y., Wong, Y. H., Pawlik, J. R., and Qian, P.-Y. (2009). Evidence for vertical transmission of bacterial symbionts from adult to embryo in the Caribbean sponge *Svenzea zeai*. *Applied and Environmental Microbiology* **75**, 6147–6156.

Lewis, S. M. (1982). Sponge-zoanthid associations: Functional interactions. In "The Atlantic Barrier Ecosystems at Carrie Bow Cay, Belize, I: Structure and Communities" (K. Rützler and I. G. Macintyre, eds), Smithsonian Contributions to Marine Sciences, vol. 12, pp. 465–474.

Macdonald, K. S., and Duffy, J. E. (2006). Two new species of sponge-dwelling snapping shrimp from the Belizean Barrier reef, with a synopsis of the *Synalpheus brooksi* species complex. *American Museum Novitates* **3543**, 1–22.

Macdonald, K. S., Ríos, R., and Duffy, J. E. (2006). Biodiversity, host specificity, and dominance by eusocial species among sponge-dwelling alpheid shrimp on the Belize Barrier Reef. *Diversity and Distributions* **12**, 165–178.

Macintyre, I. G., and Aronson, R. B. (1997). Field guidebook to the reefs of Belize. In "Proceedings of the Eighth International Coral Reef Symposium I." pp. 203–221. Smithsonian Tropical Research Institute, Balboa, Panamá.

Macintyre, I. G. and Rützler, K. (eds), (2000). Natural History of the Pelican Cays, Belize, vol. 466–480, Atoll Research Bulletin 333 pp.

Macintyre, I. G., Rützler, K., Norris, J. N., and Fauchald, K. (1982). A submarine cave near Columbus Cay, Belize: A bizarre cryptic habitat. In "The Atlantic Barrier Ecosystems at Carrie Bow Cay, Belize, I: Structure and Communities" (K. Rützler and I. G. Macintyre, eds), Smithsonian Contributions to the Marine Sciences, vol. 12, pp. 127–142.

Macintyre, I. G., Goodbody, I., Rützler, K., Littler, D. S., and Littler, M. M. (2000). A general biological and geological survey of the rims of ponds in the major mangrove islands of the Pelican Cays, Belize. *Atoll Research Bulletin* **467**, 15–46.

Macintyre, I. G., Toscano, M. A., Feller, I. C., and Faust, M. A. (2009). Decimating mangrove forests for commercial development in the Pelican Cays, Belize: Long-term ecological loss for short-term gain. In "Proceedings of the Smithsonian Marine Science Symposium." (M. A. Lang, I. G. Macintyre and K. Rützler, eds), Smithsonian Contributions to the Marine Sciences, vol. 38, pp. 281–290.

Maldonado, M., Riesgo, A., Bucci, A., and Rützler, K. (2010). Revisiting silicon budgets at a tropical continental shelf: Silica standing stocks in sponges surpass those in diatoms. *Limnology and Oceanography* **55**, 2001–2010.

Miloslavich, P., Díaz, J. M., Klein, E., Alvarado, J. J., Díaz, C., Gobin, J., Escobar-Briones, E., Cruz, J. J., Weil, E., Cortés, J., Bastidas, A. C., Robertson, D. R., Zapata, F. A., Martin, A., Castillo, J., Kazandjian, A., and Ortiz, M. (2010). Marine biodiversity in the Caribbean: Regional estimates and distribution patterns. *PLoS One* **5**, (8), 1191610.1371/journal.pone.0011916.

Moore, H. F. (1910). The commercial sponges and the sponge fisheries. *Bulletin of the Bureau of Fisheries* **28**, 399–511.

Morrison, C. L., Ríos, R., and Duffy, J. E. (2004). Phylogenetic evidence for an ancient rapid radiation of Caribbean sponge-dwelling snapping shrimps (*Synalpheus*). *Molecular Phylogenetics and Evolution* **30**, 563–581.

Muricy, G., and Minervino, J. V. (2000). A new species of *Gastrophanella* from central western Atlantic, with a discussion of the family Siphonidae (Demospongiae, Lithistida). *Journal of the Marine Biological Association of the United Kingdom* **80**, 599–605.

Musco, L., and Giangrande, A. (2005). A new sponge-associated species, *Syllis mayeri* n. sp. (Polychaeta: Syllidae), with a discussion on the status of *S. armillaris* (Müller, 1776). *Scientia Marina* **69**, (4), 467–474.

Olson, J. B., Gochfeld, D. J., and Slattery, M. (2006). *Aplysina* red band syndrome: A new threat to Caribbean sponges. *Diseases of Aquatic Organisms* **71**, 163–168.

Opishinski, T. B., Spaulding, M. L., Rützler, K., and Carpenter, M. (2001). A real time environmental data monitoring, management and analysis system for the coral reefs off the coast of Belize. In "Proceedings of the Oceans Conference 2001, Honolulu, Hawaii." pp. 1188–1197.

Parra-Velandia, F. J., Zea, W., and van Soest, R. W. M. (2012). Reef sponges of the genus *Agelas* (Porifera: Demospongiae) from the Caribbean Sea. *Zootaxa* (in press).

Priest, B. W. (1881). On an undescribed species of sponge of the genus *Polymastia*, from Honduras. *Journal of the Quekett Microscopical Club* **6**, 302–304.

Reiswig, H. M. (1976). Natural gamete release and oviparity in Caribbean Demospongiae. In "Aspects of Sponge Biology" (F. W. Harrison and R. R. Cowden, eds), pp. 99–112. Academic Press, New York.

Ríos, R., and Duffy, J. E. (1999). Description of *Synalpheus williamsi*, a new species of sponge-dwelling shrimp (Crustacea: Decapoda: Alpheidae), with remarks on its first larval stage. *Proceedings of the Biological Society of Washington* **112**, (3), 541–552.

Ríos, R., and Duffy, J. E. (2007). A review of the sponge-dwelling snapping shrimp from Carrie Bow Cay, Belize, with description of *Zuzalpheus*, new genus, and six new species (Crustacea: Decapoda: Alpheidae). *Zootaxa* **1602**, 1–89.

Ritson-Williams, R., Becerro, M. A., and Paul, V. J. (2005). Spawning of the giant barrel sponge *Xestospongia muta* in Belize. *Coral Reefs* **24**, 160.

Rützler, K. (1971). Bredin-Archbold Smithsonian biological survey of Dominica: Burrowing sponges, genus *Siphonodictyon* Bergquist, from the Caribbean. *Smithsonian Contributions to Zoology* **77**, 1–37.

Rützler, K. (1975). The role of burrowing sponges in bioerosion. *Oecologia* **19**, 203–216.

Rützler, K. (1978). Sponges in coral reefs. In "Coral Reefs: Research Methods: Monographs on Oceanographic Methodology" (D. R. Stoddart and R. E. Johannes, eds), pp. 299–313. UNESCO, Paris.

Rützler, K. (1981). An unusual bluegreen alga symbiotic with two new species of *Ulosa* (Porifera: Hymeniacidonidae) from Carrie Bow Cay, Belize. *Marine Ecology* **2**, (1), 35–50.

Rützler, K. (1987). Tetillidae (Spirophorida, Porifera): A taxonomic reevaluation. In "Taxonomy of Porifera from the N. E. Atlantic and Mediterranean Sea" (J. Vacelet and N. Boury-Esnaul, eds), NATO ASI Series G 13. pp. 189–203. Springer Verlag, Berlin.

Rützler, K. (1988). Mangrove sponge disease induced by cyanobacterial symbionts: Failure of a primitive immune system? *Diseases of Aquatic Organisms* **5**, 143–149.

Rützler, K. (1990). Association between Caribbean sponges and photosynthetic organisms. In "New Perspectives in Sponge Biology: Third International Sponge Conference." (K. Rützler, ed.), pp. 455–466. Smithsonian Institution Press, Washington, DC.

Rützler, K. (1995). Low-tide exposure of sponges in a Caribbean mangrove community. *Marine Ecology* **16**, (2), 165–179.

Rützler, K. (1996). Sponge diving: Professional but not for profit. In "Methods and Techniques of Underwater Research, Proceedings of the American Academy of Underwater Sciences 16th Annual Diving Symposium." pp. 183–204. American Academy of Underwater Sciences, Nahant, MA.

Rützler, K. (1997). The role of psammobiontic sponges in the reef community. In "Proceedings of the Eighth Coral Reef Symposium 2." pp. 1393–1398. Smithsonian Tropical Research Institute, Balboa, Panamá.

Rützler, K. (2002a). Family Clionaidae D'Orbigny, 1851. In "Systema Porifera: A Guide to the Classification of Sponges" (J. N. A. Hooper and R. W. M. van Soest, eds), pp. 173–185. Kluwer Academic/Plenum Publishers, New York.

Rützler, K. (2002b). Family Spirastrellidae Ridley & Dendy, 1886. In "Systema Porifera: A Guide to the Classification of Sponges" (J. N. A. Hooper and R. W. M. van Soest, eds), pp. 220–223. Kluwer Academic/Plenum Publishers, New York.

Rützler, K. (2002c). Impact of crustose clionid sponges on Caribbean reef corals. *Acta Geologica Hispanica* **37**, (1), 61–72.

Rützler, K. (2004). Sponges on coral reefs: A community shaped by competitive cooperation. In "Sponge Science in the New Millennium" (M. Pansini, R. Pronzato, G. Bavestrello and R. Manconi, eds), Bolletino dei Musei e degli Istituti Biologici dell' Universitá di Genova, vol. 68, pp. 85–148.

Rützler, K. (2009). Caribbean Coral Reef Ecosystems: 35 years of Smithsonian marine science in Belize. In "Proceedings of the Smithsonian Marine Science Symposium." (M. A. Lang, I. G. Macintyre and K. Rützler, eds), Smithsonian Contributions to the Marine Sciences, vol. 38, pp. 43–71.

Rützler, K., and Feller, I. C. (1988). Mangrove swamp communities. *Oceanus* **30**, (4), 16–24.

Rützler, K., and Feller, I. C. (1996). Caribbean mangrove swamps. *Scientific American* **274**, (3), 70–75.

Rützler, K., and Hooper, J. N. A. (2000). Two new genera of hadromerid sponges (Porifera, Demospongiae). *Zoosystema* **22**, (2), 337–344.

Rützler, K., and Macintyre, I. G. (1978). Siliceous sponge spicules in coral reef sediments. *Marine Biology* **49**, 147–159.

Rützler, K. and Macintyre, I. G. (eds) (1982). The Atlantic barrier ecosystems at Carrie Bow Cay, Belize, I: Structure and communities, vol. 12, Smithsonian Contributions to the Marine Sciences Smithsonian Institution Press, Washington, DC.

Rützler, K., and Macintyre, I. G. (1982b). The habitat distribution and community structure of the barrier reef complex at Carrie Bow Cay, Belize. In "The Atlantic barrier ecosystems at Carrie Bow Cay, Belize, I: Structure and Communities" (K. Rützler and I. G. Macintyre, eds), Smithsonian Contributions to the Marine Sciences, vol. 12, pp. 9–46.

Rützler, K., and Richardson, S. (1996). The Caribbean spicule tree: A sponge-imitating foraminifer (Astrorhizidae). In "Recent Advances in Sponge Biodiversity Inventory and Documentation" (P. Willenz, ed.), Bulletin de l'Institut Royal des Sciences Naturelles de Belgique, vol. 66(Suppl), pp. 143–151.

Rützler, K., and Rieger, G. (1973). Sponge burrowing: Fine structure of *Cliona lampa* penetrating calcareous substrata. *Marine Biology* **21**, 144–162.

Rützler, K., and Smith, K. P. (1992). Guide to western Atlantic species of *Cinachyrella* (Porifera: Tetillidae). *Proceedings of the Biological Society of Washington* **105**, (1), 148–164.

Rützler, K., and Smith, K. P. (1993). The genus *Terpios* (Suberitidae) and new species in the "*lobiceps*" complex. *Scientia Marina* **57**, (4), 381–393.

Rützler, K., and Vacelet, J. (2002). Family Acanthochaetetidae Fischer, 1970. In "Systema Porifera: A Guide to the Classification of Sponges" (J. N. A. Hooper and R. W. M. van Soest, eds), pp. 275–278. Kluwer Academic/Plenum Publishers, New York.

Rützler, K., Ferraris, J. D., and Larson, R. L. (1980). A new plankton sampler for coral reefs. *Marine Ecology* **1**, 65–71.

Rützler, K., Diaz, M. C., van Soest, R. W. M., Zea, S., Smith, K. P., Alvarez, B., and Wulff, J. (2000). Diversity of sponge fauna in mangrove ponds, Pelican Cays, Belize. *Atoll Research Bulletin* **476**, 229–263.

Rützler, K., van Soest, R. W. M., and Alvarez, B. (2003). *Svenzea zeai*, a Caribbean reef sponge with a giant larva, and *Scopalina ruetzleri*: A comparative fine-structural approach to classification (Demospongiae, Halichondrida, Dictyonellidae). *Invertebrate Biology* **122**, (3), 203–222.

Rützler, K., Goodbody, I., Díaz, M. C., Feller, I. C., and Macintyre, I. G. (2004). The aquatic environment of Twin Cays, Belize. *Atoll Research Bulletin* **512**, 1–49.

Rützler, K., Duran, S., and Piantoni, C. (2007a). Adaptation of reef and mangrove sponges to stress: Evidence for ecological speciation exemplified by *Chondrilla caribensis* new species (Demospongiae, Chondrosida). *Marine Ecology* **28**, (1), 95–111.

Rützler, K., Maldonado, M., Piantoni, C., and Riesgo, A. (2007b). *Iotrochota* revisited: A new sponge and review of species from the western tropical Atlantic (Poecilosclerida: Iotrochotidae). *Invertebrate Systematics* **21**, 173–185.

Rützler, K., Piantoni, C., and Diaz, M. C. (2007c). *Lissodendoryx*: rediscovered type and new tropical western Atlantic species (Porifera: Demospongiae: Poecilosclerida: Coelosphaeridae). *Journal of the Marine Biological Association of the United Kingdom* **87**, 1491–1510.

Sharp, K. H., Eam, B., Faulkner, D. J., and Haygood, M. G. (2007). Vertical transmission of diverse microbes in the tropical sponge *Corticium* sp. *Applied and Environmental Microbiology* **73**, (2), 622–629.

Sharp, K. H., Ritchie, K. B., Schupp, P. J., Ritson-Williams, R., and Paul, V. J. (2010). Bacterial acquisition in juveniles of several broadcast spawning coral species. *PLoS One* **5**, (5), e1089810.1371/journal.pone.0010898.

Shick, J. M., Lesser, M. P., and Jokiel, P. L. (1996). Effects of ultraviolet radiation on corals and other coral reef organisms. *Global Change Biology* **2**, 527–545.

Smith, F. G. W. (1939). Sponge mortality at British Honduras. *Nature* **144**, 785.

Smith, F. G. W. (1941). Sponge disease in British Honduras, and its transmission by water currents. *Ecology* **22**, (4), 415–421.

Spracklin, B. (1982). Hydroidea (Cnidaria: Hydrozoa) from Carrie Bow Cay, Belize. In "The Atlantic Barrier Ecosystems at Carrie Bow Cay, Belize, I: Structure and Communities" (K. Rützler and I. G. Macintyre, eds), Smithsonian Contributions to the Marine Sciences, vol. 12, pp. 239–252.

Steindler, L., Huchon, D., Avni, A., and Ilan, M. (2005). 16S rRNA phylogeny of sponge-associated cyanobacteria. *Applied and Environmental Microbiology* **71**, (7), 4127–4131.

Stevely, J. M., and Sweat, D. E. (1994). A preliminary evaluation of the commercial sponge resources of Belize with reference to the location of the Turneffe Islands sponge farm. *Atoll Research Bulletin* **424**, 1–21.

Stuart, A. H. (1948). World trade in sponges. *U.S. Department of Commerce. U.S. Department of Commerce Industrial Series* **82**, 1–95.

Sullivan, B. W., and Faulkner, D. J. (1990). Chemical studies of the burrowing sponge *Siphonodictyon coralliphagum*. In "New Perspectives in Sponge Biology: Third International Sponge Conference." (K. Rützler, ed.), pp. 45–50. Smithsonian Institution Press, Washington, DC.

Tavares, R., Daloze, D., Braekman, J. C., Hajdu, E., and van Soest, R. W. M. (1995). 8b-Hydroxiptilocaulin, a new guanidine alkaloid from the sponge *Monanchora arbuscula*. *Journal of Natural Products* **58**, (7), 971–1152.

Thacker, R. W., Becerro, M. A., Lumbang, W. A., and Paul, V. J. (1998). Allelopathic interactions between sponges on a tropical reef. *Ecology* **79**, (5), 1740–1750.

Thacker, R. W., Diaz, M. C., Rützler, K., Erwin, P. M., Kimble, S. J. A., Pierce, M. J., and Dillard, S. L. (2007). Phylogenetic relationships among the filamentous cyanobacterial symbionts of Caribbean sponges and a comparison of photosynthetic production between sponges hosting filamentous and unicellular cyanobacteria. In "Porifera Research: Biodiversity, Innovation, and Sustainability, Seventh International Sponge

Symposium." (M. R. Custódio, G. Lôbo-Hajdu, E. Hajdu and G. Muricy, eds), pp. 621–626. Museu Nacional, Rio de Janeiro.

Thomas, J. D., and Klebba, K. N. (2006). Studies of commensal leucothod amphipods: Two new sponge-inhabiting species from South Florida and the western Caribbean. *Journal of Crustacean Biology* **26,** (1), 13–22.

Thomas, J. D., and Klebba, K. N. (2007). New species and host associations of commensal leucothoid amphipods from coral reefs in Florida and Belize (Crustacea:Amphipoda). *Zootaxa* **1494,** 1–44.

Thomas, J. D., and Taylor, G. W. (1981). Mouthpart morphology and feeding strategies of the commensal amphipod, *Anamixis hanseni* Stebbing. *Bulletin of Marine Science* **31,** (2), 462–467.

Tóth, E., and Bauer, R. (2007). Gonopore sexing technique allows determination of sex ratios and helper composition in eusocial shrimps. *Marine Biology* **151,** (5), 1875–1886.

Tóth, E., and Duffy, J. E. (2005). Coordinated group response to nest intruders in social shrimp. *Biology Letter* **1,** 49–52. Doi:10.1098/rsbl.2004.0237.

Tóth, E., and Duffy, J. E. (2008). Influence of sociality on allometric growth and morphological differentiation in sponge-dwelling alpheid shrimp. *Biological Journal of the Linnean Society* **94,** 527–540.

Van Soest, R. W. M., and Rützler, K. (2002). Family Tetillidae Sollas, 1886. In "Systema Porifera: A Guide to the Classification of Sponges" (J. N. A. Hooper and R. W. M. van Soest, eds), pp. 85–98. Kluwer Academic/Plenum Publishers, New York.

Van Soest, R. W. M., Boury-Esnault, N., Hooper, J. N. A., Rützler, K., de Voogd, N. J., Alvarez, B., de Glasby, B., Hajdu, E., Pisera, A. B., Manconi, R., Schoenberg, C., Janussen, D., Tabachnick, K. R., Klautau, M., Picton, B., and Kelly, M. (2011). World Porifera Database. Available online at http://www.marinespecies.org/poriferaConsulted on 2011-04-29.

Vicente, V. (1989). Regional commercial sponge extinctions in the West Indies: Are recent climatic changes responsible? *Marine Ecology* **10,** (2), 179–191.

Vicente, V. P. (1990). Responses of sponges with autotrophic endosymbionts during the coral-bleaching episode in Puerto Rico. *Coral Reefs* **8,** 199–202.

Vicente, V. P., Rützler, K., and Carballeiro, N. M. (1991). Comparative morphology, ecology, and fatty acid composition of West Indian *Spheciospongia* (Demospongiae). *Marine Ecology* **12,** (3), 211–226.

Weyrer, S., Rützler, K., and Rieger, R. (1999). Serotonin in Porifera? Evidence from developing *Tedania ignis*, the Caribbean fire sponge (Demospongiae). In "Origin and Outlook, Fifth International Sponge Symposium." (J. N. A. Hooper and J. N. A. Hooper, eds), Memoirs of the Queensland Museum, vol. 44, pp. 659–665.

Wilkinson, C. R. (1987). Interocean differences in size and nutrition of coral reef sponge populations. *Science* **236,** (4809), 1654–1657.

Wilkinson, C. R., and Cheshire, A. C. (1990). Comparisons of sponge populations across the barrier reefs of Australia and Belize: Evidence for higher productivity in the Caribbean. *Marine Ecology Progress Series* **67,** 285–294.

Wulff, J. L. (2000). Sponge predators may determine differences in sponge fauna between two sets of mangrove cays, Belize Barrier Reef. *Atoll Research Bulletin* **477,** 251–266.

Wulff, J. L. (2004). Sponge on mangrove roots, Twin Cays, Belize: Early stages of community assembly. *Atoll Research Bulletin* **519,** 1–10.

Wulff, J. L. (2005). Trade-offs in resistance to competitors and predators, and their effects on the diversity of tropical marine sponges. *Journal of Animal Ecology* **74,** 313–321.

Wulff, J. L. (2006a). Sponge systematics by starfish: Predators distinguish cryptic sympatric species of Caribbean fire sponges, *Tedania ignis* and *Tedania klausi* n. sp. (Demospongiae, Poecilosclerida). *The Biological Bulletin* **211,** 83–94.

Wulff, J. L. (2006b). Ecological interactions of marine sponges. *Canadian Journal of Zoology* **84,** 146–166.

Wulff, J. L. (2008). Collaboration among sponge species increases sponge diversity and abundance in a seagrass meadow. *Marine Ecology* **29,** (2), 193–204.

Wulff, J. L. (2009). Sponge community dynamics on Caribbean mangrove roots: Significance of species idiosyncrasies. In "Proceedings of the Smithsonian Marine Science Symposium." (M. A. Lang, I. G. Macintyre and K. Rützler, eds), Smithsonian Contributions to the Marine Sciences, vol. 38, pp. 501–514.

Wulff, J. (2010). Regeneration of sponges in ecological context: Is regeneration an integral part of life history and morphological strategies? *Integrative and Comparative Biology* **50,** (4), 494–505.

Zea, S., and Weil, E. (2003). Taxonomy of the Caribbean excavating sponge species complex *Cliona caribbaea—C. aprica—C. langae* (Porifera, Hadromerida, Clionaidae). *Caribbean Journal of Science* **39,** (3), 348–370.

CHAPTER FOUR

Ecological Interactions and the Distribution, Abundance, and Diversity of Sponges

Janie Wulff[1]

Contents

Abstract

Although abiotic factors may be important first-order filters dictating which sponge species can thrive at a particular site, ecological interactions can play substantial roles influencing distribution and abundance, and thus diversity. Ecological interactions can modify the influences of abiotic factors both by further constraining distribution and abundance due to competitive or predatory interactions and by expanding habitat distribution or abundance due to beneficial interactions that ameliorate otherwise limiting circumstances. It is likely that the importance of ecological interactions has been greatly

Department of Biological Science, Florida State University, Tallahassee, FL, USA
[1]Corresponding author: Email: wulff@bio.fsu.edu

Advances in Marine Biology, Volume 61
ISSN 0065-2881, DOI: 10.1016/B978-0-12-387787-1.00003-9

underestimated because they tend to only be revealed by experiments and time-series observations in the field.

Experiments have revealed opportunistic predation to be a primary enforcer of sponge distribution boundaries that coincide with habitat boundaries in several systems. Within habitats, by contrast, dramatic effects of predators on sponge populations seem to occur primarily in cases of unusually high recruitment rates or unusually low mortality rates for the predators, which are often specialists on the sponge species affected. Competitive interactions have been demonstrated to diminish populations or exclude sponge species from a habitat in only a few cases. Cases in which competitive interactions have appeared obvious have often turned out to be neutral or even beneficial interactions when observed over time. Especially striking in this regard are sponge–sponge interactions in dense sponge-dominated communities, which may promote the continued coexistence of all participating species. Mutualistic symbioses of sponges with other animals, plants, or macroalgae have been demonstrated to increase abundance, habitat distribution, and diversity of all participants. Symbiotic microbes can enhance sponge distribution and abundance but also render their hosts more vulnerable to environmental changes. And while photosynthetic symbionts can boost growth and excavation rates for some sponge hosts, in other cases sponge growth proceeds as well or even better in diminished light.

Metrics chosen for evaluating sponge abundance make a substantial difference in interpretation of data comparing between different sites, or over time at the same site. In most cases, evaluating abundance by volume or biomass allows more ecologically meaningful interpretation of influences on distribution and abundance than does evaluating abundance by numbers of individuals or area covered. Accurate identification of species, and understanding how they are related within higher taxa, is essential. Studies in every habitat have illustrated the great power of experimental manipulations, and of time-series observations of sponge individuals, for understanding the processes underlying observed patterns; in many cases, these processes have been revealed to be ecological interactions.

Key Words: sponges; abiotic factors; ecology; interactions; predation; spongivory; competition; mutualism; abundance; diversity

1. INTRODUCTION

A surge of studies on the interactions of sponges with other organisms and with their abiotic environments has bolstered confidence in our general understanding of how sponges fit into their ecosystems. They consume the smaller sizes of particulate organic material and, in collaboration with symbiotic microbes, dissolved organics. Some sponges receive significant nutrition via photosynthetic symbionts. Sponges are in turn fed upon by a small number of charismatic animals such as angelfishes, nudibranchs, sea stars, and hawksbill turtles. Sponges are relatively successful in competition for space against non-sponge taxa, and they are masters of asexual propagation and regeneration after partial mortality. Many of their interactions are moderated by chemistry, produced either by the sponges or by their symbionts. Sponges can have profound effects, both positive and negative, on substratum stability and suitability for other organisms. Sponges are especially adept at striking up collaborative associations with organisms of all types, including other sponge species.

Rapidly changing conditions in coastal marine ecosystems are, however, generating questions that reveal uncertainties in our ability to predict what will happen to particular sponges under particular circumstances, and what the consequences will be for the ecosystems in which they live. Concerns have been expressed about both decreases and increases in sponges. For example, if plankton production rates increase due to increased water column nutrients, will sponges be clogged or grow faster? Conversely, if sponge abundance diminished dramatically, would the water column become murky, and sewage and mariculture effluents become embarrassingly even more evident? If marine protected areas inspire an upwards swing in populations of angelfishes and hawksbill turtles, will sponges be consumed to the point that coral reefs crumble and recovery of damaged reefs is stymied, or will corals flourish? If macroalgae suffer losses to disease, will sponges vanish also or increase? Will sponge pathogens flourish and photosynthetic symbionts flee in response to rapid warming of seawater? If sponge abundance increases, will it be at the expense of other sessile organisms or will it improve water quality and substratum stability? Can sponges perform homeostatic miracles, or will they finally be defeated by deteriorating conditions and vanish, taking with them the enormous number of species with which they have established symbiotic associations? Lurking within each of these questions are the challenging additional questions: Is sponge diversity as important as overall sponge abundance? Does it matter exactly which sponge species are involved, or can all of the sponges be lumped together in prognostications about the trajectories of coastal marine ecosystems?

Although abiotic factors are important first-order filters dictating which sponge species can thrive at a particular site, ecological interactions can play substantial roles in influencing distribution and abundance. Interactions with other organisms modify the influences of abiotic factors on distribution, abundance, and diversity in two main ways: by further constraining habitat distribution or abundance due to competitive or predatory interactions and by expanding habitat distribution or abundance due to beneficial interactions that ameliorate otherwise limiting circumstances. Because of the possibility of reciprocal evolutionary adjustments for ecological interactions, but not for abiotic factors, distinguishing the relative importance of abiotic and biotic influences on sponges is necessary for understanding the adaptive significance of sponge traits. On an ecological timescale as well, predictions cannot be made accurately unless distinctions are made between influences of abiotic and biotic factors. As abiotic factors change, sponge distributions change accordingly, but changes resulting from losses or gains of species with which sponges engage in significant ecological interactions can be much more rapid. Predators can be quickly eliminated by unsustainable fishing, competitors can be lost to disease in just a few months, and symbionts can be sufficiently perturbed to flee in a flash. In this review, I aim to gather current evidence on how ecological interactions with food, competitors, predators, pathogens and parasites, and mutualistic associates are intertwined with abiotic factors to influence distribution, abundance, and diversity of marine demosponges. Space considerations have forced me to defer consideration of many important aspects of sponge ecology (e.g. sponges as biomonitors, population biology and life history strategies, community dynamics, and ecosystem functional roles) in order to focus specifically on how both abiotic factors and ecological interactions have been demonstrated to influence distribution and abundance. Unequal allocation of space for coverage of the habitats considered reflects differences in the degree to which research has been focused on revealing processes.

Underrepresentation of the influence of ecological interactions on distribution, abundance, and diversity of sponges is likely in the literature, in large part because time-series observations and experimental manipulations tend to be required to demonstrate how interactions constrain or enhance distribution and abundance. Some habitats are not amenable to manipulative experiments, and shipboard-based studies are often constrained to a single short visit to each site. In a biogeographic comparison of sponge distribution patterns on cobbles across oceans, Bell and Carballo (2008) suggest that an apparent pattern of influence by biotic interactions in the Caribbean relative to the Indo-Pacific may simply reflect the greater degree to which sponges have been studied with experimental manipulations in the Caribbean.

Different types of ecological interactions are not equally easy to demonstrate, and this influences how frequently interactions are reported in the literature (e.g., Bergquist, 1978, 1999; Becerro, 2008). Predation can be observed straight away, if experimental design takes into account natural ability of predators to detect and react to prey. Habitat transplants that result in clear bite marks outside cages, but none inside, can give unambiguous answers; but comparisons of size changes inside and outside of cages can be difficult to interpret, as cages may alter sponge feeding. Competition takes longer to demonstrate, as it requires time-series observations of individuals of one species actually overgrowing and killing, or otherwise inhibiting, individuals of another species. Sometimes competition can be inferred if it can be seen that an apparently overgrown species is recently dead under another, although overgrowth could have occurred after death. Clear zones of inhibition can be seen in some cases (e.g. Turon *et al.*, 1996). Simple scores of apparent overgrowths are not necessarily evidence of competition for sponges, which are known for their uncanny abilities to tolerate or even thrive under epizoism. Overgrowth by sponges benefits many other species (review in Wulff, 2006d). Sponges growing right up to the edge of living tissue of corals can be either engaging in competition or increasing coral survival, and these possibilities cannot be distinguished without time-series observations (Goreau and Hartman, 1966; Wulff and Buss, 1979). Mutualism is most difficult to demonstrate, as it adds another layer of complexity. A problem must be identified that is only solved in the presence of the mutualistic partner, and thus competition, predation, or inhibition by some abiotic factor has to be demonstrated to differ with and without the mutualistic partner. Benefit must be measured in terms of increased growth, reproduction, or survival, and this requires following the same individuals over time. Sponges are known for their wide intraspecific variation in growth rates and in some cases defensive chemistry, imposing a requirement for control of genotype in experiments. Added to all this is the need to study potential mutualisms for long time periods because long-lived organisms, such as many sponges, may benefit from collaboration only during events that occur at time intervals that are long by human perception.

1.1. Evaluating distribution, abundance, and diversity of sponges

Distribution, abundance, and diversity are not simple, straightforward entities, especially for sponges. Abundance can be measured by numbers of individuals, area, or volume, and each of these can be measured or estimated in a variety of ways. The degree to which species are lumped together in estimating abundance varies from "sponges" (i.e. all sponges lumped together), to groups of sponges defined by growth form or other observable attribute, to painstakingly sorted and named species and subspecies. Likewise, influences on abundance are variously reported as applying to all sponges or only to particular species. Distribution boundaries can be considered at various scales, including microhabitat, habitat, and geographic; and for any particular species, different factors may constrain or extend distribution at each of these scales. Diversity measurement can be simply number of species or can involve an index that combines number of species with relative abundance, compounding interpretation struggles due to inappropriate choice of an abundance metric. Before discussing how ecological interactions influence distribution, abundance, and diversity of sponges, I briefly consider how these variables are evaluated.

1.1.1. Distribution and abundance

Sponges can dominate the biomass and species representation in benthic marine communities to the point that referring to "sponge communities" is apt. On coral reefs, mangrove prop roots, rocky intertidal shores, caves and crevices, subtidal hard bottoms in Antarctica and western Canadian fjords, and even some subtidal soft bottoms, sponge accumulations can be so dense that the underlying substratum appears irrelevant; but in other habitats sponges are minor members. Relative merits of evaluating sponge abundance by numbers of individuals, percentage cover, or volume, and the appropriate situations, for each metric, have been discussed at length and illustrated with examples by Rützler (1978, 2004) and Wulff (2001, 2009).

Conclusions from abundance studies are highly dependent on metrics chosen, as illustrated by the few studies that have provided more than one metric for explicit comparisons. Wilkinson's (1987) summary table of sponge abundance in terms of both number of individuals and biomass at various Caribbean and Great Barrier Reefs (GBRs) highlights how divergent conclusions about sponge biogeography and the environmental parameters influencing abundance can be, depending on the abundance metric used. A figure summarizing distribution patterns of the most prominent 27 sponge species on fore-reef slopes of the GBR illustrates the lack of coincidence of relative abundance in terms of numbers of individuals versus biomass (Wilkinson and Cheshire, 1989). For most of the 27 species they evaluated on 6 reefs, numbers of individuals and biomass do not vary together. On rocky substrata from 0 to 20 m, Preciado and Maldonado (2005) evaluated sponge abundance by both frequency in sampling quadrats and dry weight. Although the five species with the highest frequency of occurrence were all in the top 15 species (out of 85 species total) with respect to dry weight, the authors drew attention to three species with substantial biomass (ranked 11, 14, and 22 by dry weight) that were each found in only 1–3 of the 257 sampling quadrats that included sponges. Description of community composition by growth form in a shallow reef Caribbean community in Panama resulted in massive, encrusting, thick encrusting, and erect-branching sponges equally represented with respect to area. A very different picture of the community is conjured up by volume comparisons, as total volume of erect-branching sponges is 30 times that of encrusting sponges, and volume of massive sponges is 10 times that of encrusting sponges (Wulff, 2001).

Comparisons of community composition between sites also depend heavily on the metric used for abundance. For example, species composition at three mangrove sites in Belize and Panama appears very different when evaluated by numbers of individuals, but very similar by volume, with a single species, *Tedania ignis* Duchassaing and Michelotti, 1864, constituting 49–57% of the total, and the nine species found at all three sites constituting 73–89% of the total volume (Wulff, 2009). *T. ignis*, the "fire sponge", is the icon species for Caribbean mangrove prop roots, consistently standing out as present and dominant in the fauna. But the degree to which it appears to be dominant varies with the metric chosen. *T. ignis* in the above study constituted 8.4–20.4% of the individual sponges, it was recorded on 11–34% of individual roots at nearby sites (Diaz *et al.*, 2004), by photographs of root segments in the Florida Keys it covered 16.7% of the root area (Bingham and Young, 1995), and by line transects along the lengths of prop roots it covered 5–12% of area in Venezuela (Sutherland, 1980). By comparison with these abundance measures, 49–57% of total volume seems to inflate the relative abundance of *T. ignis*; yet this species contributes to the mangrove ecosystem by pumping and filtering water in proportion to its volume and provides shelter and food for inquilines and predators in proportion to volume as well.

Interpretations of community dynamics can vary from "highly stable" to "wildly fluctuating", for the same community, depending on metrics used for abundance. Censuses over 11 years on a shallow coral reef in San Blas, Panama, showed decrease by 53% by number of individuals but only 10.6% by volume (Wulff, 2001). Likewise, data from four complete censuses, at yearly intervals, of mangrove roots at a site in Belize, support a conclusion of enormous change by number of individuals, which varied by 50%. The opposite conclusion of great stability would be warranted based on volume data, which varied by only 12% (Wulff, 2009). In the Florida Keys, over a 4-year period, the opposite pattern emerged from data collected along randomly placed transects: density of sponge individuals increased, while area covered decreased (Chiappone and Sullivan, 1994).

The same site, evaluated by different researchers using different techniques, may appear to have changed quite dramatically in species composition solely due to employment of different evaluation metrics. One of the several illustrative examples gathered by Diaz *et al.* (2004) is Twin Cayes in Belize, which in three different studies was reported to host 20, 54, and 35 mangrove sponge species. Biogeographic comparisons may be misinterpreted if techniques applied differ. Apparently contrasting community dynamics on mangrove root censused in Venezuela (Sutherland, 1980) and the Florida Keys (Bingham and Young, 1995) led Bingham and Young to suggest that tropical systems are more stable than subtropical; but Sutherland included entire roots while Bingham and Young followed particular root segments the size of a camera framer. Because sponges "move" up and down the roots as they grow, it is possible for them to slip out of the spot monitored while still remaining present in the community. Sará (1970) illustrated the degree to which sponge individuals can shift the particular space they occupy while remaining in the community with time-series drawings of encrusting sponges in a Mediterranean cave that show the same individuals participating in the community, but in continuously shifting spots. Hughes (1996) followed sponges and corals in 12 1 m^2

quadrats by taking yearly photographs for 16 years. Sponges were remarkably constant in overall community structure, as measured by numbers of individuals and taxonomic distinction to genus. Because individual sponges could be followed in the time-series photos, he was able to document that apparent stasis was actually the result of very high rates of flux, with high rates of mortality, partial mortality, fragmentation, fusion, and recruitment. On a shallow Caribbean reef in Panama, disturbing losses of 20 of the original 39 species have been revealed by 5 full censuses of 16 m^2 (Wulff, 2006a). Declines in the same set of species on nearby reefs indicated that the problem was not confined to the study reef, but this is the only coral reef site in which individual sponges of all species have been followed over time, so there is no way to know if similar losses have been occurring elsewhere. By contrast, relying on random transects can leave result in unanswerable questions such as whether or not a shift to more but smaller individuals indicates (1) mortality of all residents, followed by recruitment, or (2) fragmentation resulting from partial mortality, or (3) merely chance placement of transects in subsequent monitoring periods. An advantage of censusing the same plots in time sequence is that it eliminates the lurking concern that apparent changes are merely artefacts of the combination of high species diversity and spatial heterogeneity.

Ultimately, the questions at hand must determine which abundance metric is employed. Trophic interactions, such as how much a sponge can filter from the water and how many bites can be taken from it by predators, scale with volume (e.g. Reiswig, 1974; Wulff, 1994), while area covered may be key for mutualisms involving sponges protecting their hosts from borers or consumers (review in Wulff, 2006d). Percentage of the substratum covered by sponges may indicate what space is unavailable to other sessile taxa, if the surface is homogeneous. Percentage cover has been frequently used in coral reef studies because of its appropriateness for corals, of which the live tissue is a consistently thin layer, regardless of overall growth form. But for sponges, ecological interpretation of percentage cover depends on the growth forms represented. Sponge volume can differ orders of magnitude for the same percentage cover, reflecting a range in thickness from 1 mm to over 1 m. Number of individuals can be used appropriately for evaluation of sponge species that do not fragment, and to compare disease prevalence or recruitment rates. More than one metric can be useful. For example, using solely numbers to evaluate the effects of disease may cause interpretation meltdown if fragmentation at lesions increases the number of individuals. However, numbers of individuals can be used in conjunction with volume to understand effects of fragmenting agents such as disease and storms (Wulff, 1995a for a hurricane example). Studies in which both numbers and biomass have been reported are particularly helpful for biogeographic comparisons (e.g. Wilkinson, 1987; Wilkinson and Cheshire, 1989).

1.1.2. Diversity

Taxonomic challenges, combined with high species diversity, prompt the question: Is it really important to accurately identify sponge species in ecological studies? The answer is unambiguously: Yes. Similar, closely related species are likely to share many important traits, but to differ in at least one ecologically important trait. Lumping species, even by genus, can lead to mistakes in estimation of population sizes, habitat distributions, and predicting responses to changes. For example, the common Caribbean mangrove fire sponge, *T. ignis*, was considered to be a habitat generalist that was unusual in inhabiting both mangrove roots and seagrass meadows (Diaz *et al.*, 2004). Reciprocal transplant and feeding choice experiments, followed by morphological and molecular study, revealed two species, *T. ignis* and *Tedania klausi* Wulff, 2006, that are distinguished ecologically by differences in palatability to sea stars (and therefore ability to inhabit seagrass meadows), susceptibility to disease, and ability to tolerate wide swings in temperature and salinity (Wulff, 2006c). Likewise, very similar sympatric Mediterranean *Scopalina* species were considered to be a single more variable species until molecular markers were used to distinguish them (Blanquer and Uriz, 2007). Once determined to be two species, life history differences between them could be distinguished that are sufficient to facilitate coexistence: *Scopalina blanensis* Blanquer and Uriz, 2008 responds opportunistically to seasonal environmental changes in temperature and food availability, while *Scopalina lophyropoda* Schmidt, 1862 responds in a more conservative manner, with similar behaviour and relatively low mortality throughout the year (Blanquer *et al.*, 2008). These are only two examples among many. Phenomenal sponge species diversity in many habitats motivates attempts to discern categories of sponges that are based on functional roles, intimate associations, suites of morphological characters, and differential vulnerability to hazards. Some divisions into categories can be made by inspection, as whether or not a sponge excavates solid carbonate or has an encrusting, massive, or tubular morphology. Categorization by other attributes, such as relative resistance to smothering by sediments, palatability to a particular predator, or possible benefits from microbial symbionts requires experiments. Grouping sponges as ecological or morphological units for data collection (i.e. not identifying to species) does not provide the same quality of information as grouping taxonomically identified sponges for subsequent analysis.

2. INFLUENCES OF ABIOTIC FACTORS AND ECOLOGICAL INTERACTIONS ON SPONGES IN VARIOUS HABITATS

Substratum type, stability, continuity, and depth; and environmental factors related to water quality, movement, and food availability; as well as ecological interactions have all been implicated as influencing distribution and abundance of sponge species. Factors do not vary alone, and so although abiotic factors often correlate well with the habitat distribution of particular sponge species, the underlying processes that actually curtail or enhance distribution and abundance are often not revealed without experiments and time-series observations. I have attempted to impose some linear organization on what is really a multidimensional interconnected network of causal factors, by focusing in turn on a series of habitat types, in each case seeking to illuminate what has been learned of how interactions with other organisms add to abiotic factors to influence distribution, abundance, and diversity of sponge species.

2.1. Subtidal rocky substrata—walls, plateaus, canyons

2.1.1. Abiotic factors

On subtidal rocky substrata, distribution and abundance of sponge species have been demonstrated to be influenced by water movement, depth, light, inclination, and other aspects of bottom topography, as well as the stability and continuity of the substratum.

Vigour of water movement has presented itself as a consistently important abiotic factor, decreasing overall abundance and constraining growth forms of sponges, and allowing only a stalwart few species to live at very exposed sites. For example, at Lough Hyne, Ireland, sponge faunas on cliffs differed between high- and low-energy environments, indicating the primary influence of wave energy; of the 96 species, only 25 were shared between cliffs and cobbles, indicating additional distinction by substratum stability (Bell and Barnes, 2003). Likewise, a diverse sponge fauna of 82 species on temperate rocky reefs in New South Wales, Australia, was revealed by ordination to consistently divide into distinct sets of species at exposed versus sheltered locations (Roberts et al., 2006). Even at this depth of 18–20 m sponge cover reflected differences, with at least 40% cover of sponges at the four sheltered locations but only 25% cover at four exposed locations. Sponge morphologies reflected hydrodynamic differences, with a preponderance of encrusting forms at exposed sites and erect forms at sheltered sites. The authors pointed out the impossibility of comparing solely exposure, as sheltered sites were also more influenced by pulses of freshwater runoff, as well as human activities.

In addition to directly disturbing organisms, water motion can wreak havoc by setting sediment in motion. An extremely high level of species turnover on shallow subtidal rocky shores at Mazatlán, Mexican Pacific, was caused by physical disturbance involving a combination of wind-motivated water movement and sediment movement and deposition (Carballo et al., 2008). Sand-sized sediment (coarse sand in summer months), which is only suspended by rough water, underscored the importance of the combination of factors. By frequently monitoring permanent quadrats over

6 years, Carballo *et al.* (2008) could follow fates of individual sponges, allowing them to definitively conclude that the influence of physical factors was sufficient to prevent competitive interactions from structuring the community.

A recurrent pattern in studies focused on subtidal hard-bottom sponges is inability to predict species composition of the assemblages at a particular site based on environmental attributes of that site and geographic distance from known sites. Exploration of canyons off of Victoria, SE Australia, yielded 165 sponge species, 79% of which were collected in only one of the five canyons (Schlacher *et al.*, 2007). Species turnover was high between sites within a canyon, as well as between canyons, and geographic distance between sites was a poor predictor of community similarity. The authors pointed out that distribution of rare species can be underestimated, especially by sled sampling, inflating the percentage of species that appear to inhabit only one site. Nevertheless, these data indicate large differences in sponge assemblages among sites. The generally high abundance of sponges in these hydrodynamically and topographically complex canyons was attributed to the great abundance of food for filter feeders. Species diversity decreased with depth in the 114–612 m range collected and increased with heterogeneity of substratum.

On subtidal rocky surfaces representing nine habitat types between 0 and 20 m on the northern Atlantic coast of Spain, Preciado and Maldonado (2005) found that substratum inclination best explained variation in sponge cover and diversity among sites. Sponge diversity (a total of 85 species in the 18 habitat-zones sampled) and biomass per quadrat were significantly greater on vertical than on horizontal substrata. They pointed out that, while the frequent dominance of horizontal surfaces by macroalgae may fuel the assumption that a disjunct distribution of sponges and macroalgae indicates that algae outcompete sponges, algal abundance is not the sole factor that varies with inclination. Sediment on horizontal surfaces may also impede sponges. As well, sponge abundance was higher on vertical substrata even at depths below the range of macroalgae. Lack of influence of competition with algae was also suggested by a pattern of sponges that were distributed independently of the presence or absence of algae, a pattern also found in the Cabrera Archipelago, in the Mediterranean off Majorca by Uriz *et al.* (1992). In the Gulf of Maine, Witman and Sebens (1990) also suggested that decrease in sponge cover by 2/3 between 45 m (i.e. below the lower limit of kelp depth distribution) and 60 m was due to increased sediment cover observed on horizontal surfaces.

The primary constraint on habitat distribution may not reveal itself without experimental manipulation. Focusing on individual species and explicit comparisons between species can help to clarify which processes influence sponge distribution and abundance. In Mazatlan Bay, Mexico, the most abundant organisms between 2 and 4 m, the sponge *Haliclona caerulea*

Hechtel, 1965 and its symbiotic associate, the branching calcareous red alga *Jania adhaerens*, were experimentally demonstrated to be constrained from living more shallowly by high mortality due to wave action (Carballo and Ávila, 2004). In conjunction with water movement, topography can influence the impact of sediment on sponges. Sediment can smother sponges by clogging their aquiferous systems. When growth of an undescribed Western Australia *Haliclona* sp. was compared at high and low levels of light, sediment, and water flow, in order to determine what factors confine it to the undersides of limestone ledges, only the low sediment treatment reduced weight loss of explants (Abdo *et al.*, 2006).

2.1.2. Ecological interactions

Distribution boundaries that coincide with abiotic factors are often caused by interactions with competitors, predators, and symbiotic associates. Macroalgae on continental shelf temperate zone hard bottoms can add their influence to topographical variations. Barthel (1986) described how *Halichondria panicea* Pallas, 1766 improved its ability to cope with medium to strong currents at the entrance of Kiel Fjord, Baltic Sea, by growing on red macroalgae that swayed with the current. Sponges and macroalgae have the opposite interaction in the central Gulf of Maine, where Witman and Sebens (1990) demonstrated clear zonation of sponges on subtidal hard substrata, with differences in sponge species composition with depth and also on vertical versus horizontal surfaces. Where kelps were abundant on horizontal–sloping surfaces above 40 m (the extinction depth of laminarian algae) sponge cover was low. Percentage cover increased with depth to a maximum of 20.8% at 45 m. High incidence of predation by a nudibranch and sea star was deemed to influence small-scale sponge distributions (Shield and Witman, 1993), but not large-scale zonation patterns, which were most influenced by negative interactions with the kelps.

On rocky reefs of the Investigator Group of islands, South Australia, a dense fucoid canopy with green algal understory dominated exposed surfaces, and although sponges were growing beneath the algae at low percentage cover, an especially rich and abundant sponge fauna was found in caves and under overhangs (Sorokin *et al.*, 2008). Likewise, greater sponge cover under rocks at Lough Hyne, Ireland, was attributed to macroalgae growing on the upper surfaces (Bell and Barnes, 2003), and Sará (1970) remarked on decreased persistence of individual sponges nearest the mouths of Ligurian caves, where they shared the substratum with macroalgae instead of other sponges.

Adding additional interactions can reverse sponge distribution patterns relative to macroalgae. At sites near Wollongong, New South Wales, Australia, sponge cover was six times higher among the kelps relative to the adjacent urchin barrens, in spite of physical disturbance by moving kelp fronds and lack of light. The lack of overlap in sponge species in the two habitats (no difference in diversity, with 10 species in each habitat) hinted at an additional factor, which was revealed to be sea urchin grazing on the barrens (Fig. 4.1). Only sponges that are chemically defended from urchins are able to live outside the kelp forests (Wright *et al.*, 1997). In the Mediterranean of NE Spain, experiments revealed that a similar set of taxa (i.e. urchins, macroalgae, and sponges) interacts very differently (Fig. 4.1). Urchin grazing facilitated growth of the sponge *Cliona viridis* Schmidt, 1862 by diminishing the fleshy seaward canopy that otherwise blocks access to sunlight for the zooxanthella symbionts of the sponge (Cebrian and Uriz, 2006; Rosell and Uriz, 1992). A second excavating sponge species that lacks photosynthetic symbionts may be favoured in competition between sponges in the darker environment that results from the absence of urchins. The next trophic level up must therefore be considered, because whether or not fishes that prey on the urchins are over-fished can determine which set of interactions prevails, by influencing urchin abundance (Cebrian and Uriz, 2006).

Aggressively invasive macroalgae have unfortunately offered opportunities to learn more about particular characteristics of algae that can affect sponges more dramatically. In Australia, Davis *et al.* (1997) documented decreased cover of sessile invertebrates, including sponges, from 48% to 23% in the 12 months following the arrival of *Caulerpa scalpelliformis* at Botany Bay, New South Wales. There was no change at reference sites during the same time period. Although sponges can be highly tolerant of epizoism, this tolerance was overwhelmed by the interwoven stolons and dense upright fronds of the *Caulerpa*, combined with the sediment they accumulated. In the Ionean Sea, Italy, Baldacconi and Corrierro (2009) also recorded substantially decreased sponge cover, but relatively little loss of species, in the 2 years during which *Caulerpa racemosa* var. *cylindracea* cover increased from scattered small bits confined to horizontal surfaces to a dense, continuous, sediment-trapping mat on all exposed surfaces. Cavity-dwelling sponges were unaffected, but on horizontal substrata, species diversity dropped from 18 to 11 and cover from 30.6% to 12.2%, and on vertical substrata species dropped from 36 to 26 and cover from 29.4% to 17.6%. The *Caulerpa* could actually anchor its stolons in the surfaces of the sponges, with the sole exception of the encrusting species *Crambe crambe* Schmidt, 1862 which was able to fend the alga off.

Macroalgae constitute a distinct set of spatial competitors against sponges because they are constrained to exposed surfaces, especially horizontal surfaces, by their requirement for sunlight, offering the possibility of refuges in the shade for sponges. Other potential competitors for space on subtidal rocky substrata include bryozoans, ascidians, and other sponges. The importance of spatial competition for sponges of the NW Mediterranean rocky sublittoral is well demonstrated by patterns in toxicity of *C. crambe* specimens, which were more toxic at sites dominated by other sessile animals relative to well-lit algal-dominated sites (Becerro *et al.*, 1997). *C. crambe*

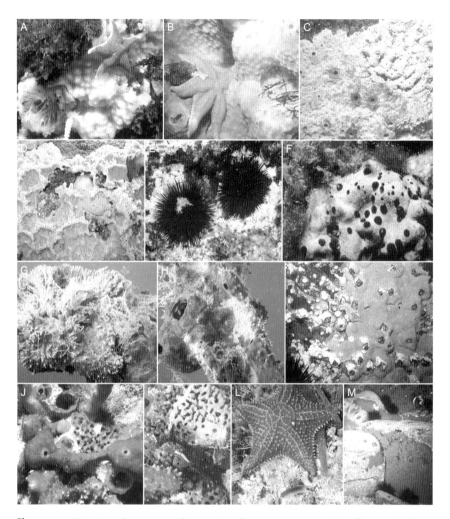

Figure 4.1 Diversity of outcomes of interactions between sponges and mollusc and echino-derm spongivores. In Antarctica, sponge-feeding sea stars may prevent *Mycale acerata*, which grows exceptionally rapidly, from overwhelming other sponges (Dayton, 1979). Photo A. Bill Baker: *Odontaster validus* eating *M. acerata*. Photo B. Bill Baker: *Perknaster fuscus* eating *Mycale acerata*. Photo C. Bill Baker: *Mycale acerata* (lower left of photo) and *Dendrilla membranosa*. In *Alaska*, an unusually dense recruitment of the dorid nudibranch *Archidoris montereyensis* elimi-nated *Halichondria panicea* from a large area of the intertidal where it had dominated the space for the previous 10 years (Knowlton and Highsmith, 2000). Photo D. Jason Hall: *Archidoris montereyensis* consuming *Halichondria* cf. *panicea* in Olympic National Park, WA. In the *Medi-terranean*, NW coast of Spain, herbivorous sea urchins facilitate growth of the zooxanthellate boring sponge *Cliona viridis* by diminishing fleshy algae that otherwise block sunlight (Cebrian and Uriz, 2006). Photo E. Enric Ballesteros *Paracentrotus lividus*. Photo F. Enric Ballesteros: The boring sponge *Cliona viridis*. In *New Zealand*, a nudibranch was two orders of magnitude more

chemistry discouraged not only regeneration of a key spatial competitor, the sponge *S. lophyropoda* Schmidt, 1862, but also the settlement of larvae of the bryozoan *Bugula neritina*. The authors point out that these results do not indicate unimportance of competition for space with seaweeds for sponge habitat distribution, but rather demonstrate the adaptive deployment of toxicity, as chemistry of an encrusting sponge is less likely to be effective against quickly growing single-holdfast seaweeds such as kelps and fucoids.

Topography can be related to sponge distribution constraints by inter-actions even in the absence of macroalgae. Off the coast of Georgia, USA, Ruzicka and Gleason (2009) related distinct sponge assemblages on vertical scarps versus plateaus to a combination of abiotic factors and predation. Of 32 species, 16 were found in only one of the habitats, and another 14 were significantly more common in one habitat. Species diversity did not differ between habitats, but density of individuals was higher on the scarps. Sponges on vertical scarps were more likely to have to withstand physical disturbance and tended to be encrusting and amorphous forms, while the sedimented surface of the plateau was handled better by erect-branching or pedunculate forms. Spongivorous fishes were more common on vertical scarps, adding a biotic component to distinguishing the sponge faunas. Nine days after four plateau species were transplanted to the scarp, signs of predation were clear, and three of the four species had lost significantly more tissue outside cages than when enclosed. In Ireland, Bell and Barnes (2003) documented another influence of topography on biotic interactions,

---

abundant on *Mycale hentscheli* grown on lines for pharmaceutical production, relative to in its natural community (Page *et al.*, 2011). Photo G. Mike Page: Severe grazing damage by *Haplodoris nodulosa* on *Mycale hentscheli*. H Mike Page: Juvenile *H. nodulosa* feeding on *Mycale hentscheli*. In *SE Australia*, a large barnacle increases recruitment success of sponges by providing a refuge from urchin grazing; different sets of sponge species live among kelps versus on urchin barrens, with only sponge species that resist sea urchin grazing in the barrens. Photo I. Andy Davis: The large barnacle *Austrobalanus imperator* and the large common urchin *Centrostephanus rodgersii* with *Tedania anhelans* (orange) and *Chondrilla australiensis* (brown). Photo M. Andy Davis: The physically defended sponge, *T. anhelans*, on vertical surfaces with *C. rodgersii*. In Belize, *Caribbean*, the massive reef sponge *Lissodendoryx colombiensis* is readily consumed by a large seagrass-dwelling sea star but is able to inhabit a seagrass meadow when sponge species that are unpalatable to the sea star overgrow it (Wulff, 2008a). Photo J. Janie Wulff: *Lissodendoyx colombiensis* overgrown by *Chondrilla caribensis* (brown), *Clathria schoenus* (yellow, branching), and *Tedania klausi* (orange-red, in bottom of photo). Photo K. Janie Wulff: The large sea star *Oreaster reticulatis* departing from where it has just consumed a portion of a large *L. colombiensis* that was not overgrown by unpalatable sponge species. Photo L. Janie Wulff: The sea star *O. reticulatis* consuming a reef species, *Mycale laevis*, that was transplanted into a cage in the seagrass, and thrived until the cage was removed, 2 h before the photo was taken. (For interpretation of the references to colour in this figure legend, the reader is referred to the Web version of this chapter.)

showing increased risk of sponges being broken off of vertical rock walls due to activities of fishes.

The possibility that predation can have substantial effects on the amorphously massive species *Mycale hentscheli* Bergquist and Fromont, 1988 in New Zealand, if normal inhibitions on the predators are absent, was discovered in the course of monitoring individuals grown on lines in aquaculture for drug production (Page *et al.*, 2011). The nudibranch *Haplodoris nodulosa* was two orders of magnitude more common on farmed sponges, causing severe depletion (Fig. 4.1). The authors suggested that the predator population explosion may have resulted from the continuous monoculture of their food supply, and possibly also the distance of the lines serving as substrata from the habitat of the natural predators of the nudibranch.

Protection of clonal invertebrates from sea urchin grazing by structure provided by large barnacles (Fig. 4.1) was suggested by a positive correlation of cover by sponges and colonial ascidians with barnacle density (Davis and Ward, 1999). The sponge *Clathria pyramida* Lendenfeld, 1888, which was known to discourage urchin grazing, stood out with a contrasting distribution negatively correlated with barnacle density, appearing to confirm the requirement of the other sponges for protection by barnacles. Unsatisfied by interpretations based solely on correlations, the authors designed experiments to test the possibility that other processes were at work. On scrubbed vertical rock faces, 4–14 m deep, they glued plaster filled barnacle tests in natural configurations and at densities spanning the natural range (Davis and Ward, 2009). As they had suspected, the barnacles influenced recruitment, and after 8 months, invertebrate cover and diversity were greater with higher barnacle density. After 56 months, clonal recruits had grown so that cover no longer increased with barnacle density, but the species diversity difference persisted. The complementary experiment of removing barnacles from the midst of established communities revealed that they were not involved in the maintenance of the sponge community, as after 22 months there were no differences between the unmanipulated and barnacles-removed sponge communities. Microhabitat protection of sponge recruits from grazers was also demonstrated to be key for the common Mediterranean sponges *C. crambe* and *S. lophyropoda*. On submerged outcrops near Blanes, sponges that settled in grooves and crevices enjoyed a reduction in mortality due to bulldozing by the sea urchin *Paracentrotus lividus* (Maldonado and Uriz, 1998).

Symbiotic associations can both enhance and constrain habitat distributions because they combine not only the abilities but also the habitat requirements of both participating species. For example, macroalgal partners in mutually beneficial associations with sponges may limit the depth distribution by their requirement for light. Restriction to depths above 4 m of the association of the coralline red alga *J. adhaerens* and the sponge

Ha. caerulea depends on the alga, which also determines the overall form of the association by growth patterns that reduce self-shading in lower-light environments (Enriquez *et al.*, 2009). Many more associations may significantly influence distribution and abundance of participating species, but establishing both the patterns and their possible adaptive significance can be challenging. For example, the brittle star *Ophiothrix fragilis* settles preferentially on sponges, especially *C. crambe*, and the very small juveniles migrate laterally to recolonize sponges if they are cleared from the sponge surfaces (Turon *et al.*, 2000). The pattern of association is extremely striking and demonstrated to be constituted purposefully, but exactly how these species influence each other is not yet determined.

Associations with microbial symbionts can also mediate the affect of abiotic variables on host sponges, and changes in abiotic variables can diminish the ability of sponges to effectively battle microbial interlopers, such as pathogens. Abundance of Mediterranean species of bath sponges, in the genera *Spongia* and *Hippospongia*, has been dramatically diminished by disease, to the point of near local extinction at some sites (e.g. Pronzato, 1999 and a recent review in Pronzato and Manconi, 2008). Sea water temperatures that were 2–4 °C above normal may have favoured enormous losses that occurred at the end of summer 1999 (e.g. Cerrano *et al.*, 2000b). Exposure time to elevated temperatures was also positively correlated with death in two mass mortality events (summers of 2008 and 2009) of *Ircinia fasciculata* Esper, 1794 in the western Mediterranean Sea (Cebrian *et al.*, 2011). Several lines of evidence implicated symbiotically associated cyanobacteria in the increased vulnerability of *I. fasciculata*: normal cyanobacteria were lost from injured individuals, photosynthetic efficiency was diminished at experimentally elevated temperatures, and the related sponge *Sarcotragus spinosulum* Schmidt, 1862, which hosts only heterotrophic bacteria, did not suffer mortality at the same times and places (Cebrian *et al.*, 2011). Precise documentation of how disease diminished population size was facilitated by conspicuous bare patches that remained on rock for several months after *I. fasciculata* individuals died of disease (Cebrian *et al.*, 2011).

Wilkinson and Vacelet (1979) made explicit comparisons of growth of several Mediterranean species in different flow and light environments by transplanting individual sponges into experimentally altered circumstances. Cyanobacteria-harbouring *Aplysina aerophoba* Nardo, 1833 grew four times as fast under a clear shield as under a black shield, a result that was not unexpected given prior examples of benefits of single-celled algae to animal hosts. Also as expected, if the algae are nutritionally helpful, growth of this species was not diminished as much by low flow conditions. Before the experiments it was less certain if cave-dwelling sponge species prefer low light, or merely accept it because caves are more suitable for some other reason; but *Aplysina cavernicola* Vacelet, 1959 and *Chondrosia reniformis* Nardo, 1847 grew better shaded from light, allowing Wilkinson and

Vacelet to designate these species as truly sciaphilous. Two individuals of the *Chondrosia* even migrated to the underside of the experimental substrata during the experiment. The experiments simultaneously demonstrated the degree to which sediment can inhibit sponges, as *A. aerophoba* control individuals grew half as much as those under clear shields. Depending on water column conditions and flow rates, sponges that gain nutritional boosts from photosynthetic symbionts may constitute a substantial portion of the fauna on subtidal rocky habitats. For example, many of the 61 demosponge species found on rocky reefs in the Investigator Group islands, South Australia, harbour cyanobacterial symbionts, reflecting the extremely clear water (Sorokin *et al.*, 2008).

Photosynthetic symbionts have also been shown to influence trophic interactions of their hosts in an unexpected way. The opisthobranch *Tylodina perversa* was inspired to feed on tissue of sponge species with cyanobacterial symbionts (preferring *A. aerophoba* over *A. cavernicola*) as well as individuals with higher densities of cyanobacteria (i.e. shallow *A. aerophoba* over deep), and even asymbiotic sponges to which cyanobacteria were added (Becerro *et al.*, 2003). In a similar example, the gastropterid opisthobranch *Sagaminopteron nigropunctatum* selects the ectosome over the choanosome of the sponge *Dysidea granulosa* (Becerro *et al.*, 2006). This feeding choice results in *S. nigropunctatum* ingesting high concentrations of cyanobacteria because *D. granulosa* has high concentrations of cyanobacteria restricted to its ectosome (Becerro and Paul, 2004; Becerro *et al.*, 2006).

2.2. Subtidal rocky substrata—cobbles and caves

Caves and cobbles are special cases of rocky substrata, with substratum discontinuity and instability added as complicating factors. Influences of disturbance regime, resource availability, colonization, and competition are so intertwined for these hard-bottom habitats that I here consider abiotic factors and ecological interactions together. Differences among individual substrata can be extreme in the case of cobbles, as substratum size can influence stability, which in turn affects disturbance rate and therefore availability of primary substratum space. An additional aspect of heterogeneity of the habitat experienced by sponges on different parts of a boulder was recently demonstrated in a study in northern France focused on *H. panicea* Pallas, 1766 and *Hymeniacidon sanguinea* Grant, 1826 (=*H. perlevis* Montagu, 1818) living on tops and bottoms of boulders (Schaal *et al.*, 2011). Stable isotope comparisons indicated that a significantly greater proportion of the food consumed by sponges living on the undersurfaces of boulders was based in decomposition of organic matter.

Substratum stability can influence species diversity through its influence on the balance between provision of new space by disturbance and overtaking of space by competitive dominants. Underlying this balance is a trade-off between competitive prowess and recruitment efficiency that directly relates provision of space to recruitment by relatively poor competitors. Rützler's (1965) demonstration that diversity of sponge communities on cobbles in the Adriatic decreased with increasing rock size, from 3 to 30 kg, was the first explicit demonstration of the influence of substratum stability (i.e. levels of disturbance) on diversity of sessile inhabitants. Subsequent studies of algae on intertidal cobbles and of corals on shallow reefs resulted in the designation of this causal relationship between intermediate levels of disturbance and peak levels of species diversity as the intermediate disturbance hypothesis (e.g. Connell, 1978).

Competitive interactions become important only on cobbles or boulders that are stable for long enough that growth of colonizers causes space to become limited. Comparing sponge faunas on boulders at a sheltered and exposed site in each of Ireland, the eastern Pacific at Mazatlan, Mexico, and Palmyra Atoll in the tropical central Pacific led Bell and Carballo (2008) to conclude that an increase in number of sponge species with cobble size in their study was due primarily to the greater area of a larger cobble receiving more larval recruits. The shapes of the curves relating species diversity to surface area varied, but all were monotonic, with no sign of a diversity decrease on larger rocks. In this case, all cobbles had surface areas of less than 3000 cm^2, and the lack of competitive exclusion as a process influencing diversity in this system was confirmed by the authors' report of bare space (30–80%) on even their largest cobbles.

The ability of sponges to profoundly influence the stability of their substrata can disconnect the relationship between cobble size and stability. Cobbles of all sizes in the shallow subtidal and lowest intertidal of the Bay of Panama, in the tropical eastern Pacific, are equally immobilized by being embedded in a colourful matrix of at least seven species of sponges, which can only be seen peeking through from spaces between cobbles. Barnacles, bryozoans, oysters, vermetids, and serpulids crowd into each other on the exposed upper surfaces, regardless of cobble size, while the sponges grip the cobbles from beneath, where they are confined by how quickly they are consumed by one of the most common fish in this habitat, the smooth puffer *Arothron hispidus* (Wulff, 1997c).

Caves are similar to cobbles in the isolation of individual substratum patches and steep gradients in availability of food and light over very small distances. Some studies of distribution and diversity patterns of sponges in caves in the Mediterranean and Ireland (e.g. Corriero *et al.*, 2000; Bell, 2002) have focused on influences of water flow and loss of sunlight-requiring potential spatial competitors. Unusual trophic interactions, as well as beneficial associations, that were first studied in Mediterranean caves have stretched our imaginations of what is possible for sponges. One striking discovery that paved the way for similar discoveries at other sites was that sponges could acquire food in such an un-sponge-like way as engaging in carnivory on small crustaceans (Vacelet and Boury-Esnault, 1995).

Indications that sponges may not necessarily abide by the same rules of ecological interaction that elegant experiments have identified for other taxa were first offered by examples of cooperation among sponge species in dense, sponge-dominated cave communities (Rützler, 1970; Sará, 1970). In shallow water caves on the Ligurian Italian coast, Sará (1970) reported more than 60 species in an area of 50 m^2 and 25 species in an area of only 2 m^2. In these particularly dense communities, with continuous sponge cover, number of species increased with increasing density. Near the mouths of the caves, an increase in diversity with decreasing density was attributed by Sará to the relative precariousness of life as a sponge in circumstances in which space must be shared with algae. He pointed out the similarity to life on small cobbles, on which space is also continuously reopening for recruitment (Rützler, 1965). But deeper in the caves, where sponges reliably abut other sponges, these communities are quite stable. By tracing outlines of encrusting sponges at monthly intervals, Sará (1970) determined that the actual location occupied by a particular sponge at a given moment was quite fluid, but that the same individuals remained in the community over the entire year. Sará presented his data relating sponge species diversity positively to density in the context of positive interactions among neighbouring sponges of different species and assembled other examples of epibiosis in situations of sponge-dominated communities. He pointed out the lack of the evidence for competitive elimination but at the same time the unavailability of bare space for recruitment of additional individuals into the community as larvae. Rützler (1970) was also attracted to dense and diverse sponge communities, and focused on a community of 34 species thriving in cavities eroded in the base of large boulders in the Adriatic. By field observations, in combination with histological sections, he revealed morphological specializations for supporting epizoic sponges, or for living as epizoic sponges, among the species inhabiting these dense communities. As creatively illustrated by these two papers published in 1970, sponges may be unique in the degree to which they engage in solving space limitation by benign or beneficial overgrowth, thereby maintaining high species diversity in extraordinarily crowded systems.

2.3. Coral reefs

2.3.1. Abiotic factors

On coral reefs, as on subtidal rocky substrata, the clearest direct abiotic influences on sponge distribution and abundance are exposure, depth, available substratum space, and details of topography, such as inclination, as well as water column productivity. Also, as on subtidal rocky substrata, the influence of light is exerted via ecological interactions, but on coral reefs, single-celled photosynthetic symbionts are more likely to be the mediators than macroalgal competitors.

Exposure to overly vigorous water movement in shallow water is the most likely cause of a consistent pattern within the wider Caribbean region and across the GBR of very low densities, biomass, and species diversity in shallow water (e.g. Wilkinson and Cheshire, 1989; Alcolado, 1990; Alvarez et al., 1990; Schmahl, 1990). Reiswig (1973) attributed the increase that he documented in total sponge volume between 20 and 50 m on the fore reef at Discovery Bay, Jamaica, to limits imposed by wave action and sedimentation in more shallow water. A critical depth, below which wave energy influence on sponge distribution and abundance drops, was suggested by Alcolado (1994) to occur between 5 and 10 m in Cuba, where diversity increases to 20 m, with a subsequent decrease between 20 and 35 m. On the Australian Barrier Reef, density, biomass, and diversity were also consistently low above 10 m and then increased to a maximum at about 20 m, reflecting a combination of light and physical disturbance (Wilkinson and Evans, 1988; Wilkinson and Cheshire, 1989). Exceptions to this pattern, where dense coral reef sponge communities are found in shallow protected areas such as leeward reefs behind algal ridges (e.g. Wulff, 2001), or at latitudes where hurricanes are rare (e.g. Wulff, 1995a), patch reefs in lagoonal systems (e.g. Schmahl, 1990), and mangrove roots (e.g. Rützler et al., 2000) lend credence to the notion that physical disturbance restricts sponges in shallow water at exposed sites.

Comprehensive regional surveys of sponge faunas have provided understanding of abiotic requirements of hundreds of individual species as well as differences among higher taxa in relationships to their environment. Every study has raised intriguing biogeography questions relating especially to faunal heterogeneity among sites. Combining data from nine expeditions allowed Reed and Pomponi (1997) to make a comprehensive analysis of distributions of nearly 300 sponge species at 417 collection sites from 0 m (but especially below 30 m) to 922 m throughout the Bahamas. Diversity was highest (206 species) in the 60–150 m zone, and although they did not quantify abundance it was clear that it peaked in this zone as well, results that concurred with other studies of deep coral reefs. Structure and diversity of sponge assemblages in the second most diverse zone, 30–60 m, strongly reflected the geomorphology, in particular, the variety of subhabitats. Many species were found only in particular depth ranges, and no species was found in all zones. Of the 3059 specimens collected, 429 were unique, a pattern found in other studies. Analysis at higher taxonomic levels revealed a striking shift in relative representation of different orders with depth, although the seven genera found in all depth zones each represented a different order. Similarly, while a geographic signature could be discerned in the species assemblages, 47 species were found in all subregions.

The importance of subhabitats defined by geomorphology was underscored by Lehnert and Fischer (1999), who applied ordination analysis to their collections, combining multiple aspects of the environment into a single analysis of distribution and abundance patterns. They collected at 102 sites at Discovery Bay, Jamaica using SCUBA, and found very clear distinctions between sponges that inhabit exposed reef surfaces versus undersurfaces of plate-shaped corals versus lagoon habitats. They pointed out the degree to which data analysis style can influence conclusions, especially the difficulty of discovering depth-related distribution patterns by using predetermined depth zones. Their collections brought the Jamaican faunal list to 157 species, of which 85% were restricted to shallow water. Of the 60 species found on the deep fore reef (using Trimix diving), only 40% were also found on the shallow reef. Statistically significant environmental variables related to substratum type included substratum inclination, back-reef, fore-reef, deep fore-reef, pinnacle, undersides of platy corals, and coral rubble. On the Bahamian slope between 91 and 531 m, substratum inclination was also a key distinguishing abiotic factor (Maldonado and Young, 1996), in this case confounded with depth because of uneven distribution of horizontal versus vertical substrata over the depth range they traversed with their submersible.

Ordination techniques were also applied to sponges of the Spermonde Archipelago of Indonesia, but on a different scale, with focus on comparing among sites spanning a large geographic area rather than microhabitat details within a set of nearby sites. Cleary and de Voogd (2007) measured a number of environmental variables for 1 day at each of 37 sites and related these to the sponge species at each site. For a total sponge fauna of 150 species, a combination of depth, exposure, and an onshore–offshore spatial component explained 56.9% of the variation in similarity among the sponge species at the sites. de Voogd and Cleary (2008) continued their Indonesian surveys with 30 patch reefs in the Thousand Islands, north of Jakarta, an area profoundly influenced by human inhabitants. Of 148 species, 43 were unique to a single site. As in other studies, the most striking distinction among faunas was related to inner versus outer sites. Faunal differences were evident at the family as well as the species level, as in the Bahamas study by Reed and Pomponi (1997).

Coral reefs and subtidal limestone rocks of the Dampier Archipelago, NW Australia, yielded 150 sponge species from 43 stations that Fromont *et al.* (2006) sorted by non-hierarchical classification. The resulting 11 groups were defined on the basis of depth, exposure, and substratum type and structure. Plotting these groups on a map illustrated a significant geographic component, but 92 of the 150 species (i.e. 61%) were found at only 1 or 2 of the 43 stations with sponges. Strikingly similar are results of Hooper and Kennedy (2002) from 22 sites on the Sunshine Coast of Southeast Queensland. Although a distinction could be made between the faunas of inner (around 2 km from shore) versus outer (around 15 km from shore) reefs, the sponge assortments on adjacent reefs were highly heterogeneous, and about 60% of the 226 species were rare or unique.

On three remote atolls of the southwestern Caribbean, Zea (2001) evaluated sponge assemblages by recording all individuals within 30 m^2 and all species within 400 m^2 at 42 stations between 2.5 and 20 m depth. Comparison with continental shelf reefs shows relatively low densities overall, likely reflecting the low concentration of suspended organic matter. Of the 96 species, 21 were found in a variety of circumstances, and the remainder were associated with circumstances described primarily by depth and exposure. As in other studies, being able to predict environmental circumstances at a site from knowing that a particular species lives there does not mean that the reverse is true. Knowing the environmental conditions at a site does not generally allow prediction of the species present. Distributions were patchy and heterogeneous on scales from tens of metres to hundreds of kilometres.

The almost unanimous finding of highly heterogeneous sponge assortments at sites that are characterized by similar abiotic factors underscores limitations on determining how abiotic factors influence distribution and abundance of individual sponge species by correlating sponge abundance with various parameters. This strong stochastic component to species present at a particular site has been an important theme for discussions of sponge distribution and abundance. History, in the forms of local species loss to disturbance, very low probability of any particular larval dispersal being successful, and enhancement of patchy distributions for some species by asexual propagation after initial recruitment by a larva (detailed discussions in, e.g. Zea, 2001; Hooper and Kennedy, 2002; Hooper et al., 1999, 2002) may play an unusually important role in determining which sponge species inhabit a particular spot.

By focusing solely on species in the order Dictyoceratida, Duckworth et al. (2008) were able to eliminate some of the variations that might result in such heterogeneous distributions and to address some of these complications. Dictyoceratids are relatively homogeneous in having larvae that are not likely to disperse far, preference for solid substrata and relatively clear water, and tough skeletons that resist fragmenting agents. Distribution and abundance patterns were strikingly like those found when sponges from all orders are included: 12 of the 23 dictyoceratid species of the Torres Straights, Australia, were only found at 1 location (4 locations, with 5–7 sites at each), and assemblages were often similar on distant reefs but very different at adjacent sites. As predicted if fragmentation is a cause of dense but widely separated patches, the one ramose species that is more likely to asexually propagate did have an especially patchy distribution.

An exception to this pattern of heterogeneous assortments of sponge species at sites characterized by similar abiotic factors may be sites that are quite extreme in at least one abiotic factor. Many species are capable of living at amenable sites, and accidents of history due to the vagaries of larval dispersal and survival of larvae after settlement cause exceptionally large differences in species composition at sites that appear to be very similar by human evaluations. But when conditions are really very poor, only a few species in a regional fauna are capable of surviving. Alcolado (1994, 2007) has pointed out this pattern in the context of sites influenced by anthropogenic pollution, and he has documented which species in the Caribbean fauna are the last to drop out in highly unfavourable sites with comprehensive time-series surveys of the Cuban sponge fauna. *Clathria venosa* Alcolado, 1984 is the species that most reliably serves as an indicator of sites affected by urban pollution in Cuba. Similarly, another thinly encrusting species in the demosponge Order Poecilosclerida, *Mycale microsigmatosa* Arndt, 1927, was the only species able to cope with all sites at Arrail do Cabo, Brazil, including sites that were affected by urban and energy-generation pollution (Muricy, 1989; Vilanova *et al.*, 2004).

A complementary approach to faunal surveys for identifying specific causes of distribution and abundance patterns is to focus on particular sponge species. Results of studies focused on single or groups of species have consistently highlighted how different the ecology of sponge species that look similar can be. Reiswig's (1973) study of factors influencing distribution and abundance of three species of large vase-shaped sponges on the north coast of Jamaica at Discovery Bay has still not been equalled for comprehensive consideration of all factors. *Verongula gigantea* Hyatt, 1875 was confined to the open, exposed habitat of the fore-reef slope platform, clearly unable to tolerate the particle-laden waters within Discovery Bay. Once established on exposed fore-reef substrata, however, *V. gigantea* individuals were undaunted by abiotic factors. Substratum collapse, caused by a combination of storm waves and bio-erosion, was the cause of the few losses from this population. On deep walls, where bases of the platy corals are readily eroded, cascading losses can constitute dispersal downslope, or if the landing spot is sediment, death by smothering. *Mycale laxissima* Duchassaing and Michelotti, 1864 was confined to reef–sand channel interfaces, a distribution coincident with flexible substrata such as gorgonians, that decreased the rate at which this narrow-stalked species was torn off by vigorous water movement. Winter storms were nevertheless the most important mortality source, and 27% of the population was lost to burial, scour, substratum collapse, and tearing loose from the substratum associated with storms in the course of a year. *Tectitethya crypta* de Laubenfels, 1949 was only found on shallow limestone ledges with relatively little sediment deposition, and the sole losses from the population, of very small individuals, were due to burial during storms.

Following in the tradition established by Henry Reiswig of studying trios of species with a common growth form, but in different orders, the erect-branching species *Iotrochota birotulata* Higgin, 1877; *Amphimedon compressa* Duchassaing and Michelotti, 1864; and *Aplysina fulva* Pallas, 1766 were scrutinized with respect to ability to cope with a variety of factors (Wulff, 1997a). They were experimentally determined to differ significantly in susceptibility to smothering by sediment, disease, and predators, as well as breakage, toppling, and pulverization by storms. Loose fragments of these species, that were generated by these factors, also differed in how well they survived, reflecting differences in reattachment success (Wulff, 1997a). In Puerto Rico, focus on *A. compressa* at sites differing in abiotic factors demonstrated increased size with depth, attributed to decreases in both growth rate and survival with increased water movement (Mercado-Molino and Yoshioka, 2009). Larger individuals were more susceptible to getting torn off by rough water, but best survival was in intermediate size classes because small individuals were eliminated by burial in sediments. After a hurricane in San Blas, Panama, this same species exhibited intermediate survival in comparison with two other erect-branching species, with relative rates of survival reflecting a balance of resistance to fragmentation and fragment survival (Wulff, 1995a). Skeletal composition strongly influences resistance to fragmentation. Among the six species with small basal attachments for which sufficient data could be collected to make statistical comparisons, the two species with skeletons solely of spongin were toppled at less than half the rate (22–24% vs. 48–60%) of the species with silica spicules as well as spongin.

In just a few hours, hurricanes can influence sponge distribution and abundance for decades afterwards. Specific effects are not readily observed, however, because they quickly become invisible as damaged sponges heal quickly or deteriorate and vanish entirely. Quantification of hurricane effects requires prior knowledge of sponges at a site and evaluation of storm damage immediately after the waves have calmed. After a major hurricane in Jamaica wrought havoc on the north coast fore reef, 5 weeks of monitoring 576 individual sponges in 67 species revealed a possible mechanism for the maintenance of a full range of growth forms among the sponges in this habitat. The immediate effect of the hurricane was serious damage to 43% of the erect-branching sponges, and less for sponges in four other growth form categories (e.g. 32% for the tough-skeletoned massive species and the least, 20%, for the encrusting species). Recovery was inversely proportional to susceptibility to damage, however, resulting in almost the same proportion of individuals lost from each of the five growth form categories after 5 weeks of either regeneration or continued deterioration (Wulff, 2006b). Curiously, the net result of damage and recovery, in terms of the proportion of the pre-hurricane populations lost, was worst for the tough massive species that were least damaged. Five years after Hurricane Allen, Wilkinson and Cheshire (1988) evaluated recolonization of a portion of the reef that had been devoid of survivors. The five species most abundant among colonists were all in the tough massive category, suggesting that, in addition to the two strategies for coping with hurricanes that were identified in the weeks following the storm (i.e. resistance to damage and recovery from damage on an individual level), a third strategy may be recovery on a population level by efficient recruitment.

Temperature can be an important constraint on latitudinal distribution of coral reef sponges, as illustrated by a geographic gradient in sponge species distributions along the Gulf Coast of Florida (Storr, 1976). Over a 230-km north–south coastline, average temperatures differed by 4 °C and mean low temperatures by 8 °C; only 10 of the 30 sponge species were distributed along the entire coastline. Along the eastern coast of Australia, an abrupt change from tropical to temperate sponge faunas in only 110 km was documented by a comprehensive geographic analysis of a total of 2324 species (Hooper et al., 2002). At the geographic edges of coral reef distribution, temperature fluctuations can veer into the unacceptably low. A January cold snap that persisted for several days in the Florida Keys resulted in sponge mortality, but death was not evenly visited upon all species. Individuals of some species suffered complete mortality, but for other species, only particular portions of each individual died, and some species appeared to be unaffected (B. Biggs et al., Florida State University, in preparation). Lower temperatures at depth were suggested to constrain reproduction and recruitment for two species of coral reef sponges, as adult sponges appeared unimpeded after they were transplanted to depths below where they were found naturally (Maldonado and Young, 1998).

Abnormally warm temperatures that motivate bleaching (i.e. loss of photosynthetic symbionts) in scleractinian corals do not necessarily cause bleaching in zooxanthellate sponges (e.g. Vicente, 1990). On Orpheus Reef, GBR, 84–87% of the corals bleached in March 1998, but all Cliona orientalis Thiele, 1900 survived (Schönberg and Wilkinson, 2001). Resistance to bleaching in clionaid boring sponges may be conferred by the ability, demonstrated in C. orientalis, to move their intracellular zooxanthellae symbionts deeper into the sponge tissue, in response to stresses (Schönberg and Suwa, 2007). Focusing on symbiotic non–photosynthetic bacteria in the Australian species Rhopaloeides odorabile Thompson, Murphy, Bergquist, and Evans, 1987, Webster et al. (2008) demonstrated loss of normal symbionts and colonization by alien microbes, including potential pathogens, when temperatures were experimentally increased to 33 °C.

Light, diminishing with depth and in cryptic spaces, plays a direct role in the lives of sponges through reactions of sponge larvae (e.g. see Maldonado, 2006 for a review). This role of light is not restricted to coral reefs, but clear water typical of healthy reefs may allow light to play a role over a much greater depth range in this habitat. Physiological sensitivity of sponges to UV light varied widely among Hawaiian sponge species (Jokiel, 1980), with the encrusting species Mycale cecilia de Laubenfels, 1936 succumbing quickly to full exposure, while *Callyspongia diffusa* Ridley, 1884 remained unhampered by UV light. A possible cost of elaborating protective pigments was suggested by the competitive exclusion of *C. diffusa* from water deeper than 3 m by the UV-sensitive *Mycale* (*Zygomycale*) *parishi* (Jokiel, 1980). Many sponges live on tropical reef flats or very shallow seagrass meadows and on intertidal shores in which no refuge from direct sunlight is available. Colour may in some cases protect sponges, and possibly pigments of photosynthetic symbionts also aid in this (discussion in Harrison and Cowden, 1976). Although habitat distribution of the common Australian reef species *R. odorabile* was positively related to light, photosynthesis could not be detected, suggesting that the apparent requirement for light reflects instead correlation of food with light or larval behaviour (Bannister *et al.*, 2011). Light may exert its greatest effects on distribution and abundance of coral reef sponges through its effects on photosynthetic symbionts, a focus of the following section.

2.3.2. Inextricable combination of abiotic factors and ecological interactions: Food for sponges

Factors that exert influence on sponge distribution and abundance through feeding by sponges cannot be readily divided into abiotic and biotic. Water column nutrients can quickly be transformed into pico plankton useful to sponges (e.g. Reiswig, 1971, 1974), and some sponges are capable of directly removing dissolved organics, in collaboration with prokaryote symbionts (e.g. Reiswig, 1981; de Goeij *et al.*, 2008; Weisz *et al.*, 2008). Light is directly transformed into ecological interactions by photosynthetic symbionts that may also feed their sponge hosts. These intertwined influences of abiotic and biotic factors involved in feeding of sponges are not confined to coral reefs, but the relative ease of *in situ* experiments has particularly promoted their study using controlled manipulations on reefs.

Nutrient enrichment of water has often been correlated with increased sponge abundance (table comparing studies in Holmes, 1997), as long as it is not combined with additional pollutants, such as inorganic particles or industrial wastes. Substantial differences in overall biomass of sponges, between the coastal and seaward portions of the GBR, and between tropical Australia and the Caribbean, have been attributed to the greater availability of nutrients near coasts. Wilkinson (1987) reported strikingly higher biomass (measured as weight) on 11 Caribbean reefs (367.5 g/m^2 at Barbados East to 2458.2 g/m^2 at Barbados West, leaving out the Jamaican site at which sponges had been recently eliminated by a major hurricane) relative to 17 reefs on the GBR, Australia (7.9 g/m^2 at Astrolabe Great to 569.9 g/m^2 at Pandora). Confining comparisons between oceans to "oceanic reefs" still yielded eight times the biomass on the Caribbean reefs. An additional related difference was the greater proportion of sponges that rely significantly on phototrophic symbionts in nutrient deficient waters of the outer GBR (Wilkinson, 1987; Wilkinson and Cheshire, 1990).

By comparison with continental shelf reefs, the sponges of three remote atolls in the southwestern Caribbean, showed relatively low densities overall, likely reflecting the meagre concentration of suspended organic matter (Zea, 2001). Zea (1994) also related sponge distribution, abundance, and diversity to a gradient in nutrients along the continental coast of Colombia that may have been natural but has been exacerbated by development of a city near the bay site. He stressed the difficulty of disentangling influences of nutrients from other things that wash off the land, such as sediment. In a survey of sponge distribution and abundance across the 230 km north-south Gulf of Mexico coastline of Florida, Storr (1976) noted that nutrient availability likely influenced the substantially greater sponge abundance and diversity near river mouths. He specifically contrasted the many enormous *Spheciospongia vesparium* Lamarck, 1815 individuals along the Gulf Coast, where nutrients from the Everglades pour into the eastern Gulf of Mexico, with the lower density and smaller individuals of this species in the relatively nutrient-poor Bahamas.

Larger sizes at increasing depth of three common Caribbean tube-shaped sponges, *Callyspongia vaginalis* Lamarck, 1814, *Agelas conifera* Schmidt, 1870, and *Aplysina fistularis* Pallas, 1766, were attributed to superior food availability by Lesser (2006). Energy budgets for *C. vaginalis* revealed a greater rate of food intake at 25 m than at 12 m at a site in the Florida Keys, reflecting significantly higher concentrations of food, especially of heterotrophic bacteria and prochlorophytes, at the deeper site (Trussell *et al.*, 2006). Respiratory costs were also higher at the shallow site, with a clear net result of significantly greater growth rates at the deeper site. Transplants between sites allowed dismissal of the possibility that genetic differences between deep and shallow sponges influenced the growth rate difference. Similarly, three of four species of typical reef sponges transplanted to mangrove roots grew significantly faster among mangroves than on the reef, suggesting response to increased levels of plankton-fuelling nutrients (Wulff, 2005). The one species that did not grow faster in the mangroves, *Desmapsamma anchorata* Carter, 1882, grows unusually rapidly on the reef (Wulff, 2008b).

Excavating sponges, of particular concern for carbonate balance on coral reefs, may be especially spurred on by water column nutrients. Infestation rates of rubble from branching corals by eroding sponge species increased with levels of coastal eutrophication in Barbados (Holmes, 1997). The particularly destructive excavating species *Cliona delitrix* Pang, 1973 was found at especially high abundances in areas influenced by sewage (e.g. Rose and Risk, 1985) and increased abundance over time was related to sewage influence (Ward-Paige *et al.*, 2005). Focusing on patterns of abundance of this voraciously excavating species with respect to a sewage outfall at seven sites in San Andrés Island (Colombia), Chaves-Fonnegra *et al.* (2007) noted that the excavator increased as they moved towards the main outfall, but when they got very close to the outfall it decreased. While increased *Escherichia coli* (the indicator of relative influence of sewage for these authors) may inspire the sponges with more food, the concomitant increase in sediment very near the source may overwhelm the benefit. Because they recorded not only the number of corals infested by *C. delitrix* but also the percentage cover of sponge and dead and live coral, they were also able to determine how the pattern with bacteria abundance was confounded by a positive correlation of sponge cover with coral (live plus dead) cover. The requirement of this species for large, recently dead corals has been identified as a confounding variable in other studies. A comparison by Chiappone *et al.* (2007) of *C. delitrix* at 181 sites in the Florida Keys revealed a distribution pattern of higher density at deeper fore-reef sites, but larger individuals on patch reefs nearer shore, possibly reflecting the need for large recently dead corals; this same requirement was manifested as greater abundance of *C. delitrix* between 12 and 20 m in Los Roques, Venezuela (Alvarez *et al.*, 1990). Deviation from the pattern of increasing boring sponges with increasing water column food at sites within the bay at Discovery Bay, Jamaica, might be due to coincident increase in sedimentation that inhibits efficient pumping of the sponges, resulting in a shift in dominant bio-eroders from sponges to worms and especially bivalves (e.g. Macdonald and Perry, 2003).

Sunlight exerts indirect influence on sponge distribution, abundance, and diversity through ecological interactions, especially by fuelling photosynthetic symbionts such as cyanobacteria, zooxanthellae, and macroalgae and by spurring the growth of photosynthetic competitors.

Photosynthetic symbionts are not just a feature of coral reefs, although clear water may allow a greater depth distribution, and the relative ease of *in situ* experiments on reefs has spurred research in this habitat. Shading sponges that harbour photosynthetic symbionts has demonstrated the potential importance of exposure, but the importance of this type of heterotroph–autotroph association is not uniform across all sponge–alga species pairs. Shading the tropical Pacific species *Lamellodysidea chlorea* de Laubenfels, 1954 for 2 weeks resulted in loss of mass, but symbiont density did not decrease, whereas shaded *Xestospongia exigua* (= *Neopetrosia exigua* Kirkpatrick, 1900) lost symbionts but not mass (Thacker, 2005). Differing reactions of these species were attributed to differences in host specificity of symbiotic cyanobacteria, with the specific association of *Oscillatoria* with *L. chlorea* being mutually beneficial, but the generalist cyanobacterium *Synechococcus spongiarum* hosted by *X. exigua* merely commensal. A pair of Caribbean species also differed in responses to shading, with *A. fulva* growing significantly less in 6 weeks under opaque canopies but *Neopetrosia subtriangularis* Duchassaing, 1850 unaffected with respect to growth, although symbiont density decreased under canopies in both sponge species (Erwin and Thacker, 2008). Significantly greater growth of symbiont-bearing *A. fulva* in lower-light mangroves than on reefs (Wulff, 2005) suggests that additional aspects of switching between feeding modes (i.e. heterotrophy vs. reliance on photosynthetic symbionts) remain to be discovered. Variation among species in response to shading may be related to how flexible they can be with respect to modes of feeding, as well as to differences among symbiont clades.

A pattern of constrained depth distribution for photosymbiont-bearing keratose sponges in the Caribbean was experimentally addressed by Maldonado and Young (1998) by transplanting individuals of the common shallow reef species *A. fistularis* Pallas, 1766 and *Ircinia felix* Duchassaing and Michelotti, 1864 from a shallow (4 m) reef to 100, 200, and 300 m. Death within 2 months of all individuals that were transplanted to 300 m was attributed to temperatures (18–9 °C) that were much lower than those of the home environment at 4 m of 26–32 °C; but transplants to 100 and 200 m did surprisingly well. Histological preparations and *in situ* photos of the sponges made before and after transplantation allowed them to conclude that symbiotic cyanobacteria remained in the same concentrations in *A. fistularis*, but the sponges lost their fistules; although cyanobacteria were lost from *I. felix* and the transplants grew unusually tall, narrow chimneys, sponges of both species grew more (although not significantly so) at depth than did controls at 4 m. The authors concluded that the absence of keratose sponges from greater depths may reflect lack of recruitment at depth, due to loss of larval viability or inability to disperse through the pycnocline. This experimental study of a distribution pattern overturned what seemed like obvious explanations of a bathymetric zonation pattern: the need of adult cyanobacterial-hosting sponges for adequate light for autotrophic symbionts, combined with an increasingly oligotrophic water column at depth.

One set of sponges that are confined to illuminated substrata are the photosymbiont-bearing clionaid boring species (e.g. López-Victoria and Zea, 2005). Distributions of three species of zooxanthellate Caribbean excavating sponge species, *Cliona aprica* Pang, 1973, *C. caribbaea* Carter, 1882, and *C. tenuis* Zea and Weil, 2003, were clearly associated with well-illuminated substrata, as well as with recently dead corals. Lack of a positive association with influence from untreated sewage, which has been demonstrated for other clionaids, suggests that zooxanthellae reliably supply their hosts in these species (López-Victoria and Zea, 2005).

The Caribbean excavating species *Cliona varians* Duchassaing and Michelotti, 1864 grows thickly over the substratum as well as excavating burrows, and zooxanthellae near the surface impart a rich golden brown colour. By manipulating light levels and pre-weighing blocks of solid carbonate substrate, Hill (1996) was able to correlate both growth rate and excavation rate with density of zooxanthellae, confirming that the sponge benefits nutritionally from the symbionts. Sunlight-fuelled symbionts also give some encrusting sponge species a significant enough boost in growth rate that they can overwhelm corals, as discussed in Section 2.3.3.2.

Curiously, while harbouring photosynthetic symbionts gives many sponges a significant boost in growth rate and appears to allow a few of them to overwhelm corals, dependence may not be as strong as for corals, and primary habitat constraints may be factors other than sunlight. In addition to the lack of sponge growth enhancement by some of these associations, harbouring photosynthetic symbionts may not be entirely beneficial even in cases of advantage. Wilkinson and Cheshire (1989) suggested that the expense of symbiont upkeep results in decreased representation of symbiont-harbouring species in coastal waters relative to oligotrophic outer reefs on the GBR. Further research into relative importance of differing flexibility in feeding mode among sponge species versus differing contributions among clades of symbionts will contribute to understanding of evolution of mutualism as well as sponge biology.

2.3.3. Ecological interactions
2.3.3.1. Symbiotic associations with macroscopic organisms
Sponges distinguish themselves by their astonishing number and variety of symbiotic associations with macroscopic organisms of all kinds. On coral reefs, many of these associations have been demonstrated to influence distribution and abundance and, in some cases, are the actual causes of distribution patterns that are correlated with abiotic factors. Alvarez *et al.* (1990) pointed out the possibility that apparent abiotic restrictions on depth for a sponge species can actually reflect the depth distribution of favoured microhabitat distributions, such as the association of *Mycale laevis* Carter, 1882 with the massive coral *Montastraea annularis* (Goreau and Hartman, 1966; Fig. 4.2). *M. laxissima*, a vase-shaped sponge with a relatively narrow basal attachment, may survive better on flexible substrata, such as gorgonians, that move with water motion, preventing the sponge from being ripped off (Reiswig, 1973).

In a very different context, overgrowth by *D. anchorata* may be facilitating the invasion of the tropical Pacific by the octocoral *Carijoa riisei*, as nudibranchs, observed on the unfouled octocoral, were absent when it was covered by the sponge (Calcinai *et al.*, 2004). In Hawaii experimental comparisons of feeding by the nudibranch, *Pyllodesmium poindemieri*, on *C. riisei* that was bare versus covered by four species of sponges, confirmed the protection afforded the octocoral by the sponges (Wagner *et al.*, 2009). How growing on *Carijoa* may benefit sponges has not been studied, but sponge associations with other colonial cnidarians on coral reefs can be mutually beneficial. Bright yellow zoanthids, *Parazoanthus swiftii*, conspicuous when embedded in dark forest green *I. birotulata*, discourage the angelfish *Holocanthus tricolor* from consuming the sponge in the Caribbean (West, 1976). Providing another caution on generalization, similar

Figure 4.2 Sponge–coral interactions, positive and negative. In the *Caribbean*, corals associated with non-excavating sponges survived an order of magnitude better than corals from
which sponges had been removed because sponges adhere corals securely to the reef frame
even if their bases are eroded (Wulff and Buss, 1979). Photo A. Janie Wulff: *Amphimedon
compressa* helping to bind *M. annularis* to the reef. Photo G. Janie Wulff: *Niphates erecta*
helping to prevent multiple portions of a *M. annularis* colony from becoming disengaged and
falling into the surrounding sediment. Photo B. Janie Wulff. The encrusting sponge, likely
Acarnus nicoleae, covering bare *M. annularis* skeleton, rendering it off limits for recruitment of
boring organisms. In the *tropical western Pacific*, the cyanobacteria-bearing encrusting sponge,
Terpios hoshinota, can rapidly overgrow living corals. Photo C. Keryea Soong: *Terpios
hoshinota* overgrowing living coral in Taiwan, where up to 30% of the corals were infested
on some reefs only a few years after *T. hoshinota* was first sighted there (Soong *et al.*, 2009). In
the *Caribbean*, *Mycale laevis* is closely associated with massive corals, which grow to

appearing associations between the zooxanthellae-hosting zoanthid, *Parazoanthus parasiticus*, in grey-blue *Niphates digitalis* Lamarck, 1814 interfered with pumping and in grey-purple *C. vaginalis* it failed to protect from angelfishes (Lewis, 1982) or sea stars (Wulff, 1995b).

Sponges associated with corals on Caribbean reefs can increase coral survival by an order of magnitude as they counter the effects of excavating organisms by gluing corals with eroded bases to the reef frame and protecting exposed skeleton from being colonized by excavating organisms. By measuring and mapping all corals on several fore-reef patch reefs, and then removing sponges from half of them, Wulff and Buss (1979) were able to confirm this benefit of sponges to corals that had been suggested by Goreau and Hartman (1966). Six months after the start of the experiment, 4% of the corals had fallen off of control reefs, but 40% had fallen off reefs from which sponges were removed. Even in cases in which some coral polyps were killed in order to allow a sponge to grip the solid carbonate skeleton, this is a small price for a coral to pay for an order of magnitude boost in colony survival.

The relationship of sponges to substratum stability on coral reefs is unusual in the degree to which the sponges themselves influence substratum

---

accommodate its large oscules, as the sponge increases survival of the coral colonies (Goreau and Hartman, 1966; Wulff and Buss, 1979). Photo D. Janie Wulff: *Mycale laevis* adhering a *Porites astreoides* colony, for which the base had been entirely eroded by boring sponges, to the reef frame. Note that the sponge does not overgrow the living coral tissue. Photo H. Janie Wulff: *M. laevis* growing to close a gap in the coverage of exposed *Montastraea annularis* skeleton. In the *Caribbean*, sponges stabilize dead coral skeletons until crustose coralline algae can cement them permanently into a stable structure, suitable for coral recruitment (Wulff, 1984). Photo E. Janie Wulff: *Aplysina cauliformis* fragments, torn from their bases in a hurricane in Jamaica, stabilizing pieces of rubble from branching corals that were also generated by the storm. On coral reefs, boring sponges that harbour zooxanthellae are major agents of destruction of solid carbonate. Photo F. Christine Schönberg: *Cliona orientalis* (lower left) and *Aka mucosa* in the same coral slab in the pavement zone in < 1 m of water, in Little Pioneer Bay, Orpheus Island, GBR, Australia. Photo L. Christine Schönberg: The boring sponge *Cliona orientalis* infesting *Platygyra daedalea* in Fig Tree Bay, Orpheus Island, Australia. Although a handful of encrusting sponges that harbour cyanobacteria are capable of rampant and rapid overgrowth of living coral, interpretations of apparent overgrowth must be based on time-series observations. Photo J. Klaus Rützler: *Chondrilla caribea* overgrowing skeletons of the corals *Acropoa cervicornis* and *Agaricia tenuifolia*, Cat Cay, Pelican Cays, Belize; note that the corals had died before the sponge covered them, and that Cat Cay hosts a particularly dense assortment of spongivorous fishes. Photo K. Klaus Rützler: *Cliona caribea* infesting *Diploria strigosa*, at the rate of 0.11–0.25 mm linear growth/day so that the distance from the tip of the knife to the leading edge of the *C. caribea* patch was covered in only 2 years. (For the colour version of this figure, the reader is referred to the Web version of this chapter.)

stability (reviews in Wulff, 2001; Bell, 2008). On one hand, as discussed in the previous section, a handful of excavating sponge species are responsible for causing disengagement from reefs of chunks of solid carbonate (e.g. Hartman, 1977, Wilkinson, 1983). Non-excavating sponges (i.e. the vast majority of species) can influence solid substrata in the opposite way, by stabilizing coral rubble, significantly improving survival of coral recruits and thereby facilitating recovery of damaged reefs (Wulff, 1984). Among the non-excavating sponges that help corals maintain their grip on the reef frame, one Caribbean species in particular, *M. laevis*, is often found in close association with large massive corals, especially in the genera *Montastraea* and *Porites* (Fig. 4.2). The sponge grows in crevices under and between colony lobes that are covered with living coral tissue, protecting the otherwise exposed coral skeletons from action of excavating organisms, and also serving to glue pieces of coral with eroded bases to the reef frame. As the corals grow, the colony shape continues to accommodate the large oscules of the sponges (Goreau and Hartman, 1966). Consumption of this species by parrotfishes, especially *Sparisoma aurofrenatum* and *Sparisoma viride*, occurred only when the surface was sliced off, and never when it was intact (Wulff, 1997b); thus in return for increasing survival of their host coral, the sponges gain both an expanding substratum and a safe place to tuck vulnerable tissue.

Some coral reef sponges augment their skeletons with macroalgae, perhaps gaining an energetic advantage by not having to expend energy on skeleton formation (examples compiled by Rützler, 1990a). On tropical eastern Pacific reefs in Panama, *Ha. caerulea* perfused with articulated coralline red algae was protected from being consumed by fishes, but when sponge pieces without algae were exposed they were quickly consumed, especially by the angelfish *Holocanthus passer*, which normally engages in planktivory on these reefs that are nearly devoid of exposed sponges (Wulff, 1997c). The Australian sponge–macroalga combination of *Haliclona cymaeformis* Esper, 1794 and *Ceratodictyon spongiosum* Zanardini may fragment more readily than most branching sponges with significant spongin content in their skeletons, because the alga serves as support; but in this case, nutritional advantages allow this symbiotic association to grow rapidly and to recover quickly after fragmenting events (Trautman *et al.*, 2000).

Playing host to photosynthetic microbes not only restricts sponge species to lighted habitats but may influence the evolution of additional associations. Symbiotic associations of sponges with zoanthid species that are obligate sponge symbionts illustrate the complex layers of interdependence resulting from symbionts that require light. Analysis of specificity of associations among 92 Caribbean sponge species and 6 zoanthid species revealed that zoanthid species that host zooxanthellae exhibit a pattern of disproportionate association with host sponges that also host photosynthetic symbionts (Swain and Wulff, 2007). An evolutionary perspective, gained by matching phylogenies, demonstrated that a host switch of a zoanthid to a sponge species without photosynthetic symbionts was accompanied by evolutionary loss of zooxanthellae from this zoanthid species, maintaining the match in requirement for sunlight between the sponge and zoanthid species (Swain, 2009).

Beneficial sponge–sponge associations, first described in temperate North American, Adriatic, and Mediterranean waters (Rützler, 1970; Sará, 1970), are also featured on coral reefs. Mutually beneficial associations among three species of erect-branching sponges, for which growth, and especially survival, is increased by adhering to a sponge of another species, are based on variation among species in susceptibility to a variety of hazards, including breakage by water movement, smothering by sediment, infection by pathogens, and consumption by a variety of predators (Wulff, 1997a). Time-series observations and a variety of experimental manipulations, comparing growth alone versus in combination with other species, confirmed that apparent overgrowths were actually mutually beneficial to participating individuals. Sponge individuals that suffer partial mortality are saved from additional mortality, that can result from being disengaged from the substratum, by firmly adhering to heterospecific sponges that are susceptible to different partial mortality sources. As in dense and diverse sponge-dominated communities in caves in the Mediterranean and Adriatic, mutually beneficial interactions between sponge species may serve to autocatalytically increase diversity by keeping all species in the community.

2.3.3.2. Competition

Competition with algae to the point that sponges are eliminated has not been reported on coral reefs. Zea (1994) was able to clarify a sequence of changes at sites on the Colombian coast near Santa Marta by repeating surveys of number of individuals and percentage cover of corals and sponges after a 2-year interval. At a site that had recently suffered considerable coral mortality due to stress by nutrients and sediment in coastal runoff, sponges had significantly increased and corals had decreased. The actual sequence was that increases in sponges followed on the heels of increases in algae, which had come at the expense of corals. Zea (1994) pointed out that some thin encrusting sponges may have been missed during the first survey when fleshy algal turfs were especially dense, but that even taking that into account, the sponge increase was significant. A recent unusual bloom of a crustose coralline alga in a small semi-enclosed bay in Bonaire has resulted in overgrowth of corals and also sponges, but whether or not sponges will be lost or be able to tolerate the overgrowth is not yet known (Eckrich et al., 2011).

Most sessile animals also appear unable to outcompete adult sponges on coral reefs, although newly recruited sponges are vulnerable. One of the rare examples is overgrowth of thinly encrusting Chondrilla caribensis forma hermatypica (Rützler et al., 2007) by the corallimorpharian Ricordea florida and the gorgonian Erythropodium caribaeorum that was documented on the fore reef in Puerto Rico (Vicente, 1990). One coral reef sponge, *Dysidea* sp., has been demonstrated to overgrow another, *Cacospongia,* in Guam. The resulting necrosis can lead to deterioration of the basal attachments of *Cacospongia* such that they lose their grip on the reef (Thacker *et al.,* 1998). Mutually beneficial associations among three branching species (Wulff, 1997a), described in the previous section, were parasitized by a fourth species that behaved conspicuously differently, overgrowing heterospecific sponges to the point of smothering them. *D. anchorata,* with a ridiculously flimsy skeleton, survived significantly better (64.3% vs. 0% after 6 months) when it grew on other erect-branching sponges than on solid carbonate substrata (Wulff, 2008b). Its role as a parasite on the sponge–sponge mutualism did not become evident until after about 12 months, when it began to get large enough to smother its hosts.

2.3.3.3. Special cases of competition: Sponges and corals
Much has been, and continues to be, made in the coral reef literature of how sponges, in general, appear to be increasing on coral reefs at the expense of corals. On the other hand, substantial losses of sponges have been reported, raising concern about loss of their important roles as water filterers, substratum stabilizers, and hosts of diverse symbionts representing every group of organisms. Variation among sponge species in their interactions with corals, ranging from rampant overgrowth of living corals to increasing coral survival by an order of magnitude, mandates that reports of sponge–coral interactions include the names of the species observed and time-series observations. Interactions that appear on initial observation to be competitive overgrowth may either fail to progress over time or actually be beneficial. Results of interactions are not merely species dependent but are also context dependent. Even the outcomes of interactions of excavating sponge species with corals, unambiguously negative for the corals, and obligate for the sponges, depend on details such as angle of encounter, temperature, and water column nutrient levels (e.g. Rützler, 2002; Schönberg, 2002, 2003; López-Victoria *et al.,* 2006). Because of these complexities, and the great importance of sponges to the existence of this threatened ecosystem, I go into much greater detail in this section.

A few sponge species have been demonstrated to be aggressively invasive and capable of overwhelming living coral, when enabled to colonize a community in which they did not evolve. In Kaneohe Bay, Hawaii, *Mycale grandis* Gray, 1867, which is native to Indonesia and Australia, has been wreaking havoc with the corals *Porites compressa* and *Montipora capitata* by smothering their living tissue since 1996. In 10 permanent photo quadrats, cover of this sponge increased 13%, while coral cover decreased 16.3% (Coles and Bolick, 2007). On the Pacific coast of Mexico, an Indo-Pacific sponge, *Chalinula nematifera* de Laubenfels, 1954 is disproportionately associated with branching corals in the genus *Pocillopora*, which it overgrows, adhering tightly to the bared skeleton. This association may be facilitating the invasion of Mexican Pacific coral reefs by *C. nematifera*, because of the relatively low light levels within the coral colonies, or perhaps the protection against predators provided by spikey *Pocillopora* branches (Ávila and Carballo, 2008). Fortunately, it had not increased its representation at the sites monitored during the study, and although the authors searched for it at 150 sites, they only found it at Isla Isabel and Cabo Pulmo.

A small handful of sponge species have been demonstrated to kill corals by chemical means. Experiments and observations of the Caribbean species *Plakortis halichondroides* Wilson, 1902 indicated that corals of 14 species were killed by contact and at distances up to 5 cm (Porter and Target, 1988); and an Australian *Haliclona* species that bears zooxanthellae and also nematocysts appears to be able to settle on and kill living *Acropora nobilis*, as necrosis has been observed within 1 cm of the sponge (Garson *et al.*, 1999; Russell *et al.*, 2003). de Voogd *et al.* (2004) recorded all neighbour interactions of four Indonesian sponge species that had been determined to be bioactive, showing that these species caused necrosis 85% of the time when they overgrew corals. The excavating species *Aka coralliphaga* Rützler, 1971 is also capable of penetrating coral covered by living tissue (Rützler, 2004).

Although only a very small number of species have been observed to consistently overwhelm corals, their effects can be significant when and where they are abundant. An example from Yemen differs from other reports of sponge–coral interactions in that the sponge, a very thinly encrusting *Clathria* species, specifically attacks the massive coral *Porites lutea* as a narrow band along the edge of the living tissue, and as it kills the coral it leaves behind bared skeleton rather than continuing to cover the substratum (Benzoni *et al.*, 2008). Although apparently a local phenomenon, at least for now, it is of concern because *P. lutea* is the primary reef building coral, constituting up to 47% of the benthic cover at the Gulf of Aden site; and *Clathria* sp. was observed in half of the coral colonies. At another Gulf of Aden site, *Clathria* sp. was noted to infest large corals that were transplanted in order to save them from destruction due to construction of a liquefied natural gas plant. Once the corals had a chance to recover from the stress of transplantation, this threat seems to have receded, as survival after over a year was 91% (Seguin *et al.*, 2008).

A small number of encrusting or excavating species, most of which harbour photosynthetic symbionts, have been demonstrated to overgrow living corals (e.g. Vicente, 1978, 1990; Rützler and Muzik, 1993). On Puerto Rican reefs, *C. caribensis* was shown to be an important aggressor against nine species of scleractinian corals (Vicente, 1990); and on coral reefs of the central and western Pacific, another encrusting sponge species that harbours cyanobacteria, *Terpios hoshinota* Rützler and Muzik, 1993, has been demonstrated to overwhelm corals. Because generalizations have frequently been made about "sponges" overwhelming corals, based on reports of these two highly unusual species; and because even these species appear to be greatly restricted in the circumstances under which they overwhelm corals, I go into considerable detail about both in the paragraphs that follow.

Terpios hoshinota (Fig. 4.2) was first reported as overrunning corals in Guam in 1973 (Bryan, 1973), and distribution and abundance studies in Okinawa 10 years later revealed a pattern of this thinly encrusting species running rampant over live corals at sites where development had increased turbidity of coastal waters (Rützler and Muzik, 1993). The possibility that *T. hoshinota* was able to gain nutritionally from the living coral tissue as it overgrew was suggested by growth rates on live coral (a mean of 6.5 mm/8 days!) that exceeded those on dead coral (Bryan, 1973). A test of this intriguing suggestion, by comparing growth on live corals versus freshly cleaned coral skeletons, revealed growth on freshly cleaned skeletons to be even faster (Plucer-Rosario, 1987), suggesting inhibition of sponge growth by the fouling community on substrata that have been exposed. The ability of *T. hoshinota* to rapidly infest large areas may be fuelled by symbiotic cyanobacteria that are so dense that the sponge can appear almost black; coupled with an unusual growth form that combines thin encrustations that quickly cover everything with narrow processes by which new substrata can be colonized asexually. On a reef in Taiwan, where an outbreak resulted in 30% of the corals being infested, Soong *et al.* (2009) positioned dark sheets over infested corals to block sunlight from fuelling the cyanobacteria hosted by *Terpios*. Blockage of sunlight caused bleaching in the coral hosts, and the sponges ceased spreading in their usual continuously encrusting growth form. But the sponges advanced thread-like processes across the shaded spaces and resumed their usual expansion once they regained lighted substrata. In spite of an unusual combination of attributes that allow *T. hoshinota* to quickly obliterate large areas of living coral, it also appears to be relatively ephemeral, vanishing at sites where it was abundant, and appearing elsewhere. Fourteen years after *T. hoshinota* had killed 87.9% of the corals at a site in Okinawa, it could not be found among the dense live corals, but it covered 50% of the substratum at a new site on a different island (Reimer *et al.*, 2010), and has now been discovered overgrowing live corals at Lizard Island, Australia (Fugii *et al.*, 2011).

Chondrilla caribensis (Fig. 4.2) also stands out as a species that is capable of overgrowing living coral under some circumstances (Vicente, 1990). When *Chondrilla* was caged with corals in the Florida Keys, it extended laterally over the corals significantly more than when it was not caged, suggesting that spongivores kept the sponge in check (Hill, 1998). Like *T. hoshinota*, *Chondrilla* has unusual abilities for spreading itself and is consequently a difficult organism to manipulate. It tends to break into pieces and migrate (e.g. Zilberberg *et al.*, 2006), entering uncertainty into interpretations of experimental results (e.g. Wilkinson and Vacelet, 1979). Experiments on growth of the closely related species, *C. nucula* Schmidt, 1862, in the Mediterranean required that the specimens be placed in small cups, and still they were able to climb out (at a rate of 2.5 cm/month) and populate the outsides of the cups (Pronzato, 2004). In the Pelican Cays, Belize, one of the sites where *C. caribensis* looks as if it has overgrown corals (Fig. 4.2), the corals had actually died as a result of bleaching before *Chondrilla* overgrew them (Macintyre, 2000), which it did in the presence of a conspicuously dense population of large spongivores, including grey angelfishes, trunk-fishes, filefishes, and spadefishes (Wulff, 2000, 2005). Consumption by fish was recorded by video when *Chondrilla* was placed on racks on sand (Dunlap and Pawlik, 1996), and Hill (1998) observed *Chondrilla* with bite marks, indicating that it was consumed, but not entirely. In Panama, it was included as one of 64 species that were consumed by naturally feeding angelfishes a few bites at a time, alternating with bites of other sponge species (i.e. smorgasbord feeding, as described by Randall and Hartman, 1968 and Wulff, 1994). *Chondrilla* ranked 28th in total volume, out of 39 sponge species in a fully censused plot, but 39th in terms of number of bites taken by angelfishes in the genus *Pomacanthus* (Wulff, 1994). It and it is consistently rejected by the large Caribbean sea star, *Oreaster reticulatis* (Wulff, 1995b). A suggestion that variation among individuals and popula-tions may account for at least some of the variation in conclusions about palatability of this species (Swearingen and Pawlik, 1998) is bolstered by the recent distinction of subspecies, one inhabiting mangrove roots, *C. caribensis* forma *caribensis*, and the other inhabiting reefs, *C. caribensis* forma *hermatypica* (Duran and Rützler, 2006; Rützler *et al.*, 2007). Thus interpretation of consumption of *Chondrilla* must take into account the source habitat. Based on literature reports of *Chondrilla* overgrowing corals in the Florida Keys and Puerto Rico (Vicente, 1990; Hill, 1998), and high frequency in hawks-bill turtle gut contents, León and Bjorndal (2002) concluded that historically much larger populations of hawksbills prevented this species from over-growing corals, but it seems that something else must be keeping this sponge uncommon on many reefs that are not currently well populated by hawks-bill turtles. For example, in Los Roques, Venezuela, *Chondrilla* was not among the 60 species in 1290 m^2 spanning a depth range from 1 to 35 m (Alvarez *et al.*, 1990); and in San Blas, Panama, it constituted only 0.085% of the total sponge volume of 33,721 cm^3 in completely censused quadrats (Wulff, 2006a). *Chondrilla* was not among the 24 most common species between 10 and 30 m depth at sites in Cuba (Alcolado, 1990); and only 5 of the 3554 (=0.14%) sponge individuals identified at shallow, medium, and deep zones in the upper Florida Keys were *Chondrilla* (Schmahl, 1990). Something besides hawksbill turtles or spongivorous fishes appears to have primary responsibility for restraining *Chondrilla*, at least on most Caribbean coral reefs, while on some reefs it appears to grow unrestrained. It is intriguing that a sponge species that can be extremely common locally, and is capable of overgrowing corals under some circumstances, has been so resistant to our developing a comprehensive understanding of what controls its distribution and abundance.

Aiming to test the hypothesis that sponges are more likely to overgrow corals on reefs that are stressed, Aerts and van Soest (1997) categorized interactions between corals and sponges at three depths at each of five sites near Santa Marta, Colombia, that varied in sedimentation rate and water column visibility. Underscoring the great importance of careful identification of sponges to species, their data analysis revealed that not only was sponge overgrowth of coral not more likely on stressed than on healthier reefs, but the chief determinant of coral overgrowth was not abiotic factors, but the presence of a particular handful of sponge species. Only 16 of the 95 sponge species at these sites engaged in overgrowth of corals at all, and of those, only *D. anchorata*, *Aplysina cauliformis* Carter, 1882, and *Callyspongia armigera* Duchassaing and Michelotti, 1864 overgrew coral in more than 10% of their contacts. Follow-up study of interactions over time revealed that apparent overgrowths often turned out to be standoffs, and that the thinly encrusting species *C. venosa* Alcolado, 1984 overgrew living coral tissue only when the coral colony had been stressed by experimental damage (Aerts, 2000).

D. anchorata, the species that cheats on mutualism among other branching sponge species by overgrowing them to the point of smothering, also bolsters its flimsy skeleton by growing over gorgonians (McLean and Yoshioka, 2008). Although it has been observed to overgrow corals, it does not survive well on rigid substrata (as detailed above). Its ability to grow at rates much faster than other Caribbean coral reef sponges is balanced by a rate of mortality that is also much higher (Wulff, 2008b). Specific growth rates after 3 months were four times those of three co-occurring branching species. The suggestion that fast growth is possible because it invests so little in its own skeleton is bolstered by biomechanical data showing that the extensibility of three other branching Caribbean reef sponges was 5–15 times as great. This flimsiness was reflected in *Desmapsamma* mortality that, after 9 months, was more than eight times that of the other species. These traits allow rapid, but relatively ephemeral, occupation of any particular site. This species that can appear at a particular moment to be overwhelming a reef can also be diminished to small scraps in an afternoon squall.

Effects of one type of ecological interaction are often moderated by another, and so segregating the effects of competition from the effects of predation is a rather futile exercise in rendering something multidimensional into the linear pattern decreed by paragraph structure. Thus many of the examples in the next section "Predation" have been, or could have been, introduced as illustrative of how competition influences sponge distribution and abundance.

2.3.3.4. Predation

Large spongivores on coral reefs include hawksbill turtles, angelfishes, and to lesser extents some trunkfishes, file fishes, puffer fishes, and rabbit fishes. Less is known of the smaller predators, such as nudibranchs and small echinoderms, and even less about inquilines that may munch on their host.

Hawksbill turtles consume a very small subset of coral reef sponge species, most of them in the orders Astrophorida, Chondrosida, and Hadromerida. Many of the species chosen are conspicuously full of silica spicules (e.g. Meylan, 1988, 1990). The large sturdy beaks of hawksbills are capable of reducing large, well-armoured sponges to nothing or to remnants, which are able to regenerate (e.g. Dam and van Diez, 1997). Data on hawksbill feeding, gained by lavage of stomach contents, provide a clear record of what has been eaten, and the relative amount of each species. Every study has confirmed the small number of species, all in a constrained representation of higher taxa, that are ingested. In the Dominican Republic, percentage cover was measured in the field for sponge and corallimorpharian species found in lavage samples in order to calculate selectivity indices. These indicated that a combination of relative abundance and preference influences turtle feeding decisions and indicated positive selection for *Spirastrella coccinea* Duchassaing and Michelotti, 1864 and *Chondrilla nucula* (= *C. caribensis*) at one site and *Myriastra kallitetilla* (= *Stelletta kallitetilla* de Laubenfels, 1936) and *C. nucula* at another site (León and Bjorndal, 2002). Unfortunately, sponge species not found in lavage samples were not included in the field survey, and so these data cannot address the question of persistent choice of these species, and a handful of others, from among the hundreds of sponge species inhabiting Caribbean coral reefs. Additional clues about hawksbill feeding choices came from Mona Island, Puerto Rico, where Dam and van Diez (1997) combined lavage sampling with observations of feeding turtles and noted that the turtles often searched under ledges and in crevices for their prey. At both cliff and reef sites, the species *Geodia neptuni* (= *Sidonops neptuni* Sollas, 1886) and *Polymastia tenax* Pulitzer-Finali, 1986 were most commonly ingested, with differences in relative rates of ingestion of these and other species reflecting differences in the sponge faunas at their sites. Strong hints that *G. neptuni* is nutritionally more valuable to the turtles were provided by reduced foraging time on the cliffs (where *Geodia* was very common), reduced variety of sponge species ingested on the cliffs, and greater growth rates of immature hawksbills living along the cliffs (Dam and van Diez, 1997). The effect on *Geodia* populations included complete consumption of some individuals, but also many individuals showed the typical healing and regeneration patterns after hawksbill bites (photo in Dam and van Diez, 1997); *Geodia* remained abundant at these sites.

Angelfishes are the other large dedicated spongivores on coral reefs. When Randall and Hartman (1968) analysed gut contents of multiple representatives of 212 species of Caribbean reef fishes, they discovered that only 11 of the species consumed sponges, and also that spongivorous fish species tended to distribute their feeding over many sponge species. Based on these data, Randall and Hartman pointed out that coral reef sponges that live on exposed surfaces were not likely to be controlled by predators. Their conclusion, inferred primarily from angelfish gut contents that in the aggregate included 46 sponge species, as well as gut contents of individual fish that had consumed as many as nine species just before the moment they were speared, has been well corroborated by extensive field observations of feeding angelfishes. Hourigan *et al.* (1989) observed feeding on over 22 sponge species in a field study of unimpeded angelfishes, and Wulff, (1994) observed two species of *Pomacanthus* consuming 64 sponge species in the course of 2285 bites. Wulff's data revealed that angelfishes fed on rarer species significantly more than predicted by their relative abundance, both by comparisons of number of bites with sponge volume and by comparisons of number of visits with number of individual sponges. Feeding sequences unambiguously confirmed that individual angelfishes took only a few (mean of 2.8) bites from each sponge, and in 92% of the cases in which they continued feeding they moved on to a sponge of another species. A different interpretation of angelfish preferences among Caribbean reef sponges was made by Pawlik *et al.* (1995) who presented pelletized sponge extracts mixed with powdered squid to wrasses in tanks and scored sponge species as deterrent if the wrasses rejected 4 or more out of 10 pellets. Because some of the species that produced extracts acceptable to wrasses also appeared frequently in Randall and Hartman's (1968) gut content data, these species were deemed preferred. It should be noted that, while gut content data can provide incontrovertible evidence that a particular species was ingested, they cannot distinguish if a prey species was consumed because it was preferred or merely because it was abundant, unless data on relative abundance of sponge species were collected at the sites where fish were collected. Thus gut content data alone cannot validate pellet assays. Some sponge species that appeared especially frequently in the gut contents analysed by Hartman, such as *C. vaginalis*, are among the most common on Caribbean reefs, suggesting that their frequency in gut contents was at least in part due to availability. This species ranked 22nd by number of bites consumed at a site where it ranked 15th by total volume (Wulff, 1994). However enticing its extracts are to wrasses, when mixed with squid powder, this species remains among the most abundant. Bite marks can sometimes be observed on the rims of *C. vaginalis* tubes, indicating the presence of spongivores capable of consuming it, but not inclined to consume very much of it at one go. In a study focused on feeding and growth rate of *C. vaginalis* at different depths, no signs of predation were ever observed, leading the authors to suggest that bottom-up control was more likely than top-down (Trussell *et al.*, 2006). For sponge species that normally live on exposed surfaces on coral reefs in the Caribbean, where most of the spongivore work has been done, predation does not appear to be a primary influence on distribution and abundance patterns.

For sponges that do not live on exposed reef surfaces, predators can play a much greater role in distribution and abundance. Some species are confined to cryptic spaces in the reef frame by spongivorous fishes that eagerly consume them if they are exposed by breaking open their cryptic habitats (e.g. Wulff, 1988, 1997b,c; Dunlap and Pawlik, 1996).

Defences against reef-dwelling spongivores that serve reef-dwelling sponge species well in their normal habitat (i.e. reefs) are not necessarily effective in other habitats. Strict boundaries of habitat distribution where reefs abut seagrass meadows may appear to be caused by inability of reef sponge species to cope with diminished solid substratum or shifting sediments. But in the Caribbean at least, the seagrass-dwelling large sea star *O. reticulatis* efficiently guarantees this habitat restriction by quickly consuming most typical coral reef sponge species if they are washed into the seagrass by a storm or moved there by underwater farmers (Wulff, 1995b). Sponge species typical of other habitats, such as mangroves, are likewise prevented from successfully colonizing coral reefs by reef-dwelling spongivorous fishes (Dunlap and Pawlik, 1996; Wulff, 2005).

Consumption of coral reef sponges by small animals, including some nudibranchs, and endosymbionts, such as shrimps and polychaetes, has also been reported, but challenges in experimental manipulation of inquilines and other small predators have made it difficult to know the extent to which these predators influence distribution and abundance of reef sponge species. Most studies have focused on preferences of the predators. Spread in geographic extent and increase in abundance of the soft coral invader *C. riisei* may be facilitated by its protection from a specialized nudibranch by association with four species of sponges in Hawaii (Wagner *et al.*, 2009). Clear consumption of sponges is demonstrated by photographs of polychaetes inhabiting *A. cauliformis* with their jaws deeply embedded in their host tissue (Tsuriumi and Reiswig, 1997), and syllid polychaetes on the surface of some sponges have been demonstrated to consume their host (e.g. Pawlik, 1983). Shrimps inhabiting canals of *Hymeniacidon caerulea* Pulitzer-Finali, 1986 clearly gain shelter, but bits of this deep royal blue sponge in their guts indicate that they are also ingesting their host (e.g. Rios and Duffy, 1999).

2.4. Coral reefs—cryptic spaces

2.4.1. Abiotic factors
Cryptic habitats, such as caves and cavities in coral reef frameworks, are inhabited by a rich diversity of sponges. Of the 92 species identified by Kobluk and van Soest (1989) in cavities in the Bonaire reef frame, only a

small proportion (14%) of the species were found over the entire 12–43 m depth range sampled, but it is unknown what might restrict depth distributions in these species. Many of the species they found in cryptic spaces also inhabit exposed surfaces, raising questions about the relative quality of life for them in these very different circumstances. Do the hidden individuals subsidize the exposed populations? Or are they merely at the fringe, barely eking out a living in cryptic spaces that may constrain size, decrease growth rates, and certainly preclude dependence on photosynthetic symbionts? Distinction must be made between cavity-filling cryptic species that entirely fill gaps between pieces of hard substrata and those, mostly thinly encrusting, species that coat the linings of cavities and crevices and undersurfaces of plate-shaped corals. These are not only morphologically, but also taxonomically distinct, and so while Kobluk and van Soest found that many cavity-inhabiting species also live on exposed surfaces, Lehnert and Fischer (1999) found that sponges inhabiting undersurfaces of platy corals constituted a highly distinct set of species.

Species typical of cryptic habitats were relatively well protected from the ravages of a major hurricane on Jamaican coral reefs; but when they were infrequently exposed by the reef frame being ripped apart, they exhibited unusually poor capacity for recovery, given the usual high regeneration capacity of sponges (Wulff, 2006b). If it is legitimate to interpret their inability to recover when exposed by reef destruction as a hint that this type of insult has not dominated their selective regime, then physical disturbance is an aspect of the abiotic environment that may be evaded in cryptic spaces on coral reefs. Sediment is the other abiotic factor from which sponges in cryptic spaces are protected. Coralline sponges, major framework builders in the Paleozoic, may now be restricted to the undersurfaces of ledges or ceilings of cavities and caves, or to the most vertical of reef walls, by their intolerance of sediment or poor ability to compete for space due to slow growth (Hartman and Goreau, 1970; Jackson et al., 1971; Willenz and Hartman, 1999). Their dense aragonite skeletons produced by slow growth are important reef framework reinforcers on deep reefs and in caves (Lang et al., 1975; Hartman, 1977).

2.4.2. Ecological interactions

As remarked in the previous section on coral reefs, at least some species have been demonstrated to be restricted to inhabiting crevices in shallow reef frames by eagerness of fishes, including some parrotfishes, to feed on them in the Caribbean (Wulff, 1988, 1997b; Dunlap and Pawlik, 1996), tropical eastern Pacific (Wulff, 1997c), and central Pacific (Backus, 1964). Two species of semi-cryptic sponges, that live with surfaces exposed, but the bulk of their tissue tucked into crevices in corals, were also consumed by parrotfishes when their surfaces were sliced off (Wulff, 1997b). However, not all species that are confined to cryptic spaces are consumed when exposed, suggesting that they fail to thrive due to surface fouling by diatoms, or abiotic factors such as sunlight or wave action (Wulff, 1997a,b).

Competition for space can be extreme in cryptic habitats. Cavity-filling species can bind disjunct pieces of solid carbonate to each other, stabilizing them until crustose coralline algae and other carbonate secreting encrusters can bind them together permanently, rendering them suitable for recruitment of coral larvae (Wulff, 1984). Internal architecture of cavity-filling cryptic sponges tends to be "cavernous", facilitating water flow in enclosed spaces and also encouraging inquilines. At least one cavity-filling species, *Hy. caerulea*, hosts eusocial shrimp (Rios and Duffy, 1999).

Encrusting sponges that line cavities or coat undersurfaces of plate-shaped corals, interact very differently, and competition for primary substratum space may dominate their lives. The ability of encrusting filter-feeding animals to evade elimination in a habitat of discontinuous substrata was explored by Buss (1976) and Buss and Jackson (1979) in a system of at least 300 species that inhabit the undersurfaces of foliaceous corals. Non-transitive competitive relationships were demonstrated by scoring overgrowth in 152 interactions among 20 species of encrusting organisms, including 8 sponge species. Buss and Jackson's resulting notion of competitive networks provides a mechanism for enhancing species diversity in habitats characterized by multiple discrete patches. A combination of position effects with multiple mechanisms of competition can increase diversity by slowing elimination from a particular discrete substratum and also by increasing the probability of different winners in each patch.

2.5. Mangroves

2.5.1. Abiotic factors

Prop roots of mangrove trees, and the associated peat banks, can support extremely dense communities dominated by sponges, at least in areas where tidal amplitude is not great. Root surface area covered by sponges can be 100%, with mean coverage reported as 31.7% in the Florida Keys (Bingham and Young, 1995), and 10–50% cover in Belize (Farnsworth and Ellison, 1996). In a comparison of diversity and abundance of sponges on mangrove roots at scales ranging from within individual roots to between cays on the Belize Barrier Reef, a striking pattern revealed by Farnsworth and Ellison (1996) was the complete lack of sponges on windward sides of cays. Luxurious sponge growth on mangrove roots in less exposed sites (e.g. 15.7 and 20.8 cm^3/cm root length on mangrove prop roots at two sites on offshore cays in Belize; Wulff, 2009) may be facilitated by relatively benign physical disturbance levels and high food availability, although extreme variations in other abiotic factors such as temperature, salinity, and turbidity can be mortal stressors.

Rützler *et al.* (2007) compiled tolerances of sediment and of temperature and salinity extremes, for 25 mangrove-dwelling sponge species, demonstrating the astonishing capacity of these fragile appearing, and uniquely porous animals to cope with environmental challenges, including temperatures of 19–41 °C, salinities of 20–38 ppt, and layers of fine sediment as thick as 5 mm. In a single tidal cycle, temperatures can vary from 18 to 32 °C in winter and from 28 to 41 °C in summer (Rützler *et al.*, 2004). Species vary widely in tolerances, and distribution patterns reflect this variation. On an offshore mangrove cay, Twin Cays Belize, differences in sponge faunas between a main channel site and a tidal creek (only 330 m away) that consisted mainly of deletions in the tidal creek (17 vs. 39 species) spurred transplant experiments to see if deletions were best explained by colonization history or by more extreme temperature and salinity fluctuations. Transplants thrived for the first 10 days, but all replicates of five of the six species had vanished 1 year later, implicating episodically unfavourable abiotic factors in constraining distribution for at least these five species (Wulff, 2004). Similarly, transplants of four species between sites in Belize that differed in abiotic factors grew during the first 16 days but then began to decline such that by 6 months later all had died (Farnsworth and Ellison, 1996). More extreme negative conditions delete more species, to the point that only a single species was found on prop roots at one Belize coastal site (Farnsworth and Ellison, 1996). At mangrove sites in the Florida Keys, where temperature and salinity fluctuate dramatically, sponge species typical of coral reefs did not fare well after transplantation, and at one of three sites, the usual mangrove species were killed as well (Pawlik *et al.*, 2007). Torrential rains in the Florida Keys entirely eliminated mangrove sponges from sites at Long Key (C. Lewis, personal communication), underscoring the degree to which episodically extreme abiotic conditions can influence sponges at sites that are vulnerable not only due to their proximity to land but also because of their geographic position at the extreme boundaries of faunal distribution.

On Caribbean mangrove roots, where low tidal amplitudes generally allow sponges to live reliably submerged very near the water line, occasional very low tides expose sponges. Even during a period of unusual tides 40 cm below MLW in the middle of sunny days, some sponges survived, providing information on variation among sponge species in ability to cope with air exposure (Rützler, 1990b). Vertical zonation patterns on mangrove roots could then be interpreted as the signature of previous low tides. Just as in rocky subtidal and coral reef habitats, negative effects on sponges of episodically unfavourable abiotic factors can be moderated or worsened through microbial symbionts in mangrove habitats. Rützler (1988) documented disease in a cyanobacteria-bearing species, *Geodia papyracea* Hechtel, 1965, in a Belizean mangrove channel, that was caused by its normal cyanobacterial symbiont multiplying at an overwhelming rate. The sponges appeared unable to control their symbionts, which may have outstripped their hosts in response to conditions, possibly temporarily warmer temperatures, that were more favourable to cyanobacteria than to sponges.

2.5.2. Ecological interactions

Heterogeneity of species distribution at spatial scales ranging from within individual roots to between geographic subregions has been the conclusion of most studies of mangrove sponge species distribution patterns (e.g. Farnsworth and Ellison, 1996). Although much variation in distribution and abundance can be attributed to variation in abiotic factors, high heterogeneity remains a characteristic even in comparisons between sites that are abiotically identical. At least some of the apparent heterogeneity reflects methods of evaluating abundance, as illustrated by three sites in Belize and Panama that appear to differ substantially in composition if evaluated with respect to numbers of individuals, but are very similar when compared in terms of volume (Wulff, 2009). Vagaries of recruitment to small separate substrata contribute to heterogeneity among roots at a particular site (e.g. Sutherland, 1980; Wulff, 2004), and current direction can influence movement of larvae within a site, with the importance of this effect strongly dependent on life histories of individual sponge species (Bingham and Young, 1991, 1995; Bingham, 1992). Understanding differences in life history strategies among sponge species that are typical of mangrove roots allows at least some of the heterogeneity among roots within a site to be interpreted as differences in successional stage (Wulff, 2009), with younger roots covered by quick recruiters and older roots tending to have accumulated poor recruiters that are adept at out-competing early successional species.

As in other communities, substratum continuity can influence the degree to which a competitively superior species can eliminate other species from a community. Rare leaps from one mangrove root to another via long flimsy extensions are possible for a few species, but in general recruitment is by larvae, as in other habitats with discrete small substrata that are separated by uninhabitable matrix. Among the sponge species in a mangrove-root system in Bahia de Buche, Venezuela, Sutherland (1980) documente an inverse relationship between efficiency of recruitment onto experimentally provided substrata and ability to acquire and hold space. This result was corroborated by Wulff (2004) in a Belizean mangrove cay where early recruitment to artificial roots (pvc pipes suspended among the roots) was disproportionately of species that were either uncommon, or not reported at all on roots.

Sutherland (1980) documented community structure and followed dynamics for 18 months on mangrove roots dominated by sponges. By explicitly comparing community development on roots versus large flat recruitment panels that he suspended among the roots, he was able to garner clues about mechanisms by which community assembly results in heterogeneous species composition from root to root. Sutherland interpreted the rapid domination of some of the panels by the single species *T. ignis*, which was not among the top recruiters, as an illustration of the extreme importance of stochastic recruitment events in this system. Because each panel had at least 10 times the surface area of an individual root, they were more likely to sample all of the larvae in the water column and, therefore, to gather species that recruit less efficiently but that may be particularly effective at gaining and maintaining control of space. Competitive elimination in this system was slowed by a disproportionately low probability that a superior competitor species recruited at all to a particular root. Just as for community dynamics on undersurfaces of coral plates, the interplay between size of continuous substrata, provision of bare space, recruitment, and competition on mangrove roots is influenced by addition of new space via growth of individual substrata.

Sutherland's (1980) remark about the Venezuelan mangrove community, "In spite of the taxonomic richness of this community, most species are extremely rare", has turned out to aptly describe other mangrove-root communities in which relative abundance has been measured. For example, nearly 54.9% of the space on Florida Keys mangrove roots was covered by just three species (Engel and Pawlik, 2005), and on mangrove props roots at three sites in Belize and Panama, 73–89% of the volume was concentrated in the nine species that all the sites had in common (Wulff, 2009). Curiously, at many Caribbean sites, the same species, *T. ignis*, dominates by whatever metric is used for abundance, even though it often does not appear on a particular root, and its presence on a root or panel does not guarantee that it will dominate that substratum (e.g. Sutherland, 1980; Wulff, 2009).

Epizoism is common among sponge species that typically inhabit mangroves, and interactions between neighbouring sponges that are overgrowing each other can be beneficial or neutral, as well as negative. An apparent competitive hierarchy of 10 species growing on mangrove prop roots in the Florida Keys was erected by Engel and Pawlik (2005) by recording whether or not sponges appeared to be growing over each other. They pointed out a counter-intuitive pattern that the most abundant species occupied a middle level in the overgrowth hierarchy, and individuals of the most basal species in the hierarchy often grew to be very large. While this pattern appears puzzling in the context of many marine systems in which competition has been demonstrated to be a key structuring process, evidence continues to accumulate that it may be normal for some sponges to not compete with sponge neighbours. Sponge-dominated communities appear to stand out in the degree to which overgrowth can increase the representation of participating species. It is important to note that this does not apply to early successional mangrove sponge species, which do get eliminated, but are also less likely be observed on roots during a one-time survey (e.g. Sutherland, 1980; Wulff, 2009). Overgrowth by non-sponge taxa, such as dense mats of the filamentous cyanobacteria genera *Lyngbia* and *Schizothrix* (Rützler *et al.*, 2004) or compound ascidians and bryozoans (personal observation), can have negative effects on sponges. Mangrove roots that are not in abiotically extreme circumstances fall into Sará's (1970) third category of sponge community types, that of "continuous sponge populations, with practically negligible intervention of other elements of the sessile macrofauna and macroflora", a situation that has consistently provided illustrations of how sponges can benefit by their neighbours also being sponges.

Differences in sponge species diversity at mangrove sites are extreme, ranging from 3 to 147 species on a list of sites compiled by Diaz et al. (2004). Very low diversity reflects abiotic variables that are simply unfavourable for most sponges, but diversity differences between sites characterized by abiotic variables that are generally favourable may be strongly influenced by ecological interactions. Mangrove prop roots in the lagoons of three mangrove-covered islands of the Pelican Cays, Belize, were inhabited by 147 sponge species, in contrast to only 57 species on roots at three sites in Twin Cays and 54 at Blue Ground Range (Rützler et al., 2000). All of these islands are far from the coast, but the Pelican Cays mangroves differ from the others in their close association with coral reefs. Spongivorous fishes that normally inhabit coral reefs, but not mangroves, were abundant among the prop roots, and many of the sponge species inhabiting Pelican Cays roots were typical of coral reefs. Reciprocal transplant experiments of sponges between the Pelican Cays (very high sponge diversity, coral reef-associated sponge fauna) and the Twin Cays (lower sponge diversity, mangrove-associated sponge fauna) demonstrated that typical mangrove species that were moved to the Pelican Cays were quickly consumed by spongivorous fishes, unless protected inside cages, and typical reef species that were moved to Twin Cays survived well until the mangrove species resident on the roots began to overgrow and ultimately eliminate them (Wulff, 2000, 2005). Reef species grown on otherwise bare pvc pipes (i.e. without competitors) that were suspended among the roots continued to thrive, and their growth rates were significantly faster than they were on a coral reef for three of four typical reef sponge species. Faster growth of coral reef species among mangrove roots highlights the possibility that ecological interactions can inhibit sponge species from living in a habitat that is otherwise superior to the one where they normally live. As in other habitats, high diversity does not necessarily indicate more favourable conditions but may result from continuous cropping of competitive dominants by consumers. In the Pelican Cays, fast-growing mangrove species that are not defended against spongivores are prevented from overwhelming slower-growing reef species by a dense assortment of angelfishes, trunkfishes, and spadefish.

Predation by fish is not routinely responsible for distribution and abundance of mangrove sponge species in typical mangrove stands that are not embedded in the tops of coral reefs. Sutherland (1980) found no differences attributable to spongivory when he compared caged versus uncaged panels situated among mangrove roots, and Bingham and Young (1995) reported no direct observations or signs of fish predation on mangrove sponges in the Florida Keys in over 270 h of diving. In a prop root community that was censused four times, at yearly intervals, abrupt decreases in relative abundance of sponge species that are known to be favoured by angelfishes were observed to result from brief and unusual residence of juvenile angelfish at one site (Wulff, 2009).

Proximity to coral reefs was the only variable that appeared to explain differences in species composition in comparisons among 8 sites (with a total of 22 sponge species) in Aruba and Curaçao that differed by the addition of typical reef species to the typical mangrove species assemblages. Within sites however, an unexpected positive association between sponge percentage cover and tannin content of mangrove roots hinted at the possibility of influence of tannins on sponge nutrition (Hunting et al., 2008) or larval settlement (Hunting et al., 2010). Mangroves may also influence at least some of the sponges inhabiting their roots by nutrient trading, carbon for nitrogen, via adventitious roots embedded in the sponges (Ellison et al., 1996). One species, Haliclona implexiformis Hechtel, 1965, grew significantly faster on roots than on pvc pipes for the 1-month duration of the experiment. In turn, the sponges increase the longevity of the roots on which they live by protecting them from boring isopods (Ellison and Farnsworth, 1990; Ellison et al., 1996).

2.6. Sediment dominated habitats, including seagrass meadows

2.6.1. Abiotic factors

A small number of marine sponge species are capable of inhabiting sediments as their primary habitat, by employing a variety of special tricks. Functional differentiation within sponge individuals is illustrated by some sediment dwellers, such as Caribbean Oceanapia spp., which have a basal portion embedded in the sediment, and an upright portion through which water flow is directed downwards through the sponge body (Werding and Sanchez, 1991), and Cervicornia cuspidifera Lamark, 1815, which not only takes in water through its upright portion but also harbours zooxanthellae there (Rützler, 1997; Rützler and Hooper, 2000). An excavating species, Cliona inconstans Dendy, 1887, also lives buried in lagoon sands in southeastern Japan; but its tall chimneys, surmounted by oscules protruding above the surface (Ise et al., 2004), indicate that water flow is in the opposite direction from that in fellow clionaid C. cuspidifera. Bubaris ammosclera Hechtel, 1969 stabilizes carbonate sand as it grows as a mat over the surface (Macintyre et al., 1968). One of the clear differences observed among the extremely diverse sponge species inhabiting the various reef-associated habitats of Tulear, Madagascar, was the incorporation of carbonate particles by the relatively few sediment-dwelling species (Vacelet and Vasseur, 1977).

Details of how this occurs have been studied for the common Caribbean seagrass-dwelling sponge, *T. crypta*, which rolls when small, organizes course sediments near its base to stabilize it when medium-sized, and then lives partially buried and incorporates unsorted coarse sediment when it is large (see Cerrano *et al.*, 2004b, 2007 for a recent review). Its ability to entirely close its osculum and to strongly contract its body may further aid its success in a habitat in which resuspended sediments might clog openings and impede water flow through aquiferous systems of the sponges (Reiswig, 1971).

In sediment-dominated habitats in which there are also scattered pieces of hard substrata, sponges may settle on the hard substrata and then grow to appear as if they are growing on the sediment (e.g. Battershill and Bergquist, 1990). A dense sponge garden on a deep reef flat in northeastern New Zealand is established on sediment-covered base rock, where the asexually generated fragments or buds, by which resident sponges propagate, first attach to other sponges, or rock and shell fragments, and ultimately to the underlying rock. They are then able to survive as they develop especially long oscular tubes (in *Polymastia* spp.) or tolerate being covered by sediment (e.g. *Cinachyra* sp., *Aaptos aaptos* Schmidt, 1864), as described by Battershill and Bergquist (1990). Similarly, a Red Sea sponge, *Biemna ehrenbergi* Keller, 1889, is attached to beach rock underlying the sediments that can bury the sponge bodies to depths of 20 cm. This species may benefit from the organic richness of the sediment as it intakes interstitial water (Ilan and Abelson, 1995). The authors point out a counterbalancing issue, which is the risk of oxygen levels diminishing too much in the interstitial water. Coral reef sponge species transplanted to a seagrass meadow inside cages (for protection from consumption by sea stars), but growing on small pieces of coral rubble, grew faster than sponges of the same genotypes and initial sizes on the reef (Wulff, in preparation), as expected if sediment-dominated habitats are richer in sponge food.

As part of a project focused on habitat for juvenile spiny lobsters, Butler *et al.* (1995) mapped all sponges of several species that are large enough to provide shelter at 27 sites in Florida Bay Unanticipated data included documentation of dramatic sponge mortality coincident with a cyanobacterial bloom. Over 80% of the individuals representing the genera *Ircinia*, *Aplysina*, *Callyspongia*, and *Hippospongia* died, as well as 40% of the *S. vesparium* Lamarck, 1815. During a second bloom in the following year, all the remaining sponges died at some sites. Although the sponges were attached to hard substrata, this habitat is predominantly seagrass meadow, and massive loss of seagrasses also occurred. Consideration of how this ecosystem might ever be fully reconstituted illustrates some of the considerable risks for sponges of living in a particularly food-rich habitat. Water that is murky with phytoplankton impedes passage of adequate sunlight for seagrasses, and loss of seagrasses eliminates binding of sediments, which are readily resuspended from shallow bottoms. Although sponges filter phytoplankton, sediment in the water column can overwhelm their canal systems. The sponge populations that were present before the mortality events have been demonstrated to account for sufficient filtering capability that subsequent blooms could be attributed to the loss of sponges (Peterson *et al.*, 2006). It is hard to see how Florida Bay can recover, unless the sponges and the seagrasses are all added back simultaneously at densities that are sufficient to clear the water and bind the sediment. Yet sponge recovery has been progressing at some sites, suggesting that sponges may contribute significant homeostatic mechanisms to this system (Stevely and Sweat 2001), although the particularly large-bodied species that had contributed much of the biomass to the community before the mortality events are taking longer to return.

Anoxic mud, the extreme case of sediment-dominated habitat and seemingly anathema to sponge species that normally inhabit hard substrata, was nevertheless tolerated by a set of mangrove-root inhabiting species. Frequent observations of large sponges of 10 species that inhabit mangrove roots partially buried in fine, organic–rich sediment, but apparently perfectly healthy, inspired Rützler *et al.* (2007) to experiment. Their experimental burials with four species confirmed that the sponges could continue to pump, and even to incorporate detritus from the sediment, even while buried for as long as 10 days.

2.6.2. Ecological interactions

Abrupt habitat distribution boundaries that happen to coincide with the edge of sediment may not necessarily indicate that sediment is the primary constraint on sponge distribution. Although reef sponge species may appear constrained by lack of continuous hard substrata where reefs abut sediment-dominated seagrass meadows, it is actually a seagrass-dwelling sea star, *O. reticulatis*, that enforces this habitat distribution boundary in the Caribbean. In feeding trials in the field, *Oreaster* consistently rejected all 6 of 6 typical seagrass meadow sponge species and 7 of 8 typical rubble bed sponge species that were offered in choice experiments in the field; but they consumed 11 of the 14 reef sponge species offered (Wulff, 1995b). While on the reef, the reef sponge species are protected from the sea star because parrotfishes and butter-flyfishes bite the sea stars if they move onto the reef; but if a storm washes reef sponges into the seagrass or a poriferologist redistributes them there, the sea stars discover and consume them within days (Wulff, 1995b).

Very different constraints shape the interaction of another seagrass meadow spongivore and its prey. In their report on unambiguous sponge-consuming equipment of the shrimp *Typton cameus* Holthuis, 1951, living in *T. klausi* in Belize, Duris *et al.* (2011) pointed out the need for relatively gentle consumption of ones' home. They suggested that a clever mechanism by which the shrimp can reduce damage to their host is to serve as effective defence against colonization by additional conspecific shrimp.

Growing on eelgrass or large bivalves and even on mobile organisms, including decorator crabs and hermit crabs, is one way in which hard substratum sponges inhabit sediment environments. Although it is a clear advantage for sponges to be on a host that has effective mechanisms for preventing burial, this is still not a simple cost-free strategy. Living substrata tend to not only be discontinuous, and sometimes mobile, but also relatively ephemeral, imposing a requirement for highly effective recruitment on sponges that utilize this option for coping with sediment. Fell and Lewandrowski (1981) demonstrated the extreme degree to which the life history of a *Halichondria* species that lives on eelgrass blades in New England estuaries is opportunistic, with high mortality balanced by very high growth rates and early, and heavy, devotion of resources to reproduction. Some sponge-living substratum associations are quite specific, such as the hermit crab sponge *Pseudospongosorites suberitoides* Diaz, van Soest, and Pomponi, 1993, which lives on shells inhabited by hermit crabs in the genus *Pagurus* on the Gulf coast of Florida (Sandford, 1994). The preference of the hermit crabs for shells of a suitable size, rather than a sponge, even if that sponge appears to confer an advantage due to its ability to grow as the crab grows, adds an additional risk for the sponge. Hermit crabs that find a shell that fits may simply discard their sponges, causing them to languish on the sediment (Sandford, 1994, 1997).

Collaborative interactions between sponges of different species can also allow hard substratum sponges to inhabit sediment-dominated habitats. Multi-species piles of sponges deposited in sand channels of a reef by a hurricane were able to stay alive by a sort of snowshoe effect (Woodley *et al.*, 1981; Wulff, personal observation), and Cerrano *et al.* (2004b) observed that sponges of different species that adhered to each other could prevent rolling on the sediment bottoms of the Belize Barrier Reef lagoon. A *Geodia* species in a seagrass meadow in Florida hosts a haplosclerid on its surface that may protect it from predators (Wilcox *et al.*, 2002). Similarly, in a seagrass meadow in Belize, a species of extremely cavernous internal architecture, which may render it relatively unaffected by shifting sediments, provides additional substratum for dense-tissued sponge species that are more vulnerable to smothering if buried. The cavernous host species, *Lissodendoryx colombiensis* Zea and van Soest, 1986, is in turn protected from being devoured by the large sea star, *O. reticulatis*, to which it is palatable (Wulff, 2008a), when it is overgrown by the dense-tissued species which are deterrent to the sea star (Fig. 4.1).

2.7. Intertidal shores

2.7.1. Abiotic factors

The extreme porosity of sponges seems as if it would render them exceptionally unsuited for inhabiting intertidal zones, but some species thrive in, and others are able to tolerate, the low intertidal, or occasional exposure by unusual tides. Even in an area of monsoon rains and intense anthropogenic pressures due to industrialization, along the NW coast of India, six sponge species were recorded at each of two sites described as rocky–muddy (Vaghela *et al.*, 2010), although the authors expressed concern for diversity loss under increasingly challenging circumstances.

In Mozambique, a diverse assemblage of sponges (33 species) dominated the sessile fauna of a wide intertidal zone in spite of extremes in temperature, salinity, and currents (Barnes, 1999). Analysis of community composition distinguished two clusters, one on exposed rocks and the other in protected caves and on boulders, reflecting the importance of current velocity to sponge distribution.

The importance of water flow for intertidal sponges is reflected in morphological alterations as well as mortality patterns that curtail distribution. McDonald *et al.* (2003) noted that intertidal *Spongia* sp. individuals in Darwin Harbour, in northern Australia, were oriented with their longest axis perpendicular to the water flow. When individuals were experimentally twisted 90°, they reoriented themselves to again have long axes facing the current. Another intertidal species in Darwin Harbour, *Cinachyrella australiensis*, produces thicker oxeas, and a higher proportion of its total mass consists of skeleton, at more disturbed sites where water flow is greater (McDonald *et al.*, 2002). Concerned about reduced current flow with the implementation of new storm surge controls at the Oosterschelde estuary in The Netherlands, Hummel *et al.* (1994) experimentally subjected *H. panicea* to different flow rates in tanks. At low flow, the sponges became covered by bacteria and died, and this effect was exacerbated by higher temperatures. Both high and low extremes of flow can make life intolerable for sponges. On the wave-dashed rocky coast of Washington, USA, *H. panicea* grows in low mounds with stiff tissues in surge channels, and thinly encrusting with less stiff tissues where water motion is less extreme. Transplants from less to more wave stress rapidly developed stiffer tissues, but transplants from more to less stress delayed switching to less stiff tissues, an adaptive choice in an unpredictable habitat (Palumbi, 1984).

Water flow in the intertidal, as in other habitats, does not have a monotonic relationship to sponge distribution and abundance, and too much water flow disrupts intertidal organisms and their communities. Sessile invertebrates, including sponges, on rubble at One Tree Island, GBR, were least abundant and less diverse at low shore levels at exposed sites. Wave action overturned rubble, exposing inhabitants to desiccation and abrasion, and significantly decreased cover in spite of the regeneration abilities of the sponges (Walker *et al.*, 2008).

2.7.2. Ecological interactions

Even in this habitat in which elegant experiments have demonstrated that upper distribution limits tend to be controlled by abiotic factors (e.g. Connell, 1961), interactions with other species have been demonstrated to either extend or constrain the distribution and abundance of sponge species. Explicit demonstrations have been made of the importance of beneficial interactions in extending the depth zone range for intertidal sponges. On wave-washed shores of the Pacific northwest of the USA, *H. panicea* inhabits the lowest intertidal but is enabled to live higher in the intertidal when associated with an erect coralline alga that protects it from desiccation (Palumbi, 1985). The association is not reliably beneficial for the sponge, as the alga is able to outcompete the sponge unless chiton grazing keeps it trimmed. In a parallel association in the tropical eastern Pacific at Mazatlan, association of *Ha. caerulea* with another erect coralline alga extends the range of the sponge by more than 1 m into the subtidal by improving resistance to waves (Carballo and Ávila, 2004).

Predation has been demonstrated to influence intertidal sponge distribution and abundance, sometimes to a great extent. In a study focused on reproductive timing of the encrusting intertidal species *Haliclona permollis* (= *Haliclona cinerea* Grant, 1826) in Oregon (Elvin, 1976), sunlight, nutrition, and tissue temperature were related to reproduction and growth, but asides about how ephemeral individuals are, due to merging with neighbours or disappearing or being broken or eaten, hint at the possible importance of interactions even in this physically challenging habitat. The intertidal sponge *H. panicea* was eliminated at a site in southcentral Alaska where the nudibranch *Archidoris montereyensis* Cooper, 1862 settled especially strongly (Figure 4.1), although the sponge had previously covered more than 50% of the substratum and had been the dominant space occupier for at least 10 years (Knowlton and Highsmith, 2000). Nudibranch numbers at the 550 m^2 site increased from the 12–42 individuals to 156 by this single recruitment event, and once the sponge was consumed, the nudibranch population plummeted as well. Underscoring the speed at which ecological interactions can alter communities, and the degree to which the alterations can be difficult to undo, the site quickly became colonized by annual macroalgae.

Cover of intertidal sponges (and also mussels and ascidians) increased when a small omnivorous grapsid crab was excluded by caging, demonstrating that preference of the crab for animal food favoured dominance of macroalgae at sites in São Paulo State, in subtropical Brazil (Christofoletti *et al.*, 2010). Restriction of sponges to spaces between and underneath cobbles in the lowest zone of the exceptionally wide intertidal on the Pacific coast of Panama is imposed by the eagerness of the smooth pufferfish *A. hispidus* to consume six species, representing five demosponge orders, if they are removed from the protection of the cobbles (Wulff, 1997c).

2.8. Antarctic hard bottoms

2.8.1. Abiotic factors

Boisterous water flow is not the only physical disturbance influencing sponge depth distribution. In the Antarctic, sponges are restricted by frequent ice scour as deep as 10–25 m (and infrequently to as deep as 600 m) and by anchor ice, which can hoist the entire bottom community to the surface with dire consequences (Dayton *et al.*, 1970; McClintock *et al.*, 2005). Below 33 m, however, sponges can cover at least 55% of the substratum (Dayton *et al.*, 1974). Species diversity increases, and then decreases, with depth. McClintock *et al.* (2005) reported 62 species in 1–100 m depth, 99 from 100 to 500 m, 25 from 500 to 1000 m, and only 16 below 1000 m in Antarctica, a pattern similar to that reported in a recently revised list of deep-sea sponges off Brazil by Hajdu and Lopes (2007), who listed 59 species in 100–200 m depth, 49 between 201 and 500, and only 7 from 1000 to 2000 m.

Temperature may seem an obvious factor influencing sponges in Antarctica, but how this abiotic variable is reflected in the sponges may not be simple. Slow growth is often a result of very low temperatures, and in general, this has been confirmed in Antarctica by time-series size measurements, in particular of common large-bodied hexactinellids. However at least two demosponges, *Homaxinella balfourensis* Ridley and Dendy, 1886 and *Mycale acerata* Kirkpatrick, 1907 can grow extremely rapidly as well as recruiting so efficiently that a *Homaxinella* population that was nearly eliminated by anchor ice quickly rebounded to 80% cover (Dayton, 1989). Thus no automatic restriction to slow growth or meagre reproduction appears to be imposed on sponges by low temperatures. Seasonal variations in available sunlight influence plankton production, and plankton can be seasonally extremely sparse to the point that it is somewhat mysterious how sponges maintain themselves (Barthel and Gutt, 1992). Sponges that are unable to at least temporarily exist under conditions of metabolic semi-quiescence may be excluded from Antarctic waters (e.g. McClintock *et al.*, 2005)

Low temperature may also have profound influences by slowing rates of larval development. Antarctica stands out in the similarity of sponge faunas among distant sites, a pattern that may reflect greater distance travelled by larvae that develop more slowly in the prevailing very low temperatures (McClintock *et al.*, 2005). Highly clumped distributions of some species suggest, however, that their larvae do not travel (Barthel and Gutt, 1992).

2.8.2. Ecological interactions

Diatoms can foul some Antarctic sponges to the point of clogging water intake pores, especially during the early summer bloom (Amsler *et al.*, 2000). Sympatric diatoms were discouraged by extracts of seven of the eight species tested. A lack of correlation of predator deterrence and diatom deterrence suggests that diatom fouling may be a significant enough issue to select for specific deterrents. Even with deterrent chemistry, sponges were thickly fouled, indicating that the diatoms are not repelled, but rather controlled by the sponges. Diatoms may even parasitize the internal tissue of sponges. From SEM analyses, Cerrano *et al.* (2000a,b) determined that diatoms of *Melosira* sp. embedded in the hexactinellid *Scolymastra joubini*

Topsent, 1916 ($=Anoxycalyx joubini$ Topsent, 1916) gain nutritionally at the expense of their hosts. By contrast, *M. acerata* appears to be able to use diatoms as food (Cerrano *et al.*, 2004a).

Antarctic sponges are consumed by spongivore and omnivore sea stars and a nudibranch (Dayton *et al.*, 1974). *M. acerata*, which grows exceptionally fast in the context of this habitat (Fig. 4.1), is preferred by the sea star *Perknaster fuscus*, which prevents it from simply taking over all the space (Dayton, 1979). Larval settlement of sponge-feeding sea stars may be moderated in turn by a filter-feeding sea star (Dayton *et al.*, 1974). Chemical defences against predators have been revealed to be at least as prevalent in Antarctic sponges as in tropical sponges (Peters *et al.*, 2009). Far from being a system in which cold temperatures slow everything to a pace at which interactions are irrelevant, Antarctic sponges are engaged in a complex web of ecological interactions strong enough to influence distribution, abundance and diversity (see McClintock *et al.*, 2005 for a comprehensive review).

3. CONCLUSIONS

Influences of abiotic and biotic effects are inextricably intertwined: details of abiotic environmental context can determine the outcomes of ecological interactions, and biotic interactions often moderate the influence of abiotic factors. Determining how these factors interact, and which are the primary influences on the distribution and abundance of particular sponge species in particular situations, is still a worthy endeavour, despite the challenges. Studies in every habitat have illustrated the great power of time-series observations of the same individuals and of experimental manipulation for understanding the processes underlying observed patterns. Influence of ecological interactions on distribution patterns may coincide with abiotic factors to the point that the causative factor is entirely obscured unless experiments are performed.

Experiments have revealed opportunistic predation to be capable of enforcing sponge distribution boundaries that coincide with habitat boundaries (e.g. coral reef–seagrass meadow) or distinct microhabitats (cryptic—exposed, kelp forest—urchin barrens). Within habitats, by contrast, dramatic effects of predators on sponge population sizes seem to occur primarily in cases of unusually high recruitment rates or unusually low mortality rates for the predators, which are often specialists on the sponge species affected (e.g. temperate rocky intertidal and subtidal, Antarctic hardbottoms). Competitive interactions have been demonstrated in a few cases to diminish populations or exclude sponge species from a habitat (e.g. reef species in mangroves, early successional species in mangroves).

Cases in which competitive interactions have appeared obvious have often turned out to be neutral or even beneficial interactions when observed over time (e.g. mangrove roots, coral reefs). Especially striking in this regard are sponge–sponge interactions in dense sponge-dominated communities, which appear to promote the continued presence of all participating species (e.g. temperate caves, coral reefs, seagrass meadows). Mutualistic symbioses with other animals, plants, or macroalgae have been demonstrated to increase abundance and habitat distribution in several habitats (e.g. coral reefs, subtidal hardbottoms, temperate, and tropical rocky intertidal). Symbiotic microbes can enhance distribution and abundance, but also render their hosts more vulnerable to environmental changes (e.g. temperate subtidal, mangroves). And while photosynthetic symbionts can boost growth and excavation rates for their sponge hosts, in other cases sponge growth proceeds as well or even better in diminished light (e.g. temperate subtidal, coral reefs).

Metrics chosen for evaluating sponge abundance make a huge difference in interpretation of data comparing between different sites or over time at the same site. In many circumstances, volume or biomass is likely to allow interpretation of influences on distribution and abundance better than numbers of individuals or area covered. Accurate identification of species and understanding how they are related within higher taxa is essential. Sponge species that look similar because they share growth form and colour can differ in attributes, such as symbionts they harbour or predators they deter, that influence their interactions and the roles they play. Even closely related species tend to differ in at least one key ecological attribute and must be distinguished for any studies relevant to conservation, as two smaller populations are not equivalent to one large one.

Predicting the outcome of ecological interactions for distribution and abundance of sponges depends on substantial understanding of details and dynamics of the ecology of the actual species involved. Apparently similar sets of species have been shown to interact completely differently. Sublittoral rocky substrata provide one good example, from among the many habitats in which sponges play major roles. In order to accurately gauge interactions outcomes in this habitat type, it must be known not only if the sea urchins are herbivores or carnivores, but also what their relative preferences among the available prey are. Whether or not predators of sea urchins control their populations must be known. Seaweed strategies must be well understood, as macroalgae can play roles ranging from outcompeting sponges to providing havens for sponges that are vulnerable to urchin grazing. Sponges may be disturbed by kelp fronds as they are whipped about by waves, and sponges with photosynthetic symbionts may suffer from diminished sunlight within algal stands; but it is the unstoppable stolons of *Caulerpa*, with their ability to spread forever asexually, accumulate sediment, and grip any sort of bottom, including sponges, that makes this green seaweed anathema for sponges. However, a sponge with a particularly effective multipurpose chemical arsenal, like *C. crambe*, can resist the stolons, perhaps paying for devoting energy to keeping its arsenal at the ready by reduced growth rates that make it vulnerable to other competitors. Knowledge of particular attributes of the sponge species in addition to resisting algal stolons is required, for example, resistance to sea urchin herbivory is a pre-requisite to thriving in urchin barrens, and dependence on photosynthetic symbionts can restrict habitat distribution even as it enhances growth rates. Striking site-specific differences in the interactions between seaweeds and sponges were only revealed by experimental manipulations, long-term observations, and application of a variety of approaches to learning about growth rates, recruitment, chemistry, and other sponge attributes, with careful attention to distinguishing individual sponge species. This rocky subtidal example is only one of many, from every type of marine habitat, that illustrate the surprises that sponges hold in store for us. The many instances in which biotic influences have been identified as important determinants of distribution and abundance of sponge species hint that many more ecological interactions of sponges await illumination.

ACKNOWLEDGEMENTS

I am grateful for extremely thoughtful and detailed manuscript reviews by James McClintock and an anonymous reviewer; for generous contributions of evocative sponge photos by Bill Baker, Emma Cebrian, Andy Davis, Ann Knowlton, Mike Page, Klaus Rützler, Christine Schönberg, and Keryea Soong; and for editorial enouragement by Mikel Becerro. I continue to be deeply grateful for my doctoral mentoring by Willard D. Hartman and G. Evelyn Hutchinson. My research is supported by the National Science Foundation under Grant No. 0550599, and by the Marine Science Network of the Smithsonian Institution, supported in part by the Hunterdon Oceanographic Research Fund. This is CCRE Contribution # 916.

REFERENCES

Abdo, D. A., Battershill, C. N., and Harvey, E. S. (2006). Manipulation of environmental variables and the effect on the growth of *Haliclona* sp.: Implications for open-water aquaculture. *Marine Biology Research* **2**, 326–332.

Aerts, L. A. M. (2000). Dynamics behind stand-off interactions in three reef sponge species and the coral *Montastrea cavernosa*. *Marine Ecology* **21**, 191–204.

Aerts, L. A. M., and van Soest, R. W. M. (1997). Quantification of sponge/coral interactions in a physically stressed reef community, NE Colombia. *Marine Ecology Progress Series* **148**, 125–134.

Alcolado, P. M. (1990). General features of Cuban sponge communities. In "New Perspectives in Sponge Biology" (K. Rützler, ed.), pp. 351–357. Smithsonian Institution Press, Washington, DC.

Alcolado, P. M. (1994). General trends in coral reef sponge communities of Cuba. In "Sponges in Time and Space: Biology, Chemistry, Paleontology" (R. W. M. van Soest, T. M. G. van Kempen and J.-C. Braekman, eds), pp. 251–255. A.A. Balkema, Rotterdam.

Alcolado, P. (2007). Reading the code of coral reef sponge community composition and structure for environmental biomonitoring: Some experiences from Cuba. In "Porifera Research: Biodiversity, Innovation and Sustainability" (M. R. Custódio, G. Lôbo-Hajdu, E. Hajdu and G. Muricy, eds), Série Livros 28. pp. 3–10. Museu Nacional, Rio de Janiero.

Alvarez, B., Diaz, C. M., and Laughlin, R. A. (1990). The sponge fauna on a fringing coral reef in Venezuela. I: Composition, distribution, abundance. In "New Perspectives in Sponge Biology" (K. Rützler, ed.), pp. 358–366. Smithsonian Institution Press, Washington, DC.

Amsler, C. D., Moeller, C. B., McClintock, J. B., Iken, K. B., and Baker, B. J. (2000). Chemical defenses against diatom fouling in Antarctic marine sponges. *Biofouling* **16,** 29–45.

Ávila, E., and Carballo, J. L. (2008). A preliminary assessment of the invasiveness of the Indo-Pacific sponge *Chalinula nematifera* on coral communities from the tropical Eastern Pacific. *Biological Invasions* **11,** 257–264.

Backus, G. (1964). The effects of fish-grazing on invertebrate evolution in shallow tropical waters. *Allan Hancock Foundation Occasional Paper* **27,** 1–29.

Baldacconi, R., and Corrierro, G. (2009). Effects of the spread of the alga *Caulerpa racemosa* var. *cylindracea* on the sponge assemblage from coralligenous concretions of the Apulian coast (Ionian Sea, Italy). *Marine Ecology—An Evolutionary Perspective* **30,** 337–345.

Bannister, R. J., Hoogenboom, M. O., Anthony, K. R. N., Battershill, C. N., Whalan, S., Webster, N. S., and de Nys, R. (2011). Incongruence between the distribution of a common coral reef sponge and photosynthesis. *Marine Ecology Progress Series* **423,** 95–100.

Barnes, D. K. A. (1999). High diversity of tropical intertidal zone sponges in temperature, salinity, and current extremes. *African Journal of Ecology* **37,** 424–434.

Barthel, D. (1986). On the ecophysiology of the sponge *Halichondria panicea* in Kiel Bight. I. Substrate specificity, growth and reproduction. *Marine Biology* **32,** 291–298.

Barthel, D., and Gutt, J. (1992). Sponge associations in the Weddell Sea. *Antarctic Science* **4,** 137–150.

Battershill, C. N., and Bergquist, P. R. (1990). The influence of storms on asexual reproduction, recruitment, and survivorship of sponges. In "New Perspectives in Sponge Biology" (K. Rützler, ed.), pp. 396–403. Smithsonian Institution Press, Washington, DC.

Becerro, M. A. (2008). Quantitative trends in sponge ecology research. *Marine Ecology* **29,** 167–177.

Becerro, M. A., and Paul, V. J. (2004). Effects of depth and light on secondary metabolites and cyanobacterial symbionts of *Dysidea granulosa*. *Marine Ecology Progress Series* **280,** 115–128.

Becerro, M. A., Uriz, M. J., and Turon, X. (1997). Chemically-mediated interactions in benthic organisms: The chemical ecology of *Crambe crambe* (Porifera, Poecilosclerida). *Hydrobiologia* **355,** 77–89.

Becerro, M. A., Turon, X., Uriz, M. J., and Templado, J. (2003). Can a sponge feeder be a herbivore? *Tylodina perversa* (Gastropoda) feeding on *Aplysina aerophoba* (Demospongiae). *Biological Journal of the Linnean Society* **78,** 429–438.

Becerro, M. A., Starmer, J. A., and Paul, V. J. (2006). Chemical defenses of cryptic and aposematic gastropterid molluscs feeding on their host sponge *Dysidea granulosa*. *Journal of Chemical Ecology* **32,** 1491–1500.

Bell, J. J. (2002). The sponge community in a semi-submerged temperate sea cave: Density, diversity and richness. *P.S.Z.N.I. Marine Ecology* **23,** 297–311.

Bell, J. J. (2008). The functional roles of marine sponges. *Estuarine, Coastal and Shelf Science* **79,** 341–353.

Bell, J. J., and Barnes, D. K. A. (2003). Effect of disturbance on assemblages: An example using Porifera. *Biological Bulletin* **205,** 144–159.

Bell, J. J., and Carballo, J. L. (2008). Patterns of sponge biodiversity and abundance across different biogeographic regions. *Marine Biology* **155,** 563–570.

Benzoni, F., Calcinai, B., Eisinger, M., and Klaus, R. (2008). Coral disease mimic: Sponge attacks *Porites lutea* in Yemen. *Coral Reefs* **27,** 695.

Bergquist, P. R. (1978). Sponges. University of California Press, Berkeley and Los Angeles.

Bergquist, P. R. (1999). The present state of sponge science. *Memoirs of the Queensland Museum* **44,** 23–26.

Biggs, B., Strimaitis, A., Wulff, J. (in prep). The effect of a cold water shock on the coral reef sponge fauna of the Florida Keys, USA.

Bingham, B. L. (1992). Life histories in an epifaunal community: Coupling of adult and larval processes. *Ecology* **73,** 2244–2259.

Bingham, B. L., and Young, C. M. (1991). The influence of sponges on invertebrate recruitment - A field-test of allelopathy. *Marine Biology* **109,** 19–26.

Bingham, B. L., and Young, C. M. (1995). Stochastic events and dynamics of a mangrove root epifaunal community. *P.S.Z.N.I. Marine Ecology* **16,** 145–163.

Blanquer, A., and Uriz, M. J. (2007). Sponge cryptic species revealed by mitochondrial and ribosomal genes: A phylogenetic approach. *Molecular Phylogenetics and Evolution* **45,** 392–397.

Blanquer, A., Uriz, M. J., and Agell, G. (2008). Hidden diversity in sympatric sponges: Adjusting life-history dynamics to share substrate. *Marine Ecology Progress Series* **371,** 109–115.

Bryan, P. G. (1973). Growth rate, toxicity and distribution of the encrusting sponge *Terpios* sp. (Hadromerida: Suberitidae) in Guam, Mariana Islands. *Micronesica* **9,** 237–242.

Buss, L. W. (1976). Better living through chemistry: The relationship between allelochemical interactions and competitive networks. In "Aspects of Sponge Biology" (F. W. Harrison and R. R. Cowden, eds), pp. 315–327. Academic Press, New York.

Buss, L. W., and Jackson, J. B. C. (1979). Competitive networks—Non-transitive competitive relationships in cryptic coral-reef environments. *American Naturalist* **113,** 223–234.

Butler, M. J., Hunt, J. H., Herrnkind, W. F., Childress, M. J., Bertelsen, R., Sharp, W., Matthews, T., Field, J. M., and Marshall, H. G. (1995). Cascading disturbances in Florida Bay, USA: Cyanobacteria blooms, sponge mortality and implications for juvenile spiny lobsters *Panulirus argus*. *Marine Ecology Progress Series* **129,** 119–125.

Calcinai, B., Bavestrello, G., and Cerrano, C. (2004). Dispersal and association of two alien species in the Indonesian coral reefs: The octocoral *Carijoa riisei* and the demosponge *Desmapsamma anchorata*. *Journal of the Marine Biological Association of the United Kingdom* **84,** 937–941.

Carballo, J. L., and Ávila, E. (2004). Population dynamics of a mutualistic interaction between the sponge *Haliclona caerulea* and the red alga *Jania adhaerens*. *Marine Ecology Progress Series* **279,** 93–104.

Carballo, J. L., Vega, C., Cruz-Barraza, J. A., Yañez, B., Nava, H., Ávila, E., and Wilson, M. (2008). Short- and long-term patterns of sponge diversity on a rocky tropical coast: Evidence of large-scale structuring factors. *Marine Ecology* **29,** 216–236.

Cebrian, E., and Uriz, M. J. (2006). Grazing on fleshy seaweeds by sea urchins facilitates sponge *Cliona viridis* growth. *Marine Ecology Progress Series* **323,** 83–89.

Cebrian, E., Uriz, M. J., Garrabou, J., and Ballesteros, E. (2011). Sponge mass mortalities in a warming Mediterranean Sea: Are cyanobacteria-harboring species worse off? *PloS One* **6,** e20211.

Cerrano, C., Arillo, A., Bavestrello, G., Calcinai, B., Catteneo-Vietti, R., Penna, A., Sarà, M., and Totti, C. (2000a). Diatom invasion in the antarctic hexactinellid sponge *Scolymastra joubini*. *Polar Biology* **23,** 441–444.

Cerrano, C., Bavestrello, G., Bianchi, G. N., Catteneo-Vietti, R., Bava, S., Morganti, C., Morri, C., Picco, P., Sara, G., Schiaparelli, S., Sicarrd, A., and Sponga, G. (2000b). A catastrophic mass-mortality of gorgonians and other organisms in the Ligurean Sea (North-western Mediterranean), summer 1999. *Ecology Letters* **3**, 284–293.

Cerrano, C., Calcinai, B., Cucchiari, E., De Camillo, C., Totti, C., and Bavestrello, G. (2004a). The diversity of relationships between Antarctic sponges and diatoms: The case of *Mycale acerata* Kirkpatrick, 1907 (Porifera, Demospongiae). *Polar Biology* **27**, 231–237.

Cerrano, C., Pansini, M., Valisano, L., Calcinai, B., Sarà, M., and Bavestrello, G. (2004b). Lagoon sponges from Carrie Bow Cay (Belize): Ecological benefits of selective sediment incorporation. (M. Pansini, R. Pronsato, G. Bavestrello and R. Manconi, eds), Sponge Science in the New Millennium, vol. 68, pp. 239–252. Bollettino dei Musei Istituti Biologici, Universitá di Genova.

Cerrano, C., Calcinai, B., Di Camillo, C. G., Valisano, L., and Bavestrello, G. (2007). How and why do sponges incorporate foreign material? Strategies in Porifera. In "Porifera Research: Biodiversity, Innovation and Sustainability" (M. R. Custódio, G. Lôbo-Hajdu, E. Hajdu and G. Muricy, eds), Série Livros 28. pp. 239–246. Museu Nacional, Rio de Janiero.

Chaves-Fonnegra, A., Zea, S., and Gómez, M. L. (2007). Abundance of the excavating sponge *Cliona delitrix* in relation to sewage discharge at San Andrés Island, sw Caribbean, Colombia. *Boletín de Investigaciones Marinas y Costeras* **36**, 63–78.

Chiappone, M., and Sullivan, K. M. (1994). Ecological structure and dynamics of near-shore hard-bottom communities in the Florida Keys. *Bulletin of Marine Science* **54**, 747–756.

Chiappone, M., Rutten, L. M., Miller, S. L., and Swanson, D. W. (2007). Large-scale distributional patterns of the encrusting and excavating sponge *Cliona delitrix* Pang on Florida Keys coral substrates. In "Porifera Research: Biodiversity, Innovation and Sustainability" (M. R. Custódio, G. Lôbo-Hajdu, E. Hajdu and G. Muricy, eds), Série Livros 28. pp. 255–263. Museu Nacional, Rio de Janiero.

Christofoletti, R. A., Murakami, V. A., Oliveira, D. N., Barreto, R. E., and Flores, A. A. V. (2010). Foraging by the omnivorous crab *Pachygrapsus transversus* affects the structure of assemblages on sub-tropical rocky shores. *Marine Ecology Progress Series* **420**, 125–134.

Cleary, D. F. R., and de Voogd, N. J. (2007). Environmental associations of sponges in the Spermonde Archipelago, Indonesia. *Journal of the Marine Biological Association of the United Kingdom* **87**, 1669–1676.

Coles, S. L., and Bolick, H. (2007). Invasive introduced sponge *Mycale grandis* overgrows reef corals in Kaneohe Bay, Oahu, Hawaii. *Coral Reefs* **26**, 911.

Connell, J. H. (1961). The influence of interspecific competition and other factors on the distribution of the barnacle *Cthamalus stellatus*. *Ecology* **42**, 710–723.

Connell, J. H. (1978). Diversity in the tropical rain forests and coral reefs. *Science* **199**, 1302–1310.

Corriero, G., Liaci, L. S., Ruggiero, D., and Pansini, M. (2000). The sponge community of a semi-submerged Mediterranean cave. *P.S.Z.N.I. Marine Ecology* **21**, 85–96.

Dam, R., and van Diez, C. E. (1997). Predation by hawksbill turtles at Mona Island, Puerto Rico. Proceedings of the 8th International Coral Reef Symposium, Panama, vol. 2, pp. 1421–1426.

Davis, A. R., and Ward, D. W. (1999). Does the large barnacle *Austrobalanus imperator* (Darwin, 1954) structure benthic invertebrate communities in SE Australia? *Memoirs of the Queensland Museum* **44**, 125–130.

Davis, A. R., and Ward, D. W. (2009). Establishment and persistence of species–rich patches in a species–poor landscape: Role of a structure-forming subtidal barnacle. *Marine Ecology Progress Series* **380**, 187–198.

Davis, A. R., Roberts, D. E., and Cummins, S. P. (1997). Rapid invasion of a sponge-dominated deep-reef by *Caulerpa scalpelliformis* (Chlorophyta) in Botany Bay, New South Wales. *Australian Journal of Ecology* **22**, 146–150.

Dayton, P. K. (1979). Observations of growth, dispersal and population dynamics of some sponges in McMurdo Sound, Antarctica. Colloques Internationaux du CNRS 291. pp. 271–282.

Dayton, P. K. (1989). Interdecadal variation in an Antarctic sponge and its predators from oceanographic climate shifts. *Science* **245**, 1484–1486.

Dayton, P. K., Robilliard, G. A., and Paine, R. T. (1970). Benthic faunal zonation as a result of anchor ice at McMurdo Sound, Antarctica. (M. Holgate, ed.), Antarctic Ecology, vol. 1, pp. 244–258. Academic Press, London.

Dayton, P. K., Robilliard, G. A., Paine, R. T., and Dayton, L. B. (1974). Biological accommodation in the benthic community at McMurdo Sound, Antarctica. *Ecological Monographs* **44**, 105–128.

De Goeij, J. M., van den Berg, H., van Oostveen, M. M., Epping, E. H. G., and van Duyl, F. C. (2008). Major bulk dissolved organic carbon (DOC) removal by encrusting coral reef cavity sponges. *Marine Ecology Progress Series* **357**, 139–151.

de Voogd, N. J., and Cleary, D. F. R. (2008). An analysis of sponge distribution and diversity at three taxonomic levels in the Thousand Islands/Jakarta Bay reef complex, West Java, Indonesia. *Marine Ecology* **29**, 205–215.

de Voogd, N. J., Necking, L. E., Hoeksema, B. W., Noor, A., and van Soest, R. W. M. (2004). Sponge interactions with spatial competitors in the Spermonde Archipelago. *Bollettino dei Musei Istituti Biologici, Università di Genova* **68**, 253–261.

Diaz, M. C., Smith, K. P., and Rützler, K. (2004). Sponge species richness and abundance as indicators of mangrove epibenthic community health. *Atoll Research Bulletin* **518**, 1–17.

Duckworth, A. R., Wolff, C., Evans-Illidge, E., Whalan, S., and Lui, S. (2008). Spatial variability in community structure of Dictyoceratid sponges across Torres Strait, Australia. *Continental Shelf Research* **28**, 2168–2173.

Dunlap, M., and Pawlik, J. R. (1996). Video-monitored predation by Caribbean reef fishes on an array of mangrove and reef sponges. *Marine Biology* **126**, 117–123.

Duran, S., and Rützler, K. (2006). Ecological speciation in a Caribbean marine sponge. *Molecular Phylogenetics and Evolution* **40**, 292–297.

Duris, Z., Horká, I., Juracka, P. J., Petrusek, A., and Sandford, F. (2011). These squatters are not innocent: The evidence of parasitism in sponge-inhabiting shrimps. *PloS One* **6**, e21987.

Eckrich, C. E., Peachey, R. B. J., and Engel, M. S. (2011). Crustose, calcareous algal bloom (*Ramicusta* sp.) overgrowing scleractinian corals, gorgonians, a hydrocoral, sponges, and other algae in Lac Bay, Bonaire, Dutch Caribbean. *Coral Reefs* **30**, 131.

Ellison, A. M., and Farnsworth, E. J. (1990). The ecology of Belizean mangrove-root fouling communities. I. Epibenthic fauna are barriers to isopod attack of red mangrove roots. *Journal of Experimental Marine Biology and Ecology* **142**, 91–104.

Ellison, A. M., Farnsworth, E. J., and Twilley, R. R. (1996). Facultative mutualism between red mangroves and root-fouling sponges in Belizean mangal. *Ecology* **77**, 2431–2444.

Elvin, D. W. (1976). Seasonal growth and reproduction of an intertidal sponge, *Haliclona permollis* (Bowerbank). *Biological Bulletin* **151**, 108–125.

Engel, S., and Pawlik, J. (2005). Interactions among Florida sponges. II. Mangrove habitats. *Marine Ecology Progress Series* **303**, 145–152.

Enriquez, S., Ávila, E., and Carballo, J. L. (2009). Phenotypic plasticity induced in transplant experiments in a mutualistic association between the red alga *Jania adhaerens* (Rhodophyta, Corallinales) and the sponge *Haliclona caerulea* (Porifera: Haplosclerida): Morphological responses of the alga. *Journal of Phycology* **45**, 81–90.

Erwin, P. M., and Thacker, R. W. (2008). Phototrophic nutrition and symbiont diversity of two Caribbean sponge-cyanobacteria symbioses. *Marine Ecology Progress Series* **362**, 139–147.

Farnsworth, E. J., and Ellison, A. M. (1996). Scale dependent spatial and temporal variability in biogeography of mangrove-root epibiont communities. *Ecological Monographs* **66,** 45–66.

Fell, P. E., and Lewandrowski, K. B. (1981). Population dynamics of the estuarine sponge, Halichondria sp., within a New England eelgrass community. *Journal of Experimental Marine Biology and Ecology* **55,** 49–63.

Fromont, J., Vanderklift, M. A., and Kendrick, G. A. (2006). Marine sponges of the Dampier Archipelago, Western Australia: Patterns of species distributions, abundance and diversity. *Biodiversity and Conservation* **15,** 3731–3750.

Fugii, T., Hirose, E., Keshavmurthy, S., Chen, C. A., Zhou, W., and Reimer, J. D. (2011). Coral-killing cyanobacteriosponge (*Terpios hoshinota*) on the Great Barrier Reef. *Coral Reefs* **30,** 483.

Garson, M. J., Clark, R. J., Webb, R. I., Field, K. L., Charan, R. D., and McCaffrey, E. J. (1999). Ecological role of cytotoxic alkaloids: *Haliclona* n. sp., an unusual sponge/dinoflagellate association. *Memoirs of the Queensland Museum* **44,** 205–213.

Goreau, T. F., and Hartman, W. D. (1966). Sponge: Effect on the form of reef corals. *Science* **151,** 343–344.

Hajdu, E., and Lopes, D. A. (2007). Checklist of Brazilian deep-sea sponges. In "Porifera Research: Biodiversity, Innovation and Sustainability" (M. R. Custódio, G. Lôbo-Hajdu, E. Hajdu and G. Muricy, eds), Série Livros 28. pp. 353–359. Museu Nacional, Rio de Janiero.

Harrison, F. W. and Cowden, R. R. (eds) (1976). Aspects of Sponge Biology. Academic Press, New York, 354pp.

Hartman, W. D. (1977). Sponges as reef builders and shapers. *Studies in Geology* **4,** 127–134.

Hartman, W. D., and Goreau, T. F. (1970). Jamaican coralline sponges: Their morphology, ecology, and fossil relatives. In "The Biology of the Porifera." Zoological Society London Symposium 25. (W. C. Fry, ed.), pp. 205–243.

Hill, M. S. (1996). Symbiotic zooxanthellae enhance boring and growth rates of the tropical sponge *Anthosigmella varians* forma *varians*. *Marine Biology* **125,** 649–654.

Hill, M. S. (1998). Spongivory on Caribbean reefs releases corals from competition with sponges. *Oecologia* **117,** 143–150.

Holmes, K. E. (1997). Eutrophication and its effect on bioeroding sponge communities. Proceedings of the 8th International Coral Reef Symposium 2. pp. 1411–1416.

Hooper, J. N. A., and Kennedy, J. A. (2002). Small-scale patterns of sponge biodiversity (Porifera) on Sunshine Coast reefs, eastern Australia. *Invertebrate Systematics* **16,** 637–653.

Hooper, J. N. A., Kennedy, J. A., List-Armitage, S. E., Cook, S. D., and Quinn, R. (1999). Biodiversity, species composition and distribution of marine sponges in northeast Australia. *Memoirs of the Queensland Museum* **44,** 263–274.

Hooper, J. N. A., Kennedy, J. A., and Quinn, R. J. (2002). Biodiversity "hotspots", patterns of richness and endemism, and taxonomic affinities of tropical Australian sponges. *Biodiversity and Conservation* **11,** 851–885.

Hourigan, T. F., Stanton, F. G., Motta, P. J., Kelley, C. D., and Carlson, B. (1989). The feeding ecology of three species of Caribbean angelfishes (family Pomacanthidae). *Environmental Biology of Fishes* **24,** 105–116.

Hughes, T. P. (1996). Demographic approaches to community dynamics: A coral reef example. *Ecology* **77,** 2256–2260.

Hummel, H., Fortuin, A. W., Bogaards, R. H., Meijboom, A., and DeWolf, L. (1994). The effects of prolonged emersion and submersion by tidal manipulation on marine macrobenthos. *Hydrobiologia* **283,** 219–234.

Hunting, E. R., van Soest, R. W. M., van der Geest, H. G., Vos, A., and Debrot, A. O. (2008). Diversity and spatial heterogeneity of mangrove associated sponges of Curaçao and Aruba. *Contributions to Zoology* **77,** 205–215.

Hunting, E. R., van der Geest, H. G., Krieg, A. J., van Mierlo, M. B. L., and van Soest, R. W. M. (2010). Mangrove-sponge associations. *Aquatic Ecology* **44,** 679–684.

Ilan, M., and Abelson, A. (1995). The life of a sponge in a sandy lagoon. *Biological Bulletin* **189,** 363–369.

Ise, Y., Takeda, M., and Watanabe, Y. (2004). Psammobiontic Clionidae (Demospongiae: Hadromerida) in lagoons of the Ryukyu Islands, southwestern Japan. *Bollettino dei Musei Istituti Biologici, Universitá di Genova* **68,** 381–389.

Jackson, J. B. C., Goreau, T. F., and Hartman, W. D. (1971). Recent brachiopod-coralline sponge communities and their paleoecological significance. *Science* **173,** 623–625.

Jokiel, P. L. (1980). Solar ultraviolet radiation and coral reef epifauna. *Science* **207,** 1069–1071.

Knowlton, A. L., and Highsmith, R. C. (2000). Convergence in the time-space continuum: A predator-prey interaction. *Marine Ecology Progress Series* **195,** 285–291.

Kobluk, D. R., and van Soest, R. W. M. (1989). Cavity-dwelling sponges in a southern Caribbean coral reef and their paleontological implications. *Bulletin of Marine Science* **44,** 1207–1235.

Lang, J. C., Hartman, W. D., and Goreau, T. F. (1975). Sclerosponges: Primary framework constructors on the Jamaican deep forereef. *Journal of Marine Research* **33,** 223–231.

Lehnert, H., and Fischer, H. (1999). Distribution patterns of sponges and corals down to 107 m off North Jamaica. *Memoirs of the Queensland Museum* **44,** 307–316.

León, Y. M., and Bjorndal, K. A. (2002). Selective feeding in the hawksbill turtle, an important predator in coral reef ecosystems. *Marine Ecology Progress Series* **245,** 249–258.

Lesser, M. P. (2006). Benthic-pelagic coupling on coral reefs: Feeding and growth of Caribbean sponges. *Journal of Experimental Marine Biology and Ecology* **328,** 277–288.

Lewis, S. M. (1982). Sponge-zoanthid associations: Functional interactions. *Smithsonian Contributions to the Marine Sciences* **12,** 465–474.

López-Victoria, M., and Zea, S. (2005). Current trends of space occupation by encrusting excavating sponges on Colombian coral reefs. *Marine Ecology—An Evolutionary Perspective* **26,** 33–41.

López-Victoria, M., Zea, S., and Weil, E. (2006). Competition for space between encrusting excavating Caribbean sponges and other coral reef organisms. *Marine Ecology Progress Series* **312,** 113–121.

Macdonald, I. A., and Perry, C. T. (2003). Biological degradation of coral framework in a turbid lagoon environment, Discovery Bay, north Jamaica. *Coral Reefs* **22,** 523–535.

Macintyre, I. G. (2000). Status of scleractinian corals in some Pelican Cays ponds following the 1998 bleaching event. *Atoll Research Bulletin* **466,** 35–37.

Macintyre, I. G., Mountjoy, E. W., and D'Anglejan, B. F. (1968). An occurrence of submarine cementation of carbonate sediments off the west coast of Barbados, WI. *Journal of Sedimentary Petrology* **38,** 660–664.

Maldonado, M. (2006). The ecology of sponge larva. *Canadian Journal of Zoology* **84,** 175–194.

Maldonado, M., and Uriz, M. J. (1998). Microrefuge exploitation by subtidal encrusting sponges: Patterns of settlement and post-settlement survival. *Marine Ecology Progress Series* **174,** 141–150.

Maldonado, M., and Young, C. M. (1996). Bathymetric patterns of sponge distribution on the Bahamian slope. *Deep-Sea Research* **43,** 897–915.

Maldonado, M., and Young, C. M. (1998). Limits on the bathymetric distribution of keratose sponges: A field test in deep water. *Marine Ecology Progress Series* **174,** 123–139.

McClintock, J. B., Amsler, C. D., Baker, B. J., and van Soest, R. W. M. (2005). Ecology of Antarctic marine sponges: An overview. *Integrative and Comparative Biology* **45,** 359–368.

McDonald, J. I., Hooper, J. N. A., and McGuiness, K. A. (2002). Environmentally influenced variability in the morphology of *Cinachyrella australiensis* (Carter 1886) (Porifera: Spirophorida: Tetillidae). *Marine and Freshwater Research* **53**, 79–84.

McDonald, J. I., McGuiness, K. A., and Hooper, J. N. A. (2003). Influence of re-orientation on alignment to flow and tissue production in a *Spongia* sp. (Porifera: Demospongiae: Dictyoceratida). *Journal of Experimental Marine Biology and Ecology* **296**, 13–22.

McLean, E. L., and Yoshioka, P. M. (2008). Substratum effects on the growth and survivorship of the sponge *Desmapsamma anchorata*. *Caribbean Journal of Science* **44**, 83–89.

Mercado-Molino, A. E., and Yoshioka, P. M. (2009). Relationships between water motion and size-specific survivorship and growth of the demosponge *Amphimedon compressa*. *Journal of Experimental Marine Biology and Ecology* **375**, 51–56.

Meylan, A. (1988). Spongivory in hawksbill turtles – A diet of glass. *Science* **239**, 393–395.

Meylan, A. (1990). Nutritional characteristics of sponges in the diet of the hawksbill turtle. *Eretmochelys imbricata*. In "New Perspectives in Sponge Biology" (K. Rützler, ed.), pp. 472–477. Smithsonian Institution Press, Washington, DC.

Muricy, G. (1989). Sponges as pollution bio-monitors at Arrail do Cabo, Southeastern Brazil. *Revista Brasileira de Biologia* **49**, 347–354.

Page, M. J., Handley, S. J., Northcote, P. T., Cairney, D., and Willan, R. C. (2011). Successes and pitfalls in the aquaculture of the sponge *Mycale hentscheli*. *Aquaculture* **312**, 52–61.

Palumbi, S. R. (1984). Tactics of acclimation: Morphological changes of sponges in an unpredictable environment. *Science* **225**, 1478–1480.

Palumbi, S. R. (1985). Spatial variation in an alga-sponge commensalism and the evolution of ecological interactions. *American Naturalist* **126**, 267–274.

Pawlik, J. R. (1983). A sponge-eating worm from Bermuda: *Branchiosyllis oculata* (Polychaeta, Syllidae). *P.S.Z.N.I. Marine Ecology* **4**, 65–79.

Pawlik, J. R., Chanas, B., Toonen, R. J., and Fenical, W. (1995). Defenses of Caribbean sponges against predatory reef fish. I. Chemical deterrency. *Marine Ecology Progress Series* **127**, 183–194.

Pawlik, J. R., McMurray, S. E., and Henkel, T. P. (2007). Abiotic factors control sponge ecology in Florida mangroves. *Marine Ecology Progress Series* **339**, 93–98.

Peters, K. J., Amsler, C. D., McClintock, J. B., van Soest, R. W. M., and Baker, B. (2009). Palatability and chemical defenses of sponges from the western Antarctic Penninsula. *Marine Ecology Progress Series* **385**, 77–85.

Peterson, B. J., Chester, C. M., Jochem, F. J., and Fourqurean, J. W. (2006). Potential role of sponge communities in controlling phytoplankton blooms in Florida Bay. *Marine Ecology Progress Series* **328**, 93–103.

Plucer-Rosario, G. (1987). The effect of substratum on the growth of *Terpios*, an encrusting sponge which kills corals. *Coral Reefs* **5**, 197–200.

Porter, J. W., and Target, N. (1988). Allelochemical interactions between sponges and corals. *Biological Bulletin* **175**, 230–239.

Preciado, I., and Maldonado, M. (2005). Reassessing the spatial relationship between sponges and macroalgae in sublittoral rocky bottoms: A descriptive approach. *Helgoland Marine Research* **59**, 141–150.

Pronzato, R. (1999). Sponge-fishing, disease and farming in the Mediterranean Sea. *Aquatic Conservation: Marine and Freshwater Ecosystems* **9**, 485–493.

Pronzato, R. (2004). A climber sponge. *Bollettino dei Musei Istituti Biologici, Universitá di Genova* **68**, 549–552.

Pronzato, R., and Manconi, R. (2008). Mediterranean commercial sponges: Over 5000 years of natural history and cultural heritage. *Marine Ecology* **29**, 146–166.

Randall, J. E., and Hartman, W. D. (1968). Sponge-feeding fishes of the West Indies. *Marine Biology* **1**, 216–225.

Reed, J. K., and Pomponi, S. A. (1997). Biodiversity and distribution of deep and shallow water sponges in the Bahamas. *Proceedings of the 8th International Coral Reef Symposium, Panama,* vol. 2, pp. 1687–1692.

Reimer, J. D., Nozawa, Y., and Hirose, E. (2010). Domination and disappearance of the black sponge: A quarter century after the initial Terpios outbreak in southern Japan. *Zoological Studies* **50,** 394.

Reiswig, H. M. (1971). Particle feeding in natural populations of three marine demosponges. *Biological Bulletin* **141,** 568–591.

Reiswig, H. M. (1973). Population dynamics of three Jamaican Demospongiae. *Bulletin of Marine Science* **23,** 191–226.

Reiswig, H. M. (1974). Water transport, respiration, and energetics of three tropical marine sponges. *Journal of Experimental Marine Biology and Ecology* **14,** 231–249.

Reiswig, H. M. (1981). Partial carbon and energy budgets of the bacteriosponge *Verongia fistularis* (Porifera: Demospongiae) in Barbados. *Marine Ecology* **2,** 273–293.

Rios, R., and Duffy, J. E. (1999). Description of *Synalpheus williamsi,* a new species of sponge-dwelling shrimp (Crustacea: Decapoda: Alpheidae), with remarks on its first larval stage. *Proceedings of the Biological Society of Washington* **112,** 541–552.

Roberts, D. E., Cummins, S. P., Davis, A. R., and Chapman, M. G. (2006). Structure and dynamics of sponge-dominated assemblage on exposed and sheltered temperate reefs. *Marine Ecology Progress Series* **321,** 19–30.

Rose, C. S., and Risk, M. J. (1985). Increase in *Cliona delitrix* infestation of *Montastrea cavernosa* heads on organically polluted portions of the Grand Cayman fringing reef. *Marine Ecology* **6,** 345–363.

Rosell, D., and Uriz, M. J. (1992). Do associated zooxanthellae and the nature of the substratum affect survival, attachment, and growth of *Cliona viridis* (Porifera: Hadromerida)? An experimental approach. *Marine Biology* **114,** 503–507.

Russell, B. D., Degnan, B. M., Garson, M. J., and Skilleter, G. A. (2003). Distribution of a nematocyst-bearing sponge in relation to potential coral donors. *Coral Reefs* **22,** 11–16.

Rützler, K. (1965). Substratstabilität im marinen Benthos als ökologishcher Faktor, dargestellt am Beispiel adriatischer Porifera. *International Review of Hydrobiology* **50,** 281–292.

Rützler, K. (1970). Spatial competition among Porifera: Solution by epizoism. *Oecologia* **5,** 85–95.

Rützler, K. (1978). Sponges in coral reefs. In "Coral Reefs: Research Methods" (D. R. Stoddart and R. E. Johannes, eds), Monographs on Oceanographic Methodology. **5,** pp. 299–313. UNESCO, Paris.

Rützler, K. (1988). Mangrove sponge disease induced by cyanobacterial symbionts: Failure of a primitive immune system? *Diseases of Aquatic Organisms* **5,** 143–149.

Rützler, K. (1990a). Associations between Caribbean sponges and photosynthetic organisms. In "New Perspectives in Sponge Biology" (K. Rützler, ed.), pp. 455–466. Smithsonian Institution Press, Washington, DC.

Rützler, K. (1990b). Low-tide exposure of sponges in a Caribbean mangrove community. *P.S.Z.N. I. Marine Ecology* **16,** 165–179.

Rützler, K. (1997). The role of psammobiontic sponges in the reef community. *Proceedings of the 8th International Coral Reef Symposium, Panama,* vol. 2, pp. 1393–1398.

Rützler, K. (2002). Impact of crustose clionid sponges on Caribbean coral reefs. *Acta Geologica Hispanica* **37,** 61–72.

Rützler, K. (2004). Sponges on coral reefs: A community shaped by competitive cooperation. *Bollettino dei Musei Istituti Biologici, Universitá di Genova* **68,** 85–148.

Rützler, K., and Hooper, J. N. A. (2000). Two new genera of hadromerid sponges (Porifera, Demospongiae). *Zoosystema* **22,** 337–344.

Rützler, K., and Muzik, K. (1993). *Terpios hoshinota,* a new cyanobacteriosponge threatening Pacific reefs. *Sciencias Marinas* **57,** 395–403.

Rützler, K., Diaz, M. C., van Soest, R. W. M., Zea, S., Smith, K. P., Alvarez, B., and Wulff, J. L. (2000). Diversity of sponge fauna in mangrove ponds, Pelican Cays, Belize. *Atoll Research Bulletin* **477,** 231–250.

Rützler, K., Goodbody, I., Diaz, M. C., Feller, I. C., and Macintyre, I. G. (2004). The aquatic environment of Twin Cays, Belize. *Atoll Research Bulletin* **512,** 1–49.

Rützler, K., Duran, S., and Piantoni, C. (2007). Adaptation of reef and mangrove sponges to stress: Evidence for ecological speciation exemplified by *Chondrilla caribensis* new species (Demospongiae, Chondrosida). *Marine Ecology* **28,** 95–111.

Ruzicka, R., and Gleason, D. F. (2009). Sponge community structure and anti-predator defenses on temperate reefs of the South Atlantic Bight. *Journal of Experimental Marine Biology and Ecology* **380,** 36–46.

Sandford, F. (1994). The Florida hermit-crab sponge, a little known 'mobile' sponge from the NE corner of the Gulf of Mexico, and its hermit crab associates. In "Sponges in Time and Space: Biology, Chemistry, Paleontology" (R. W. M. van Soest, T. M. G. van Kempen and J.-C. Braekman, eds), pp. 273–278. A.A. Balkema, Rotterdam.

Sandford, F. (1997). Sponge/shell switching by the hermit crab *Pagurus impressus. Invertebrate Biology* **114,** 73–78.

Sará, M. (1970). Competition and cooperation in sponge populations. In "Symposia of the Zoological Society of London 25," pp. 273–284.

Schaal, G., Riera, P., and Leroux, C. (2011). Microscale variations of food web functioning within a rocky shore invertebrate community. *Marine Biology* **158,** 623–630.

Schlacher, T. A., Schlacher-Hoehlinger, M. A., Williams, A., Althaus, F., Hooper, J., and Kloser, R. (2007). Richness and distribution of sponge megabenthos in continental margin canyons off southeastern Australia. *Marine Ecology Progress Series* **340,** 73–88.

Schmahl, G. P. (1990). Community structure and ecology of sponges associated with four southern Florida coral reefs. In "New Perspectives in Sponge Biology" (K. Rützler, ed.), pp. 376–383. Smithsonian Institution Press, Washington, DC.

Schönberg, C. H. L. (2002). Substrate effects on the bioeroding demosponge *Cliona orientalis*. 1. Bioerosion rates. *P.S.Z.N.I. Marine Ecology* **23,** 313–326.

Schönberg, C. H. L. (2003). Substrate effects on the bioeroding demosponge *Cliona orientalis*. 2. Substrate colonization and tissue growth. *P.S.Z.N.I. Marine Ecology* **24,** 59–74.

Schönberg, C. H. L., and Suwa, R. (2007). Why bioeroding sponges may be better hosts for symbiotic dinoflagellates than many corals. In "Porifera Research: Biodiversity, Innovation and Sustainability" (M. R. Custódio, G. Lôbo-Hajdu, E. Hajdu and G. Muricy, eds), Série Livros 28. pp. 569–580. Museu Nacional, Rio de Janeiro.

Schönberg, C. H. L., and Wilkinson, C. R. (2001). Induced colonization of corals by a clionid bioeroding sponge. *Coral Reefs* **20,** 69–76.

Seguin, F., Le Brun, O., Hirst, R., Al-Thary, I., and Dutrieux, E. (2008). Large coral transplantation in Bal Haf (Yemen): An opportunity to save corals during the construction of a liquefied natural gas plant using innovative techniques. Proceedings of the 11th International Coral Reef Symposium, Ft. Lauderdale.

Shield, C. J., and Witman, J. D. (1993). The impact of *Henricia sanguinolenta* (O.F. Müller) (Echinodermata:Asteroidea) predation on the finger sponges, *Isodictya* spp. *Journal of Experimental Marine Biology and Ecology* **166,** 107–133.

Soong, K., Yang, S.-L., and Chen, C. A. (2009). A novel dispersal mechanism of a coral-threatening sponge, *Terios hoshinota* (Suberitidae, Porifera). *Zoological Studies* **48,** 596.

Sorokin, S. J., Laperousaz, T. C. D., and Collings, G. J. (2008). Investigator Group Expedition 2006: Sponges (Porifera). *Transactions of the Royal Society of South Australia* **132,** 163–172.

Stevely, J. M., and Sweat, D. E. (2001). The recovery of sponge populations in Florida Bay and Upper Keys following a widespread sponge mortality. Final Report, Florida Fish and Wildlife Conservation Commission.

Storr, J. F. (1976). Ecological factors controlling sponge distribution in the Gulf of Mexico and the resulting zonation. In "Aspects of Sponge Biology" (F. W. Harrison and R. R. Cowden, eds), pp. 261–276. Academic Press, New York.

Sutherland, J. P. (1980). Dynamics of the epibenthic community on roots of the mangrove *Rhizophora mangle*, at Bahia de Buche, Venezuela. *Marine Biology* **58**, 75–84.

Swain, T. D. (2009). Phylogeny-based species delimitations and the evolution of host associations in symbiotic zoanthids (Anthozoa, Zoanthidea) of the wider Caribbean region. *Zoological Journal of the Linnean Society* **156**, 223–238.

Swain, T. D., and Wulff, J. L. (2007). Diversity and specificity of Caribbean sponge-zoanthid symbioses: A foundation for understanding the adaptive significance of symbioses and generating hypotheses about higher-order systematics. *Biological Journal of the Linnean Society* **92**, 695–711.

Swearingen, D. C., and Pawlik, J. R. (1998). Variability in the chemical defense of the sponge *Chondrilla nucula* against predatory reef fishes. *Marine Biology* **131**, 619–627.

Thacker, R. W. (2005). Impacts of shading on sponge-cyanobacteria symbioses: A comparison between host-specific and generalist associations. *Integrative and Comparative Biology* **45**, 369–376.

Thacker, R. W., Becerro, M. A., Lumbang, W. A., and Paul, V. J. (1998). Allelopathic interactions between sponges on a tropical reef. *Ecology* **79**, 1740–1750.

Trautman, D. A., Hinde, R., and Borowitzka, M. A. (2000). Population dynamics of an association between a coral reef sponge and a red macroalga. *Journal of Experimental Marine Biology and Ecology* **244**, 87–105.

Trussell, G. C., Lesser, M. P., Patterson, M. R., and Genovese, S. J. (2006). Depth-specific differences in growth of the reef sponge *Callyspongia vaginalis*: Role of bottom-up effects. *Marine Ecology Progress Series* **323**, 149–158.

Tsuriumi, M., and Reiswig, H. M. (1997). Sexual vs. asexual reproduction in an oviparous rope-form sponge, *Aplysina cauliformis* (Porifera: Verongida). *Invertebrate Reproduction and Development* **32**, 1–9.

Turon, X., Becerro, M. A., Uriz, M. J., and Llopis, J. (1996). Small scale association measures in epibenthic communities as a clue for allelochemical interactions. *Oecologia* **108**, 351–360.

Turon, X., Codina, M., Tarjuelo, I., Uriz, M. J., and Becerro, M. A. (2000). Mass recruitment of *Ophiothrix fragilis* (Ophiuroidea) on sponges: Settlement patterns and post-settlement dynamics. *Marine Ecology Progress Series* **200**, 201–212.

Uriz, M. J., Rosell, D., and Martin, D. (1992). The sponge populations of the Cabrera Archipelago (Belearic Islands): Characteristics, distribution and abundance of the most representative species. *P.S.Z.N.I. Marine Ecology* **13**, 101–117.

Vacelet, J., and Boury-Esnault, N. (1995). Carnivorous sponges. *Nature* **373**, 333–335.

Vacelet, J., and Vasseur, P. (1977). Sponge distribution in coral reefs and related areas in the vicinity of Tuléar (Madagascar). Proceedings of the 3rd International Coral Reef Symposium, Miami, vol. 1, pp. 113–117.

Vaghela, A., Bhadja, P., Ramoliya, J., Patel, N., and Kundu, R. (2010). Seasonal variations in the water quality, diversity and population ecology of intertidal macrofauna at an industrially influenced coast. *Water Science and Technology* **61**, 1505–1514.

Vicente, V. P. (1978). An ecological evaluation of the West Indian demosponge *Anthosigmella varians* (Hadromerida, Spirastrellida). *Bulletin of Marine Science* **28**, 771–777.

Vicente, V. P. (1990). Overgrowth activity by the encrusting sponge *Chondrilla nucula* on a coral reef in Puerto Rico. In "New Perspectives in Sponge Biology" (K. Rützler, ed.), pp. 436–443. Smithsonian Institution Press, Washington, DC.

Vilanova, E., Mayer-Pinto, M., Curbelo-Fernandez, M. P., and Da Silva, S. H. G. (2004). The impact of a nuclear power plant discharge on the sponge community of a tropical bay (se Brazil). *Bollettino dei Musei Istituti Biologici, Universitá di Genova* **68**, 647–654.

Wagner, D., Kahng, S. E., and Toonen, R. J. (2009). Observations on the life history and feeding ecology of a specialized nudibranch predator (*Phyllodesmium poindemieri*), with implications for biocontrol of an invasive octocoral (*Carijoa riisei*) in Hawaii. *Journal of Experimental Marine Biology and Ecology* **372,** 64–74.

Walker, S. J., Degnan, B. M., Hooper, J. N. A., and Skilleter, G. A. (2008). Will increased storm disturbance affect the biodiversity of intertidal, nonscleractinian sessile fauna on coral reefs? *Global Change Biology* **14,** 2755–2770.

Ward-Paige, C. A., Risk, M. J., Sherwood, O. A., and Jaap, W. C. (2005). Clionid sponge surveys on the Florida reef tract suggest land-based nutrient inputs. *Marine Pollution Bulletin* **51,** 570–579.

Webster, N. S., Cobb, R. E., and Negri, A. P. (2008). Temperature thresholds for bacterial symbiosis with a sponge. *ISME Journal* **8,** 830–842.

Weisz, J. B., Lindquist, N., and Martens, C. S. (2008). Do associated microbial abundances impact marine demosponge pumping rates and tissue densities? *Oecologia* **155,** 367–376.

Werding, B., and Sanchez, H. (1991). Life habits and functional morphology of the sediment infaunal sponges *Oceanapia oleracea* and *Oceanapia peltata* (Porifera: Haplosclerida). *Zoomorphology* **110,** 203–208.

West, D. A. (1976). Aposematic coloration and mutualism in sponge-dwelling tropical zoanthids. In "Coelenterate Biology and Behavior" (G. O. Mackie, ed.), pp. 443–452. Plenum Press, New York.

Wilcox, T. P., Hill, M., and DeMeo, K. (2002). Observations on a new two-sponge symbiosis in the Florida Keys. *Coral Reefs* **21,** 198–204.

Wilkinson, C. R. (1983). Role of sponges in coral reef structural processes. In "Perspectives on Coral Reefs" (D. J. Barnes, ed.), pp. 263–274. Brian Clouston Publisher, Manuka, Australia.

Wilkinson, C. R. (1987). Interocean differences in size and nutrition of coral reef sponge populations. *Science* **236,** 1654–1657.

Wilkinson, C. R., and Cheshire, A. C. (1988). Growth rate of Jamaican coral reef sponges after Hurricane Allen. *Biological Bulletin* **175,** 175–179.

Wilkinson, C. R., and Cheshire, A. C. (1989). Patterns in the distribution of sponge populations across the central Great Barrier Reef. *Coral Reefs* **8,** 127–134.

Wilkinson, C. R., and Cheshire, A. C. (1990). Comparisons of sponge populations across the barrier reefs of Australia and Belize: Evidence for higher productivity in the Caribbean. *Marine Ecology Progress Series* **67,** 285–294.

Wilkinson, C. R., and Evans, E. (1988). Sponge distribution across Davies Reef, Great Barrier Reef, relative to location, depth, and water movement. *Coral Reefs* **8,** 1–7.

Wilkinson, C. R., and Vacelet, J. (1979). Transplantation of marine sponges to different conditions of light and current. *Journal of Experimental Marine Biology and Ecology* **17,** 91–104.

Willenz, P., and Hartman, W. D. (1999). Growth and regeneration rates of the calcareous skeleton of the Caribbean coralline sponge *Ceratoporella nicholsoni*: A long term survey. *Memoirs of the Queensland Museum* **44,** 675–685.

Witman, J. D., and Sebens, K. P. (1990). Distribution and ecology of sponges at a subtidal rock ledge in the central Gulf of Maine. In "New Perspectives in Sponge Biology" (K. Rützler, ed.), pp. 391–396. Smithsonian Institution Press, Washington, DC.

Woodley, J. D., Chornesky, E. A., Clifford, P. A., Jackson, J. B. C., Kaufman, L. S., Lang, J. C., Pearson, M. P., Porter, J. W., Rooney, M. C., Rylaarsdam, K. W., Tunnicliffe, V. J.Wahle, C. W. *et al.* (1981). Hurricane Allen's impact on Jamaican coral reefs. *Science* **214,** 749–755.

Wright, J. T., Benkendorff, K., and Davis, A. R. (1997). Habitat associated differences in temperate sponge assemblages: The importance of chemical defense. *Journal of Experimental Marine Biology and Ecology* **213,** 199–213.

Wulff, J. L. (1984). Sponge-mediated coral reef growth and rejuvenation. *Coral Reefs* **3,** 157–163.

Wulff, J. L. (1988). Fish predation on cryptic sponges of Caribbean coral reefs. *American Zoologist* **28,** A166.

Wulff, J. L. (1994). Sponge-feeding by Caribbean angelfishes, trunkfishes, and filefishes. In "Sponges in Time and Space: Biology, Chemistry, Paleontology" (R. W. M. van Soest, T. M. G. van Kempen and J.-C. Braekman, eds), pp. 265–271. A.A. Balkema, Rotterdam.

Wulff, J. L. (1995a). Effects of a hurricane on survival and orientation of large, erect coral reef sponges. *Coral Reefs* **14,** 55–61.

Wulff, J. L. (1995b). Sponge-feeding by the Caribbean starfish *Oreaster reticulatus*. *Marine Biology* **123,** 313–325.

Wulff, J. L. (1997a). Mutually beneficial associations among species of coral reef sponges. *Ecology* **78,** 146–159.

Wulff, J. L. (1997b). Parrotfish predation on cryptic sponges of Caribbean coral reefs. *Marine Biology* **129,** 41–52.

Wulff, J. L. (1997c). Causes and consequences of differences in sponges diversity and abundance between the Caribbean and eastern Pacific at Panama. Proceedings of the 8th International Coral Reef Symposium, Panama 2. pp. 1377–1382.

Wulff, J. L. (2000). Sponge predators may determine differences in sponge fauna between two sets of mangrove cays, Belize Barrier Reef. *Atoll Research Bulletin* **477,** 251–263.

Wulff, J. L. (2001). Assessing and monitoring coral reef sponges: Why and how? *Bulletin of Marine Science* **69,** 831–846.

Wulff, J. L. (2004). Sponges on mangrove roots, Twin Cays, Belize: Early stages of community assembly. *Atoll Research Bulletin* **519,** 1–10.

Wulff, J. L. (2005). Trade-offs in resistance to competitors and predators, and their effects on the diversity of tropical marine sponges. *Journal of Animal Ecology* **74,** 313–321.

Wulff, J. L. (2006a). Rapid diversity and abundance decline in a Caribbean coral reef sponge community. *Biological Conservation* **127,** 167–176.

Wulff, J. L. (2006b). Resistance vs. recovery: Morphological strategies of coral reef sponges. *Functional Ecology* **20,** 699–708.

Wulff, J. L. (2006c). Sponge systematics by starfish: Predators distinguish cryptic sympatric species of Caribbean fire sponges, *Tedania ignis* and *Tedania klausi* n. sp. (Demospongiae, Poecilosclerida). *Biological Bulletin* **211,** 83–94.

Wulff, J. L. (2006d). Ecological interactions of marine sponges. *Canadian Journal of Zoology* **84,** 146–166.

Wulff, J. L. (2008a). Collaboration among sponge species increases sponge diversity and abundance in a seagrass meadow. *Marine Ecology: An Evolutionary Perspective* **29,** 193–204.

Wulff, J. L. (2008b). Life history differences among coral reef sponges promote mutualism or exploitation of mutualism by influencing partner fidelity feedback. *American Naturalist* **171,** 597–609.

Wulff, J. L. (2009). Sponge community dynamics on Caribbean mangrove roots: Significance of species idiosyncrasies. In "Smithsonian Contributions to Marine Science 38," pp. 501–514. Smithsonian Institution Scholarly Press, Washington, DC.

Wulff, J. L. (in prep). Context-dependency of growth rate in tropical marine sponges.

Wulff, J. L., and Buss, L. W. (1979). Do sponges help hold coral reefs together? *Nature* **281,** 474–475.

Zea, S. (1994). Patterns of coral and sponge abundance in stressed coral reefs Santa Marta, Colombian Caribbean. In "Sponges in Time and Space: Biology, Chemistry, Paleontology" (R. W. M. van Soest, T. M. G. van Kempen and J.-C. Braekman, eds), pp. 257–264. A.A. Balkema, Rotterdam.

Zea, S. (2001). Patterns of sponge (Porifera, Demospongiae) distribution in remote, oceanic reef complexes of the southwestern Caribbean. *Revista de al Academia Colombian de Ciencias Exactas, Físicas y Naturales* **25,** 579–592.

Zilberberg, C., Solé-Cava, A. M., and Klautau, M. (2006). The extent of asexual reproduction in sponges of the genus *Chondrilla* (Demospongiae: Chondrosida) from the Caribbean and Brazilian coasts. *Journal of Experimental Marine Biology and Ecology* **336,** 211–220.

CHAPTER FIVE

SPONGE ECOLOGY IN THE MOLECULAR ERA

Maria J. Uriz[1] and Xavier Turon

Contents

Department of Marine Ecology, Centre d'Estudis Avançats de Blanes (CEAB-CSIC), Blanes, Girona, Spain
[1]Corresponding author: Email: iosune@ceab.csic.es

Advances in Marine Biology, Volume 61
ISSN 0065-2881, DOI: 10.1016/B978-0-12-387787-1.00006-4

Abstract

Knowledge of the functioning, health state, and capacity for recovery of marine benthic organisms and assemblages has become essential to adequately manage and preserve marine biodiversity. Molecular tools have allowed an entirely new way to tackle old and new questions in conservation biology and ecology, and sponge science is following this lead. In this review, we discuss the biological and ecological studies of sponges that have used molecular markers during the past 20 years and present an outlook for expected trends in the molecular ecology of sponges in the near future. We go from (1) the interface between inter- and intraspecies studies, to (2) phylogeography and population level analyses, (3) intra-population features such as clonality and chimerism, and (4) environmentally modulated gene expression. A range of molecular markers has been assayed with contrasting success to reveal cryptic species and to assess the genetic diversity and connectivity of sponge populations, as well as their capacity to respond to environmental changes. We discuss the pros and cons of the molecular gene partitions used to date and the prospects of a plentiful supply of new markers for sponge ecological studies in the near future, in light of recently available molecular technologies. We predict that molecular ecology studies of sponges will move from genetics (the use of one or some genes) to genomics (extensive genome or transcriptome sequencing) in the forthcoming years and that sponge ecologists will take advantage of this research trend to answer ecological and biological questions that would have been impossible to address a few years ago.

Key Words: Phylogeography; population genetics; cryptic speciation; clonality; chimerism; gene expression; microsatellites; Porifera

1. INTRODUCTION

Coastal benthic ecosystems are in danger worldwide as a result of human activities. Consequently, assessment of benthic biodiversity and population vulnerability is a crucial ecological concern for marine biologists. Studies of marine biodiversity and vulnerability benefited massively from the incorporation of molecular tools, and, as a result, knowledge of the functioning, health state, and capacity for recovery of marine benthic organisms and assemblages has improved vastly during the past decades (Haig, 1998; Sweijd et al., 2000; DeSalle and Amato, 2004; Bickford et al., 2006).

Molecular ecology is a relatively new discipline, which has resulted from applying molecular tools to both traditional and new ecological issues. These tools allowed an entirely new way to interrogate organisms and to tackle old and new questions in marine ecology. The health state and capacity for adaptation to environmental changes of benthic invertebrates have been estimated by analyzing several genetic descriptors such as gene diversity, gene flow, departures from Hardy–Weinberg equilibrium, inbreeding, and changes in effective population size, among others (Grosberg and Cunningham, 2001; Hellberg *et al.*, 2002; Pearse and Crandall, 2004; Charlesworth, 2009). These descriptors are calculated from data on allele frequencies or sequence differences, and from the partitioning of their variation within and among populations.

For the accurate assessment of population descriptors, neutral molecular markers with an adequate degree of polymorphism are necessary. At the beginning of the molecular ecology era, the markers commonly used for ecological issues were polymorphic enzymatic proteins (allozymes), which proved suitable for assessing population differentiation and adaptation to particular environmental conditions. While allozymes have been very useful in many sponge studies (reviewed in Solé-Cava and Boury-Esnault, 1999; Borchiellini *et al.*, 2000; Van Oppen *et al.*, 2002), they present important practical problems such as the requirement for fresh tissue, troubles with the interpretation of the electrophoresis gels, and the difficulty to compare across studies. These drawbacks, together with new technological developments such as polymerase chain reaction (PCR), led researchers in the field of molecular ecology to move from analyses of proteins to genes, and sponge molecular ecology followed this trend.

Sequence data from several gene partitions of mitochondrial and nuclear DNA have been used for ecological issues involving marine benthic invertebrates. In addition to mere allele frequency data, sequences have the advantage of containing useful phylogenetic information. Mitochondrial DNA in particular has been and still is of prime importance in phylogeography and population genetics (Avise, 2000, 2009). However, sponges feature a low level of intraspecies variability in mtDNA that has hindered the application of this marker (Duran *et al.*, 2004a; Wörheide *et al.*, 2005). It remains unclear why poriferan mtDNA displays low rates of evolution (Lavrov *et al.*, 2005), but this fact restricts the applicability of some of the most popular markers for studies of sponge population genetics. Nevertheless, only a restricted subset of mitochondrial genes (*COI* in particular) has been assayed so far and more research on other genes is necessary.

Finding a nuclear substitute for mtDNA is problematic because of technical hitches such as allele resolution, the prevalence of paralogy, recombination, and longer coalescent times compared to mitochondrial genes (Palumbi *et al.*, 2001; Zhang and Hewitt, 2003). Internal transcribed spacers (ITSs) separating conserved regions of the *rRNA* genes have been used for studies at phylogenetic and population levels (e.g. Wörheide *et al.*, 2002a,b; Duran *et al.*, 2004b). However, it has been recognized that intragenomic polymorphisms (IGP), due to a lack of homogenization of the multiple copies of the *rRNA* gene clusters, can greatly limit the application of this marker to population genetics of sponges, and that levels of IGP should be determined and taken into account in any study (Wörheide *et al.*, 2004; Duran *et al.*, 2004b). Clearly, in sponges, development of new nuclear markers, preferably single copy genes with high variability (introns), is necessary for advancement in the fields of population genetics and demography (Wörheide *et al.*, 2005). New technologies will surely increase our ability to develop large numbers of markers in non-model organisms such as sponges (Thompson *et al.*, 2010).

Microsatellites, also known as simple sequence repeats (SSRs) or short tandem repeats (STRs), are among the most variable and ubiquitous types of DNA sequence in the genome (Li *et al.*, 2002). Given their high mutation rate, they make possible fine-scale analysis of the genetic relationships among populations (Bowcock *et al.*, 1994) and, coupled with new analytical tools (e.g. Csilléry *et al.*, 2010), they appear to be the best choice for studies on population differentiation, gene flow, and clonality in sponges. They can provide useful demographic parameters to answer ecological questions and can help make risk assessment and predictions on the fate of sponge populations submitted to exploitation or to harmful conditions (e.g. Dailianis *et al.*, 2011). However, the species-specific nature of microsatellites makes it necessary to develop them *de novo* for each new target species, in a time-consuming procedure involving the preparation and screening of genomic libraries. This drawback is likely responsible for the scarce number of sponge species (8) for which microsatellite markers have been developed so far: *Crambe crambe* (Schmidt) (Duran *et al.*, 2002), *Halichondria panicea* (Pallas) (Knowlton *et al.*, 2003), *Scopalina lophyropoda* Schmidt (Blanquer *et al.*, 2005), *Hymeniacidon sinapium* de Laubenfels (Hoshino and Fujita, 2006), *Spongia lamella* (= *S. agaricina*) (Schulze) (Noyer *et al.*, 2009), *Spongia officinalis* Linnaeus (Dailianis and Tsigenopoulos, 2010), *Ephydatia fluviatilis* Topsent (Cigliarelli *et al.*, 2008), and *Paraleucilla magna* Klautau, Monteiro and Borojevic (Guardiola *et al.*, 2011).

The microsatellite markers developed for sponges have allowed a number of recent studies at the intra- and interpopulation level, uncovering patterns of genetic structure at several scales, and allowing the study of clonality and chimerism (Duran *et al.*, 2004c; Calderón *et al.*, 2007; Hoshino *et al.*, 2008; Blanquer *et al.*, 2009; Blanquer and Uriz, 2010, 2011; Noyer, 2010; Dailianis *et al.*, 2011; Guardiola *et al.*, 2011). Doubtlessly, in recent years, the application of microsatellites has revitalized the field of sponge molecular ecology. With the new technologies of massive sequencing, loci containing tandem repeats (microsatellites) can be easily obtained and optimized by sequencing a small part of the genome (Agell and Uriz, 2010; Jennings *et al.*, 2011). These new technologies will surely fuel the development of microsatellite markers in the forthcoming years so that we expect an explosive increase of research on molecular ecology and biodiversity of non-model organisms such as sponges.

Undoubtedly, many sponge ecological studies can benefit from using molecular approaches to strengthen their conclusions. In particular, one of the key issues in the field of the sponge molecular ecology is the assessment of how sponges respond to environmental changes, whether weak or strong, cyclic or stochastic. The responses of individuals and populations to a changing environment and the causes underlying adaptation are major topics in the field of molecular ecology (Carroll *et al.*, 2007). Mobile organisms can respond to suboptimal environmental conditions by migrating to a more favourable area. However, sponges live fixed to the sea bottom as adults and cannot migrate; sponges must respond to new conditions only physiologically, by acclimating their metabolism for survival. The ability of sponges to persist in a given area is determined by their genetic constitution. Under adverse conditions, selective mortality may occur in sponge populations resulting in local genetic adaptation, or even populations may become locally extinct. In the two cases, genetic markers can assist in evaluating the impact of the environmental changes on sponge assemblages (Hutchings *et al.*, 2007).

Although there is little doubt about the benefits of applying molecular techniques to improve and expand the field of sponge ecology, classical ecological approaches have been decisive for the formulation of appealing questions that can be approached by using molecular tools. Ecological and biological issues related to reproduction (e.g. McKinnon *et al.*, 2004), growth, resistance/vulnerability to man-induced perturbations, responses to natural changes, chemically or physically mediated interactions, competition, facilitation, commensalism, trophic and larval ecology (Palumbi *et al.*, 2008), are still poorly known in the Porifera. Thus, classical physiological and ecological approaches continue to be necessary in sponge studies as a source of new questions and hypotheses that can nowadays be addressed with an array of molecular tools.

In this review, we will consider the main aspects where molecular tools have contributed to the advancement of our knowledge on sponge ecology. We will go from (1) the interface between inter- and intraspecies studies to (2) phylogeography and population level analyses, (3) intra-population features such as clonality and chimerism, and (4) environmentally modulated gene expression. We leave out of this review, on purpose, the extensive recent literature on molecular markers applied to studies of sponge symbiont assemblages, which will be dealt with elsewhere (Thacker and Freeman, 2012). We will end this retrospective with a prospect of future directions and developments that we foresee field will witness in the forthcoming years.

2. WHERE MOLECULAR MARKERS ALERT US ABOUT HIDDEN SPONGE DIVERSITY: CRYPTIC SPECIATION AND ITS ECOLOGICAL REPERCUSSIONS

2.1. Role of molecular markers in the discovery of cryptic sponge species

We are in a time of accelerated biodiversity loss, with many species disappearing even before they are identified (Williams and Hilbert, 2006) while, at the same time, other species are moved from their native areas to new ranges where they can threaten local biota (Kaiser and Gallagher, 1997). Assessment of local and introduced biodiversity is urgent and mandatory before issues of conservation and protection can be addressed. Yet this is a difficult task, in particular in groups that lack sufficient morphological characters and whose taxonomy is difficult, such as sponges. This is the first aspect in which genetic tools have come to the rescue in recent years.

Molecular tools are especially powerful in disclosing previously undetected taxonomic diversity, such as that represented by the so-called cryptic species, which cannot usually be resolved efficiently with only morphological characters. The message stemming from the application of molecular tools is that diversity has been grossly underestimated in benthic invertebrates in general (Knowlton, 2000) and in sponges in particular (Wörheide *et al.*, 2005). The finding of cryptic species is a common outcome whenever sponges have been investigated using genetic markers. Often, studies that focused on population genetics of a single species have revealed a previously misperceived species complex whose status needed clarification as a first step (e.g. Miller *et al.*, 2001; Zilberberg *et al.*, 2006a; Blanquer and Uriz, 2007; Xavier *et al.*, 2010a).

The use of molecular markers has proved that populations of marine sessile invertebrates in general (e.g. Palumbi *et al.*, 1997; Bierne *et al.*, 2003) and sponges in particular (e.g. Duran *et al.*, 2004b; Nichols and Barnes, 2005; Blanquer *et al.*, 2009; Blanquer and Uriz, 2010) have strong spatial structure and restricted gene flow (see section 3) which are favourable conditions for reproductive isolation (Knowlton, 1993) and, thus, speciation. Cryptic species, besides hindering biodiversity assessment, create problems for non-taxonomic research. Even if they are very close morphologically, cryptic species often show contrasting physiological, reproductive, and/or other biological traits (Blanquer *et al.*, 2008) so that their misidentification may cause serious inconsistencies in biological and ecological studies.

Widespread geographic distributions have been often reported in the old literature for many sponge species because of the lack of clear morphological differences (e.g. in spicule shape and size) among individuals inhabiting distant areas. The idea of "cosmopolitan" sponge species, however, is at odds with what we know about the limited dispersal capacity of most sponge larvae (e.g. Boury-Esnault *et al.*, 1993; Uriz *et al.*, 1998; Mariani *et al.*, 2005; Uriz *et al.*, 2008). Molecular analyses soon made claims of cosmopolitanism fall into disrepute as examples of overconservative systematics (Klautau *et al.*, 1999; Lazoski *et al.*, 2001; Miller *et al.*, 2001; Wörheide *et al.*, 2002b; Boury-Esnault and Solé-Cava, 2004), and the existence of several cryptic sibling species was demonstrated in most cases. Thanks to the extensive use of molecular approaches to improve species identification, the number of recognized cryptic sponge species is steadily increasing (e.g. Solé-Cava *et al.*, 1991a,b; Klautau *et al.*, 1999; Solé-Cava and Boury-Esnault, 1999; Blanquer and Uriz, 2007; Pérez *et al.*, 2011).

Several markers have proved useful for assessing the taxonomic status of sponge morphotypes or species complexes (see Cárdenas et al., 2012) and thus to detect cryptic species. Allozymes, besides being suitable markers for studies of population genetics in sponges, also resulted useful for establishing species boundaries (e.g. Solé-Cava and Thorpe, 1986, 1994; Solé-Cava et al., 1991a,b, 1992; Bavestrello and Sarà, 1992; Boury-Esnault et al., 1992; Sarà et al., 1993; Klautau et al., 1994; Barbieri et al., 1995; Muricy et al., 1996a,b; reviewed in Solé-Cava and Boury-Esnault, 1999). But both mitochondrial and nuclear sequences have replaced allozymes for species delimitation in the past years. *ITS* sequences were used to study taxa purportedly widely distributed, uncovering the existence of several species (e.g. Wörheide et al., 2002b). The 5′end or Folmer partition (Folmer et al., 1994) of the mitochondrial gene cytochrome oxidase subunit 1 (*COI*), which has been proposed as the standard marker for DNA barcoding (e.g. Hebert et al., 2003), has proved to be suitable to discriminate species in many cases (e.g. Blanquer and Uriz, 2007) because of its extraordinarily low intraspecies variability in sponges as compared to other groups (e.g. Duran et al., 2004a; López-Legentil and Pawlik, 2009). A slightly longer fragment of the same gene, including the I3-M11 partition, was claimed to improve resolution and phylogenetic signal compared to the Folmer partition (Erpenbeck et al., 2006a; López-Legentil and Pawlik, 2009). Of course, more solid results are obtained when several genes are used and congruent patterns appear. Thus, multiple nuclear and mitochondrial markers have been used in some studies to detect or confirm species differentiation (Zilberberg et al., 2006a; Blanquer and Uriz, 2007; Xavier et al., 2010a; Reveillaud et al., 2011a,b). Occasionally, cryptic species were first detected during studies of population genetics, using allozymes or microsatellites, and then confirmed by mitochondrial and nuclear sequences (e.g. Blanquer and Uriz, 2007). In retrospect, morphological characters matching the new species boundaries could be found in some cases (Muricy et al., 1996a; Blanquer and Uriz, 2008).

2.2. Representative case examples

Cryptic speciation in sponges has been uncovered using molecular markers in at least 23 species complexes resulting in ca. 50 cryptic species (Table 5.1). Solé-Cava and coworkers first applied allozymes to establish the species boundaries of Calcarea and Demospongiae across geographical clines. Solé-Cava and Thorpe (1986) described two new species within the *Suberites ficus* (Johnston) complex while Solé-Cava et al. (1991a) studied two geographically distant populations of the allegedly cosmopolitan species, *Clathrina clathrus* (Schmidt) and *C. cerebrum* (Haeckel). In both cases, populations of the two species from the South West Atlantic (Brazil) and the Mediterranean Sea showed high levels of genetic divergence, which allowed the

Table 5.1 Selected cases of sponge species complexes resulting in the determination of cryptic sibling species after a molecular study

Original species or species complex	Molecular marker	Resulting cryptic species	Reference
Suberites ficus	Allozymes	*S. pagurorum* *S. rubrus*	Solé-Cava and Thorpe (1986)
Thethya aurantium	Allozymes	*T. aurantium* *T. citrine* *Tethya* sp.	Sarà et al. (1989)
Axinella damicornis/A. verrucosa	Allozymes	*A. damicornis* *A. verrucosa* *Axinella* sp.	Solé-Cava et al. (1991b)
Clathrina clathrus	Allozymes	*C. aurea* South–West Atlantic *C. clathrus* sp. Mediterranean	Solé-Cava et al. (1991a)
Clathrina cerebrum	Allozymes	*C. cerebrum* Mediterranean *C. brasiliensis* North Atlantic	Solé-Cava et al. (1991a)
Oscarella lobularis	Allozymes	*O. lobularis* Mediterranean *O. tuberculata* Mediterranean	Boury-Esnault et al. (1992)
Petrosia ficiformis	Allozymes	*P. ficiformis* Mediterranean *P. clavata* Mediterranean	Bavestrello and Sarà (1992)
Plakina trilopha	Allozymes	*P. trilopha* Widespread Mediterranean *P. endoumensis* Cave1 Mediterranean *P. jani*	Muricy et al. (1996b)

Species	Markers/methods	Taxa/localities	Reference
Latrunculia spp.		Cave2 Mediterranean *Plakina* sp3 Cave3 Mediterranean	Miller et al. (2001)
Astrosclera willeyana	Allozymes ITS1-5.8S-ITS2 region	8 *Latrunculia* species *Astrosclera* sp1 Red Sea *Astrosclera* sp2 Great Barrier Reef *A. willeyana* Fiji/Vanuatu	Wörheide et al. (2002a)
Pachymatisma normani	*COI* and *ITS1-5.8S ITS2*	*P. normani* North Atlantic *P. johnstoni* North Atlantic	Cárdenas et al. (2007)
Chondrosia reniformis	Allozymes	*C. reniformis* Mediterranean *Chondrosia* sp. Atlantic (from Bermuda to Brazil)	Lazoski et al. (2001)
Chondrilla cf. *nucula*	5′-end *COI*	*C.* cf *nucula* Caribbean mangroves *C.* cf *nucula* Caribbean reefs	Duran & Rützler (2006)
Halichondria panicea	5′-end *COI*	*H. panicea* sp1 Alaska *H. panicea* sp2 NE Atlantic	Erpenbeck et al. (2004)
Scopalina lophyropoda	5′-end *COI, 28S, 16S*	*S. lophyropoda* Mediterranean *S. blanensis* Mediterranean *S. ceutensis* Mediterranean, North Africa *S. canariensis* Atlantic, Canary Islands	Blanquer and Uriz (2007, 2008)
Cliona celata	5′-end *COI, ATPase8,* and *28S*	*C. celata* sp1 North Atlantic *C. celata* sp2 North Atlantic *C. celata* sp3	Xavier et al. (2010a)

(continued)

Table 5.1 (continued)

Original species or species complex	Molecular marker	Resulting cryptic species	Reference
Hexadella spp.	COI, 28S, ATPS-intron	Western Mediterranean	Reveillaud et al. (2011a)
		C. celata sp4	
		Western Mediterranean	
		H. detritifera NW Atlantic (Irish, Scottish, and Norvegian coasts and Greenland Sea)	
		H. cf. detritifera Ionian Sea, Irish coasts, Bay of Biscay	
		Hexadella sp.	
		Mediterranean deep-sea	
		Hexadella cf. pruvoti	
		Mediterranean, shallow sea	
Plocamionida spp.	MIM6 and 13M11partitions of 28S & COI	P. tylotata North Atlantic	Reveillaud et al. (2011b)
		P. grandichela North Atlantic	
		P. ambigua (tornata)	
		North Atlantic	
		P. microcionides	
		North Atlantic	
Halisarca spp.	Mitochondrial genome	H. dujardini	Ereskovsky et al. (2011)
		Atlanto-Mediterranean	
		H. harmelini	
		Mediterranean	

authors to consider them different species with a disjointed geographical distribution (Table 5.1). Allozymes proved also useful to discover a new cryptic species phenotypically intermediate between the well-known *Aaxinella verrucosa* Brøndsted and *A. damicornis* (Esper) (Solé-Cava *et al.*, 1991b).

Genetic (allozyme electrophoresis) and cytological studies of the aspiculate homoscleromorph *Oscarella lobularis* (Schmidt) from the Mediterranean revealed the existence of several species, including two cryptic polymorphic species: *Oscarella tuberculata* (Schmidt) and *O. lobularis* (Boury-Esnault *et al.*, 1992). The phenotypic differences between these two species are hard to find, but an accurate histological study (Muricy *et al.*, 1996a) found several cytological diagnostic characters such as amount of collagen, types of vacuolar cells, and types of symbiotic bacteria.

Muricy *et al.* (1996b), by using eleven allozyme loci, distinguished four species within the "*Plakina trilopha*" complex, which were difficult to identify on the basis of their spicule characteristics. Indeed, the values of the *I* index of genetic identity (Nei, 1978) and the presence of diagnostic alleles for each of the several morphotypes provided evidence of the presence of four species (Muricy *et al.*, 1996b). The "true" *P. trilopha* Schulze was widely distributed in the Mediterranean, but the other three species in the complex were found only in certain caves and vertical walls at single sites (each species) around Marseilles (France). These caves represent exceptional habitats (Vacelet *et al.*, 1994) that may have kept sponge populations isolated. The authors hypothesized that these *Plakina* species likely evolved by independent colonization events in the different caves of the region, followed by reproductive isolation of the subpopulations due to restricted water circulation in the caves and low dispersal capabilities of their larvae (Muricy *et al.*, 1996b).

Perhaps the most paradigmatic instance of cryptic speciation in sponges has been the *Chondrilla* "*nucula*" complex. The type material is from the Adriatic and it was considered cosmopolitan until it was used precisely to illustrate the problems with overconservative systematics. Klautau *et al.* (1999) using 10 allozyme loci showed the existence of five genetic clades in samples from the Atlantic and the Mediterranean. The authors tentatively retained the original name for the Mediterranean clade and assigned letters to the other four groups. Morphological characters (spicule sizes) did not correlate with species boundaries defined genetically. Zilberberg *et al.* (2006a) further analyzed several forms belonging to what was formally *Chondrilla* "*nucula*" Schmidt in the Caribbean and Brazilian coasts using allozymes. Although focused on the assessment of asexual reproduction, this study found two of the species defined by Klautau *et al.* (1999) and two more species belonging to this complex, one in the Caribbean and one in Brazil. Duran and Rützler (2006) analyzed the 5′end partition of *COI* in individuals of what they called *C*. cf. *nucula* inhabiting two contrasting ecosystems in the Caribbean: mangal swamps and coral reefs. Each habitat was occupied by a distinct morphotype, which differed among them in colour and general shape: the lighter coloured, thinner morphotype from the coral reefs and the darker and thicker morphotype from the mangal swamps. Five out of 12 haplotypes found were specific of mangrove habitats, while another five were exclusive of coral reefs (Duran and Rützler, 2006). An AMOVA based on haplotype frequencies statistically supported high genetic isolation between the two habitats present in the same localities. Moreover, populations from the same habitat (either mangrove or reef) separated by more than 1000 km had a similar haplotype composition. This system represented the first instance of ecological speciation in sponges. The 5′end partition of *COI* also allowed differentiating between the Alaskan populations of *Halichondria panicea* and those from the NE Atlantic. The resulting phylogenetic tree clearly showed two clades, which represented two separate species (Erpenbeck *et al.*, 2004).

Complete mitochondrial genomes have been recently used to identify cryptic species of *Halisarca* Johnston (Ereskovsky *et al.*, 2011). *Halisarca dujardini* Johnston, and *H. harmelini* Ereskovsky, Lavrov, Boury-Esnault, and Vacelet, had identical mitochondrial genomes as for their gene content and gene arrangement but differed in size by ∼1300 bp (6.8%). The overall genetic distance between coding sequences of the two species was much greater than previously reported for species of non-bilaterian animals (Ereskovsky *et al.*, 2011). This genetic difference calls for caution about claims of highly conserved mitochondrial genomes in sponges, based only on the cytochrome oxidase gene.

Other gene fragments have also been assayed with different success. Wörheide *et al.* (2002a) used the *ITS1-5.8S-ITS2* region to analyze several populations of the purportedly circum-Pacific coralline demosponge *Astrosclera willeyana* Lister. Despite the small number of differences between sequences, these authors concluded that, under a strict version of the phylogenetic species concept, the populations from the Red Sea, the GBR, and Fiji/Vanuatu represented distinct species, of which the Fiji/Vanuatu species would correspond to *A. willeyana s.s.* (but see Wörheide, 2006, for a lack of differentiation among two of the species using another marker). The species status for these three populations according to the molecular markers used corroborated the three previously detected groups based on morphological characters (Wörheide, 1998). However, Nichols and Barnes (2005) using *ITS* markers failed to resolve the phylogenetic relations among representatives of the genus *Placospongia* Gray from West Pacific, Caribbean, and Indo-Pacific populations. Although discrete lineages were found in the several geographical regions, cryptic species could not be established. These authors stated that, because of the intragenomic variation of *ITS*, the phylogenetic structure in their dataset reflected duplication events rather than relationships among individuals.

Using a partial sequence of the *COI* and a *ITS1-5.8S-ITS2* nuclear fragment, Cárdenas *et al.* (2007) showed that *Pachymatisma johnstonia* (Bowerbank) and *P. normani* Sollas, two astrophorid species that had been considered, synonymous were indeed good species, undistinguishable on the basis of spicule shapes and sizes.

Molecular markers can confirm species boundaries but sometimes reveal lack of speciation. Lazoski *et al.* (2001) using 13 allozymes discovered that the Atlantic populations of *Chondrosia reniformis* Nardo from Bermuda and Brazil were indeed a separate species from the Mediterranean *C. reniformis*. However, populations of *C. reniformis* from the Atlantic, separated by up to 8600 km of distance, showed remarkable genetic similarity, albeit with significant population structure (see section 3).

Cryptic speciation was unexpected for the only known Mediterranean *Scopalina* (*S. lophyropoda*) despite the few diagnostic characters available for the species (Blanquer and Uriz, 2008). However, microsatellites developed for *S. lophyropoda* failed to amplify some populations along the distribution area reported for the species (from the Adriatic to the Canary Islands), pointing to the possible presence of a species complex (Blanquer and Uriz, 2007). Mitochondrial *COI* (5′-end, Folmer partition) and 16S rRNA gene, together with sequences of the nuclear *28S* rRNA gene, confirmed that the populations that did not amplify were indeed three new species (Blanquer and Uriz, 2007, 2008). One of them (*S. blanensis* Blanquer and Uriz) shared habitat and could even be found in contact with the true *S. lophyropoda* (Fig. 5.1). The other two (*S. ceutensis* Blanquer and Uriz and *S. canariensis* Blanquer and Uriz), however, seemed to be restricted to geographically distant areas: North Africa (Mediterranean) and Canary Islands (North Atlantic), respectively.

Sometimes the cryptic species resulting from a species complex show a disjointed distribution. Several Atlanto-Mediterranean populations of the sponge *Cliona celata* Grant, until then considered a cosmopolitan species, were analyzed using mitochondrial *COI* and *ATP8* synthetase, as well as the nuclear *28S* rRNA gene (Xavier *et al.*, 2010a). The phylogenetic reconstructions indicated the existence of four well-supported clades with a clear gap between intra- and inter-clade divergences. Consequently, *C. celata* represents a complex of four cryptic species, with contrasting distributions: two species occurring along the Atlantic European coasts and the other two in the Mediterranean and Macaronesian islands. These results confirmed previous findings obtained with allozyme markers, which led to the suggestion of splitting the Mediterranean "*C. celata*" into two species (Barbieri *et al.*, 1995). These results also showed that the boring and massive growth forms of this excavating sponge are truly different growth stages or ecological phenotypes of the same species.

Reveillaud *et al.* (2011a) used the Folmer partition of the mitochondrial *COI* gene, the D3–D5 region of the nuclear large ribosomal subunit

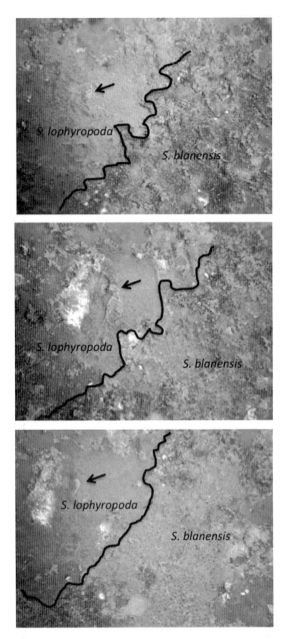

Figure 5.1 An example of highly dynamic encrusting sponges: three sequencial snapshots showing fissions, fusions, growth, and shrinkage of the cryptic Mediterranean species *Scopalina lophyropoda* and *S. blanensis*, along the year (arrows point of a reference object). (For colour version of this figure, the reader is referred to the Web version of this chapter.)

(*28S rRNA* gene) and the second intron of the nuclear *ATP* synthetase, beta subunit gene (*ATPS*) to establish the boundaries of the three Atlanto-Mediterranean species of *Hexadella* Topsent and to investigate the presence of cryptic species within this genus. Phylogenetic analyses revealed several divergent clades for the deep-sea sponges, congruent across the mitochondrial and nuclear markers. One clade contained specimens from the Irish, the Scottish, and Norwegian margins and from the Greenland Sea (*H. dedritifera* Topsent), another clade contained specimens from the Ionian Sea, the Bay of Biscay, and the Irish margin (*H. cf. dedritifera*), and a third clade corresponded to a new Mediterranean deep-sea species (*Hexadella* sp.). Furthermore, another cryptic shallow-water species (*H cf. pruvoti* Topsent) was also revealed in the Mediterranean Sea and in the Gorringe Bank (North Atlantic). The *ATPS* marker, first applied to sponges by Bentlage and Wörheide (2007), proved its applicability for species delimitation in this group in representatives of the genus *Hexadella*. Reveillaud *et al.* (2011b) using the I3M11 and M1M6 partitions of the *COI*, and *28S rRNA* gene sequences, in combination with sponge morphology, detected an underestimated biodiversity of the genus *Plocamionida* Topsent along 3000 km of European margins, with three additional valid species besides *P. ambigua* (Bowerbank).

Although most molecular studies have detected genetic variation in sponges associated with even subtle morphological differences that had been considered without diagnostic value in traditional sponge taxonomy (e.g. colour: Boury-Esnault *et al.*, 1992; Klautau *et al.*, 1999; Knowlton, 2000; Miller *et al.*, 2001; Blanquer and Uriz, 2008), a few studies have reported the lack of genetic variability in sympatric morphotypes that differed in colour and shape (e.g. Solé-Cava and Thorpe, 1986; Boury-Esnault *et al.*, 1992; López-Legentil and Pawlik, 2009). For instance, the Caribbean sponge, *Xestospongia muta* (Schmidt), has three main morphotypes, which are characterized by their digitate, rough, or smooth external surface, respectively. The haplotype network of their populations based on the I3-M11 partition of *COI* indicated that the high degree of morphological differentiation did not reflect genetic boundaries in *X. muta*, and that gene flow occurs between these morphotypes (López-Legentil and Pawlik, 2009). The genus *Xestospongia* De Laubenfels seems to harbour species with wide intraspecies phenotypic variation. No correlation between external morphology and sterol chemotypes (Kerr and Kelly-Borges, 1994), which had been considered to have chemotaxonomic value (Bergquist *et al.*, 1990), has been reported for the Indo-Pacific species of *Xestospongia*. This seems to represent a particular case where environmental conditions (mainly currents) may determine the shape of the outer surface of the individuals, while populations living in contrasting habitats are genetically connected may be because of a relatively wide dispersion of gametes in these oviparous sponges.

Another example of high intraspecies phenotypic plasticity unrelated with genetic differentiation is *Callyspongia vaginalis* (Lamarck), which shows three morphotypes differing in colour and external surface in the Caribbean. The genetic distances among these three morphotypes and the close species *Callyspongia fallax* Duchassaing and Michelotti (all of them with oxeas as the only spicule type) were assessed by partitions of two mitochondrial (*COI* and *16S*) and two nuclear (*18S* and *28S* rRNA) genes (López-Legentil *et al.*, 2010). None of these genetic markers provided evidence for differentiation among the morphotypes of *C. vaginalis* or between these and the congeneric *C. fallax*. Morphological characters (spicule sizes and spongin fibre characteristics) showed differences not linked to genetic patterns. As in *Xestospongia muta*, *C. vaginalis* seems to maintain a high degree of phenotypic plasticity, and their morphological characteristics did not indicate reproductive boundaries.

2.3. Conclusions

Molecular markers can reveal hidden sponge biodiversity and, thus, improve the consistency of ecological studies. They also assist taxonomists in assessing the taxonomic value of phenotypic characters, distinguishing those that are evolutionarily fixed from those that result from environmental plasticity. Morphological characters that have been traditionally disregarded for species identification can turn out to be adequate diagnostic characters for discriminating some species complexes, and the reverse is also true. In some cases (e.g. *Chondrilla* Schmidt, *Plakina* Schulze, and *Scopalina* Schmidt), an "*a posteriori*" scrutiny of the phenotypic characters of the cryptic species revealed differences in colour, skeletal arrangement, amount of spongin surrounding the skeletal tracks, or sponge surface features. At the same time, although the main tale is one of hidden diversity, suspected cryptic species may result in mere morphotypes, as revealed by genetic divergence below the "between–species" differentiation threshold.

It should also be noted that the lack of genetic differences derived from a single marker must be interpreted with care, since the same gene may show contrasting mutation rates in different lineages of the Porifera. Hence, the choice of a suitable marker strongly depends on the evolutionary context of each single taxon (Heim *et al.*, 2007) and, as a consequence, several markers should be assayed before a decision on the taxonomic status of a given species is taken. A foreseeable trend in the near future is the incorporation of more markers and the use of a combined, multilocus approach to assess sponge biodiversity.

According to the many examples of cryptic sponge species with disjointed distributions reported in the literature, sponge speciation seems to have occurred mainly in allopatry even at small geographical scales (hundreds of kilometres). Allopatric speciation has been proposed for the two species of the

Clathrina clathrus complex (Solé-Cava and Boury-Esnault, 1999), as well as for *Scopalina lophyropoda*, *S. ceutensis*, and *S. canariensis* (Blanquer and Uriz, 2007). In these cases, the most plausible scenario for their speciation is that derived from isolation by distance, as a result of a strong larval philopatry, which is a shared trait of most sponge species (e.g. Uriz *et al.*, 1998; Uriz *et al.*, 2008). Speciation of *S. lophyropoda* and *S. blanensis*, despite their currently over-lapping habitat, might have also originated in allopatry, with *S. blanensis* diverging at the Central Mediterranean and then recolonizing the western Mediterranean coasts (Blanquer and Uriz, 2007, 2008).

Interrupted gene flow among populations due to physical barriers produces genetic divergence and the consequent speciation. However, ecological reproductive barriers may also bring about speciation. Understanding the role of ecological factors in speciation will require an integrated knowledge on ecological, evolutionary, and behavioural aspects, as well as on the selective pressures operating in natural populations. Until now, there has been little evidence of limited interbreeding within sponge populations related to ecological niche differentiation. The cryptic speciation within the *Chondrilla* cf. *nucula* complex in two disjointed habitats, mangal swamps and coral reefs, illustrates the only example so far of ecologically driven speciation in sponges (Duran and Rützler, 2006).

A combination of physical and ecological barriers may be underlying the intense species radiation of the Homoscleromorpha *Plakina* in caves of a small geographical area of the western Mediterranean (Muricy *et al.*, 1996b). The physical isolation of caves, which in some respects can be considered as islands (e.g. Vacelet *et al.*, 1994) and the particular ecological conditions of these biotopes such as trophic depletion, reduced water movement and light (e.g. Martí *et al.*, 2004) adds to the poor larval dispersal of sponges in general (e.g. Mariani *et al.*, 2005) and can boost sponge speciation.

The new clades that are arising using genetic techniques represent a substantial increase in the number of sponge species currently known. The detection of cryptic species by molecular methods will continue to improve our knowledge of the true diversity in the world oceans in the forthcoming years. There is an active dispute about the relative merits of molecular-based and traditional descriptions of species (e.g. Schlick-Steiner *et al.*, 2007; Packer *et al.*, 2009; Cook *et al.*, 2010; Bucklin *et al.*, 2011). In our view, phenotypic descriptions should complement molecular-based species descriptions to make them practically available in ecological studies. Moreover, studies on the biological, biochemical, and ecological aspects of cryptic species are also recommended to approach a multidisciplinary "species concept" (Manuel *et al.*, 2003; Erpenbeck *et al.*, 2004, 2006b; Loukaci *et al.*, 2004; Blanquer *et al.*, 2009; Pérez *et al.*, 2011; Cárdenas *et al.*, 2012). Several biological aspects, such as the extent and timing of the reproductive period, investment in reproduction, recruitment success, growth, and competitive abilities, can be different among cryptic species (e.g. Blanquer *et al.*, 2008). Disregarding true sponge biodiversity by ignoring the presence of cryptic species in an ecosystem will unavoidably lead to incorrect conclusions in ecological studies.

3. POPULATION GENETICS AND PHYLOGEOGRAPHY

Population genetics focuses on the genetic characteristics that shape populations and influence their success or failure at ecological and historical time scales. The health state of populations and their capacity to adapt to environmental changes can be estimated by several genetic descriptors such as gene diversity, effective population size, gene flow, kinship, inbreeding, and the extent of asexual reproduction, among others. Predictions obtained from theory are compared with empirical results obtained from actual populations to make inferences about patterns, processes, and cause-and-effect relationships in the biological world (Hamilton, 2009). On the other hand, understanding the historical events that have contributed to the current geographical distributions of populations is the main goal of phylogeography (Avise, 2000). The topics of population genetics merge with those of phylogeography and intraspecific phylogeny (Avise, 2004) as both address genetic differentiation among populations, but they focus on processes that occur at contrasting time scales (i.e. present-day recurrent processes, historical events, and evolutionary time scales) and, consequently, the molecular markers that are suitable to provide the necessary information for each issue may not be the same. Studies of population genetics of sponges lag behind other marine taxa, especially regarding the development and application of updated molecular markers. It has been suggested that sponge populations may not reach equilibrium rapidly and, consequently, highly variable molecular markers are required, together with particular evolutionary models (Wörheide *et al.*, 2004).

3.1. Choice of variable molecular markers at the intraspecies level

A range of molecular markers has been assayed with contrasting success in genetic studies of sponge populations. The divergent results of the several studies draw attention on the contrasting behaviour that a gene may show in different species and made it clear that new markers are still needed for answering the diverse ecological questions.

3.1.1. Allozymes

Allozymes proved to be markers with high intraspecies variability in sponges and were thus the molecular marker of choice for studying population structure and differentiation in sponges until the end of the past century, (e.g. Thorpe and Solé-Cava, 1994; Lazoski *et al.*, 2001; Miller *et al.*, 2001) and continue to be used today (Whalan *et al.*, 2005, 2008). Allozyme electrophoresis has also proven very useful to detect cryptic species and false cosmopolitanism (reviewed in Solé-Cava and Boury-Esnault, 1999; Borchiellini *et al.*, 2000; see section 2).

3.1.2. Gene sequences
3.1.2.1. Nuclear gene sequences
The *ITS* regions (*ITS1* and *ITS2*) have been widely used for phylogeographic studies of sponges (Lopez *et al.*, 2002; Wörheide *et al.*, 2002a,b, 2004; Duran *et al.*, 2004b; Nichols and Barnes, 2005) likely because of the availability of PCR primers, which can be used across a range of taxa. Conserved and variable regions alternate in these genes, which facilitates the design of PCR primers in conserved regions, flanking more variable sequences (Nichols and Barnes, 2005). However, these genes were not sufficiently variable in all the targeted species. For instance, individuals of *Placospongia* from both sides of the Isthmus of Panama show little divergence in the targeted *ITS* region (Nichols and Barnes, 2005) as did individuals of the coralline sponge *Astrosclera willeyana* from the Red Sea, the Great Barrier reef, and Fiji/Vanuatu (Wörheide *et al.*, 2002a). Moreover, *ITS*s may show IGP in sponges (Van Oppen *et al.*, 2002; Wörheide *et al.*, 2004; Duran *et al.*, 2004b; Hoshino *et al.*, 2008), which complicates the interpretation of phylogeographic results using these markers. Single-strand conformation polymorphism methods (e.g. Lôbo-Hadju *et al.*, 2004) or cloning must be used to screen IGP in sponges before using *ITS* as markers for population genetics, because moderate ITS paralogy can be tolerable for phylogenetic studies, but less so for population level studies (Wörheide *et al.*, 2004).

Other gene partitions have also been assayed. An intron of the *ATP beta* synthetase gene was used (in combination with *ITS*) for a phylogeographic and population genetic study of two calcareous sponges in the Pacific and Indo-Pacific regions (Bentlage and Wörheide, 2007; Wörheide *et al.*, 2008) and seems to be a promising tool.

3.1.2.2. Mitochondrial gene sequences
Mitochondrial genes are widely used in population genetics and phylogeographic studies of marine organisms because they are maternally inherited without recombination, have shorter coalescence times, and are expected to undergo lineage sorting three times faster than nuclear markers (Avise *et al.*, 1987; Palumbi *et al.*, 2001). However, the mtDNA gene partitions that have been used with success for other invertebrates (e.g. *COI*) are extremely conserved in sponges (e.g. Knowlton, 2000) and they have been rather proposed for use as markers for medium and low-level phylogenies in lineages that diverged up to 200 MYA (e.g. Erpenbeck *et al.*, 2002).

The several studies that have used mitochondrial markers, in particular, the 5′ region (Folmer *et al.*, 1994) of the cytochrome *c* oxidase subunit I (*COI*), to determine the genetic structure of sponge populations, revealed very low levels of variability of this gene even over broad geographic scales (tens of thousands of kilometres, Wörheide, 2006), although population structure could be demonstrated in general (Duran *et al.*, 2004a; Duran and Rützler, 2006; DeBiasse *et al.*, 2010). The I3-M11 partition of the same gene seems to show a slightly higher intraspecies variability in sponges than the Folmer partition (Erpenbeck *et al.*, 2006a; López-legentil and Pawlik, 2009), and thus it seems more suitable for sponge population level studies. In some cases, where *COI* sequences indicated significant differences among targeted "populations", they were in fact revealing cryptic species (e.g. Xavier *et al.*, 2010a).

Another mitochondrial gene, the *NADH* dehydrogenase subunit 5 (nad5), has recently been assayed for intraspecific genetic diversity in two sponge species (Hoshino *et al.*, 2008). As in the case of *COI*, the *nad5* gene showed very low genetic diversity that hinders its applicability in this group.

3.1.2.3. Amplified fragment length polymorphisms

Amplified fragment length polymorphisms (AFLPs) are reliable and relatively easy to obtain markers used for studies of population genetics in several invertebrate groups (Mueller and Wolfenbarger, 1999). In sponges, however, AFLPs have only been used to differentiate *Ephydatia mülleri* and *E. fluviatilis* when gemmules, which harbour the diagnostic spicules, are not present. AFLPs have been proposed to use in combination with other markers in population level studies in those areas where the two species are in sympatry, to avoid overestimation of genetic differentiation among populations of one species due to erroneous species attribution (Gigliarelli *et al.*, 2008).

3.1.2.4. Microsatellites

Because of the intra-genomic variability of *ITS*, the low resolution of sponge-mtDNA sequences even over long geographical distances, and the methodological constraints of allozymes, microsatellites appear to be the best choice among the currently available markers for sponge studies of population differentiation, genetic diversity, gene flow, clonality, and other population genetic descriptors. Microsatellites or SSRs are small DNA stretches consisting of a repeated core sequence of a few base pairs. Because they are highly polymorphic in the number of repeats (and thus in length), and co-dominant, they have been extensively used since the 1990s in population assignments, paternity analyses, and fine-scale dispersal analyses of terrestrial organisms (Webster and Reichart, 2005). The major drawback of this technique has been the cumbersome effort needed to generate a statistically relevant number of such polymorphic loci in non-model organisms (Zane *et al.*, 2002). Consequently, they have been only occasionally used up to now in sponge studies. There are at present eight sponge species for which microsatellites have been developed (listed in section 1).

3.2. Genetic differentiation at large and regional geographical scales

Genetic differentiation among sponge populations seems to be the rule in studies at broad scales. Benzie *et al.* (1994) studied allozyme variation at six polymorphic loci in four dictyoceratid species (*Phyllospongia lamellosa* (Esper), *P. alcicornis* (Esper), *Carteriospongia flabellifera* (Bowerbank), and *Collospongia auris* (Bergquist, Cambie, and Kernan)) in the western Coral Sea (Pacific Ocean). The allele frequencies showed that the populations of these species were in Hardy–Weinberg equilibrium, presumably as a result of random mating in local populations. Genetic differentiation was found for all the populations of all studied species, two of which followed the isolation-by-distance model (Table 5.2). The study allowed the authors to detect a barrier to gene flow between some populations caused by the South Equatorial Current since the genetic divergence found among populations North and South of this current was higher than expected from the geographic distances between them.

As said above (see section 2), Klautau *et al.* (1999) demonstrated with the use of 10 allozyme loci the false cosmopolitanism and sibling speciation in the *Chondrilla nucula* complex. In the same work, they analyzed the differentiation of seven populations of one of the genetic forms (*Chondrilla* sp. B) along 2700 km of Brazilian coastline and found that they were highly structured, indicating low gene flow along the coast studied.

Lazoski *et al.* (2001) analyzed both inter- and intraspecies variation of 13 allozyme polymorphic loci in Mediterranean and Atlantic (Bermuda and Brazil) populations of the purportedly cosmopolitan species *Chondrosia reniformis*. The low genetic identities of Atlantic and Mediterranean sponges were compatible with the presence of two cryptic species. However, the West Atlantic populations of *C. reniformis* were genetically similar over a distance of > 8000 km, and a high gene flow scenario was suggested in what was a completely atypical result for sponges. The interpretation of this result, however, should be taken with caution. Genetic homogeneity was calculated based on Nei's I index (Nei, 1978), and it was high for both Mediterranean (>0.96) and West Atlantic populations (>0.88). These values, however, are well within the range of those found in intraspecies comparisons in sponges (Solé-Cava and Boury-Esnault, 1999). On the other hand, analyses of genetic differentiation (Fst) showed that the populations have a significant genetic structure in both areas ($p < 0.0001$) in spite of high genetic similarity.

Table 5.2 Collation of works reporting values of population differentiation (DIFF, using Fst or analogous measures) in sponges

Species	Marker	Range	Diff.	IBD	Observations	Reference
Phyllospongia lamellosa	Allozymes	10s–550 km	YES	NO	HWE except 2 populations	Benzie *et al.* (1994)
Phyllospongia alcicornis	Allozymes	10s–700 km	YES	YES	HWE	Benzie *et al.* (1994)
Carterospongia flabellifera	Allozymes	10s–500 km	YES	YES	HWE	Benzie *et al.* (1994)
Collospongia auris	Allozymes	10s–250 km	YES	NA	HWE except one population	Benzie *et al.* (1994)
Chondrilla 'nucula' B	Allozymes	2–2700 km	YES	NA	Evidence for cryptic speciation	Klautau *et al.* (1999)
Chondrosia sp.	Allozymes	Up to 8600 km	YES	NO	HWE. High genetic identity but significant Fst	Lazoski *et al.* (2001)
Crambe crambe	COI	10s–2700 km	YES	NO	Only 2 haplotypes	Duran *et al.* (2004a)
Crambe crambe	ITS	10s–3000 km	YES	NA	Intra-genomic variation detected	Duran *et al.* (2004b)
Crambe crambe	Microsats	10s–3000 km	YES	YES	Heterozygote deficiency, clonality present	Duran *et al.* (2004c)
Haliclona sp.	Allozymes	10s m	NO	NA	Heterozygote deficiency	Whalan *et al.* (2005)
		100s m	YES	NA	Heterozygote deficiency	
		100s km	YES	NA	Heterozygote deficiency	
Chondrilla cf. *nucula*	COI	4–18 km	YES	NA	Suspected ecological speciation process	Duran and Rützler (2006)
Crambe crambe	Microsats	1000–1700 km	YES	NA	Suspected ecological speciation process	Calderón *et al.* (2007)
		0–7 m	YES	NA	Heterozygote deficiency, clonality present	
Pericharax heteroraphis	*ATPSβ-iII*	Up to 2000 km	YES	NA	HWE in general. Differentiation between pooled regional populations	Bentlage and Wörheide (2007)

Species	Markers	Geographic range	Differentiation	HWE	Notes	Reference
Hymeniacidon flavia	*ITS, nad5*	10s–1000 km	YES	NA	Only 2 nad5 haplotypes	Hoshino et al. (2008)
Hymeniacidon sanguinea	*ITS, nad5*	10s–1600 km	NO	NA	Only 2 nad5 haplotypes	Hoshino et al. (2008)
Leucetta chagosensis	*ATPSβ-iII*	10s–4500 km	YES	YES	SW Pacific subset of populations	Wörheide et al. (2008)
Scopalina lophyropoda	Microsats	0–100 m	YES	NA	Heterozygote excess, little clonality	Blanquer et al. (2009)
Rhopaloeides odorabile	Allozymes & COI	1–10s of km	NO	NA	No differentiation in general, but some localized divergent populations	Whalan et al. (2008)
Xestospongia muta	COI	150–1800 km	YES	NO	I3-M11 COI partition	López-Legentil and Pawlik (2009)
Spongia lamella	Microsats	7–1800 km	YES	YES	Heterozygte deficiency	Noyer (2010)
Phorbas fictitius	COI	10s–100s km	NO	NO	I3-M11 COI partition	Xavier et al. (2010b)
		100s–1000s km	YES	NO		
Callyspongia vaginalis	COI	10s–100s km	YES	NO	Possible occasional long distance dispersal	DeBiasse et al. (2010)
Scopalina lophyropoda	Microsats	10s m–10s km	YES	YES	Heterozygote excess. Little clonality	Blanquer and Uriz (2010)
		10s–100s km	YES	YES	Heterozygote excess. Little clonality	
Paraleucilla magna	Microsats	10–50 m	YES	YES	Heterozygote deficiency. Introduced species	Guardiola et al. (2011)
Spongia officinalis	*COI and microsats*	7–1900 km	YES	YES (but not within basins)	Heterozygote deficiency	Dailianis et al. (2011)

Approximate geographic ranges are listed (taken from Google Earth when not indicated by the authors). The differentiation criterion "YES" or "NO" is not absolute; it is based on the appreciation that most population comparisons did or did not indicate significant differentiation. Results of analyses of isolation by distance (IBD) are also listed when reported (otherwise NA, not applicable). HWE, Hardy–Weinberg equilibrium.

In a study of a species of *Haliclona* Grant with allozymes, Whalan *et al.* (2005) investigated the genetic structure of populations at several spatial scales. Although only two loci were included in the study, the results were consistent with panmictic populations within reefs, but significant differentiation was found at intermediate (between reefs separated by hundreds of metres) and large (between areas 400 km apart) scales.

Nuclear ribosomal *ITS*s have often been used for sponge phylogeographic studies. Duran *et al.* (2004b) analyzed sequence variation in the nuclear ribosomal *ITS*s (*ITS-1* and *ITS-2*) in 11 populations of the sponge *Crambe crambe* across the species distribution range in the western Mediterranean and Atlantic Ocean. They reported the first confirmed instance of intra-genomic variation of *ITS*s in sponges. Phylogeographic, nested clade, and population genetic analyses revealed highly structured populations affected by restricted gene flow and isolation by distance. The authors speculated about a recent expansion of the species distribution range to the Macaronesian region from the Mediterranean and stated that the pattern observed was not likely to be the result of a natural biogeographic relationship between these zones but of a man-mediated introduction.

ITS sequences were much more variable in populations of two species of *Hymeniacidon* Bowerbank (*H. flavia* Sim and Lee and *H. sinapium* de Laubenfels) from Japan than *NADH* dehydrogenase subunit 5 (*nad5*) mitochondrial DNA sequences (Hoshino *et al.*, 2008), which only showed two haplotypes per species along their respective distribution ranges. Several significant genetic structures were detected in the nested clade analysis for *H. flavia*, indicating restricted gene flow with isolation by distance, while *H. sinapium* showed very little genetic variation. The authors speculated about a recent introduction via natural or man-mediated processes to the Western Pacific of *H. sinapium*, which may have experienced a bottleneck as a result of founder effects during introduction, or a severe population decline followed by rapid range expansion. The geographic genetic structure of *H. flavia* suggests low dispersal ability of its larvae, whereas higher larval dispersal was suggested for *H. sinapium*.

Bentlage and Wörheide (2007) developed a new nuclear marker for sponges (the second intron of the nuclear *ATP* synthetase beta subunit gene, *ATPSβ-iII*) and analyzed it together with *ITS* sequences to uncover phylogeographic patterns of the coral reef sponge *Pericharax heteroraphis* Poléjaeff in the southwest Pacific. Variation among *ITS* sequences was low in contrast to *ATPSβ*-iII, indicating a better performance of the newly developed marker for population studies. A statistical parsimony network suggested a past population subdivision with subsequent range expansion for GBR alleles. The authors expressed concern about the small sizes of most sampled populations but, based on the pairwise Fst values among pooled regional populations, they reported a high degree of differentiation between Indonesia and the GBR, Queensland Plateau and Vanuatu. Moreover, Vanuatu was strongly differentiated from the Queensland Plateau, central and southern GBR, whereas the differentiation between Vanuatu and the northern GBR was considerably smaller.

Wörheide *et al.* (2008) studied the genetic divergence among Indo-Pacific populations of the calcareous sponge *Leucetta chagosensis* Dendy by using two nuclear markers (ITS 1 and 2) and the same intron (*ATPSβ-iII*), used in the previous study (Bentlage and Wörheide, 2007). A deep phylogeographic structure was found, congruent across the *ITS* and *ATPSβ-iII* markers. One phylogeographic clade contained specimens from the Indian Ocean and Red Sea, another clade was composed of individuals from the Philippines, and two other clades consisted of sponges from NW Pacific and SW Pacific with an area of overlap in the Great Barrier Reef/Coral Sea. Gene flow was low among most regional populations, which showed isolation by distance along the Equatorial Current in the South-western Pacific. Overall, the results pointed towards stepping-stone dispersal with some putative long distance exchange, consistent with expectations from low dispersal capabilities. Both founder and vicariance events during the late Pliocene and Pleistocene were speculated to be partially responsible for generating the deep phylogeographic structure found.

Duran *et al.* (2004a) performed the first study of population structure in sponges using *COI* sequence data (5′-end or Folmer partition). Eight populations of the poecilosclerid *C. crambe*, separated by distances from 20 to 3000 km, were analyzed. As mentioned, low variability of this gene was found (only two haplotypes). Nevertheless, the different frequencies of these haplotypes revealed genetic structure and low gene flow between populations separated by tens of kilometres.

The phylogeographic study of the purportedly circum-Pacific species *Astrosclera willeyana* across the Indo-Pacific using the Folmer partition of *COI* (Wörheide, 2006) is a paradigmatic example of the low variability of the sponge mtDNA. Only three *COI* haplotypes with a maximum *p*-distance of 0.42% were identified across the Indo-Pacific populations spanning more than 20,000 km. The haplotype distribution, however, was uneven, as all Pacific individuals had one of the haplotypes with the exception of a single population featuring a second haplotype. The Red Sea population consisted of individuals with the third haplotype found.

Whalan *et al.* (2008) used the Folmer partition of *COI* for analyzing the population structure of the species *Rhopaloeides odorabile* Thompson, Murphy, Bergquist, and Evans in the central GBR. Sampling distances ranged between 100 m and 140 km. Moreover, they analyzed the same samples with three polymorphic allozyme loci and compared the results with both markers. Populations did not show structure for any of the two markers and no evidence for genetic differentiation between inner- and mid-reef sites was revealed. Nuclear and mtDNA markers indicate large-scale genetic admixture in this species, although there was some evidence for small, localized, genetic differences between some populations, which the authors attributed to reef-specific hydrodynamics.

López-Legentil and Pawlik (2009) compared the two above-mentioned partitions of *COI* in seven populations of *Xestospongia muta* from Florida, the Bahamas and Belize and found higher nucleotide diversity in the I3-M11 partition than in the 5′-end partition. Pairwise tests of genetic differentiation among geographic locations based on Fst values showed significant genetic differentiation between most populations, but this genetic differentiation did not follow the isolation-by-distance model. These authors explained the differentiation found by the patterns of ocean currents, although they did not discard that the limited dispersal of larvae contributed to the differentiation found. The authors advised to consider local hydrological features in future plans for management and conservation of sponges in coral reefs.

The genetic population structure of the common branching sponge, *Callyspongia vaginalis*, along more than 450 km of the Florida reef system, from Palm Beach to the Dry Tortugas, was assessed by using sequences of the Folmer partition of the *COI* gene (DeBiasse *et al.*, 2010). No clear pattern of genetic differentiation was revealed. The strong structure of populations from most sampling locations was attributed to larval philopatry as in other sponge species. However, in a few cases, non-significant pairwise Fst values were found between relatively distant sampling sites. The genetic connectivity between populations far away from each other led the authors to suggest that some long distance larval dispersal may occur via ocean currents or larval transport within sponge fragments as reported for the Mediterranean *Scopalina lophyropoda* (Maldonado and Uriz, 1999).

Sequence variation in the I3-M11 partition of the mtDNA *COI* gene was analyzed in 10 populations of the Atlanto-Mediterranean demosponge *Phorbas fictitius* (Bowerbank) (Porifera: Poecilosclerida) at a regional scale comparing mainland (Iberian) and insular (Macaronesian) populations, and at a local scale focusing on different islands of the Azores archipelago (Xavier *et al.*, 2010b). Genetic differentiation based of Fst estimates was found among most populations at both scales revealing highly structured populations. This confirms the presumably low dispersal potential of this species and the geographical isolation of the studied populations. However, the authors found evidence of long distance dispersal events between some populations. Only two haplotypes were shared by mainland and insular localities. Phylogenetic and network analyses indicate a separation of insular (Macaronesian) and mainland (Iberian) populations. The phylogenetic analysis pointed to the Macaronesian Islands as the species origin area with posterior expansion to mainland locations via current-mediated dispersal of larvae or sponge fragments. This study adds to the growing evidence of structured populations in the marine realm and highlights the importance of the Macaronesian islands on the evolutionary history of the Northeast Atlantic marine biota (Xavier *et al.*, 2010b).

The levels of genetic divergence among populations of *Phorbas fictitius* using the I3-M11 partition (Xavier *et al.*, 2010b) were of the same order of magnitude than those of *Xestospongia muta* populations using the same *COI* partition (López-Legentil and Pawlik, 2009), and much higher than the values found in Folmer's *COI* partition in several species at similar and even larger spatial scales (Duran *et al.*, 2004a; Wörheide, 2006). These studies therefore support that this alternative partition of the *COI* gene is more suitable than the 5'-end partition to infer intraspecific patterns in sponges, as already shown for interspecies relationships by Erpenbeck *et al.* (2006a,b). However, the finding of suitable polymorphic gene partitions for sponge population studies is not resolved and new genes need to be explored. Recently, partial sequences of the *ATP* synthase 6 (*ATP6*) and the cytochrome oxidase 2 (*CO2*) genes and two spacers: one located between *ATP6* and *CO2* and the other between the *NADH* dehydrogenase subunit 5 *(nad5)* and the small subunit ribosomal RNA genes have been assayed simultaneously in taxonomical distant sponges (Rua *et al.*, 2011) with contrasting success for alpha-level systematics, phylogeography, and population genetics.

The use of microsatellites allowed detection of marked population structure of the species *Crambe crambe* at local and regional scales (Duran *et al.*, 2004c) with more accuracy than sequence data (Duran *et al.*, 2004a, 2004b). Eleven populations were analyzed at six loci in locations placed along the Atlanto-Mediterranean distribution range of the species. High levels of between-population structure were found, and a significant isolation-by-distance pattern was observed. A strong genetic structure was also found within sampled sites. Patterns of allelic distribution between populations suggest the possibility of a recent colonization of the Atlantic range from the Mediterranean Sea as already proposed by Duran *et al.* (2004b) using sequence data.

The genetic structure of the Mediterranean sponge *Scopalina lophyropoda* (Schmidt) was analyzed at several spatial scales (from tens of metres to thousands of kilometres) by using seven specific microsatellite loci (Blanquer and Uriz, 2010). The genetic diversity of *S. lophyropoda* was structured at the three spatial scales studied: within populations, between populations of a geographic region, and between isolated geographic regions, although some stochastic gene flow might occur among populations within a region. The genetic structure followed an isolation-by-distance pattern according to the Mantel test. However, several of the genetic descriptors gave unexpected results. Despite philopatric larval dispersal (Uriz *et al.*, 1998) and fission events in the species (Blanquer and Uriz, 2010), heterozygote excess was found in many populations, and the contribution of clonality to the population genetic make-up was minor. The heterozygote excess and the lack of inbreeding were envisaged to be the result of either sperm dispersal, a strong selection against mating between relatives to avoid inbreeding depression or a high longevity of genets combined with recruitment events by allopatric larvae.

The population genetics of two emblematic Mediterranean bath sponges have been studied recently by microsatellites. Seven populations along the western Mediterranean and the Portugal coasts of *Spongia lamella* were analyzed by using seven microsatellite loci (Noyer, 2010). Inbreeding was the main characteristic for all loci and populations, which was attributed to mating among relatives or to the existence of breeding subunits within populations. Although the results should be taken with care because of the high rate of null and unsized alleles and the low number of individuals in some populations, partitioning of the molecular variance (AMOVA) showed that genetic data were spatially structured with significant differences within populations and among populations of each region. Genetic structure was found in all the populations examined, which followed an isolation-by-distance model.

In contrast, a genetic study on the Mediterranean bath sponge *Spongia officinalis* (Dailianis *et al.*, 2011) reports a high genetic diversity in most populations despite the species' harvesting and the recurrent massive mortality episodes (Pérez *et al.*, 2000) that decimate its populations. Population genetic analysis along the species distribution range (from eastern Mediterranean to the Strait of Gibraltar) using eight microsatellite loci showed low levels of genetic structure, not correlated to geographic distance, inside geographic sectors (western and eastern Mediterranean). Anthropogenic and natural mechanisms were speculated to be involved in enhancing larval dispersal, resulting in an unusual connectivity among sponge populations at a regional scale. Specimens were also analyzed using the $5'$-end partition of *COI* to verify whether the several morphotypes of the species described (Vacelet, 1959) were indeed cryptic species. *COI* sequences indicated that only one species is present throughout the Mediterranean, except in the Gibraltar zone where another cryptic *Spongia* sp. could be present (Dailianis *et al.*, 2011).

3.3. Small-scale genetic structure

The few studies that assessed the genetic structure of sponges at small spatial scales (i.e. among populations separated by tens of metres or within populations at the scale of a few metres) have been performed using microsatellite markers. A small-scale study of the population structure of the sponge *Crambe crambe* from a single rocky wall (inter-individual distances from 0 to 7 m) was done using six microsatellite markers and autocorrelation analysis on mapped individuals (Calderón *et al.*, 2007). The results showed a strong genetic similarity of sponges separated by less than 100 cm. Even when the effect of clonality was removed from the analysis, the trend of genetic relatedness was significant within the first distance classes (30–40 cm). On the contrary, genetic similarities in sponges 2–7 m apart were within the same range as sponges from other walls of the same locality, or from other Mediterranean localities. Estimated mean dispersal distances per generation were ca. 35 cm, and neighbourhood sizes were estimated at ca. 33 sponges. This indicated that, although some or many of the larvae could disperse away from the population of origin, enough propagules settled in the close vicinity of their mother sponges so as to build a marked genetic structure at very small spatial scales. Interestingly, the results strongly pointed to the existence of some degree of self-fertilization in this population

The sponge larval philopatry reported from behavioural studies (e.g. Uriz *et al.*, 1998) seems to be reflected in the inability of larvae to overcome subtle barriers such as unidirectional currents or small submarine walls (e.g. Blanquer *et al.*, 2009; Guardiola *et al.*, 2011). The Mediterranean sponge *Scopalina lophyropoda* is a clear instance of strongly restricted gene flow even among populations separated by tens of metres. Blanquer *et al.* (2009) mapped and characterized genetically all the individuals of three populations placed on three vertical walls separated ca. 100 m each and analyzed the contribution of sexual and asexual reproduction, and the breeding and mating system, to the spatial genetic structure (SGS) using seven microsatellites. SGS was analyzed at increasing distances by autocorrelation analysis. Significant autocorrelation and thus SGS was found at the smaller distances analyzed (from 1 to 10 m), which underpins the larval philopatry reported in behavioural studies (Uriz *et al.*, 1998). All these patterns, however, contrasted with the lack of inbreeding detected in the populations, which is in agreement with data on other marine modular invertebrates (McFadden, 1997; Ayre and Miller, 2006) and confirms that strong SGS does not necessarily imply inbreeding.

Microsatellites resulted informative markers to establish the genetic structure of very close subpopulations of the allochthonous calcareous sponge *Paraleucilla magna* at the Blanes littoral (NE of the Iberian Peninsula). This is the only genetic study so far of an introduced sponge species, which has colonized the Mediterranean from the Atlantic in recent years. Low but statistically significant genetic differentiation was found among the three subpopulations of *P. magna* established the first study year despite they were placed less than 100 m apart, and the short time after their establishment (Guardiola *et al.*, 2011). Several estimators of genetic differentiation allowed to confirm the differentiation among populations. F*st* values were similar to those reported for *Scopalina lophyropoda* in the same localities. Populations showed a heterozygote deficit, attributable to inbreeding, which is in agreement with the species' patchy distribution and an extreme philopatry of their larvae (Lanna *et al.*, 2007; Frotscher and Uriz, 2008). The authors speculated about a founder effect as the cause of this genetic pattern since these small populations were recently established in the study area (ca. 10 years ago).

3.4. Temporal genetic structure

The only study of temporal genetic differentiation across three consecutive years has been performed on the allochthonous calcareous sponge *Paraleucilla magna* (Guardiola *et al.*, 2011). The species is annual (Frotscher and

Uriz, 2008) so that the building of the yearly populations relied exclusively on recruits resulting from those larvae released from the previous cohort. The population genetic features allowed the authors to predict either high genetic variation in populations across time, if allopatric recruitment occurs, or low genetic differentiation across years if the yearly cohorts result from philopatric larvae. Low but statistically significant differentiation of the three populations occurred across years. These results also showed hetero-zygote deficit and allele instability in the populations over the 3 years, which are consistent with a recent establishment of these populations in the study area. However, one population disappeared in the second study year, likely as a result of recruitment failure, and the other two remained slightly differentiated, while the three populations were in place again in the third study year, and showed significant F*st* values. Overall, the population descriptors pointed to this species as a good opportunistic colonizer but highly sensitive to stochastic events affecting recruitment and thus with low–medium predicted impact on native communities.

3.5. Conclusions

The studies considered in this review highlighted some shortcomings of gene sequence data used, thus far, for the assessment of sponge population genetics. Mitochondrial genes (in particular, the Folmer partition of the *COI* gene) are too conserved for reliably detecting intraspecies genetic patterns, and even the I3-M11 partition of this same gene is not variable enough for detecting microevolutionary processes of sponge populations. On the other hand, *ITS* sequences have moderate intraspecies variability but can show intra-genomic variation that must be investigated before these markers can be used. New sequence markers for sponges, such as the *ATPSβ-iII* intron, proved useful for revealing differentiation at large geographical distances and deserve further attention. Microsatellites have now replaced the use of allozymes, but there are only a few studies using microsatellites in sponges as yet. Nevertheless, these studies proved that microsatellite markers were highly polymorphic in sponges and are a good choice to analyze geographic and temporal patterns at a wide range of scales, including the intra-population level. However, the high variability associated with microsatellite loci means that sample sizes must be large (Ruzzante, 1998). Microsatellites in sponges also tend to consist of imperfect repeats and to show amplification failures or null alleles that complicate the interpretation of results. More and better microsatellite markers need to be developed, although the initial phase of screening and optimizing markers for each study species represents an important cost in terms of time and money.

Overall, our screening of works that formally analyzed population differentiation at different scales and with different markers (Table 5.2) shows an overwhelming majority of positive results. In general, as Boury-Esnault and Solé-Cava (2004) stated, we can conclude that sponge populations are genetically structured, which is in accordance with the short-lived type of larva they feature. Isolation by distance has not been substantiated in all studies that have analyzed it, though (Table 5.2). This points to the importance of sporadic phenomena such as episodes of long-range dispersal or to the relevance of hydrological features in shaping the distribution of sponge populations. It is also noticeable that, with intriguing exceptions (e.g. *Scopalina lophyropoda*), the use of microsatellite markers has always shown a deficit of heterozygotes in the populations, while allozymes tended to indicate that populations are in Hardy–Weinberg equilibrium (but see Whalan *et al.*, 2005). More research is necessary on more species to ascertain whether inbreeding is indeed a common feature of sponge populations or whether the results obtained are an artefact of the marker used (e.g. null alleles in microsatellites).

Despite the drawbacks associated with the selected marker or sampling procedures, the information gathered from studies of population genetics of key species is widely applicable in conservation issues. The genetic fitness of populations and their connectivity should become pivotal aspects in management and design of marine protected areas and should also be considered in regulation policies for exploited natural resources (Palumbi, 2003; Cognetti and Maltagliati, 2004; DeSalle and Amato, 2004; Bell, 2008). Selective protection or removal of target key species (often emblematic animals or plants) often produces ecosystem disequilibria (VanBlaricom and Estes, 1988). Thus, protection of whole habitats or, at least, of as many species as possible, is desirable. Consequently, studies of population genetics of "not-so-emblematic" species such as sponges and other invertebrates are increasingly taken into consideration in conservation policies.

4. CONTRIBUTION OF ASEXUAL PHENOMENA TO POPULATION STRUCTURE: CLONALITY AND CHIMERISM IN SPONGES

Sponges, as other benthic invertebrates, can combine sexual and asexual reproduction. The former implies the production of small dispersive propagules (larvae) that potentially allow colonization of new habitats. In contrast, asexual reproduction often results in the production of larger propagules with lower dispersal abilities. However, long dispersing propagules, such as floating buds, drifting fragments, and other external gemmules, can also be produced asexually. The ecological and evolutionary significance of these diverse reproductive mechanisms is far reaching. Asexual reproduction allows rapid colonization of an area or rapid recovery after disturbance, at the cost of genetic diversity. Sexual reproduction is often restricted temporally (e.g. seasonally) and is a slower process, but it allows new adaptive combinations of the genetic pool; long distance dispersal of larvae may allow colonization of new areas.

In other groups of invertebrates such as cnidarians, asexual reproduction generally dominates population structure locally, while sexual propagules ensure gene flow between populations (McFadden, 1997). However, this pattern cannot be extrapolated to sponges and it remains unclear under which circumstances one or another strategy should be favoured. Likely, the pattern may change with the particular biological characteristics of the species and their habitats.

General ecological theory predicts that asexual reproduction may be adaptive in extreme and disturbed habitats, allowing the population to grow quickly while environmental conditions are favourable and to repopulate quickly after disturbance, while sex will be favoured in biotically complex and undisturbed environments (Bell, 1982; Trivers, 1985). However, asexual propagation can also be a mechanism for maintaining or increasing the abundance of well-adapted genotypes in a relatively unchanging habitat (Maynard-Smith, 1978; Carvalho, 1994). This seems to be the case of Antarctic sponges, which exhibit a high production of asexual buds independently of their taxonomy (Sarà et al., 2002; Teixidó et al., 2006). On the other hand, sexual reproduction with its associated genetic recombination is advantageous in moderately changing environments where a substrate of genetic variability, upon which selection could act, is necessary (Williams, 1975; Sebens and Thorne, 1985; Charlesworth, 1993; Bürger, 1999; Bengtsson and Ceplitis, 2000). In benthic cnidarians, asexual reproduction has been related to stable habitats and little predation pressure (Ayre, 1984; Karlson, 1991). However, Coffroth and Lasker (1998) found in a gorgonian the highest amount of reproduction by fragmentation at intermediate levels of disturbance. In fact, sexual and asexual reproductive strategies can successfully coexist in time and space in the same sponge individuals (Sarà and Vacelet, 1973; Johnson, 1978; Bautista-Guerrero et al., 2010; Leong and Pawlik, 2010), or occur in different seasons (Ereskovsky and Tokina, 2007), thus reflecting that these two strategies are evolutionarily stable in a range of environmental situations.

4.1. Sponge asexual propagules

Asexual propagules have been reported in representatives of all classes of Porifera (Ereskovsky and Tokina, 2007). They are diverse in shape, location within the sponge, and function (Fig. 5.2) and in most cases contribute to increase the clonality of sponge populations. Ecologically, they can be divided into dispersive (those that can travel to some distance) and non-dispersive propagules (basically recruiting in the same population that

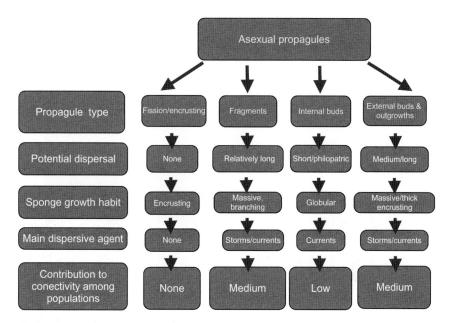

Figure 5.2 Schematic representation of the potential contribution of the main types of asexual propagules to the connectivity among sponge populations. (For colour version of this figure, the reader is referred to the Web version of this chapter.)

originates them). Among the former, external buds, commonly produced as projections at the sponge surface, are the most widespread (Merejkowsky, 1879). The calcareous *Sycon* spp. (Connes, 1964) and *Clathrina blanca* (Miklucho-Maclay) (Johnson, 1978), the hadromerids *Tethya* spp. (Connes, 1967; Gaino *et al.*, 2006) and *Polymastia penicillus* (Montagu) (Plotkin and Ereskovsky, 1997), the axinellid *Axinella damicornis* (Esper) (Boury-Esnault, 1970), the poecilosclerid *Mycale contarenii* (Martens) (De Vos, 1965; Corriero *et al.*, 1998), the astrophorid *Thenea muricata* (Bowerbank) (Uriz, 1983), and the hexactinellid *Lophocalyx* spp. (Schulze, 1887) all are examples that often show external small, roundish buds at the end of spicule bundles. In other cases (e.g. *Aplysina* Nardo, *Anoxycalyx* Kirkparick, and Rosella spp.), the propagules consist of irregular outgrowths of large size (from millimetres to centimetres) whose stalks progressively constrict off, the propagule finally breaking free, moving to some distance, and attaching to the substrate as new individuals (Vacelet, 1959; Sánchez, 1984; Teixidó *et al.*, 2006). In stable environments, these abundant propagula can result in sponge stands dominated by the asexually reproduced species (Fig. 5.3B).

A particular process is found in sponges without rigid skeletons but with high amounts of collagen such as some chondrosids (e.g. *Chondrosia reniformis*, Bonasoro *et al.*, 2001; *Chondrilla "nucula"* complex, Zilberberg *et al.*, 2006a)

Figure 5.3 Asexual reproduction in Antarctic sponges with patchy distributions. (A) External budding of a *Rossella* (Hecactinellida) species. (B) *Rossella* bed. (C) Internal gemmules in a Tetillidae. (D) Tetillidae bed. Photgraphs by Thomas Lundälv, UGOT. (For colour version of this figure, the reader is referred to the Web version of this chapter.)

and homosclerophorids (e.g. *Oscarella lobularis*, Sarà and Vacelet, 1973) which, when placed in overhangs, produce outgrowths that elongate towards the floor, resulting in teardrop-shaped fragments tethered by thin strands of tissue that finally break off, the fragment attaching to the underlying substrate.

External bubble-like buds have been recently reported for the homosclerophorids *Oscarella lobularis* and *O. tuberculata* Schmidt (Ereskovsky and *Tokina*, 2007). These buds float in the water column and can develop very fast into functional sponges, so they could disperse farther than denser external buds that will tend to sink faster. Bubble-like buds are also seasonally found in *Prosuberites epiphytum* Lamarck from the Mediterranean Sea (authors' personal observation) and might have gone unnoticed in other sponge species.

Internal buds (gemmules) were first described in freshwater sponges (Brien, 1967) but have also been reported for marine demosponges. They have been studied in detail in the hadromerid *Stylocordyla borealis* (Loven) (= *S. chupachups* Uriz, Gili, Orejas, and Pérez-Porro) (Sarà *et al.*, 2002), but they are particularly characteristic of the spirophorid genera *Craniella* Schmidt, *Cinachyra* Sollas, *Cinachyrella* Wilson, and *Tetilla* Schmidt (e.g. Chen *et al.*, 1997). These gemmules (Fig. 5.3C) are small and appear to be released through the sponge canals (e.g. Sarà *et al.*, 2002). Their dispersal capabilities are expected to be, in general, lower than those of sexually produced larvae. Moreover, sponge buds have been rarely found in the water column so that their contribution to the species dispersal can only be assumed except where genetic studies of the species populations are performed.

Among asexual processes that *a priori* do not result in dispersal, individual fissions are frequent in encrusting species (Fig. 5.1). In contrast, casting off fragments in massive and branching sponges, in particular under rough sea conditions (Wulff, 1985, 1986, 1991, 1995; Battershill and Bergquist, 1990; Tsurumi and Reiswig, 1997), may contribute to a certain extent to the species dispersal. These easy-to-fragment species can exploit this condition, for example, to survive after predator attacks or physical disturbances (Leong and Pawlik, 2010). Spontaneous fission is common in thinly encrusting sponges (Ayling, 1983) and has been analyzed in several species, such as *Crambe crambe* (Turon *et al.*, 1998; Garrabou and Zabala, 2001; Teixidó *et al.*, 2009), *Hemimycale columella* (Bowerbank) (Garrabou and Zabala, 2001), *Halichondria okadai* (Kadota) (Tanaka, 2002), *Scopalina lophyropoda* (Blanquer *et al.*, 2008), *Hymedesmia* spp., and *Microciona* spp. (authors' unpublished research).

The internal gemmules attached to the substrate of some hadromerids such as *Suberites domuncula* and *S. ficus* (Johnston) (Connes, 1977) do not contribute either to the species dispersal. These gemmules are covered by spongin and only develop in direct contact with the water once the adult sponge dies. The internal gemmules of the excavating clionaid genera *Aka* de Laubenfels, *Pione* Gray (Rosell and Uriz, 2002; Schönberg, 2002), *Cliona* Grant (Rützler, 1974; Rosell and Uriz, 2002), and *Thoosa* Hancock (Volz, 1939; Bautista-Guerrero *et al.*, 2010) are reinforced by collagen and spicules and seem to play a primary role similar to that of *Suberites* gemmules, that is, to ensure the genotype persistence. These gemmules are comparable to the statoblasts of the freshwater Spongillids, which only develop once the producer sponge disappears (Brien, 1967; Saller, 1990).

4.2. Clonality in sponge populations

Field monitoring studies have assessed the importance of asexual phenomena (fission, fragmentation, budding) in the population dynamics and life history of sponges (e.g. Ayling, 1983; Wulff, 1985, 1986, 1991; Pansini and Pronzato, 1990; Turon *et al.*, 1998; Teixidó *et al.*, 2006; Blanquer *et al.*, 2008; Leong and Pawlik, 2010). However, they usually cover a restricted time frame and may lead to wrong interpretations of the relative importance of sexual and asexual reproduction on the long term (see below). Determining the extent of clonality is necessary not only because of its biological implications but also because clonality causes biases in population genetic parameters (genetic diversity, linkage disequilibrium, inbreeding coefficients, effective population sizes). While some works used histocompatibility techniques to define clones in sponges (Neigel and Avise, 1983;

Neigel and Schmahl, 1984), recent studies have tackled the issue using genetic tools. In some cases, data on clonality are obtained in population genetic studies (e.g. Davis *et al.*, 1996; Miller *et al.*, 2001; Duran *et al.*, 2004c; Whalan *et al.*, 2005), but a few works addressed specifically the analysis of clonality at low spatial scales (Zilberberg *et al.*, 2006a,b; Calderón *et al.*, 2007; Blanquer *et al.*, 2009).

For the genotyping of clones, hypervariable markers are needed. In this sense, allozymes and microsatellites have been the markers of choice in sponges. When identical multilocus genotypes (MLGs) are found, the probability that they derive from sexual reproduction can be calculated (e.g. Stenberg *et al.*, 2003) and, if negligibly small, the groups of shared MLG are considered the result of asexual processes. Care should be taken, however, in the presence of inbreeding, as the probability of finding identical MLG by chance as a result of sexual reproduction increases significantly with respect to a randomly mating population with the same allele frequencies (Duran *et al.*, 2004b). The contribution of asexual reproduction (C_{asex}) can be measured as the complement (in percentage) of the ratio of different genotypes in a population, $(1 - (N_c/N)) \times 100$, where N_c is the number of different MLGs, and N is the total number of sponges analyzed (adapted from Ellstrand and Roose, 1987). C_{asex} represents the maximal possible contribution of asexuality, as differences at loci not genotyped and repeated generation of the same MLG through sexual reproduction (even if unlikely) would render the true clonality smaller than the one estimated by C_{asex}.

Davis *et al.* (1996) analyzed the genetic composition of sponges (*Halisarca laxus* (Lendenfeld)) epibiont on clump-forming solitary ascidians (*Pyura spinifera* (Quoy and Gaimard)) using six allozymes. They found that, in most cases, a single sponge clone is found on each ascidian clump, implicating an asexual colonization mechanism. Only 63 different genotypes were found among 172 sponge samples ($C_{asex} = 63\%$).

The extent of clonality in sponge populations can vary widely among closely related species. Miller *et al.* (2001), in an allozyme study of several morphs belonging to the genus *Latrunculia* du Bocage in New Zealand, found extensive cryptic speciation (with up to eight different species) and evidence for asexual reproduction in each of these eight species, although the contribution of clonality to the population structure was variable among species (C_{asex} indices from 18% to 63%). Identical MLGs (nine allozymes) were found only within sites, mostly comprising groups of two to four individuals, but up to 20 sponges shared a single MLG in Wellington.

Duran *et al.* (2004c) studied 11 populations of the poecilosclerid *Crambe crambe* in its Atlanto-Mediterranean range using 6 microsatellite loci. The work was primarily addressed at studying interpopulation structure but, in spite of a sampling planned to avoid clonemates (individuals collected at least 5 m apart), genotypically identical individuals for the six loci were found in three of the populations. A total of 13 groups of MLG, each comprising 2 or 3 sponges, were found. In most cases, the pattern found was highly unlikely to be the result of sexual reproduction even taking into account the heterozygote deficit in the calculation of clonal probabilities. The authors suggested that fragmentation with subsequent dispersal of fragments, rather than simple fission, could explain the finding of clones more than 5 m apart.

The extent of asexual reproduction was highly variable in four species of the genus *Chondrilla* Schmidt, between species and among populations of the same species, in a study along the Caribbean and Brazilian shores using allozyme markers (Zilberberg *et al.*, 2006a). The C_{asex} index ranged from 27% to 52% in three of the species, while in the fourth, it was 7% in one population and 39% in another, which was interpreted as reflecting differences in stability of the respective habitats. Most clonemates were found at short distances, indicating fission as the likely mechanism of clone production, and that asexuality has a role in filling in the available habitat rather than in dispersal. In addition, some individuals in overhangs shared MLG with those just below, indicating a role of the production of teardrop fragments in the clonal structure of the species. In another study with allozyme markers, Zilberberg *et al.* (2006b) found that 52 coral-excavating *Cliona delitrix* Pang sponges from 13 coral heads separated from a few to several hundred metres had all unique genotypes ($C_{asex}=0$). The starting hypothesis was that coral breakage and associated sponge fragmentation would be a mechanism for the spread of this bioeroding species. However, the results indicated that population maintenance is due to sexual reproduction alone in the area surveyed.

Calderón *et al.* (2007) genotyped 6 microsatellite loci in 177 specimens of the sponge *Crambe crambe* located in a single wall (all inter-individual distances <7 m) to ascertain small-scale genetic structure and clonality in this species. Given previous knowledge on the discrete fission rates (Turon *et al.*, 1998) and the larval behaviour of this species (Uriz *et al.*, 1998; Mariani *et al.*, 2006), it was expected that clonality would have a minor role in the establishment of this population and that little genetic structure would be found at the scale of a few metres. The results were not coherent with expectations, 76 of the individuals had non-unique MLG, forming 24 clones of size 2–8 sponges, and this pattern could not be explained by sexual reproduction. The contribution of asexual reproduction (C_{asex}) was, therefore, ca. 30%. The inter-clone distances were less than 1 m, with a mean of ca. 20 cm. In addition, strong genetic similarity of sponges located less than 50 cm apart was found even after the effect of clonality was removed, indicating that both sexual and asexual processes acted at small scales and contributed to the establishment of the species.

In another microsatellite-based study, Blanquer *et al.* (2009) analyzed small-scale genetic structure in *S. lophyropoda* in the NW Mediterranean. In this case, field studies have revealed a dynamic system of fissions and fusions (Blanquer *et al.*, 2008) and, consequently, an important contribution of clonality to the make-up of the populations was expected. However, the analysis of seven microsatellites in three populations revealed that the extent of clonality was minor ($C_{asex} = 11\%$), and only two clones of four and three individuals, respectively, were proven to be the result of asexual phenomena. Nevertheless, strong genetic structure, attributable to larval philopatry, was found at small scales (< 5 m). The lack of clonality detected can be in part explained by the finding (Blanquer and Uriz, 2011) that this sponge had a prevalence of chimeric individuals (see below). Thus, if there are different MLGs in a single individual, the assessment of fission events using genetic markers becomes unreliable.

No clones were reported to influence the genetic structure of the populations of several species belonging to the taxonomically distant taxa Calcaronea (within the class Calcarea) and Haplosclerida and Dictyoceratida (within Demospongiae). In a study of *Haliclona* sp. at spatial scales ranging from metres to hundreds of kilometres, Whalan *et al.* (2005) found no evidence that asexual reproduction is important for the maintenance of populations. Although limited to only two polymorphic allozyme markers, the levels of genetic diversity were high, and the number of identical MLGs found was coherent with chance expectation in sexually reproducing populations given the allele frequencies found. In the Mediterranean populations of the introduced calcareous sponge *Paraleucilla magna*, clonality was found to be almost inexistent ($C_{asex} = 1.5\%$, Guardiola *et al.*, 2011). No clonality was reported in the genetic studies of the Mediterranean bath sponges *Spongia officinalis* (Dailianis *et al.*, 2011) and *S. lamella* ($= S.$ *agaricina*) (Noyer, 2010), which is accordance with the lack of asexual buds and the unbreakable (elastic) consistence of the two species, which makes fragmentation unlikely.

In conclusion, from the scarce number of works that applied genetic tools to the characterization of clonality in sponges, a wide variability in the contribution of asexual reproduction to the structure of populations can be concluded. A second lesson that can be gleaned from these studies is that, in many cases, predictions from field studies and conventional population dynamics were not supported by genetic data (Zilberberg *et al.*, 2006b; Calderón *et al.*, 2007; Blanquer *et al.*, 2009). Field studies are, in general, restricted temporally and miss the historical perspective, which is particularly important in long-lived, slow-growing species whose population structure is built up over the years. In this sense, molecular techniques allow an analysis of clonality that keeps the history of past events, providing an assessment of the long-term contribution of sexual and asexual processes to the demography of the populations under study, rather than a temporally restricted snapshot (Calderón *et al.*, 2007). More studies on species with different morphologies and under different environmental conditions are called for to reveal the full extent of the importance of clonality in sponges.

4.3. Sponge chimerism

The existence of fusion processes between sponges (e.g. Turon *et al.*, 1998; Blanquer *et al.*, 2008) raises the interesting possibility of chimera formation. Histocompatibility assays and natural observations (Neigel and Avise, 1983; Neigel and Schmahl, 1984) indicated that contact between genetically distinct (allogeneic) sponges results in a rejection reaction, while contact between clonally derived fragments (isogeneic) often leads to fusion. However, fusion between sibling larvae has been reported (e.g. Van de Vyver, 1970; Maldonado, 1998), but this capability to fuse between allogeneics is reported to be lost during ontogeny (Ilan and Loya, 1990; Maldonado, 1998; Gauthier and Degnan, 2008). The genetic mechanism responsible for allorecognition is not known at present in sponges and is an important field for future research.

The biological and ecological implications of chimerism have been widely discussed (e.g. Grosberg, 1988a; Pineda-Krch and Lehtila, 2004; Rinkevich, 2005). Among the selective advantages, escape in size from predation and exploitation of joint genomic fitness have been proposed. On the other hand, somatic and germ cell parasitism can occur, sometimes leading to resorption of one of the partners. While these phenomena have been intensely studied in other invertebrate groups such as ascidians (reviewed in Rinkevich, 2005), much less is known in sponges, although takeover of the germ cell population of chimeric sponges by one of the fusion partners has been proposed (Gauthier and Degnan, 2008).

Only two studies have addressed the study of chimerism using genetic tools. In a grafting experiment with *Niphates erecta* Duchanssaing and Michelotti, Neigel and Avise (1985) compared the responses (either fusion or non-fusion) with genotypic composition at three allozyme loci. Although compatibility and genetic results were largely concordant, in the sense that most acceptances occurred in individuals with identical genotypes (and the converse was true for non-fusions), a few grafts between sponges with different allozyme genotypes resulted in allogeneic fusion. In the only genetic study of chimerism in a natural population of a sponge, Blanquer and Uriz (2011) using seven microsatellite loci found a high incidence of multichimerism in *Scopalina lophyropoda*, with a total of 36 MLGs in 13 sponges sampled at four points of their bodies. In some cases, each intra-individual sample had a different MLG. Interestingly, larvae produced by the colonies had genotypes compatible with the transmission of different MLG, so no germ line dominance seemed to occur.

4.4. Conclusions

The few studies on sponge clonality using molecular tools indicated that the contribution of asexual reproduction to the genetic structure of the sponge populations is lower than previously assumed given the many instances of asexual propagules reported for these animals. In particular, no true clones were found in genetic studies of massive sponges such as *Spongia officinalis*, *S. lamella*, and *Paraleucilla magna* or in the excavating *Cliona delitrix*. Populations of encrusting species would seem more prone to harbour clones resulting from fission, because the released fragments remain attached to the substrate and thus they do not leave the population of origin (e.g. *Crambe crambe*). However, this cannot be generalized to all the encrusting species, since an example exists of a thinly encrusting, highly fissionable species with minimal numbers of clonemates in their populations (*Scopalina lophyropoda*). On the other hand, sponge clonality is expected to be higher in relative stable habitats such as deep bottoms or the Antarctic shelf, where sponge distribution use to be very patchy (Rice *et al.*, 1990) and some producers of asexual propagules dominate (Teixidó *et al.*, 2006).

All the instances so far studied address the role of clonality at small scales (within populations), and no evidence has been found for the potential role of widely dispersing asexual propagules (e.g. floating buds). These studies are of course more challenging methodologically, as the probability of encountering asexual propagules far for the source is very small. However, a convenient starting point could be the analysis with genetic markers of population structure of actively fragmenting sponges with potential for relatively long levels of dispersion, such as some branching forms (Wulff, 1985, 1991).

Natural occurrence of sponge chimerism has been but recently proved unambiguously in a sponge species (Blanquer and Uriz, 2011). Whether this characteristic is exclusive of *Scopalina lophyropoda* or it exists in other species remains to be explored. Sponge chimerism could help to understand the long-term success of relatively small (as for the number of individuals) sponge populations, which are highly structured and patchily distributed, by minimizing genetic drift. The study of chimerism in sponges is in its infancy, and no doubt more work using molecular markers will cast light on the prevalence and ecological relevance or this mechanism in natural populations.

5. DIFFERENTIAL EXPRESSION OF FUNCTIONAL GENES

One of the main topics in marine ecology is the study of the responses of individuals, populations and assemblages to environmental changes. These changes exert selective pressures at ecological time scales. The outcome will depend on the species ability to adjust its physiology and biology to new scenarios.

The identification of cellular changes associated with the acclimatization of species to particular habitats and environmental changes can be assessed by analyzing differential expression of selected genes and can help in predicting the organisms' ability to survive perturbations. Differential gene expression in individuals under contrasting environmental conditions or experimental treatments has been analyzed through protein quantification, Northern blot, real-time qPCR, microarrays, or, recently, massive sequencing tools. In sponges, nevertheless, only the first three approaches have been used so far. As in other organisms, gene expression studies allow detecting potential adaptive traits in sponge individuals living in contrasting habitats or under adverse environmental conditions. Moreover, these studies have also contributed to the identification of particular genes involved in basic biological functions such as biomineralization, mating, or development, which have decisive implications in the ecology of the species.

Studies of gene and protein expression in sponges have proliferated since the 1990s (Fig. 5.4). The identification of functional genes from clone libraries was the first approach to discover new genes in sponges (e.g. Urgarković et al., 1991; Schröder et al., 2000). Subsequently, the expression of the identified genes was analyzed under experimental conditions, mainly in laboratory settings (e.g. Krasko et al., 1999; 2000; Schröder et al., 2000), and only rarely in the field. The biological function of the expressed genes was speculated based on sequence similarity with genes of known function in model organisms.

5.1. Genetic and environmental regulators of sponge biomineralization

Biomineralization refers to the biologically controlled process of precipitation of mineral salts. The skeletal elements of animals are the immediate result of biomineralization. The process involves expression of genes that code for protein/enzymes facilitating mineral precipitation and skeleton shaping and greatly depends on environmental factors such as silicon or calcium concentration, temperature, and water alkalinity. Sponge skeletons can be siliceous (in the form of hydrated silica) and calcareous (crystallized as calcite or aragonite). However, one of the most pursued aim in the studies of sponge gene expression has been to find out the mechanisms involved in the formation of sponge siliceous skeletons: the proteins/genes responsible, the associated genes, the environmental conditions inducing gene expression, and the localization of the process, whether intra-, extracellular, or both (e.g. Uriz et al., 2000; Schröder et al., 2005b; Müller et al., 2008a,b). Sponge mineral skeletons play important biological and ecological roles in both siliceous and calcareous sponges (Uriz et al., 2003; Uriz, 2006). They allow sponges to organize their aquiferous system in a three-dimensional plan, thus enhancing sponge accessibility to the resources of the water column and diminishing the negative potential effects of sedimentation. It has been speculated that the calcareous skeletons of sponges and other

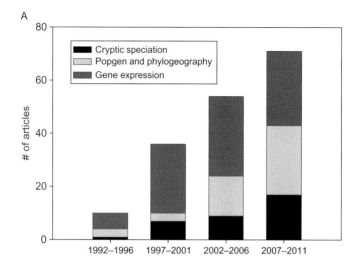

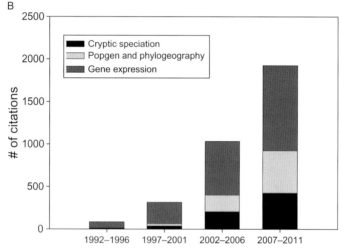

Figure 5.4 Bibliometric data of the main fields of sponge research covered in this review during the past 20 years (in 5-year intervals). Data were obtained from searches in the Science Citation Index Expanded through the Web of Science (Thomson Reuters©). Search criteria (in the "Topic" field) were as follows: For cryptic speciation: "(sponge or porifera) and ("cryptic speci★")"; for phylogeography and population genetics: "(sponge or porifera) and (phylogeog★ or "population genetic★")"; for gene expression: "(sponge or porifera) and ("gene expression")". Searches were refined manually to meet the criteria of this review (f.i., non-molecular-based studies of cryptic speciation were filtered out). (A) Number of publications since 1990. (B) Number of citations since 1990.

organisms might have arisen to detoxify cells from an excess of calcium ions incompatible with life (Lowenstam and Margulis, 1980). Thus, the current sponge skeletons, which play decisive structural and defensive roles in present-day sponges, may be an example of an exaptation mechanism (i.e. fixation of a character that had one function in an ancestral form and a new function in a descendant form) (Uriz, 2006). The genetic basis and the environmental control of the sponge biomineralization have received a great deal of attention in the past 20 years, particularly in siliceous sponges.

Shimizu *et al.* (1998) first elucidated the structure of proteins involved in silicification of the sponge *Tethya aurantium* (Pallas) (silicateins). The cDNA sequence of the most abundant silicatein (alpha) revealed that these proteins were highly similar to members of the cathepsin L and papain family of proteases. Later, Cha *et al.* (1999) experimentally demonstrated the enzymatic activity of sponge silicateins and their involvement in directing silica polymerization *in vitro*. Subsequent research, however, demonstrated that the process is far for simple and that external factors influenced considerably the process.

Schröder *et al.* (2005b) studied spicule formation in cell aggregates (primmorphs) of *Suberites domuncula*. These authors showed that spicule formation initiates intracellularly independently of the Si concentration in the water, but silicatein expression strongly increased in sponge primmorphs after cultivation at high concentrations of silicate (60 μmol/L). This agreed with the higher spicule thickness and strength of sponges living in silicate-rich environments such as bathyal depths, upwelling areas, Antarctic waters, and particular silicon-rich habitats from the Pacific and North Atlantic (e.g. Maldonado *et al.*, 1999). They also reported that a cluster of three genes, which contained the silicatein-α gene, underwent coordinated expression after treatment with silicic acid. Since the expression of these genes occurred synchronously, it was suggested that the three were involved in the formation of spicules (Schröder *et al.*, 2005b). Although the first studies on sponge silicification focused on representatives of the Class Demospongiae, a silicatein-related protein has also been found in the spicules of the hexactinellids *Monorhaphis chuni* Schulze (Müller *et al.*, 2008a) and *Crateromorpha meyeri* Gray (Müller *et al.*, 2008b). Thus, in both Demospongiae and Hexactinellida, cathepsin-related silicatein enzymes appear to form part of the siliceous matrix of spicules.

Mn-sulphate in the water induced formation of weak spicules with irregular porous surfaces in *Suberites domuncula* primmorphs by downregulating aquaporin-8 gene expression (Müller *et al.*, 2011). Aquaporin has been reported to be involved in dehydration and hardening of bio-silica following the bio-silica polycondensation reaction.

Other mineral elements in the ambient water appear to induce the expression of genes related with structural proteins. Both silicatein and collagen, which appear to be co-involved in spicule formation, were up-regulated after silicic acid exposure (Krasko *et al.*, 2000) or ferritin exposure (Krasko *et al.*, 2002). Conversely, myotrophin seems to increase collagen gene expression but not silicatein expression (Krasko *et al.*, 2000).

5.2. Pollutants, stress, and gene expression

One of the most extensively dealt issues related to protein/gene expression in sponges is the assessment the potential effects of pollution. Differential gene expression in sponges submitted to several kinds of pollutants and stressors has been studied trying to detect early effects of noxious substances on sponge populations before they become lethal. Several studies analyzed the up- or downregulation of the expression (mRNA levels) of functional genes under various concentrations of metal and PCB pollutants, while others reported on protein expression (e.g. Wiens *et al.*, 1998; Agell *et al.*, 2001; Cebrian *et al.*, 2006; López-Legentil *et al.*, 2008) or, more rarely, protein activity (Saby *et al.*, 2009).

Regeneration after partial mortality is at the basis of the sponge ecological success. Urgarković *et al.* (1991) determined, by densitometric scanning of the autoradiograms from dot-blot analyses, that the expression of *RAS* gene involved in regeneration in *Geodia cydonium* (Jameson) was strongly induced by genotoxic xenobiotics. Reaggregation of dissociated *G. cydoniun* cells was induced by an aggregation factor that caused expression of *RAS* gene in the cells, and this expression was inhibited by detergents at concentrations within a pollution–realistic range.

Polychlorinated biphenyl PCB 118 pollutants induced the expression of two chaperone proteins in the marine sponge *Geodia cydoniun* (Wiens *et al.*, 1998): the 14-3-3 protein and the heat shock protein HSP70 (primarily involved in folding of proteins). Using the cDNAs coding for these two proteins and Northern blot electrophoresis, the authors demonstrated that none of the two chaperones were detected in the absence of PCB. In contrast, after incubation of sponge tissue with PCB 118 during 12 h, the transcripts of the two chaperones were detectable, and the corresponding proteins appeared after 24 h. Due to the broad cross-reactivity of their antibodies throughout the Metazoa, these chaperones were proposed as useful biomarkers for monitoring environmental PCBs. The expression of both HSP proteins and corresponding genes have been widely assessed in sponges thanks to the extremely conserved sequences of these genes, which allowed the use of universal primers for cross-amplification in sponges (Müller *et al.*, 1995; Koziol *et al.*, 1996; Koziol *et al.*, 1997; Krasko *et al.*, 1997; Wiens *et al.*, 1998). The expression of stress proteins was also quantified in natural populations of the sponge *Crambe crambe*, under both polluted and non-polluted conditions, both at the laboratory and in a field experiment, which demonstrated than HSP70 expression was induced under metal pollution, although only semi-quantitative methods (Western blot) were used for protein determination (Agell *et al.*, 2001). In contrast, no differential HSP protein expression was found, either in the laboratory or in a transplant experiment, in *Chondrosia reniformis*, under high copper concentrations (Cebrian *et al.*, 2006) compared to low copper controls.

Episodes of mass diseases and mortalities have been often reported for sponges inhabiting both tropical and temperate Seas (Galstoff, 1942; Webster, 2007; Cebrian *et al.*, 2011). The giant barrel sponge *Xestospongia muta*, which is one of the largest and most conspicuous components of Caribbean coral reef communities, increased expression of the HSP70 gene (as measured with real-time qPCR) when it underwent fatal bleaching in the field and in response to thermal and salinity variation in the laboratory, while HSP expression remained constant during the cyclic sublethal bleaching (López-Legentil *et al.*, 2008).

Müller *et al.* (1995) and Bachinski *et al.* (1997) studied the effects of thermal stress on the sponges *Suberites domuncula* and *Ephydatia fluviatilis*, respectively, by analyzing either HSP protein or gene expression, which was up-regulated under the high temperature treatment. Krasko *et al.* (1999) reported significant increase of the intracellular Ca^{++} concentration and reduction of the starvation-induced apoptosis in sponge aggregates of the demosponge *S. domuncula* submitted to 5 mM ethylene. The expression of two genes, a *SDERR* encoding for a potential ethylene-responsive protein, termed *ERR_SUBDO*, and a Ca^{++}/calmodulin-dependent protein kinase, were up-regulated after exposure to ethylene. Database searches revealed that the sponge polypeptide shared high similarity (82% at the amino acid level) with the ethylene-inducible protein from plants. No other ethylene-responsive proteins had been identified in Metazoa previously. The expression of both genes in primmorphs of *S. domuncula* increased fivefold after a 3-day incubation period with ethylene. Ethylene is one major alkene in seawater resulting from oceanic dissolved organic carbon by photochemical reactions. The authors proposed that, because sponges are efficient benthic filter feeders, they are likely to take up large amounts of ethylene from the surrounding water, and this organic molecule could play a positive role in calcium homeostasis and apoptosis reduction of the sponge cells.

Krasko *et al.* (2001) identified sponge genes involved in cell protection in front of environmental threats. Besides the previously identified efficient defence systems such as chaperones and the P-170/multidrug resistance pump system (Müller *et al.*, 1996), they reported a further multidrug resistance pathway that is related to the pad one homologue (*POHL*) mechanism that had also been identified in humans. It is suggested that proteolysis is involved in the inactivation of xenobiotics by the *POHL* system. Two cDNAs were cloned, one from the demosponge *Geodia cydonium* and a second from the hexactinellid *Aphrocallistes vastus* Schulze. The two sponge cDNAs were highly similar to each other as well as to the known sequences from fungi and other Metazoa. Under controlled laboratory conditions, the expression of the potential multidrug resistance gene

POHL was strongly up-regulated in response to the natural toxins staurosporine or taxol in *G. cydonium*. The relevance of the expression pattern of the *G. cydonium* gene *POHL* for the assessment of pollution in the field was determined at differentially polluted environments.

Differential gene expression related with immune reactions in response to the bacterial endotoxine LSP in *Suberites domuncula* was analyzed in populations from a confined, polluted habitat (i.e. Limski Canal in Croatia: Mediterranean Sea) and from the open coast in the same area (Schröder *et al.*, 2005a). Most of the differentially expressed transcripts coded for the allograft inflammatory factor-1 (*AIF-1*), a molecule involved in self/non-self recognition in *S. domuncula*. The level of gene expression of the *AIF-1* gene was determined by Northern blotting and real-time qPCR and resulted in much higher values in the sponges living in the open Sea. These results pointed to an immuno-depression in sponges inhabiting contaminated areas. Other potential indicators of sponge immuno-depression, such as the inhibition of the 2-5′-oligoadenylate synthetase (OAS) in sponges exposed to metal pollution, have been reported (Saby *et al.*, 2009). Sponge OAS activity was quantified as the amount of reaction products (2-5A oligomers) after the incubation in a PCR machine of the sponge extracts with ATP, using a C18 reverse-phase column on HPLC. *In vitro*, activity of OAS from *Geodia cydonium* and *Crella elegans* (Schmidt) was inhibited to a variable extent by Cu, Mn, Zn, and Fe ions.

5.3. Symbioses and horizontal gene transfer

Symbiosis with cyanobacteria is widespread in sponges inhabiting shallow waters, in particular, in the Indo-Pacific ocean shelf (Wilkinson and Fay, 1979; Wilkinson, 1987; Lemloh *et al.*, 2009) but also in other tropical and temperate seas (Vacelet, 1975; Erwin and Thacker, 2008). Cyanobacteria influence sponge growth and competition abilities (Erwin and Thacker, 2008) and thus greatly account for the sponge ecological success (see Thacker and Freeman, 2012). Steindler *et al.* (2007) analyzed for the first time gene expression in relation to the presence of symbiotic cyanobacteria. The target species was the common Mediterranean sponge *Petrosia ficiformis* (Poiret) that can harbour or not cyanobacteria as a function of the amount of irradiance arriving to its habitat. After suppression substractive hybridization to identify uniquely expressed genes, and separation and cloning of the PCR products resulting from symbiotic and aposymbiotic individuals, the authors isolated two novel genes (named *PfSym1* and *PfSym2*). These genes were screened for differential expression in two fragments of the same individual (clones), which harboured or not the symbiotic cyanobacteria (after a medium term transplantation experiment), via Northern blotting. Both genes were up-regulated in sponges harbouring the cyanobacterial symbionts.

Despite the recent blooming literature on the identification of sponge-associated microbes by pyrosequencing techniques (e.g. Turque *et al.*, 2010;

Webster *et al.*, 2010; Lee *et al.*, 2011), the molecular basis of the sponge-microorganism symbioses and the exchange of metabolites between sponges and their associated microbes remain poorly understood. Gene expression studies may help understand the interactions between the symbiotic partners. Müller *et al.* (2004) reported that the demosponge *Suberites domuncula* expressed, under optimal aeration conditions, the enzyme tyrosinase, which synthesizes diphenols from monophenolic compounds. The authors assumed that these products serve as carbon source for symbiotic bacteria to grow.

Bacteria and fungi are common symbionts of sponges (Taylor *et al.*, 2007; Erwin *et al.*, 2012). Horizontal gene transfer from bacteria (Jackson *et al.*, 2011) and fungi (Rot *et al.*, 2006) to sponges has been suggested. Horizontal gene transfer is considered to be a major evolutionary force in eukaryotes (Kozo-Polyansky *et al.*, 2010). Jackson *et al.* (2011) have recently identified a new protein (spherulin) from the spherulites that form the calcareous skeleton of the sponge *Astrosclera willeyana* and speculated on a horizontally transferred gene from the symbiotic bacteria to the sponge. Rot *et al.* (2006) showed some lines of evidence demonstrating that introns can be found in the mitochondria of sponges (Porifera) based on the sequencing of a 2349 bp fragment of the mitochondrial *COI* gene from the sponge *Tetilla* sp. (Spirophorida). They report the presence of an intron with group I intron characteristics as in Cnidarians. The intron encodes a putative homing endonuclease. A phylogenetic analysis of the *COI* protein sequence and of the intron open reading frame suggests that the intron may have been transmitted horizontally from a fungus donor. The authors suggested that the horizontal gene transfer of a mitochondrial intron was facilitated by a symbiotic relationship between fungus and sponge. Once more, ecological (symbiotic) relationships seem to have implications at the genomic level. To better understand the mode of transmission of mitochondrial introns in sponges, Szitenberg *et al.* (2010) have studied *COI* intron distribution among representatives of Tetillidae. Out of 17 *COI* sequences of Tetillidae examined, only 6 were found to possess group I introns. These different forms of introns had distinct secondary structures. Since sponges harbouring the same intron type did not always form monophyletic groups, the authors suggested that sponge introns might have been transferred horizontally.

5.4. Expression of other sponge genes with ecological implications

Most sponges are long living organisms, which confers stability to the sponge-dominated assemblages. As a consequence of their permanence across years in the same habitat, many sponges are keystone species in structuring benthic assemblages, where they play a significant role in species interactions such as competition, facilitation, or symbiosis, among others (Sarà and Vacelet, 1973). The sponge longevity has been associated to the reported capacity of sponge cells to proliferate almost indefinitely due to the presence of high telomerase activity (Koziol *et al.*, 1998). To this regard, Schröder *et al.* (2000) identified genes in the marine sponge *Suberites domuncula*, which were differentially expressed in proliferating and non-proliferating cells. Moreover, they sequenced the complete cDNA corresponding to the transcript of the *SDLAGL* gene, which encoded a polypeptide (named putative longevity assurance-like polypeptide LAGL_SUBDO), which showed a high sequence similarity to the longevity assurance genes from other Metazoa. This gene was highly expressed in aggregated proliferating cells but not in isolated single cells.

5.5. Conclusions

The publication of the complete genome of *Amphimedon queenslandica* (Srivastava *et al.*, 2010) has accelerated the identification of new sponge genes, shedding light on the sponge development and phylogenetic relationships with other animals. These studies have mainly focused on the identification and expression of genes related to functional aspects of sponges, such as the origin of the animal body plans and the developmental patterning processes (e.g. Larroux *et al.*, 2006; Adamska *et al.*, 2007; Lapébie *et al.*, 2009). Enumeration of these works is out of the scope of this ecologically oriented review. However, the publication of a sponge genome has been undoubtedly a major breakthrough in sponge science. Further exploitation of this invaluable resource will likely lead, in the next years, to the characterization of many genes of ecological relevance. Moreover, protein expression studies should be conducted in parallel to gene expression studies to validate the true repercussion of the observed differential gene expression in the cell biology of the sponges.

Coupled with the generation of genomic information, the application of high-throughput sequencing technologies to gene expression quantification, the so-called RNA-seq techniques (Marioni *et al.*, 2008; Wang *et al.*, 2009), is expected to boost the number of studies on sponges in the forthcoming years. This technique is being employed to quantify differential expression of stress-related genes of the Mediterranean sponge *Crella elegans* submitted to temperature stress, as well as during its reproductive cycle and the resistance phases posterior to larval release (authors, current research).

6. FORTHCOMING TRENDS IN SPONGE MOLECULAR ECOLOGY: HOPES AND PROSPECTS

A simple literature perusal shows that the main topics covered in this review have blossomed in the past two decades, with a markedly increasing trend in number of publications and citations (Fig. 5.4). We expect that the ongoing trend of fruitful research in the interface between molecular systematics and molecular ecology will continue to progress significantly in the near future. In particular, we foresee that more and more instances of cryptic speciation will be uncovered and will require modification of existing taxonomic arrangements. This field will likely progress along two ways: deciphering the mechanisms of formation of sibling species (vicariance, isolation by distance, founder effects, etc.) and improving the formal incorporation of molecular (genomic, proteomic, and metabolomic) knowledge in species definition (Ivanišević et al., 2011; Cárdenas et al., 2012). The use of functional genes related to reproduction such as gamete recognition proteins (Swanson and Vacquier, 2002; Turner and Hoekstra, 2008) in parallel with neutral markers and metabolic fingerprinting will greatly aid in the first of these goals.

At the same time, we anticipate that new and better markers will be developed for sponge population genetic studies. Next-generation sequencing methods allow faster and more efficient development of microsatellites, as well as identification of new candidate genes for population level studies. These data can ease the incorporation of other kinds of markers requiring extensive genomic knowledge, such as single nucleotide polymorphisms (SNP; Kim and Misra, 2007). SNP have never been used in sponges but are nevertheless promising tools judging from their results in other groups (reviewed in Clark et al., 2010) and may outcompete microsatellites in the future (Morin et al., 2004).

According to the consistent results reported in previous studies, we can predict that the present trend of high population structure and little gene flow will continue to prove its generality in sponge populations. However, we hope to be able to adequately track the (occasional) long-distance dispersal events that have been suggested for some species. Better sampling designs and analyses with more informative markers, coupled with knowledge of regional current patterns and particle dispersal models, will undoubtedly contribute to this goal (Siegel et al., 2008).

Patterns of temporal genetic variation are increasingly being studied in benthic invertebrates, and they add another dimension to the picture of the distribution of genetic variability (e.g. Lee and Boulding, 2007; Calderón et al., 2009, 2012). These studies are more challenging in sponges, as they cannot be reliably aged in general. However, studies on annual species (Guardiola et al., 2011) or monitoring of yearly recruits in populations can shed light on the dynamic mechanisms that result in the spatial distribution of genetic variation that we observe.

We also expect to see more studies that analyze the relative role of sexual and asexual reproduction in the building up of the populations. The studies so far seem to indicate that the importance of asexual reproduction is less than foreseen, but the picture can change as more species and particular habitats are investigated. Furthermore, the studies should move from intra-population processes to the assessment of the importance of asexual propagules in the colonization process and the gene flow among populations. As for the other ecological fields mentioned, we await the development of highly variable markers, which is a prerequisite for this kind of studies (Arnaud-Haond *et al.*, 2005).

Hopefully, molecular studies will unearth whether chimerism, only found up to now in natural populations of *Scopalina lophyropoda* (Blanquer and Uriz, 2011), is widespread in other sponge species, which will raise intriguing questions as to its generation, maintenance, and ecological relevance. We also foresee that the genetic basis of self/non-self recognition (Grosberg, 1988b) in sponges will be fully described and the genes implicated will be incorporated to the field of sponge molecular ecology.

At the same time, the range of questions addressed will widen with the application of new techniques of gene expression analyses (Steindler *et al.*, 2007; Marioni *et al.*, 2008; Wang *et al.*, 2009). Differential gene expression will allow us to gain insights into the actual functioning and responses of individuals and populations rather than just assessing demographic parameters and connectivity. Further research likely will also focus on micro-RNA genes regulating gene expression. Only recently, the first paper reporting the application of interference RNA techniques for gene silencing in sponges has been published (Rivera *et al.*, 2011). These techniques open a new avenue of possibilities for reverse genetics studies on functional aspects of sponges.

Sponges are paradigmatic invertebrates as for their ancestral symbiosis with an array of prokaryotic communities (Taylor *et al.*, 2007). Horizontal gene transfer from their symbiotic prokaryotes, which has been recently reported in a couple of cases (Rot *et al.*, 2006; Jackson *et al.*, 2011), may prove to be prevalent in sponges, where studied with the appropriated molecular tools.

All together, we urge researchers to turn sponges into a data-rich group in terms of the molecular patterns and mechanisms underlying their ecological distribution and vulnerability (Carvalho *et al.*, 2010). If we succeed, sponges can make a pivotal group for studies on conservation and management of the marine coastal environment. Too often, this research focuses on some "emblematic" groups or "flagship" species, sometimes more "conspicuous" but less relevant ecologically than sponges. Because of their diversity, abundance and roles in the trophic networks, and their plentiful interactions with prokaryotes and eukaryotes, sponges must be in the limelight in all studies aiming at setting up conservation policies (Cowen *et al.*, 2007; Arrieta *et al.*, 2010).

We are at present in the brink of a revolution in molecular ecology (Tautz *et al.*, 2010), and sponge science will follow this lead in parallel with (and hopefully not behind) marine science in general. We can predict that studies will move from genetics (the use of one or a few genes) to genomics (extensive genome or transcriptome sequencing)(Wilson *et al.*, 2005; Bouck and Vision, 2007; Dupont *et al.*, 2007; Mittard-Runte *et al.*,). Conceptual and methodological developments made in model organisms will be extended to non-model species, and sponges should benefit from this trend. We are moving from a field that was rich in theory and poor in data, to a data-rich world, which should be matched with new theoretical and analytical tools (Hamilton, 2009). Proteomics and metabolomics will join in, and together they will transform many aspects of sponge research, allowing a more complete ecological approach involving the simultaneous study of many components of the interactions of organisms with the environment (Johnson and Browman, 2007). Thus, sponge ecologists will soon have the tools to answer ecological and biological questions that would have been impossible to address a few years ago (Clark *et al.*, 2010).

ACKNOWLEDGEMENTS

This study has been funded by projects BENTHOMICS ref. CTM2010-22218 from the Spanish Ministry of Science and Innovation, ECIMAR from ANR, France COCONET project ref. 287844, EU (FP7/2007-2013), and 2009-SGR655 and 2009-SGR484 from the Generalitat of Catalunya.

REFERENCES

Adamska, M., Degnan, S. M., Green, K. M., Adamski, M., Craigie, A., Larroux, C., and Degnan, B. M. (2007). Wnt and TGF-b expression in the sponge *Amphimedon queenslandica* and the origin of metazoan embryonic patterning. *PLoS One* **2**, e1031 doi:10.1371/journal.pone.0001031.

Agell, G., and Uriz, M. J. (2010). Molecular markers for sponge population studies in the genomic era. Are microsatellites still the best choice? In "Proceedings VIII World Sponge Conference." 128.

Agell, G., Uriz, M. J., Cebrian, E., and Martí, R. (2001). Does stress protein induction by copper modify natural toxicity in sponges? *Environmental Toxicology and Chemistry* **20**, 2588–2593.

Arnaud-Haond, S., Alberto, F., Teixeira, S., Procaccini, G., Serrao, E. A., and Duarte, C. M. (2005). Assessing genetic diversity in clonal organisms: Low diversity or low resolution? Combining power and cost efficiency in selecting markers. *The Journal of Heredity* **96**, 434–440.

Arrieta, J. M., Arnaud-Haond, S., and Duarte, C. M. (2010). What lies underneath: Conserving the oceans' genetic resources. *Proceedings of the National Academy of Sciences of the United States of America* **107**, 18318–18324.

Avise, J. C. (2000). Phylogeography. The History and Formation of Species. Harvard University Press, Cambridge.

Avise, J. C. (2004). Natural Markers, Natural History, and Evolution. (2nd edn.) Sinauer Associates, Sunderland.

Avise, J. C. (2009). Phylogeography: Retrospect and prospect. *Journal of Biogeography* **36**, 3–15.

Avise, J. C., Arnold, J., Ball, R. M., Bermingham, E., Lamb, T., Neigel, J. E., Reeb, C. A., and Saunders, N. C. (1987). Intraspecific phylogeography: The mitochondrial DNA bridge between population genetics and systematics. *Annual Review of Ecology and Systematics* **18**, 489–522.

Ayling, A. M. (1983). Growth and regeneration rates in thinly encrusting demospongiae from temperate waters. *The Biological Bulletin* **165,** 343–352.

Ayre, D. J. (1984). The effects of sexual and asexual reproduction on geographic variation in the sea anemone *Actinia tenebrosa. Oecologia (Berlin)* **62,** 222–229.

Ayre, D. J., and Miller, K. (2006). Random mating in the brooding coral *Acropora palifera. Marine Ecology Progress Series* **3007,** 155–160.

Bachinski, N., Koziol, C., Batel, R., Labura, Z., Schröder, H. C., and Müller, W. E. G. (1997). Immediate early response of the marine sponge *Suberites domuncula* to heat stress: Reduction of trehalose and glutathione concentrations and glutathione S-transferase activity. *Journal of Experimental Marine Biology and Ecology* **201,** 129–141.

Barbieri, M., Bavestrello, G., and Sarà, M. (1995). Morphological and ecological differences in two electrophoretically detected species of *Cliona* (Porifera, Demospongiae). *Biological Journal of the Linnean Society* **54,** 193–200.

Battershill, C. N., and Bergquist, P. R. (1990). The influence of storms on asexual reproduction, recruitment, and survivorship of sponges. In "New Perspectives in Sponge Biology" (K. Rützler, ed.), pp. 397–403. Smithsonian Institution Press, Washington.

Bautista-Guerrero, E., Carballo, J. L., and Maldonado, M. (2010). Reproductive cycle of the coral-excavating sponge *Thoosa mismalolli* (Clionaidae) from Mexican Pacific coral reefs. *Invertebrate Biology* **129,** 285–296.

Bavestrello, G., and Sarà, M. (1992). Morphological and genetic differences in ecologically distinct populations of *Petrosia* (Porifera, Demospongiae). *Biological Journal of the Linnean Society* **47,** 49–60.

Bell, G. (1982). The masterpiece of nature: The evolution and genetics of sexuality. University of California Press, Berkeley.

Bell, J. J. (2008). Connectivity between island marine protected areas and the mainland. *Biological Conservation* **141,** 2807–2820.

Bengtsson, B. O., and Ceplitis, A. (2000). The balance between sexual and asexual reproduction in plants living in variable environments. *Journal of Evolutionary Biology* **13,** 415–422.

Bentlage, B., and Wörheide, G. (2007). Low genetic structuring among *Pericharax heteroraphis* (Porifera: Calcarea) populations from the Great Barrier Reef (Australia), revealed by analysis of nrDNA and nuclear intron sequences. *Coral Reefs* **26,** 807–816.

Benzie, J. A. H., Sandusky, C., and Wilkinson, C. R. (1994). Genetic structure of dictyoceratid sponge populations on the western Coral Sea reefs. *Marine Biology* **119,** 335–345.

Bergquist, P. R., Karuso, P., and Cambie, R. C. (1990). Taxonomic relationships within the Dendroceratida: A biological and chemotaxonomical appraisal. In "New Perspectives in Sponge Biology" (K. Rützler, ed.), pp. 72–78. Smithsonian Institution Press, Washington.

Bickford, D., Lohman, D. J., Sodhi, N. S., Ng, P. K. L., Meier, R., Winker, K., Ingram, K. K., and Das, I. (2006). Cryptic species as a window on diversity and conservation. *Trends in Ecology & Evolution* **22,** 148–155.

Bierne, N., Borsa, P., Daguin, C., Jollivet, D., Viard, F., Bonhomme, F., and David, P. (2003). Introgression patterns in the mosaic hybrid zone between *Mytilus edulis* and *M. galloprovincialis. Molecular Ecology* **12,** 447–462.

Blanquer, A., and Uriz, M. J. (2007). Cryptic speciation in marine sponges evidenced by mitochondrial and nuclear genes: A phylogenetic approach. *Molecular Phylogenetics and Evolution* **45,** 392–397.

Blanquer, A., and Uriz, M. J. (2008). 'A posteriori' searching for phenotypic characters to describe new cryptic species of sponges revealed by molecular markers (Dictyonellidae: *Scopalina*). *Invertebrate Systematics* **22,** 489–502.

Blanquer, A., Agell, G., and Uriz, M. J. (2008). Hidden diversity in sympatric sponges: Adjusting life-history dynamics to share substrate. *Marine Ecology Progress Series* **371,** 109–115.

Blanquer, A., and Uriz, M. J. (2010). Population genetics at three spatial scales of a rare sponge living in fragmented habitats. *BMC Evolutionary Biology* **10**, 13.

Blanquer, A., and Uriz, M. J. (2011). "Living together apart": The hidden genetic diversity of sponge populations. *Molecular Biology and Evolution* **28**, 2435–2438.

Blanquer, A., Uriz, M. J., and Pascual, M. (2005). Polymorphic microsatellite loci isolated from the marine sponge *Scopalina lophyropoda* (Demospongiae: Halichondrida). *Molecular Ecology Notes* **5**, 466–468.

Blanquer, A., Uriz, M. J., and Caujapé-Castells, J. (2009). Small-scale spatial genetic structure in *Scopalina lophyropoda*, an encrusting sponge with philopatric larval dispersal and frequent fission and fusion events. *Marine Ecology Progress Series* **380**, 95–102.

Bonasoro, F., Wilkie, I. C., Bavestrello, G., Cerrano, C., and Candia Carnevali, M. D. (2001). Dynamic structure of the mesohyl in the sponge *Chondrosia reniformis* (Porifera, Demospongiae). *Zoomorphology* **121**, 109–121.

Borchiellini, C., Chombard, C., Lafay, B., and Boury-Esnault, N. (2000). Molecular systematics of sponges (Porifera). *Hydrobiologia* **420**, 15–27.

Bouck, A., and Vision, T. (2007). The molecular ecologist's guide to expressed sequence tags. *Molecular Ecology* **16**, 907–924.

Boury-Esnault, N. (1970). Un phénomène de bourgeonnement externe chez l'éponge *Axinella damicornis* (Esper). *Cahiers de Biologie Marine* **11**, 491–496.

Boury-Esnault, N., and Solé-Cava, A. (2004). Recent contribution of genetics to the study of sponge systematics and biology. *Bollettino dei Musei e degli Istituti Biologici dell'Università di Genova* **68**, 3–18.

Boury-Esnault, N., Solé-Cava, A. M., and Thorpe, J. P. (1992). Genetic and cytological divergence between colour morphs of the Mediterranean sponge *Oscarella lobularis* Schmidt (Porifera, Demospongiae, Oscarellidae). *Journal of Natural History* **26**, 271–284.

Boury-Esnault, N., Pansini, M., and Uriz, M. J. (1993). Cosmopolitism in sponges: The "complex" *Guitarra fimbriata* with description of a new species of *Guitarra* from the northeast Atlantic. *Scientia Marina* **57**, 367–373.

Bowcock, A. M., Ruiz-Linares, A., Tomfohrde, J., Minch, E., Kidd, J. R., and Cavalli-Sforza, L. L. (1994). High resolution of human evolutionary trees with polymorphic microsatellites. *Nature* **368**, 455–457.

Brien, P. (1967). Un nouveau mode de statoblastogenèse chez une éponge d'eau douce africaine (*Potamolepis stendelli*, Jaffe). *Bulletins de l'Académie Royale des Sciences, des Lettres et des Beaux-Arts de Belgique* **53**, 552–571.

Bucklin, A., Steinke, D., and Blanco-Bercial, L. (2011). DNA barcoding of marine Metazoa. *Annual Review of Marine Science* **3**, 471–508.

Bürger, R. (1999). Evolution of genetic variability and the advantage of sex and recombination in changing environments. *Genetics* **153**, 1055–1069.

Calderón, I., Ortega, N., Duran, S., Becerro, M. A., Pascual, M., and Turon, X. (2007). Finding the relevant scale: Clonality and genetic structure in a marine invertebrate (*Crambe crambe*, Porifera). *Molecular Ecology* **16**, 1799–1810.

Calderón, I., Palacín, C., and Turon, X. (2009). Microsatellite markers reveal shallow genetic differentiation between cohorts of the common sea urchin *Paracentrotus lividus* (Lamarck) in northwest Mediterranean. *Molecular Ecology* **18**, 3036–3049.

Calderón, I., Pita, L., Brusciotti, S., Palacín, C., and Turon, X. (2012). Time and space: Genetic structure of the cohorts of the common sea urchin *Paracentrotus lividus* in Western Mediterranean. *Marine Biology* **159**, 187–197.

Cárdenas, P., Xavier, J., Tendal, O. S., Schander, C., and Rapp, H. T. (2007). Redescription and resurrection of *Pachymatisma normani* (Demospongiae: Geodiidae), with remarks on the genus *Pachymatisma* Bowerbank. *Journal of the Marine Biological Association of the United Kingdom* **87**, 1511–1525.

Cárdenas, P., Pérez, T., and Boury-Esnault, N. (2012). Sponge systematics facing new challenges. *Advances in Marine Biology* **61,** 81–209.

Carroll, S. P., Hendry, A. P., Reznick, D. N., and Fox, C. W. (2007). Evolution on ecological time-scales. *Functional Ecology* **21,** 387–393.

Carvalho, G. R. (1994). Evolutionary genetics of aquatic clonal invertebrates: Concepts, problems and prospects. In "Genetics and Evolution of Aquatic Organisms" (A. R. Beaumont, ed.), pp. 291–322. Chapman & Hall, London.

Carvalho, G. R., Creer, S., Allen, M. J., Costa, F. O., Tsigenopoulos, C. S., Le Goff-Vitry, M., Magoulas, A., Medlin, L., and Metfies, K. (2010). Genomics in the discovery and monitoring of marine biodiversity. In "Advances in Marine Genomics" (J. M. Cok, K. Tessmar-Raible, C. Boven and F. Viard, eds), pp. 1–32. Springer, New York.

Cebrian, E., Agell, G., Martí, R., and Uriz, M. J. (2006). Response of the Mediterranean sponge *Chondrosia reniformis* Nardo to copper pollution. *Environmental Pollution* **141,** 452–458.

Cebrian, E., Uriz, M. J., Garrabou, J., and Ballesteros, E. (2011). Sponge mass mortalities in a warming Mediterranean Sea: Are cyanobacteria-harboring species worse off? *PLoS One* **6,** e20211 doi:10.1371/journal.pone.0020211.

Cha, J. N., Shimizu, K., Zhou, Y., Christianssen, S. C., Chmelka, B. F., Stucky, G. D., and Morse, D. E. (1999). Silicatein filaments and subunits from a marine sponge direct the polymerization of silica and silicones in vitro. *Proceedings of the National Academy of Sciences of the United States of America* **96,** 361–365.

Charlesworth, B. (1993). The evolution of sex and recombination in a varying environment. *The Journal of Heredity* **84,** 345–350.

Charlesworth, B. (2009). Effective population size and patterns of molecular evolution and variation. *Nature Reviews. Genetics* **10,** 205.

Chen, Y. H., Chen, C. P., and Chang, K. H. (1997). Budding cycle and bud morphology of the globe-shaped sponge *Cinachyra australiensis*. *Zoological Studies* **36,** 194–200.

Clark, M. A., Tanguy, A., Jollivet, D., Bonhomme, F., Guinand, B., and Viard, F. (2010). Populations and pathways: Genomic approaches to understanding population structure and environmental adaptation. In "Introduction to Marine Genomics, Chapter 3" (J. M. Cock, K. Tessmar-Raible, C. Boyen and F. Viard, eds), pp. 73–118. Springer, London.

Coffroth, M. A., and Lasker, H. R. (1998). Population structure of a clonal gorgonian coral: The interplay between clonal reproduction and disturbance. *Evolution* **52,** 379–393.

Cognetti, G., and Maltagliati, F. (2004). Strategies of genetic biodiversity conservation in the marine environment. *Marine Pollution Bulletin* **48,** 811–812.

Connes, R. (1964). Contribution à l'étude de la proliferation par voie asexuée chez le *Sycon*. *Bulletin de la Société Zoologique de France* **89,** 188–195.

Connes, R. (1967). Structure et development des bourgeons chez l'éponge siliceuse *Tethya lyncurium* Lamarck. Recherches experimentales et cytologiques. *Archives de Zoologie Experimentale et Générale* **108,** 157–195.

Connes, R. (1977). Contribution à l'étude de la gemmulogènese chez la démosponge marine *Suberites domuncula* (Olivi) Nardo. *Archives de Zoologie Experimentale et Générale* **118,** 391–407.

Cook, L. G., Edwards, R. D., Crisp, M. D., and Hardy, N. B. (2010). Need taxonomy always be required for new species descriptions? *Invertebrate Systematics* **24,** 322–326.

Corriero, G., Liaci, L. S., Marzano, C. N., and Gaino, E. (1998). Reproductive strategies of *Mycale contarenii* (Porifera, Demospongiae). *Marine Biology* **131,** 319–327.

Cowen, R. K., Gawarkiewicz, G., Pineda, J., Thorrold, S. S., and Werner, F. E. (2007). Population connectivity in marine systems: An overview. *Oceanography* **20,** 14–21.

Csilléry, K., Blum, M. G. B., Gaggiotti, O. E., and François, O. (2010). Approximate Bayesian Computation (ABC) in practice. *Trends in Ecology & Evolution* **25,** 410–418.

Dailianis, T., and Tsigenopoulos, C. S. (2010). Characterization of polymorphic microsatellite markers for the endangered Mediterranean bath sponge *Spongia officinalis* L. *Conservation Genetics* **11,** 1155–1158.

Dailianis, T., Tsigenopoulos, C. S., Dounas, C., and Voultsiadou, E. (2011). Genetic diversity of the imperilled bath sponge *Spongia officinalis* Linnaeus, 1759 across the Mediterranean Sea: Patterns of population differentiation and implications for taxonomy and conservation. *Molecular Ecology* **20**, 3757–3772.

Davis, A. R., Ayre, D. J., Billingham, M. R., Styan, C. A., and White, G. A. (1996). The encrusting sponge *Halisarca laxus*: Population genetics and association with ascidian *Pyura spinifera*. *Marine Biology* **126**, 27–33.

De Vos, C. (1965). Le bourgeonnement externe de l'éponge *Mycale contarenii* (Martens) (Démosponges). *Bulletin du Muséum national d'Histoire Naturelle, Paris* **37**, 548–555.

DeBiasse, M. B., Richards, V. P., and Shivji, M. S. (2010). Genetic assessment of connectivity in the common reef sponge, *Callyspongia vaginalis* (Demospongiae: Haplosclerida) reveals high population structure along the Florida reef tract. *Coral Reefs* **29**, 47–55.

DeSalle, R., and Amato, G. (2004). The expansion of conservation genetics. *Nature Reviews. Genetics* **5**, 702–712.

Dupont, S., Wilson, K., Obst, M., Sköld, H., Nakano, H., and Thorndyke, M. C. (2007). Marine ecological genomics: When genomics meets marine ecology. *Marine Ecology Progress Series* **332**, 257–273.

Duran, S., and Rützler, K. (2006). Ecological speciation in a Caribbean marine sponge. *Molecular Phylogenetics and Evolution* **40**, 292–297.

Duran, S., Pascual, M., Estoup, A., and Turon, X. (2002). Polymorphic microsatellite loci in the sponge *Crambe crambe* (Porifera: Poecilosclerida) and their variation in two distant populations. *Molecular Ecology Notes* **2**, 478–480.

Duran, S., Pascual, M., and Turon, X. (2004a). Low levels of genetic variation in mtDNA sequences over the western Mediterranean and Atlantic range of the sponge *Crambe crambe* (Poecilosclerida). *Marine Biology* **144**, 31–35.

Duran, S., Giribet, G., and Turon, X. (2004b). Phylogeographical history of the sponge *Crambe crambe* (Porifera, Poecilosclerida): Range expansion and recent invasion of the Macaronesian islands from the Mediterranean Sea. *Molecular Ecology* **13**, 109–122.

Duran, S., Pascual, M., Estoup, A., and Turon, X. (2004c). Strong population structure in the marine sponge *Crambe crambe* (Poecilosclerida) as revealed by microsatellite markers. *Molecular Ecology* **13**, 511–522.

Ellstrand, N. C., and Roose, M. L. (1987). Patterns of genotypic diversity in clonal plant species. *American Journal of Botany* **74**, 123–131.

Ereskovsky, A. V., and Tokina, D. B. (2007). Asexual reproduction in homoscleromorph sponges (Porifera; Homoscleromorpha). *Marine Biology* **151**, 425–434.

Ereskovsky, A. V., Lavrov, D. V., Boury-Esnault, N., and Vacelet, J. (2011). Molecular and morphological description of a new species of *Halisarca* (Demospongiae: Halisarcida) from Mediterranean Sea with redescription of the type species *Halisarca dujardini*. *Zootaxa* **2768**, 5–31.

Erpenbeck, D., Breeuwer, J. A. J., Van der Velde, H. C., and Van Soest, R. W. M. (2002). Unravelling host and symbiont phylogenies of halichondrid sponges (Demospongiae, Porifera) using a mitochondrial marker. *Marine Biology* **141**, 377–386.

Erpenbeck, D., Knowlton, A. L., Talbot, S. L., Highsmith, R. C., and Van Soest, R. W. M. (2004). A molecular comparison of Alaska and northeast Atlantic *Halichondria panicea* (Pallas 1776) (Porifera: Demospongiae) populations. Proceedings VI International Sponge Conference. *Bollettino dei Musei e degli Istituti Biologici dell'Università di Genova* **68**, 319–325.

Erpenbeck, J., Hooper, J. N. A., and Wörheide, G. (2006a). CO1 phylogenies in diploblasts and the "Barcoding of Life"—Are we sequencing a suboptimal partition? *Molecular Ecology Notes* **6**, 550–553.

Erpenbeck, D., Breeuwer, J. A. J., Parra-Velandia, F. J., and van Soest, R. W. M. (2006b). Speculation with spiculation?—Three independent gene fragments and biochemical characters versus morphology in demosponge higher classification. *Molecular Phylogenetics and Evolution* **38,** 293–305.

Erwin, P. M., and Thacker, R. W. (2008). Phototrophic nutrition and symbiont diversity of two Caribbean sponge-cyanobacteria symbioses. *Marine Ecology Progress Series* **362,** 139–147.

Erwin, P. M., López-Legentil, S., González-Perch, R., and Turon, X. (2012). A specific mix of generalists: Bacterial symbionts in Mediterranean Ircinia spp.. *FEMS Microbiology Ecology* **79,** 619–637. doi:10.1111/j.1574-6941.2011.01243.x.

Folmer, O., Black, M., Hoech, W., Lutz, R., and Vrijenhoeck, R. (1994). DNA primers for amplification of mitochondrial cytochrome c oxidase subunit 1 from diverse metazoan invertebrates. *Molecular Marine Biology and Biotechnology* **3,** 294–299.

Frotscher, P. J., and Uriz, M. J. (2008). Reproduction and life cycle of the calcareous sponge *Paraleucilla magna* in the Mediterranean Sea. In "XV Symposium Ibérico de Estudios de Bentos Marino, Funchal, Madeira." Book of Abstracts.

Gaino, E., Scalera Liaci, L., Sciscioli, M., and Corriero, G. (2006). Investigation of the budding process in *Tethya citrina* and *Tethya aurantium* (Porifera, Demospongiae). *Zoomorphology* **125,** 87–97.

Galstoff, P. S. (1942). Wasting diseases causing mortality of sponges in the West Indies and Gulf of Mexico. *Proceedings 8th American Scientific Congress* **3,** 411–421.

Garrabou, J., and Zabala, M. (2001). Growth dynamics in four Mediterranean demosponges. *Estuarine, Coastal and Shelf Science* **52,** 293–303.

Gauthier, M., and Degnan, B. M. (2008). Partitioning of genetically distinct cell populations in chimeric juveniles of the sponge *Amphimedon queenslandica*. *Developmental and Comparative Immunology* **32,** 1270–1280.

Gigliarelli, L., Lucentini, L., Palomba, A., Sgaravizzi, G., Lancioni, H., Lanfaloni, L., Willenz, P., Gaino, E., and Panara, F. (2008). Applications of PCR-RFLPs for differentiating two freshwater sponges: *Ephydatia fluviatilis* and *Ephydatia mülleri*. *Hydrobiologia* **605,** 265–269.

Grosberg, R. K. (1988a). Life-history variation within a population of the colonial ascidian *Botryllus schlosseri*. I. The genetic and environmental control of seasonal variation. *Evolution* **42,** 900–920.

Grosberg, R. K. (1988b). The evolution of allorecognition specificity in clonal invertebrates. *The Quarterly Review of Biology* **63,** 377–412.

Grosberg, R., and Cunningham, C. W. (2001). Genetic structure in the sea. From populations to communities. In "Marine Community Ecology" (M. D. Bertness, S. D. Gaines and M. E. Hay, eds), pp. 61–84. Sinauer Associates, Sunderland, Massachusetts.

Guardiola, M., Frotscher, J., and Uriz, M. J. (2011). Genetic structure and differentiation at a short-time scale of the introduced calcareous sponge *Paraleucilla magna* to the western Mediterranean. *Hydrobiologia* **687,** doi:10.1007/s10750-011-0948-1.

Haig, S. M. (1998). Molecular contributions to conservation. *Ecology* **79,** 413–425.

Hamilton, M. B. (2009). Population Genetics. Wiley-Blackwell, Oxford.

Hebert, P. D. N., Cywinska, A., Ball, S. L., and deWaard, J. R. (2003). Biological identifications through DNA barcodes. *Proceedings of the Royal Society of London, Series B: Biological Science* **270,** 313–332.

Heim, I., Nickel, N., and Brümmer, F. (2007). Molecular markers for species discrimination in poriferans: A case study on species of the genus *Aplysina*. In "Porifera Research: Biodiversity, Innovation and Sustainability" (M. R. Custódio, G. Lôbo-Hajdu, E. Hajdu and G. Muricy, eds), pp. 361–371. Universidade do Rio de Janeiro, Brazil.

Hellberg, M. E., Burton, R. S., Neigel, J. E., and Palumbi, S. R. (2002). Genetic assessment of connectivity among marine populations. *Bulletin of Marine Science* **70,** 273–290.

Hoshino, S., and Fujita, T. (2006). Isolation of polymorphic microsatellite markers from *Hymeniacidon sinapium* (Porifera: Demospongiae:Halichondrida). *Molecular Ecology Notes* **6,** 829–831.

Hoshino, S., Saito, D. S., and Fujita, T. (2008). Contrasting genetic structure of two Pacific *Hymeniacidon* species. *Hydrobiologia* **603,** 313–326.

Hutchings, P., Ahyong, S., Byrne, M., Przeslawski, R., and Wörheide, G. (2007). Vulnerability of benthic invertebrates of the Great Barrier Reef to climate change. In "Climate Change and the Great Barrier Reef. A Vulnerability Assessment" (J. E. Johnson and P. A. Marshall, eds), pp. 309–356. Great Barrier Reef Marine Park Authority and Australian Greenhouse Office, Townsville.

Ilan, M., and Loya, Y. (1990). Ontogenetic variation in sponge histocompatibility responses. *The Biological Bulletin* **179,** 279–286.

Ivanišević, J., Thomas, O. P., Lejeusne, C., Chevaldonné, P., and Pérez, T. (2011). Metabolic fingerprinting as an indicator of biodiversity: Towards understanding inter-specific relationships among Homoscleromorpha sponges. *Metabolomics* **7,** 289–304.

Jackson, D. J., Macis, L., Reitner, J., and Wörheide, G. (2011). A horizontal gene transfer supported the evolution of an early metazoan biomineralization strategy. *BMC Evolutionary Biology* **11,** 238.

Jennings, T. N., Knaus, B. J., Mullins, T. D., Haig, S. M., and Cronn, R. C. (2011). Multiplexed microsatellite recovery using massively parallel sequencing. *Molecular Ecology Resources* **11,** 1060–1067 doi:10.1111/j.1755-0998.2011.03033.x.

Johnson, M. F. (1978). Studies on the reproductive cycles of the calcareous sponges *Clathrina coriacea* and *C. blanca*. *Marine Biology* **50,** 73–79.

Johnson, S. C., and Browman, H. I. (2007). Introducing genomics, proteomics and metabolomics in marine ecology. *Marine Ecology Progress Series* **332,** 247–248.

Kaiser, J., and Gallagher, R. (1997). Does diversity lure invaders? *Science* **277,** 1204–1205.

Karlson, R. H. (1991). Fission and the dynamics of genets and ramets in clonal Cnidarian populations. *Hydrobiologia* **2016,** 235–240.

Kerr, R. G., and Kelly-Borges, M. (1994). Biochemical and morphological heterogeneity in the Caribbean sponge *Xestospongia muta* (Petrosida: Petrosiidae). In "Sponges in Time and Space" (R. W. M. van Soest, Th.M. G. van Kempen and J. C. Braekman, eds), pp. 65–73. Balkema, Rotterdam.

Kim, A., and Misra, A. (2007). SNP genotyping: Technologies and biomedical applications. *Annual Review of Biomedical Engineering* **9,** 289–320.

Klautau, M., Solé-Cava, A. M., and Borojevic, R. (1994). Biochemical systematics of sibling sympatric species of *Clathrina* (Porifera: Calcarea). *Biochemical Systematics and Ecology* **33,** 367–375.

Klautau, M., Russo, C. A. M., Lazoski, C., Boury-Esnault, N., Thorpe, J. P., and Solé-Cava, A. M. (1999). Does cosmopolitanism result from overconservative systematics? A case study using the marine sponge *Chondrilla nucula*. *Evolution* **53,** 1414–1422.

Knowlton, N. (1993). Sibling species in the sea. *Annual Review of Ecology and Systematics* **24,** 189–216.

Knowlton, N. (2000). Molecular genetics analyses of species boundaries in the sea. *Hydrobiologia* **420,** 73–90.

Knowlton, A. L., Pierson, B. J., Talbott, S. L., and Highsmith, R. C. (2003). Isolation and characterization of microsatellite loci in the intertidal sponge *Halichondria panicea*. *Molecular Ecology Notes* **3,** 560–562.

Koziol, C., Wagner-Hüslsmann, C., Mikoc, A., Gamulin, V., Kruse, M., Pancer, Z., Schäcke, H., and Müller, W. E. G. (1996). Cloning of heat-inducible biomarker, the cDNA encoding the 70 kDa heat shock protein, from the marine sponge *Geodia cydonium*: Response to natural stressors. *Marine Ecology Progress Series* **136,** 153–161.

Koziol, C., Batel, R., Arinc, E., Schröder, H. C., and Müller, W. E. G. (1997). Expression of the potential biomarker heat shock protein 70 and its regulator DnaJ homolog, by temperature stress in the sponge *Geodia cydonium*. *Marine Ecology Progress Series* **154,** 261–268.

Koziol, C., Borojevic, R., Steffen, R., and Müller, W. E. G. (1998). Sponges (Porifera) model systems to study the shift from immortality to senescent somatic cells: The telomerase activity in somatic cells. *Mechanisms of Ageing and Development* **100,** 107–120.

Kozo-Polyansky, B. M., Fet, V., and Margulis, L. (2010). Rediscovering Symbiogenesis. Harvard University Press, Harvard.

Krasko, A., Scheffer, U., Koziol, C., Pancer, Z., Batel, R., Badria, F. A., and Müller, W. E. G. (1997). Diagnosis of sublethal stress in the marine sponge *Geodia cydonium*: Application of the 70 kDa heat-shock protein and a novel biomarker, the Rab GDP dissociation inhibitor, as probes. *Aquatic Toxicology* **37,** 157–168.

Krasko, A., Schröder, H. C., Perovic, S., Steffen, R., Kruse, M., Reichert, W., Müller, I. M., and Müller, W. E. G. (1999). Ethylene modulates gene expression in cells of the marine sponge *Suberites domuncula* and reduces the degree of apoptosis. *The Journal of Biological Chemistry* **274,** 31524–31530.

Krasko, A., Batel, R., Schröder, H. C., Müller, I. M., and Müller, W. E. G. (2000). Expression of silicatein and collagen genes in the marine sponge *Suberites domuncula* is controlled by silicate and myotrophin. *European Journal of Biochemistry* **267,** 4878–4887.

Krasko, A., Kurelec, B., Batel, R., Müller, I. M., and Müller, W. E. G. (2001). Potential multidrug resistance gene POHL: an ecologically relevant indicator in marine sponges. *Environmental Toxicology and Chemistry* **20,** 198–204.

Krasko, A., Schröder, H. C., Batel, R., Grebenjuk, V. A., Steffen, R., Müller, I. M., and Müller, W. E. G. (2002). Iron induces proliferation and morphogenesis in primmorphs from the marine sponge *Suberites domuncula*. *DNA and Cell Biology* **21,** 67–80.

Lanna, E., Monteiro, L. C., and Klautau, M. (2007). Life cycle of *Paraleucilla magna* Klautau, Monteiro and Borojevic, 2004 (Porifera, Calcarea). In "Porifera Research," pp. 413–418.

Lapébie, P., Gazave, E., Ereskovsky, A., Derelle, R., Bézac, C., Renard, R., Houliston, E., and Borchiellini, C. (2009). WNT/b-catenin signalling and epithelial patterning in the homoscleromorph sponge *Oscarella*. *PLoS One* **4**(6), e5823. doi:101371/journal.pone.0005823.

Larroux, C., Fahey, B., Liubicich, D., Hinman, V. F., Gauthier, M., Gongora, M., Green, K., Wörheide, G., Leys, S. P., and Degnan, B. M. (2006). Developmental expression of transcription factor genes in a demosponge: Insights into the origin of metazoan multicellularity. *Evolution and Development* **8,** 150–173.

Lavrov, D. V., Forget, L., Kelly, M., and Lang, B. F. (2005). Mitochondrial genomes of two demosponges provide insights into an early stage of animal evolution. *Molecular Biology and Evolution* **22,** 1231–1239.

Lazoski, C., Solé-Cava, A. M., Boury-Esnault, N., Klautau, M., and Russo, C. A. (2001). Cryptic speciation in a high gene flow scenario in the oviparous marine sponge *Chondrosia reniformis*. *Marine Biology* **139,** 421–429.

Lee, H. J., and Boulding, E. G. (2007). Mitochondrial DNA variation in space and time in the northeastern Pacific gastropod, *Littorina keenae*. *Molecular Ecology* **16,** 3084–3103.

Lee, O. O., Wang, Y., Yang, J., Lafi, F. F., Al-Suwailem, A., and Qian, P.-Y. (2011). Pyrosequencing reveals highly diverse and species-specific microbial communities in sponges from the Red Sea. *The ISME Journal* **5,** 650–664.

Lemloh, M. L., Fromont, J., Brümmer, F., and Usher, K. M. (2009). Diversity and abundance of photosynthetic sponges in temperate Western Australia. *BMC Ecology* **9,** 1–13.

Leong, W., and Pawlik, J. R. (2010). Fragments of propagules? Reproductive tradeoffs among *Callyspongia* spp. from Florida coral reefs. *Oikos* **119,** 1417–1422.

Li, Y. C., Korol, A. B., Fahima, T., Beiles, A., and Nevo, E. (2002). Microsatellites: Genomic distribution, putative functions and mutational mechanisms: A review. *Molecular Ecology* **11,** 2453–2465.

Lôbo-Hadju, G., Guimarães, A. C. R., Salgado, A., Lamarao, F. R. M., Vieiralves, T., Mansure, J. J., and Albano, R. M. (2004). Intragenomic, intra-and interspecific variation in the rDNA ITS of Porifera revealed by PCR-single strand conformation polymorphism (PCR-SSCP). *Bollettino dei Musei e degli Istitute Biologici dell' Università di Genova* **68,** 413–423.

Lopez, J. V., Peterson, C. L., Willoughby, R., Wright, A. E., Enright, E., Zoladz, S., Reed, J. K., and Pomponi, S. (2002). Characterization of genetic markers for *in vitro* dell line identification of the marine sponge *Axinella corrugata. The Journal of Heredity* **94,** 27–36.

López-Legentil, S., and Pawlik, J. R. (2009). Genetic structure of the Caribbean giant barrel sponge *Xestospongia muta* using the I3-M11 partition of COI. *Coral Reefs* **28,** 157–165.

López-Legentil, S., Song, B., Mcmurray, S. E., and Pawlik, J. R. (2008). Bleaching and stress in coral reef ecosystems: hsp70 expression by the giant barrel sponge *Xestospongia muta. Molecular Ecology* **17,** 1840–1849.

López-Legentil, S., Erwin, P. M., Henkel, T. P., Loh, T.-K., and Pawlik, J. R. (2010). Phenotypic plasticity in the Caribbean sponge *Callyspongia vaginalis* (Porifera: Haplosclerida). *Scientia Marina* **74,** 445–453.

Loukaci, A., Muricy, G., Brouard, J. P., Guyot, M., Vacelet, J., and Boury-Esnault, N. (2004). Chemical divergence between two sibling species of *Oscarella* (Porifera) from the Mediterranean Sea. *Biochemical Systematics and Ecology* **32,** 893–899.

Lowenstam, H. A., and Margulis, L. (1980). Evolutionary prerequisites for early phanerozoic calcareous skeletons. *Bio Systems* **12,** 27–41.

Maldonado, M. (1998). Do chimeric sponges have improved chances of survival? *Marine Ecology Progress Series* **164,** 301–306.

Maldonado, M., and Uriz, M. J. (1999). Sexual propagation by sponge fragments. *Nature* **398,** 476.

Maldonado, M., Carmona, C., Uriz, M. J., and Cruzado, A. (1999). Decline in Mesozic reed-building sponges explained by silicon limitation. *Nature* **401,** 765–788.

Manuel, M., Borchiellini, C., Alivon, E., Le Parco, Y., Vacelet, J., and Boury-Esnault, N. (2003). Phylogeny and evolution of calcareous sponges: Monophyly of Calcinea and Calcaronea, high level of morphological homoplasy, and the primitive nature of axial symmetry. *Systematic Biology* **52,** 311–333.

Mariani, S., Uriz, M., and Turon, X. (2005). The dynamics of sponge larvae assemblages from northwestern Mediterranean nearshore bottoms. *Journal of Plankton Research* **27,** 249–262.

Mariani, S., Uriz, M., Turon, X., and Alcoverro, T. (2006). Dispersal strategies in sponge larvae: Integrating the life history of larvae and the hydrologic component. *Oecologia* **149,** 174–184.

Marioni, J. C., Mason, C. E., Mane, S. M., Stephens, M., and Gilad, Y. (2008). RNA-seq: An assessment of technical reproducibility and comparison with gene expression arrays. *Genome Research* **18,** 1509–1517.

Martí, R., Uriz, M. J., Ballesteros, E., and Turon, X. (2004). Benthic assemblages in two Mediterranean caves: Species diversity and coverage as a function of abiotic parameters and geographic distance. *Journal of the Marine Biological Association of the United Kingdom* **84,** 557–572.

Maynard-Smith, J. (1978). The Evolution of Sex. Cambridge University Press, Cambridge.

McFadden, C. S. (1997). Contributions of sexual and asexual reproduction to population structure in the clonal soft coral, *Alcyonium rudyi. Evolution* **51,** 112–126.

McKinnon, J. S., Mori, S., Blackman, B. K., David, L., Kingsley, D. M., Jamieson, L., Chou, J., and Schluter, D. (2004). Evidence for ecology's role in speciation. *Nature* **429,** 294–298.

Merejkowsky, C. D. (1879). Reproduction des éponges par bourgeonnement extérieur. *Archives de Zoologie Expérimentale et Générale* **8,** 417–433.

Miller, K., Alvarez, B., Battershill, C., Northcote, P., and Parthasarathy, H. (2001). Genetic, morphological, and chemical divergence in the sponge genus *Latrunculia* (Porifera: Demospongiae) from New Zealand. *Marine Biology* **139,** 235–250.

Mittard-Runte, V., Bekel, T., Blom, J., Dondrup, M., Henckel, K., Jaenicke, S., Krause, L., Linke, B., Neuweger, H., Schneiker-Bekel, S., and Goesmann, A. (2010). Practical guide: Genomic techniques and how to apply them to marine questions. In "Introduction to Marine Genomics, Chapter 3" (J. M. Cock, K. Tessmar-Raible, C. Boyen and F. Viard, eds), pp. 314–378. Springer, London.

Morin, P. A., Luikart, G., Wayne, R. K., and the SNP workshop group, R. K. (2004). SNPs in ecology, evolution and conservation. *Trends in Ecology & Evolution* **19,** 208–216.

Mueller, U. G., and Wolfenbarger, L. L. (1999). AFLP genotyping and fingerprinting. *Trends in Ecology & Evolution* **14,** 389–394.

Müller, W. E. G., Koziol, C., Kurelec, B., Dapper, J., Batel, R., and Rinkevich, B. (1995). Combinatory effects of temperature stress and nonionic organic pollutants on stress protein (hsp70) gene expression in the freshwater sponge *Ephydatia fluviatilis*. *Archives of Environmental Contamination and Toxicology* **14,** 1203–1208.

Müller, W. E. G., Steffen, R., Rinkevich, B., Matranga, V., and Kurelec, B. (1996). The multixenobiotic resistance mechanism in the marine sponge *Suberites domuncula*: Its potential applicability for the evaluation of *environmental pollution* by toxic compounds. *Marine Biology* **125,** 165–170.

Müller, W. E. G., Perovic, S., Schröder, H. C., and Breter, H. J. (2004). Oxygen as a morphogenic factor in sponges: Expression of a tyrosinase gene in the sponge *Suberites domuncula*. *Micron* **35,** 87–88.

Müller, W. E. G., Boreiko, A., Schloßmacher, U., Wang, X., Eckert, C., Kropf, K., Li, J., and Schröder, H. C. (2008a). Identification of a silicatein (-related) protease in the giant spicules of the deep-sea hexactinellid *Monorhaphis chuni*. *The Journal of Experimental Biology* **211,** 300–309.

Müller, W. E. G., Wang, X., Kropf, K., Boreiko, A., Schloßmacher, U., Brandt, D., Schröder, H. C., and Wiens, M. (2008b). Silicatein expression in the hexactinellid *Crateromorpha meyeri*: The lead marker gene restricted to siliceous sponges. *Cell and Tissue Research* **333,** 339–351.

Müller, W. E. G., Wang, X., Wiens, M., Schloßmacher, U., Jochum, K. P., and Schröder, H. C. (2011). Hardening of bio-silica in sponge spicules involves an aging process that alters its enzymatic polycondensation: Evidence for an aquaporin-mediated water absorption. *Biochimica et Biophysica Acta* **1810,** 713–726.

Muricy, G., Boury-Esnault, N., Bézac, C., and Vacelet, J. (1996a). Cytological evidence for cryptic speciation in Mediterranean *Oscarella* species (Porifera: Homoscleromorpha). *Canadian Journal of Zoology* **74,** 881–896.

Muricy, G., Solé-Cava, A. M., Thorpe, J. P., and Boury-Esnault, N. (1996b). Genetic evidence for extensive cryptic speciation in the subtidal sponge *Plakina trilopha* (Porifera: Demospongiae: Homoscleromorpha). *Marine Ecology Progress Series* **138,** 181–187.

Nei, M. (1978). Estimation of average heterozygosity and genetic distance from a small number of individuals. *Genetics* **89,** 583–590.

Neigel, J. E., and Avise, J. C. (1983). Histocompatibility bioassays of population structure in marine sponges. Clonal structure in Verongia longissima and Iotrochota birotulata. *The Journal of Heredity* **74,** 134–140.

Neigel, J. E., and Avise, J. C. (1985). The precision of histocompatibility response in clonal recognition in tropical marine sponges. *Evolution* **39,** 724–732.

Neigel, J. E., and Schmahl, G. P. (1984). Phenotipic variation within histocompatibility-defined clones of marine sponges. *Science* **224,** 413–415.

Nichols, S. A., and Barnes, P. A. G. (2005). A molecular phylogeny and historical biogeography of the marine sponge genus *Placospongia* (Phylum Porifera) indicate low dispersal capabilities and widespread crypsis. *Journal of Experimental Marine Biology and Ecology* **323,** 1–15.

Noyer, C. (2010). Relationship between genetic, bacterial and chemical diversity of the Mediterranean sponge *Spongia agaricina*. PhD Thesis, University of Barcelona.

Noyer, C., Agell, G., Pascual, M., and Becerro, M. A. (2009). Isolation and characterization of microsatellite loci from the endangered Mediterranean sponge *Spongia agaricina* (Demospongiae: Dictyoceratida). *Conservation Genetics* **10,** 1895–1898.

Packer, L., Gibbs, J., Sheffield, C., and Hanner, R. (2009). DNA barcoding and the mediocrity of morphology. *Molecular Ecology Resources* **9,** 42–50.

Palumbi, S. R. (2003). Population genetics, demographic connectivity, and the design of marine reserves. *Ecological Applications* **13,** S146–S158.

Palumbi, S. R., Grabowsky, G., Duda, T., Geyer, L., and Tachino, N. (1997). Speciation and population genetic structure in tropical Pacific sea urchins. *Evolution* **51,** 1506–1517.

Palumbi, S. R., Cipriano, F., and Hare, M. P. (2001). Predicting nuclear gene coalescence from mitochondrial data: The three-times rule. *Evolution* **55,** 859–868.

Palumbi, S. R., McLeod, K. L., and Grunbaum, D. (2008). Ecosystems in action: Lessons from marine ecology about recovery, resistance, and reversibility. *BioScience* **58,** 1.

Pansini, M., and Pronzato, R. (1990). Observations on the dynamics of a Mediterranean sponge community. In "New Perspectives in Sponge Biology" (K. Rützler, ed.), pp. 405–415. Smithsonian Institution Press, Washington, DC.

Pearse, D. E., and Crandall, K. A. (2004). Beyond Fst: Analysis of population genetic data for conservation. *Conservation Genetics* **5,** 585–602.

Pérez, T., Garrabou, J., Sartoretto, S., Harmelin, J.-G., Francour, P., and Vacelet, J. (2000). Mortalité massive d'invertébrés marins: un événement sans précédent en Méditerranée nord-occidentale. *Life Sciences* **323,** 853–865.

Pérez, T., Ivanišević, J., Dubois, M., Thomas, O. P., Tokina, D., and Ereskovsky, A. V. (2011). Oscarella balibaloi, a new sponge species (Homoscleromorpha: Plakinidae) from the western Mediterranean Sea: Cytological description, reproductive cycle, and ecology. *Marine Ecology* **23,** 174–187.

Pineda-Krch, M., and Lehtila, K. (2004). Costs and benefits of genetic heterogeneity within organisms. *Journal of Evolutionary Biology* **17,** 1167–1177.

Plotkin, A. S., and Ereskovsky, A. V. (1997). Ecological features of asexual reproduction of the White Sea sponge *Polymastia mammillaris* (Demospongiae, Tetractinomorpha) in Kandalaksha Bay. In "Modern Problems of Poriferan Biology" (A. V. Ereskovsky, H. Keupp and R. Kohring, eds), pp. 127–132Selb Fach Geowiss E20, Berlin.

Reveillaud, J., Remerie, T., Van Soest, R. W. M., Erpenbeck, D., Cárdenas, P., Derycke, S., and Xavier, J. R. (2011a). Species boundaries and phylogenetic relationships between Atlanto-Mediterranean shallow-water and deep-sea coral associated *Hexadella* species (Porifera, Ianthellidae). *Molecular Phylogenetics and Evolution* **56**(1), 104–114.

Reveillaud, J., Remerie, T., Van Soest, R. W. M., Erpenbeck, D., Cárdenas, P., Derycke, S., and Xavier, J. R. (2011b). Phylogenetic relationships among NE Atlantic *Plocamionida* Topsent (1927) (Porifera, Poecilosclerida): Under-estimated diversity in reef ecosystems. *PLoS One* **6,** e16533 doi:10.1371/journal.pone.0016533.

Rice, A. L., Thurston, M. H., and New, A. L. (1990). Dense aggregations of a hexactinellid sponge, *Pheronema carpenteri*, in the Porcupine Seabight (northeast Atlantic Ocean), and possible causes. *Progress in Oceanography* **24,** 179–196.

Rinkevich, B. (2005). Natural chimerism in colonial urochordates. *Journal of Experimental Marine Biology and Ecology* **322,** 93–109.

Rivera, A. S., Hammel, J. U., Haen, K. M., Danka, E. S., Brandon Cieniewicz, B., Winters, I. P., Posfai, D., Wörheide, G., Lavrov, D. V., Knight, S. W., Hill, M. S., Hill, A. L., and Nickel, M. (2011). RNA interference in marine and freshwater sponges: Actin knockdown in *Tethya wilhelma* and *Ephydatia muelleri* by ingested dsRNA expressing bacteria. *BMC Biotechnology* **2011,** 67.

Rosell, D., and Uriz, M. J. (2002). Excavating and endolithic sponge species (Porifera) from the Mediterranean: Species descriptions and identification key. *Organisms, Diversity and Evolution* **1,** 1–31.

Rot, C., Goldfarb, I., Ilan, M., and Huchon, D. (2006). Putative cross-kingdom horizontal gene transfer in sponge (Porifera) mitochondria. *BMC Evolutionary Biology* **6,** 71.

Rua, C. P. J., Zilberberg, C., and Solé-Cava, A. (2011). New polymorphic mitochondrial markers for sponge phylogeography. *Journal of the Marine Biological Association of the United Kingdom* **91,** 1015–1022. doi:10.1017/S0025315410002122.

Rützler, K. (1974). The burrowing sponges of Bermuda. *Smithsonian Contributions to Zoology* **165,** 1–32.

Ruzzante, D. E. (1998). A comparison of several measures of genetic distance and population structure with microsatellite data: Bias and sampling variance. *Canadian Journal of Fisheries and Aquatic Sciences* **55,** 1–14.

Saby, E., Justesen, J., Kelve, M., and Uriz, M. J. (2009). In vitro effects of metal pollution on Mediterranean sponges: Species-specific inhibition of $2',5'$ oligoadenylate synthetase. *Aquatic Toxicology* **94,** 204–210.

Saller, U. (1990). Formation and construction of asexual buds of the freshwater sponge *Radiospongilla cerebellata* (Porifera, Spongillidae). *Zoomorphology* **109**(6), 295–301.

Sánchez, H. (1984). Cultivo experimental de esponjas en el laboratorio. *Anales del Instituto de Investigaciones Marinas Punta de Betin* **14,** 17–28.

Sarà, M., and Vacelet, J. (1973). Écologie des démosponges. In "Traité de Zoologie. Anatomie, Systématique, Biologie. III Spongiares" (P.-P. Grassé, ed.), pp. 462–576. Masson and Cie, Paris.

Sarà, M., Corriero, G., and Bavestrello, G. (1993). *Tethya* (Porifera, Demospogiae) species coexisting in a Maldivian coral reef lagoon: Taxonomical, genetic and ecological data. *Marine Ecology* **14,** 341–355.

Sarà, A., Cerrano, C., and Sarà, M. (2002). Viviparous development in the Antarctic sponge *Stylocordyla borealis* Lovén, 1868. *Polar Biology* **25,** 425–431.

Sarà, A., Mensi, P., Manconi, R., Bavestrello, G., and Barletto, E. (1989). Genetic variability in Mediterranean populations of Tethya (Porifera: Demospongiae). In "Reproduction, Genetics, and Distributions of Marine Organisms" (J. S. Ryland and P. A. Tyler, eds), pp. 293–298. Olsen and Olsen, Fredensborg, Denmark.

Schlick-Steiner, B. C., Seifert, B., Stauffer, C., Christian, E., Crozier, R. H., and Steiner, F. M. (2007). Without morphology, cryptic species stay in taxonomic crypsis following discovery. *Trends in Ecology & Evolution* **22,** 391–392.

Schönberg, C. H. L. (2002). *Pione lampa*, a bioeroding sponge in a worm reef. *Hydrobiologia* **482,** 49–68.

Schröder, H. C., Kruse, M., Batel, R., Müller, I. M., and Müller, W. E. G. (2000). Cloning and expression of the sponge longevity gene SDLAGL. *Mechanisms of Development* **95,** 219–220.

Schröder, H. C., Grebenjuk, V. A., Binder, M., Skorokhod, A., Batel, R., Hassanein, H., and Müller, W. E. G. (2005a). Functional molecular biodiversity: assessing the immune status of two sponge populations (*Suberites domuncula*) on the molecular level. *Marine Ecology* **25,** 93–108.

Schröder, H. C., Perović-Ottstadt, S., Grebenjuk, V. A., Engel, S., Müller, I. M., and Müller, W. E. G. (2005b). Biosilica formation in spicules of the sponge *Suberites domuncula*: Synchronous expression of a gene cluster. *Genomics* **85,** 666–678.

Schulze, F. E. (1887). Report on the Hexactinellida collected by H.M.S. 'Challenger' during the years 1873-1876. Report on the Scientific Results of the Voyage of H.M.S. 'Challenger', 1873-1876 *Zoology* **21**, 1–514.

Sebens, K. P., and Thorne, B. L. (1985). Coexistence of clones, clonal diversity, and the effects of disturbance. In "Population Biology and Evolution of Clonal Organisms" (J. B. C. Jackson, L. W. Buss and R. E. Cook, eds), pp. 357–398. Yale University Press, New Haven and London.

Shimizu, K., Cha, J., Stucky, G. D., and Morse, D. E. (1998). Silicatein alpha: Cathepsin L-like protein in sponge biosilica. *Proceedings of the National Academy of Sciences of the United States of America* **95**, 6234–6238.

Siegel, D. A., Mitarai, S., Costello, C. J., Gaines, S. D., Kendall, B. E., Warner, R. R., and Winters, K. B. (2008). The stochastic nature of larval connectivity among nearshore marine populations. *Proceedings of the National Academy of Sciences of the United States of America* **105**, 8974–8979.

Solé-Cava, A. M., and Boury-Esnault, N. (1999). Patterns of intra- and interspecific divergence in marine sponges. *Memoirs of the Queensland Museum* **44**, 591–602.

Solé-Cava, A. M., and Thorpe, J. P. (1986). Genetic differentiation between morphotypes of the marine sponge *Suberites ficus* (Demospongiae: Hadromerida). *Marine Biology* **93**, 247–253.

Solé-Cava, A. M., and Thorpe, J. P. (1994). Evolutionary genetics of marine sponges. In "Sponges in Time and Space" (R. W. M. Van Soest, T. M. G. Van Kempen and J. C. Braekman, eds), pp. 55–62. Balkema, Rotterdam.

Solé-Cava, A. M., Klautau, M., Boury-Esnault, N., Borojevic, R., and Thorpe, J. P. (1991a). Genetic evidence for cryptic speciation in allopatric populations of two cosmopolitan species of the calcareous sponge genus *Clathrina*. *Marine Biology* **111**, 381–386.

Solé-Cava, A. M., Thorpe, J. P., and Manconi, R. (1991b). A new Mediterranean species of *Axinella* detected by biochemical genetic methods. In "Fossil and Recent Sponges" (H. Keupp and J. Reitner, eds), pp. 313–321. Springer, Berlin.

Solé-Cava, A. M., Boury-Esnault, N., Vacelet, J., and Thorpe, J. P. (1992). Biochemical genetic divergence in sponges of the genera *Corticium* and *Oscarella* (Demospongiae: Homoscleromorpha) in the Mediterranean Sea. *Marine Biology* **113**, 299–304.

Srivastava, M., Simakov, O., Chapman, J., Fahey, B., Gauthier, E. A., Mitros, T., Richards, G. S., Conaco, C., Dacre, M., Hellsten, U., Larroux, C., Putnam, N. H., Stanke, M., Adamska, M., Darling, A., Degnan, S. M., Oakley, T., Plachetzki, D., Zhai, Y., Adamski, M., Calcino, A., Cummins, S. F., Goodstein, D. M., Harris, C., Jackson, D., Leys, S. P., Shu, S., Woodcroft, B. J., Vervoort, M., Kosik, K. S., Manning, G., Degnan, B. M., and Rokhsar, D. S. (2010). The *Amphimedon queenslandica* genome and the evolution of animal complexity. *Nature* **466**, 720–726.

Steindler, L., Schuster, S., Ilan, M., Avni, A., Cerrano, C., and Beer, S. (2007). Differential gene expression in a marine sponge in relation to its symbiotic state. *Marine Biotechnology* **9**, 543–549.

Stenberg, P., Lundmark, M., and Saura, A. (2003). MLGsim: A program for detecting clones using a simulation approach. *Molecular Ecology Notes* **3**, 329–331.

Swanson, W. J., and Vacquier, V. D. (2002). The rapid evolution of reproductive proteins. *Nature Reviews. Genetics* **3**, 137–144.

Sweijd, N. A., Bowie, R. C. K., Evans, B. S., and Lopata, A. L. (2000). Molecular genetics and the management and conservation of marine organisms. *Hydrobiologia* **420**, 153–164.

Szitenberg, A., Rot, C., Ilan, M., and Huchon, D. (2010). Diversity of sponge mitochondrial introns revealed by cox 1 sequences of Tetillidae. *BMC Evolutionary Biology* **10**, 288.

Tanaka, K. (2002). Growth dynamics and mortality of the intertidal encrusting sponge *Halichondria okadai* (Demospongiae, Halichondrida). *Marine Biology* **140**, 383–389.

Tautz, D., Ellegren, H., and Weigel, D. (2010). Next generation molecular ecology. *Molecular Ecology* **19,** 1–3.

Taylor, M. W., Radax, R., Steger, D., and Wagner, M. (2007). Sponge-associated microorganisms: Evolution, ecology, and biotechnological potential. *Microbiology and Molecular Biology Reviews* **71,** 295–347.

Teixidó, N., Gili, J. M., Uriz, M. J., Gutt, J., and Arntz, W. E. (2006). Observations of asexual reproductive strategies in Antarctic hexactinellid sponges from ROV video records. *Deep-Sea Research II* **53,** 972–984.

Teixidó, N., Pineda, M. C., and Garrabou, J. (2009). Decadal demographic trends of a long-lived temperate encrusting sponge. *Marine Ecology Progress Series* **275,** 113–124.

Thacker, R. W., and Freeman, C. J. (2012). Sponge-microbe symbioses: recent advances and new directions. *Advances in Marine Biology* **62**.

Thompson, R. C., Wang, I. J., and Johnson, J. R. (2010). Genome-enabled development of DNA markers for ecology, evolution and conservation. *Molecular Ecology* **19,** 2184–2195.

Thorpe, J. P., and Solé-Cava, A. M. (1994). The use of allozyme electrophoresis in invertebrate systematics. *Zoologica Scripta* **23,** 3–18.

Trivers, R. L. (1985). Social Evolution. Benjamin/Cummings Publishing, Menlo Park.

Tsurumi, M., and Reiswig, H. M. (1997). Sexual *vs* asexual reproduction in an oviparous rope-forming sponge, *Aplysina cauliformis* (Porifera: Verongida). *Invertebrate Reproduction and Development* **32,** 1–9.

Turner, L. M., and Hoekstra, H. E. (2008). The evolution of reproductive proteins: Causes and consequences. *The International Journal of Developmental Biology* **52,** 769–780.

Turon, X., Tarjuelo, I., and Uriz, M. J. (1998). Growth dynamics and mortality of the encrusting sponge Crambe crambe (Poecilosclerida) in contrasting habitats: Correlation with population structure and investment in defence. *Functional Ecology* **12,** 631–639.

Turque, A. S., Batista, D., Silveira, C. B., Cardoso, A. M., Vieira, R. P., Moraes, F. C., Clementino, M. M., Albano, R. M., Paranhos, R., Martins, O. B., and Muricy, G. (2010). Environmental shaping of sponge associated archaeal communities. *PLoS One* **5**(12), e15774.

Urgarković, D., Kurelec, B., Robitzki, A., Müller, W. E. G., and Schröder, H. C. (1991). Inhibition of aggregation-factor-induced ras gene expression in the sponge *Geodia cydonium* by detergent-polluted seawater: A sensitive biological assay for low-level detergent pollution. *Marine Ecology Progress Series* **71,** 253–258.

Uriz, M. J. (1983). Contributión a la fauna de esponjas (Demospongiae) de Cataluña. *Annales de la Sección de Ciencias del Colegio Universitario de Gerona. Monografías* **1,** 1–220.

Uriz, M. J. (2006). Mineral spiculogenesis in sponges. *Canadian Journal of Zoology* **84,** 322–356.

Uriz, M. J., Maldonado, M., Turon, X., and Marti, R. (1998). How do reproductive output, larval behaviour, and recruitment contribute to adult spatial patterns in Mediterranean encrusting sponges? *Marine Ecology Progress Series* **167,** 137–148.

Uriz, M. J., Turon, X., and Becerro, M. (2000). Silica deposition in Demosponges: Spiculogenesis in *Crambe crambe*. *Cell and Tissue Research* **301,** 299–309.

Uriz, M. J., Turon, X., Becerro, M. A., and Agell, G. (2003). Siliceous spicules and skeleton frameworks in sponges: Origin, diversity, ultrastructural patterns, biological functions. *Microscopy Research and Technique* **62,** 279–299.

Uriz, M. J., Turon, X., and Mariani, S. (2008). Ultrastructure and dispersal potential of sponge larvae: Tufted versus evenly ciliated parenchymellae. *Marine Ecology* **29,** 280–297.

Vacelet, J. (1959). Répartition générale des éponges et systématique des éponges cornées de la region de Marseille et de quelques stations méditérranéennes. *Recueil des Travaux de la Station marine d' Endoume* **16,** 39–101.

Vacelet, J. (1975). Étude en microscopie électronique de l'association entre bactéries et spongiaires du Genre *Verongia* (Dictyoceratida). *Journal de Microscopie et de Biologie Cellulaire* **23**, 271–288.

Vacelet, J., Boury-Esnault, N., and Harmelin, J. G. (1994). Hexactinellid cave, a unique deep-sea habitat in the scuba zone. *Deep Sea Research* **41**, 965–973.

Van de Vyver, G. (1970). La non confluence intraspécifique chez les Spongiares et la notion d'individu. *Annales d'Embryologie et de Morphogénèse* **3**, 251–262.

Van Oppen, M. J. H., Wörheide, G., and Takabayashi, M. (2002). Nuclear markers in evolutionary and population genetic studies of scleractinian corals and sponges. Proceedings 9th International Coral Reef Symposium. **1**, pp. 23–27.

VanBlaricom, G. R., and Estes, J. A. (1988). The Community Ecology of Sea Otters. Springer-Verlag, New York.

Volz, P. (1939). Die Bohrschwämme (Clioniden) der Adria. *Thalassia* **3**, 1–64.

Wang, Z., Gerstein, M., and Snyder, M. (2009). RNA-Seq: A revolutionary tool for transcriptomics. *Nature Reviews. Genetics* **10**, 57–63.

Webster, N. S. (2007). Sponge disease: A global threat? *Environmental Microbiology* **9**, 1363–1375.

Webster, M. S., and Reichart, L. (2005). Use of microsatellites for parentage and kinship analyses in animals. *Methods in Enzymology* **395**, 222–238.

Webster, N. S., Taylor, M. W., Behnam, F., Lücker, S., Rattei, T., Whalan, S., Horn, M., and Wagner, M. (2010). Deep sequencing reveals exceptional diversity and modes of transmission for bacterial sponge symbionts. *Environmental Microbiology* **12**, 2070–2082.

Whalan, S., Johnson, M. S., Harvey, E., and Battershill, C. (2005). Mode of reproduction, recruitment, and genetic subdivision in the brooding sponge *Haliclona* sp.. *Marine Biology* **146**, 425–433.

Whalan, S., De Nys, R., Smith-Keune, C., Evans, B. S., Battershill, C., and Jerry, D. R. (2008). Low genetic variability within and among populations of the brooding sponge Rhopaloeides odorabile on the central Barrier Reef. *Aquatic Biology* **3**, 111–119.

Wiens, M., Koziol, C., Hassanein, H. M. A., Batel, R., Schröder, H. C., and Müller, W. E. G. (1998). Induction of gene expression of the chaperones 14-3-3 and HSP70 by PCB118 (2,3',4,4',5- pentachlorobiphenyl) in the marine sponge *Geodia cydonium*: Novel biomarkers for polychlorinated biphenyls. *Marine Ecology Progress Series* **165**, 247–257.

Wilkinson, C. R. (1987). Productivity and abundance of large sponge population on Flinders Reef flats, Coral Sea. *Coral Reefs* **5**, 183–188.

Wilkinson, C. R., and Fay, P. (1979). Nitrogen fixation in coral reef sponges with symbiotic cyanobacteria. *Nature* **279**, 527–529.

Williams, G. C. (1975). Sex and Evolution. Princeton University, Princeton, NJ.

Williams, S. E., and Hilbert, D. (2006). Climate change threats to the biodiversity of tropical rainforests in Australia. In "Emerging Threats to Tropical Forests" (W. F. Laurance and C. Peres, eds), pp. 33–52. Chicago University Press, Chicago.

Wilson, K., Thorndyke, M. C., Nilsen, F., Rogers, A., and Martinez, P. (2005). Marine systems: Moving into the genomics era. *Marine Ecology* **26**, 3–16.

Wörheide, G. (1998). The reef cave dwelling ultra-conservative coralline demosponge *Astrosclera willeyana* Lister 1900 from the Indo-Pacific. Micromorphology, ultrastructure, biocalcification, isotope record, taxonomy, biogeography, phylogeny. *Facies* **38**, 1–88.

Wörheide, G. (2006). Low variation in partial cytochrome oxidase subunit 1 (CO1) mitochondrial sequences in the coralline demosponge *Astrosclera willeyana* across the Indo-Pacific. *Marine Biology* **148**, 907–912.

Wörheide, G., Degnan, B. M., Hooper, J. N. A., and Reitner, J. (2002a). Biogeography and taxonomy of the Indo-Pacific reef cave dwelling coralline demosponge *Astrosclera willeyana*—New data from nuclear internal transcribed spacer sequences. In "Proceedings 9th International Coral Reef Symposium" (K. M. Moosa, S. Soemodihardjo, A. Soegiarto, K. Romimohtarto, A. Nontji, Soekarno and Suharsono, eds), pp. 339–346. Ministry for Environment, Indonesian Institute of Sciences, International Society for Reef Studies, Yakarta.

Wörheide, G., Hooper, J. N. A., and Degnan, B. M. (2002b). Phylogeography of western Pacific *Leucetta chagonensis* (Porifera: Calcarea) from ribosomal DNA sequences: Implications for population history and conservation of the Great Barrier Reef World Heritage Area (Australia). *Molecular Ecology* **11**, 1753–1768.

Wörheide, G., Fromont, J., and Solé-Cava, A. (2004). Population genetics and phylogeography of sponges—A workshop synthesis. *Bollettino dei Musei e degli Istituti biologici dell Università di Genova* **68**, 683–688.

Wörheide, G., Solé-Cava, A. M., and Hooper, J. N. A. (2005). Biodiversity, molecular ecology and phylogeography of marine sponges: Patterns, implications and outlooks. *Integrative and Comparative Biology* **45**, 377–385 doi:10.1093/icb/45.2.377.

Wörheide, G., Epp, L., and Macis, L. (2008). Deep genetic divergence among Indo-Pacific populations of the coral reef sponge *Leucetta chagosensis* (Leucettidae): Founder effects, vicariance or both? *BMC Evolutionary Biology* **8**, 24.

Wulff, J. L. (1985). Dispersal and survival of fragments of coral reef sponges. *Proceedings of the Fifth International Coral Reef Congress* **5**, 119–124.

Wulff, J. L. (1986). Variation in clone structure of fragmenting coral reef sponges. *Biological Journal of the Linnean Society* **27**, 211–230.

Wulff, J. L. (1991). Asexual fragmentation, genotype success, and population dynamics of the erect branching sponges. *Journal of Experimental Marine Biology and Ecology* **149**, 227–247.

Wulff, J. L. (1995). Effects of a hurricane on survival and orientation of large, erect coral reef sponges. *Coral Reefs* **14**, 55–61.

Xavier, J. R., Rachello-Dolmen, P. G., Parra-Velandia, F., Schönberg, C. H. L., Breeuwer, J. A. J., and Van Soest, R. W. M. (2010a). Molecular evidence of cryptic speciation in the "cosmopolitan" excavating sponge Cliona celata (Porifera, Clionaidae). *Molecular Phylogenetics and Evolution* **56**, 13–20.

Xavier, J. R., Van Soest, R. W. M., Breeuwer, J. A. J., Martins, A. M. F., and Menken, S. B. J. (2010b). Phylogeography, genetic diversity and structure of the poecilosclerid sponge *Phorbas fictitius* at oceanic islands. *Contributions to Zoology* **79**, 119–129.

Zane, L., Bargelloni, L., and Patarnello, T. (2002). Strategies for microsatellite isolation: A review. *Molecular Ecology* **11**, 1–16.

Zhang, D. X., and Hewitt, G. M. (2003). Nuclear DNA analyses in genetic studies of populations: Practice, problems and prospects. *Molecular Ecology* **12**, 563–584.

Zilberberg, C., Solé-Cava, A. M., and Klautau, M. (2006a). The extent of asexual reproduction in sponges of the genus *Chondrilla* (Demospongiae: Chondrosida) from the Caribbean and the Brazilian coasts. *Journal of Experimental Marine Biology and Ecology* **336**, 211–220.

Zilberberg, C., Maldonado, M., and Solé-Cava, A. M. (2006b). Assessment of the relative contribution of asexual propagation in a population of the coral-excavating sponge *Cliona delitrix* from the Bahamas. *Coral Reefs* **25**, 297–301.

Subject Index

Note: Page numbers followed by *f* and *t* indicate figures and tables, respectively.

Taxonomic Index

Note: Page numbers followed by *f* and *t* indicate figures and tables, respectively.